Springer Biographies

The books published in the Springer Biographies tell of the life and work of scholars, innovators, and pioneers in all fields of learning and throughout the ages. Prominent scientists and philosophers will feature, but so too will lesser known personalities whose significant contributions deserve greater recognition and whose remarkable life stories will stir and motivate readers. Authored by historians and other academic writers, the volumes describe and analyse the main achievements of their subjects in manner accessible to nonspecialists, interweaving these with salient aspects of the protagonists' personal lives. Autobiographies and memoirs also fall into the scope of the series.

More information about this series at http://www.springer.com/series/13617

Mario Bunge

Between Two Worlds

Memoirs of a Philosopher-Scientist

Springer

Mario Bunge
Department of Philosophy
McGill University
Montreal, Canada

Springer Biographies
ISBN 978-3-319-29250-2 ISBN 978-3-319-29251-9 (eBook)
DOI 10.1007/978-3-319-29251-9

Library of Congress Control Number: 2016931313

Printed on acid-free paper

This Springer imprint is published by Springer Nature
The registered company is Springer International Publishing AG Switzerland

Preface

I decided long ago never to write my memoirs, because I knew that episodic memory is rather creative and, therefore, unreliable. But I changed my opinion when I read some strange biographies of mine, where even my mother's maiden name had been altered. Besides, some relatives and friends put pressure on me: they thought that, as both a philosopher-scientist and a hyphenated Canadian, I had something interesting to tell. Let them bear some of my guilt.

Some readers may wonder which are the worlds mentioned in the title of this glimpse into my life and work. They will find out if they read up to the middle of the book. And then they will realize that the worlds in question are not two but four. I put "two" in the title so as not to alienate potential readers rightly suspicious of talk about alternative worlds. I tell my life as a sequence of episodic recalls of the form what-what for who-where-when. And I describe my work and times by outlining only a few landmarks. When I started writing this book, I could not stop: contrary to my expectations, my memories poured out effortlessly – perhaps a sign that I had enjoyed living even in hard times.

I must have unfairly forgotten many people who have been good to me. It is also certain that I have underrated or even rubbished some people who deserved better. I will be the first to deplore all that. In any case, look out!

Montreal, Canada
Fall of 2015

Mario Bunge

Contents

List of Illustrations

Chapter 1
Childhood

Family Background

I was one of the myriad unintended by-products of the First World War (1914–1918). Indeed, I would guess that I was conceived during a celebration of the armistice that put end of that carnage – the most global, long, bloody, absurd, and unpopular war in history. I surmise that my parents, though from very different backgrounds, met by chance at the Eden Hotel in La Falda, a hill resort in Córdoba, at the center of Argentina, during one of the celebrations of that longawaited event.

My mother, Marie Müser – whom everybody called Mariechen, or Little Mary – was a tall and stunningly beautiful 36 years old German who had immigrated 6 years before and worked as a nurse in the German Hospital in Rosario. This was then Argentina's second city, on the awesome Paraná, the third-longest river in the world. My father, Augusto Bunge, was 5 years older, handsome, elegant, cultivated, a physician and recently elected national congressman. Some years earlier, the national government had commissioned him to visit Germany and France to study their public health care systems and report on them – which he did in two large volumes full of statistics. Augusto had earned a reputation as an expert in public health and labor law. And he belonged to one of the country's so-called patrician families – in contrast to my mother, who had humble origins.

My parents were good-looking, well-mannered, well-traveled, spoke "high" German, and loved Goethe and Schiller. They were moderately gregarious, avid readers, nature lovers, and feared germs, though not air drafts – then often blamed for colds. Both were passionate about health care, individual and public. And the war, though very distant, had moved and disgusted them. It had also made them oscillate between their admiration for Germany and their hatred for German militarism.

Besides, both Mariechen and Augusto were available: she was single, and he was legally separated from his wife, Belén Holmberg. (Divorce came to Argentina only in 1987.) When I asked my father once why he had broken up with his first wife, he

M. Bunge, *Between Two Worlds*, Springer Biographies,
DOI 10.1007/978-3-319-29251-9_1

told me that she did not accompany him well on the piano while he played the violin. On another occasion he said that she refused to have children, whom he adored. Belén, for her part, was said to hold that Augusto was impatient, haughty, and irascible. The truth of this matter is hard to know and no longer relevant.

At 16, Mariechen joined the German Red Cross to train as a nurse. Subsequently, she was sent to work at hospitals in the two German colonies in China, Tsingtau and Kiaochu, where cholera and typhus were endemic. She took the Trans Siberian Railway from Moscow to Harbin, a trip that lasted 2 weeks, and made the rest of the journey by train and coach in a territory infested with bandits. Mariechen had been happy in the German compounds in China, because she felt that her work was useful, she liked the Chinese, and was treated as an equal by doctors and colonial officers. One of her patients, a humble Chinese peasant, thanked her for having put in his armpit the most wonderful and effective of her remedies: that little stick with numerals marked on it.

In 1912, upon returning from China accompanying convalescing colonial officials and soldiers, Mariechen learned that the German imperial government had just decreed general mobilization. She realized that the drums of war were announcing what the great powers – especially Germany, Austria, France, and Britain – had been preparing for decades with the sole end of expanding their respective empires. Then and there, my mother resolved to sail, along with a sister, on the first ship bound for America.

Mariechen in Argentina

The first available boat was bound for Rosario in Argentina. The two sisters, certified nurses, were immediately employed by the British German Hospital, which split into two when the war broke out. At that time, nursing was almost as prestigious and well paid as teaching, and the Europeans, who had started to value skill over pedigree, admitted nurses in society. In particular, the Eden Hotel, where Mariechen and her sister stayed, welcomed lower-class professionals, like teachers and nurses, as long as they were presentable and spoke "high" German. Besides, Ida Eichhorn, a co-owner of the hotel, who had been a patient of my mother's, was her friend and did not charge her. Ida and Mariechen kept up their friendship until 1933, when the hotel presented Hitler with an aircraft.

On winning the war at a staggering cost, the Allies forced the Treaty of Versailles on the central powers. That treaty crushed the losers with war reparations, and presented Japan with the German colonies in China and the Pacific. Italy, a last-minute ally, was given some Greek islands that had been under Ottoman occupation. The talk about the right of peoples to self-determination was thus shown to be mere window-dressing.

My father and the great economist John Maynard Keynes warned that the said Treaty, advertised as securing peace for all time, was the seed of a second world war. Of course, no statesmen listened to that warning: they were more worried by the Bolshevik revolution of 1917 and its echoes abroad than about the future of civilization. The German right-wing exploited the popular resentment caused by the Treaty, as well as the despair caused by the Great Depression and the fear of Communism, to arm well-organized militias that specialized in provoking street riots. President Hindenburg, already senile, was invited to watch a parade of one of them, the SA brownshirts, and told that they were soldiers back from the front. He commented: "They march very well. Must have taken many prisoners."

These street disorders undermined the weak democratic Weimar Republic, and prepared the *putsch* that put Hitler and his thugs in power, although they had gotten only 37 % of the votes. The complacency of the academic community is well known (see Shirer 1860). The support of the legal positivists Hans Kelsen and Carl Schmitt to the legal justification of that coup is seldom remembered, but has been well documented (see Dyzenhaus 1997). The Weimar Republic has been unfairly maligned: while it was a political powder keg, it was also a cultural wonder, for it outshined all the other European nations in mathematics, the natural sciences, and cinematography, and was the equal of some in technology, literature, music, and the plastic arts. One reason for this superiority is that the Germans had kept their best brains far from the front.

Repercussion of Nazism in Argentina

Germany changed gears in January, 1933, and only 6 years later started a world war that was to be even more murderous than the first. This event was decisive for the German community in Argentina, which split almost overnight into two. Each of them had its own daily newspaper. The democratic *Argentinisches Tageblatt* belonged to the Alemann brothers, friends of my father's. Neither this daily nor its pro-Nazi counterpart had any influence outside the German colony. By contrast, *El pampero*, the Spanish language evening paper, had a wide readership, for fascism was rather popular since the military coup of 1930. This was largely due to hatred of the British, who owned the main railways and meatpacking companies, and were known to interfere strongly in political affairs. In particular, the conservative and corrupt but anti-German president Roberto M. Ortiz (1938–1940) was said to have been appointed at a London corporate boardroom.

President Ortiz was blinded by diabetes in mid-1940, and Congress debated his impeachment for incompetence, for being unable to do his job. We the democrats, both the genuine and the fake, took to downtown streets to march and shout, loudly but peacefully. The mounted police, popularly called "the Cossacks," chased us, and I got a mild stroke with a sabre. Ortiz was sacked and his Vice President, an obscure provincial lawyer accused of representing the landowners and ranchers, favored the fascists and harassed the anti-fascists, my father among them, for the crime of helping the Allied cause. A socialist congressman quipped that, while the ranchers looked up to Germany, their cows looked down on Britain. As usual since the fascist coup of 1930, most of the intellectuals and writers kept silent.

Hitler's ascent to power did not affect my parents' marriage, for both were socialists. But it disconcerted and saddened them, for they admired German literature and medicine. It was disconcerting that a people noted for their poets, musicians and scientists, had fallen into the hands of a gang of murderers to the applause of most German professors and students. The situation was so absurd that many thought the Nazi regime would not last. This helps to explain why so many anti-Nazis and Jews tried to emigrate only weeks before the war.

My Parent's Ancestors

As I indicated before, my parents were socially very uneven. My mother's family belonged to the upper working class (or lower middle class), as her father, Wilhelm Müser (b. 1841), had been a stationmaster, and his wife, Amalie Schroeder (b. 1850), a housewife. Marie Herminie Ida was born in the small town of Hardechsen, near Hannover. This was the great Leibniz's birthplace and also, according to my mother, the place where the purest German (*Hochdeutsch*) was spoken – which sounds melodious by comparison with the harsh Prussian barking and the soft but garbled southern dialects. In 1714 Georg Ludwig, Duke of Hannover, was made king of Great Britain under the name George I. He moved to London along with his entire court, that included great musicians like Händel and great scientists like Herschel. My mother would not have been received at either the Hannover or the Windsor courts, for her ancestors had not been officers or even pastors, and she had attended only primary school and nursing college.

The Bunge family had originated in Gøtland, a wooded island off the Swedish coast. There is still a village called Bunge, renowned for its medieval sculptures, but its inhabitants are not in agreement over the correct pronunciation of their village's name. Early in the seventeenth century, some Bunges migrated to the Rhineland and settled there, whereas others migrated to Russia and produced the prime minister who wrote the decree liberating the serfs – a feat that cost him his job.

In the succeeding century other Bunges left for Holland, Belgium, Canada, and the USA. In Belgium they founded the multinational cereal import firm that is still operating in all continents from the *Bungehuis* in Antwerp. The Argentine branch, called Bunge & Born, is so powerful, that it used to be known as "The Octopus," but there have not been any Bunges in it since 1912. See the following diagram.

Abridged genealogy of the founder Carl August (from Cárdenas & Payá 1985a,b)

Johann Bunge

|

Johann Heinrich & Gertrude Tilemann (1679)

|

David Caspar & Isabel Voss von Lunem

|

Johann Heinrich (1710) & Clara Büdemann (1744)

Diederich David (1745–1814) & Clara Krupp (1743–1813)
|
Henriette Caroline (1772–1828) & Adam Scheuten
|
Carl August (1804–1849) & Genara Peña Lezica (1809–1857)
|
Raimundo Octavio (1844–1910) & María Luisa Arteaga (1853–1934)
|

Carlos Octavio (1875–1918), Augusto (1877–1943) & Marie Müser (1882–1982) and 6 siblings
|
Mario Augusto (b.1919)
|

&1. Julia Molina y Vedia (1912–1985)	&2. Marta Irene Cavallo (b. 1938)
|	|
Carlos F. (b. 1941), Mario Jr. (b. 1943),	Eric R. (b. 1967), Silvia A. (b. 1973)

My branch of the family goes back to Carl August Bunge (1804–1849), an enterprising and handsome businessman who came in 1827 to Buenos Aires from Unna, in the Rhineland-Palatinate. Seven years later he married the young and childless widow María Genara Peña de Lezica. She belonged to a family of wealthy merchants and smugglers who claimed to be descended from Pelagius, or Don Pelayo. This was the Visigothic warrior who is believed to have started the *Reconquista*, or war for the political independence of Spain from Arab rule, a process that lasted nearly eight centuries. The Peña-Lezicas also claimed to be descended from the Spanish captain Francisco de Ampuero, whom Francisco Pizarro – the pig herdsman who headed the conquest and looting of golden Perú – forced to marry the Inca princess christened as Inés Yupanqui, who gave him five children. Another disreputable branch of my family tree is the Krupp family, which amassed a fabulous fortune manufacturing the best and largest guns.

A more distinguished if distant relative was Ernesto Guevara, aka Che. His father, whom I met briefly three decades ago, told me that I am related to both himself and to Celia de la Serna, his first wife and Che's mother. Her sister Carmen and her husband, the journalist C. Córdova Iturburu, aka Policho, were close friends of my parents. I never met the Che but, along with millions, I admire his selfless dedication to the cause of the downtrodden, while regretting that he tried to export the Cuban revolution to Bolivia without first inquiring whether the people whom he wished to help were ready to join him. The cause of social justice has benefited more from politics informed by social science and grassroots movements, as in Scandinavia, than from martyrdom and Jacobin elitism. However, I do not have a recipe for bloodlessly unseating brutal dictators who enjoy the support of powerful empires.

Finally, there was William Bunge, an American mathematical geographer who came to Montreal chased by Senator McCarthy's hounds. At my university's Allan Memorial Institute of Mental Health, William was an unwitting guinea pig in the infamous experiment, carried out by famous McGill professors funded by the CIA, designed to see whether it was possible to control minds by means of massive LSD doses. This brutal treatment left William handicapped for life, and he lived the rest of it on a mean pension in a distant and cold Quebec village. William is one of the millions of forgotten lives destroyed by the Cold War. He is also a reminder of the need to honor the Hippocratic oath, perhaps the earliest humanistic manifesto.

In any event, it is absurd to be proud or ashamed of one's ancestors, since one does not choose them. When Columbus started his "discovery" voyages, Spain had about six million inhabitants, one third of whom were landless nobles, most of them hungry and ragged, and yet being forbidden to work with their hands under penalty of being erased from the nobility registry. So far as I know, none of those beggars made their mark. Few if any of them ventured into the Americas: all the Spaniards who travelled to the "new" continent to do useful work had been landless and uneducated peasants. Instead of gold, they found the green and fence-less pampas, where it was not uncommon for a handful of men to kill a cow for a single meal.

The Offspring of Carl August and Genara

With the help his kid brother Heinrich August, and presumably with that of his rich in-laws as well, Carl August built a gigantic business, which included land and finance. Thus, he fully integrated into his new land, though without severing his European ties. Indeed he was the consul of Prussia and Holland, and helped organize the first German school, hospital, and Lutheran temple. Apparently, Carl August was not bothered by the tyranny of the wealthy rancher Juan Manuel de Rosas (1829–1852), under which he lived nearly all his years in Argentina. The first rebellious Bunge was his grandson Augusto (1877–1943), my father – a socialist since his student days. The million or so immigrants who settled later in Argentina behaved similarly. Some of the most remarkable success stories were those of the Arabs and Jews, in particular those who succeeded in farming and in the so-called liberal professions. Their children did not care for their parent's language, religion, or customs. I first saw skullcaps in New York. But in some rural villages I saw taciturn Jews wearing gaucho-style shirts, pantaloons, and boots; and once I was forced to shake the hand of the first Arab President of the Republic. However, let us go back to the Argentine Bunges.

Carl August and Genara had eight children. All of them, except Octavio Raimundo, got rich buying and exploiting large tracts of fertile land that the General Roca's government had stolen from the Araucanian Indians and sold in installments. My grandfather Octavio was the exception because he held that "judges should not owe money." He earned a doctorate in Law, was appointed a judge, and eventually became the Chief Justice of the nation. Octavio married María Luisa Arteaga, a Uruguayan of Basque descent. In the Basque Country there is a village and an ugly fake medieval castle, the Arteaga Tower. María Luisa was an unsmiling and devout Catholic, but her husband was anticlerical, as his travel memoir show. It is likely that he was also a Freemason. Such a combination of a skeptical man with a pious wife was then frequent. All the founding fathers of Argentina had been Freemasons and excommunicated by Rome, but they knew the power of religion as a tool of social control. Thus, Manuel Belgrano, the first Argentinian economist and one of the Founding Fathers, was not a churchgoer but forced his troops to attend mass before battle.

María Luisa and Octavio had eight live children, most of whom distinguished themselves. The first-born, Carlos Octavio, was a judge, a professor of social psychology, a prolific writer, and a public intellectual. Augusto, his closest friend and my father, had studied at the Jesuit school, where he won all the medals, which filled a shoe box. He kept winning them even after telling his chagrined theology professor that he had lost his faith. He studied medicine and almost failed an exam for mentioning Pasteur's findings on germs, but graduated in 1900 with the gold medal. His doctoral dissertation dealt with tuberculosis – the most dreaded sickness at the time – as a social disease. Thus from the start he joined the ranks of the "social hygiene" movement, which waged a strong campaign for free universal health care – a movement that won out nearly a century ago in nearly all the advanced countries (Sánchez 2007).

Augusto's brothers Alejandro and Jorge earned their engineering doctorates in Germany. Alejandro became a top economist and organizer of statistical bureaus in several countries. He also founded the Argentine economic review, as well as an independent center of economic studies. He rightly criticized free trade policies for preventing the development of native industries. He was also very devout, and ended up supporting fascism and breaking with my father. I liked his sons Rafo, Max and Fritz, all very out-going and sportive.

Jorge Bunge, my father's youngest sibling, became a successful modernist architect. He also designed and managed Pinamar, a large sea resort built on sand dunes, which he transformed into pine forests – one of the few triumphs of alchemy. In 1935, Jorge invited my father to join him in this enterprise, warning him: "Pinamar will last, whereas nothing will remain of your political efforts." How prescient! He visited us once in a blue moon, and advised me when my father died. Jorge had been widowed as a young man, and never remarried because he did not wish to impose a foster mother on his only child, my cousin Cecilia, on whom my father and I doted. She told me once that her father was not worried about the afterlife: "If I am sent to Hell, I will have interesting conversations with Augusto; and if I end up in Paradise, I will enjoy the company of Sor [Sister] Alejandro."

(For the Bunge brothers, see Cárdenas and Payá 1995, 1997).

Mariechen and the Bunges

Obviously, given her humble origins and her having a manual profession, Mariechen did not fit into a family proud of their ancestry.

María Luisa Arteaga, my father's mother – dry, haughty, pious and always in black – refused to meet my mother because she was not married. She called her "Augusto's concubine." O the mysteries of Christian charity and family values! In 1924, when my sister Eva was born, my parents got married by proxy in Uruguay, but this marriage was not valid in Argentina. Divorce came to this country only in 1987, after a fierce battle with the Catholic Church. I was therefore a natural, bastard, or adulterine child, as the law had it, even though my father recognized me the day I was born, September 21, 1919. For that reason, or rather folly, after my father's death my mother could not collect the pension she was entitled to as the widow of a national congressman who had served his country with distinction for 20 years. And because I was an "adulterous" child, I had to pay exorbitant death duties, which forced me to put our property on auction and sell part of it at a cut rate. That was only the first of half a dozen failed business ventures of mine.

Despite my low legal status, my Argentine relatives treated me as family. After all, I was blonde and they said I had the Bunge looks, although others saw me as similar to my mother. I got along well with Uncle Jorge and Aunt Julia, whom we visited with some frequency. We visited them; they did not visit us.

We did not visit my aunt Delfina and her husband, Manuel Gálvez, because they practiced a somber and exclusive Catholicism, on top of which they sympathized with fascism. However, I found Gálvez, Uncle Manolo, the most interesting of my uncles because he was a first-rank novelist, and his novels about the infamous Paraguay war had moved me to tears as a child. He was very hard of hearing, and used a hearing aid activated by a battery in his handkerchief pocket. He turned it on when intending to speak, and off when one was about to answer him. Then he pretended to listen while contemplating askance a sty on one eye.

My cousins, nearly all of them older than I, treated me affectionately. I remember in particular Luis María, Roberto's son, as well as Alejandro's sons Rafo, Max, and Fritz. Years later I befriended Max, a failed businessman who wore a bowler hat. He and his bride had spent their first night in a tent on one of the wooded islands in the enormous Paraná delta. At midnight shooting started that lasted until daylight. Max and his wife thought it was a fight between rival smuggler gangs, but the reports turned out to be caused by bamboo canes that had been burning in a bonfire. My cousin Delfina Gálvez is an architect and fiction writer who is now 102 years old and alert as ever. She had married Amancio Williams, my favorite architect and one of the earliest plane pilots in the country. His highly original work, a sort of aerial Bauhaus, is better known at Harvard than in Argentina. Despite his ecclesiastic connections, Amancio got very few commissions, for he was too advanced for our conservative society.

Cecilia, my favorite cousin as well as my father's favorite niece, was a couple of years younger than me. I found her sweet and pretty, and loved to play with her in the presence of her English nanny during our weekly visit to our grandmother. Once, to gain her admiration, I made my horse caracole and stand on its hind legs. Another time we got into an ethical dispute: Cecilia admired our aunt Julia's charitable work, while I thought that charity only covered up poverty. My beloved Cecilia broke into tears, and I did not know where to hide. Of course, now I understand that Aunt Julia's selfless work among lepers was filling a hole left by the state.

My relation with Cecilia ceased when our grandmother died, but we rekindled it half a century later, when my name started to appear in newspapers and she retired from the business emporium managed by her beloved husband Enrique Shaw, who died from leukemia. Enrique claimed that, although he had lived as a bourgeois, he was dying as a worker, for his employees had donated all his blood. The Argentine society of Catholic businessmen regards those generous blood transfusions as miraculous, and has initiated Enrique's canonization. I regard it as even more miraculous that his widow, a housewife and mother of eight, took overnight the reins of the Shaw-Bunge emporium – which includes a bank, a glass factory, and the Pinamar sea resort. Cecilia made it flourish for 35 years without a college degree or previous business experience – all that in a notoriously sexist and crooked business environment. Maybe character and controlled despair compensated for her lack of an MBA. To be sure, her huge initial endowment helped, but many others – myself included – squandered their inheritances.

The Shack

My father decided that nature should be my main teacher. This is why, shortly after my birth, he bought in installments an old adobe shack with half an acre of land, which he slowly expanded to twice its original size. This property was situated in semirural West Florida, four blocks from a small train station, half an hour from Retiro, a railway hub in downtown Buenos Aires.

Our neighborhood was modest and sparsely populated. Many of our neighbors were hard-working Italians who cultivated vegetables, fruits and flowers that they sold in the city. Our original dwelling was a small and primitive adobe shack with a smelly outhouse – all my father could afford at the time. But we enjoyed the shade of a majestic tree, the *ombú* that graces the Argentine *pampas*, or plains – whence the name our parents gave our place: *El ombú*. We also owned the tallest windmill in the place, which extracted water from a deep well – until it died up three decades later. My father had the house improved in stages, starting with protecting the windows with iron bars. Since the walls were made of dried mud, however, the thiefs were able to pull out the bars effortlessly.

We were surrounded by farms of between 5 and 10 ha. The largest of them, Bottini's, had a hedge of seena-seena, a thorny and semi-transparent bush. Our own property was protected by maclura, also thorny but denser, though not enough to prevent some ladies from complaining to the police that my father was sun-bathing in the nude.

I had the run of the neighborhood from age 3, and spent long hours watching the labors of our neighbors – who called me Testa Bianca (White Head) – as well as the roadmen. These took a short lunch break and ate a loaf of bread and a tin of sardines for 20 cents (the equivalent of an American nickel). They were paid 4 pesos a day, the equivalent of 1 dollar – one-half the wage of an American industrial worker and one-fifth of what the Ford workers earned in 1914. They were frugal because some supported their families in Europe, and others paid in 120 installments for the lot of land they had bought, and where they hoped to build their own house with the help of friends. Some of them visited their vacant lots every Sunday.

I felt so free and confident in that sparsely populated region with hardly any traffic, that at age 4, following a row with my mother, I decided to leave home. I packed a few belongings in a small bag, and started walking on a newly opened road with no map or even goal in mind. After an hour or so, some neighbors, took me back home. A few years later, following another fight with my mother, I wrote to my father, who was in Europe on professional business, complaining about a punishment that "your wife" had given me. That was only the first of the many letters I mailed abroad, first in Spanish and later in Esperanto, on stamps, politics, and other subjects. Our political police did not waste time opening letters with steam and resealing them.

Shopping

We bought clothes and other durables in the city, preferably at the socialist cooperative or at the English-owned department stores Harrods and Gath and Chávez, famous for their summer cloth caps. My favorite shop sold Hornby toy trains and exotic foodstuff, such as English dog biscuits and American breakfast cereal – which we tried for a while just to humour my father. American wares were rare, while British-made men's winter clothes were popular among the upper middle-class. By contrast, no lady would dream of following the British women's fashion, regarded as ugly and vulgar compared with the French one.

American cars, with the exception of the early Fords, were appreciated only by farmers because they were cheap, easy to drive and repair, as well as light and high above the ground, and thus capable of braving muddy country roads. Only the heavy and expensive but elegant European cars were considered classy. Rational city folks would not have cars at all. My father never had one, while I bought two ruinous nth-hand cars in succession, a British Ford and a Packard, as soon as I thought I could afford them. They spent as much time at the mechanic's as at our garage, and their tyres made from reconditioned ones, had to be repaired every 100 km or so. But the Packard, huffing and puffing, took us to distant places such as the Calchaquí valleys, which seemed frozen in colonial times.

Back to daily life in *El Ombú*. We grew in our own orchard, or bought from peddlers, most of the vegetables, fruits, and flowers we consumed. Among them were some fancy vegetables, such as New Zealand spinach, salsify, and watercress;

fruits, such as blackberries, kakis, and raspberries; and flowers, such as anemones, Paraguayan jasmine, and rare rose varieties. The rest, from coal and meat to medicines and school utensils, were bought at the few shops in the village, only four blocks away.

At that time, long before the emergence of super-markets, shopkeepers were kept very busy. The hardware-store owner had to cut to measure every piece of wood or pane of glass, and weigh every bunch of nails. The pharmacist had to prepare nearly every prescription, using a mortar and scales. The grocer had to cut and wrap up 10 cents' worth of sugar or four slices of cheese. The butcher had to cut, weigh, and wrap 50 cents' worth of thin slices of meat. Many customers bought on credit, and the shop keeper had to wait until his customers could pay for his monthly bill. I was often in arrears because I had no regular source of income. Thus, all such transactions involved trust – a vanishing social virtue.

Once in a while we bought from a horse cart loaded with things of a single kind offered at reasonable prices. When a Dutch ship arrived, we were visited by a cart loaded with spherical cheeses coated in red wax, or with Bols gin in earthenware bottles that, once emptied, could be used as foot warmers. At other times one could buy cheap fresh fish, melons, or fruit from orchards in the immense and beautiful delta of the Paraná river that originates in Amazonia. Hunger and shanty towns were unknown in the country before the Great Depression that came 1 year after it started on Wall Street as a result of a process that still baffles economists blinded by theories that postulate equilibrium and do not involve time, the universal parameter.

San Martín Avenue, the only paved street in the region, reached the brown Río de la Plata, 4 km away, where I used to swim in summer. At the beginning we took the only bus in the region, but after a while the potholes grew to the point of rendering car traffic impossible, so I walked 8 km – an exercise that gave me strong legs. The potholes came with a silver lining: after a heavy rainfall, my friends and I used them as miniature pools. The other streets, all dirt, were roamed by horses, cows, and bulls. Once I was chased by a bull and, to run faster, got rid of the steel contraptions designed to straighten my legs bent from calcium deficiency.

The mail was delivered by Señor Reyes, a serious man who came on horseback. When the streets were paved, he rode a tricycle. And when he retired, Reyes bought a huge old car, which he drove as a taxi at a constant speed, ignoring all intersections. The street lights came several decades later, but I had taught myself to drive long before, so I still drive Reyes-style.

Even professional burglars rode on horseback. But some of them were sophisticated in other ways. When they ransacked our house, they chloroformed us and took everything portable. My father had to walk barefoot and in pajamas to borrow clothes for himself and my mother from our nearest friends, Roberto and Beatriz Giusti. The political policemen who emptied the residences of suspects in the 1970s had it much easier: they loaded the stolen goods on the sinister green Ford Falcons and vans of the police force.

Development of El Ombú

Our land was initially covered in weeds taller than myself, and had only two trees: the largest *ombú* I have ever seen, and a ceiba tree only slightly smaller than the one in Sydney's botanical garden. Our toilet was a primitive latrine inhabited by a swarm of buzzing blue-green flies. I was terrified at the prospect of falling into that seemingly bottomless pit, a fear may have accelerated my toilet training, although my mother claimed that she had me trained by the age of 4 months. This must have been a wild exaggeration. What is true is that in the developing countries learning that skill takes half the time it takes in the advanced ones, where incontinence in everything is big business.

We washed in a large tin vat, near the coal stove, which was also used to wash our laundry. Our rooms were lighted by kerosene lamps. One night Bruja, or Witch, our friendly but clumsy black bitch, overturned a lamp, and the flames spread quickly. Alarmed, my father shot her then and there with his revolver – an action that I could not forgive him. I became familiar with death, and with the idea that there is nothing beyond it, at a very early age.

Our cottage became more comfortable every year with the help of mortgages and of the master builder Egidio and his men. Every morning, before leaving for his downtown office, my father reminded him: "Egidio, do not forget to add Ceresita [the most popular brand of moisture-proofing] to the cement mix." Eventually we got a wall telephone, a modern bathroom, electric energy, a manual hot-water washing machine, and even a clever Swedish gas refrigerator.

At the same time, my father transformed the wilderness into a beautiful combination of orchard and garden, where artichokes rubbed leaves with rose bushes and fruit trees. We had 130 rose varieties, my father's pride and joy. I have seen only one comparable combination of beauty with utility: the garden-orchard tended by Nikos Tsilibaris, our best friend in Corfu.

Our Gardeners

My father did all the gardening with his own hands with my help. I kept so close to him that once he crushed one of my little toes with his rummer, and was much more afflicted than I. Eventually we were able to hire gardeners, among them Job, the cheerful Dutchman who had sailed the Malaysian straits, exhibited on his back a beautiful peacock tattooed in colors on his back, stuck a knitting needle in his body without drawing blood, and got drunk every weekend; the lazy German who had been a provincial congressman and addressed my father as *Genosse* (Comrade), and cared only for mushrooms; the Japanese who started by drawing a plan of the property, cut down a thorny tree that reminded my father of his childhood, and left a letter of resignation admitting his ignorance of the local flora; the Spaniard Eduardo Rey López, who had lost an eye while serving in the Spanish navy and used to upbraid the roosters as "Effeminates!" when the chickens failed to lay enough eggs. We also had two competent Micheles from northern Italy. The first of them deserves a separate paragraph.

Michele Buontempo was educated and skillful, hard-working and handsome. One day, our faithful if ugly maid/cook Rosa, whom we adored, gave us notice. She accused Michele of flying on his wings and entering her room through the nonexistent chimney to rape her. Her accusation was all the more ridiculous given their differences in looks and ages. So, we were forced to accept Rosa's resignation. Some time later she visited us, to the joy of all of us, including my pets.

Rosa was replaced by Clotilde, just out of a Catholic convent. She had been taught only how to pray and scrub, as well as the minutest details of Palestinian geography. She could not even place Argentina, let alone Buenos Aires, in my atlas. Clotilde was ignorant but not blind: she noticed Michele, and spent all her wages on clothes and cosmetics to woo him.

Very soon, Michele asked for my help in writing her love letters and in conveying them to her. He proposed her, and of course Clotilde accepted him. My parents, chronic matchmakers, proceeded to help them. My father located a lot of land in the fertile Río Negro valley, in upper Patagonia, suitable for fruit growing, and the newly weds settled there. A couple of years later, Michele visited us. He was distraught and weeping: Clotilde, who had produced a child, had turned mad and could no longer work on the farm or even in the house. He had left his wife interned at the local hospital, and confided his baby to a friendly neighbor. He came to us for advice and help I lived until 20 in that paradise, accompanied by several dogs, the cat Mustapha, a pair of peacocks, oven-birds, seagulls, and songbirds, and a lovely little titi monkey who jumped to my father's shoulder, pressed his cheek against my father's, and fell asleep while the alpha male read in his favorite armchair. Tití was murdered by our senior dog, Dacky, who could not tolerate competition any more than the free-marketeers who plot the ruin of their rivals while praising competition.

Our Neighbors

Our neighbors, most of them descended from Spaniards or Italians, tended small farms, from 5 to 10 acres, or minded small shops. We were on excellent terms with most of them, and were invited to their weddings and funerals. The wakes were solemnized by choirs of professional weepers, all clad in black, who looked freshly extracted from Greek tragedies.

The largest farm, Bottini's, produced wine in addition to vegetables. When vintage came, the year's crop of Concord grapes was thrown into a huge wooden vat. Those grapes were popularly known as "bedbug grapes" because it was not uncommon to crush such an evil tasting insect when eating them straight from the vine, as I liked to do.

No sooner was the vat full, than all the members of the Bottini family, along with their hired hands, entered the vat with their bare unwashed feet, and, tucking up their pants or skirts, stepped vigorously on the grapes, bugs and all. The work was easy but hard, and it was encouraged by monotonous Italian chants. Finally, the grape juice was poured into large oak barrels, where it was allowed to ferment.

One summer morning, Bottini senior ordered his two farm hands to take his four dogs, elderly but still vigorous, to the sidewalk and attach them with wire to a post. At sunset, when the dogs were exhausted from heat and lack of water, they were beaten to death, one after the other and slowly. This was the cruelest scene I have ever witnessed. Another cruelty I watched every year was the breaking of draught horses destined to drag the municipal garbage carts. The horse-breaker, helped by two men, would bring down his nervous victim and, with his *facón* – a silver-handled dagger – would make incisions around the animal's eyes until, terrified and exhausted, it would stop kicking.

Old Bottini's Offspring

Bottini senior had a daughter and a son with whom we had friendly relations. The young woman was a teacher and lived in the village. My mother asked her to teach me to write. At that time, learning to write started with drawing hundreds of meaningless strokes. I pressed the pencil so hard that the patient and sweet Miss Bottini complained that I ploughed the exercise book. I never acquired a decent hand, except for mathematical symbols, which my classmates found elegant. Learning to type at age 7 on the discarded typewriter of my father's office did not help. Nor did it help that my father's own handwriting was hardly legible, as was *de rigueur* among physicians.

Miss Bottini's bother, Liborio, was tall, thin, placid and amiable, and had no known occupation. In 1945, when General Perón founded the Labor Party, supposedly the unionist wing of his movement, Liborio joined it. This event surprised us all, because he had never worked, hence was not a member of any trade union. Liborio's gait became firmer and his weak voice a tad louder. But his days in the limelight were few, for 2 years later the General ordered the dissolution of his hasty creation, as well as the incarceration of Cipriano Reyes, its main leader.

Reyes had been the strongman of the meat-packers union, which he had wrestled from the Communists at pistol point. A movie was made about the gunfights in the cold rooms of the big meat-packing plants, all of them in American or British hands, which, according to the Communist leadership, were "friendly imperialists", and hence the union should drop its grievances – exactly what the Peronists needed to win the workers over.

Spruille Braden, the American ambassador, encouraged us, the opposition, to what we wrongly judged to be the local version of fascism, while in fact it was an original product that in the end introduced a few good changes, such as the feminine vote, along with many bad ones, such as the union mafias. Even I, a mere foot soldier of the opposition, got a congratulatory letter from the Ambassador, as well as an invitation to join the small group of people who dared gather one winter morning to mourn the passing of President Roosevelt. The General, a brilliant tactician, devised a winning slogan: "¡Braden o Perón!"

Apparently, Cipriano Reyes had political ambitions of his own, which The Leader could not tolerate. Liborio, a loyal friend, visited him every week in his jail. When we met at the train station we discussed the rumor that his erstwhile boss had been subjected to a lobotomy to be tamed. But this rumor, like most of the misinformation that circulates when there is no freedom of the press, turned out to be false. Indeed, Reyes was freed when Perón was ousted in 1955, he agitated against Peronism, and lived to the age of 96 years.

Most Interesting Neighbors

Our dearest neighbors were Roberto and Beatriz Giusti and their five little girls. I was especially close to Leonor, and even closer to her sister Liliana, 1 year younger than me. Leonor, only 1 year my senior but far more mature than I, bossed and organized us all. My father, who doted on her, wrote a poem, "Silver bell", that she recited on the school stage at a national holiday celebration. But in the middle of her performance she got so nervous, that she broke into tears and worse. Leonor and I attended the same school, and my desk was right behind hers. Once, when one her long golden tresses invaded my space, I could not resist the temptation to dip it into my inkwell. I paid for this mischief, which I hope was out of character.

Another notable in our neighborhood was the popular radio journalist Juan José de Soiza Reilly, famous for speaking at a high speed, who solicited help for individuals in need, such as the fat lady who needed a corset. Equally respected was the solicitor Luis Bivort, who kept the minutes of the local municipal association that my father had founded and met monthly at home. Then there was the pharmacist who was said to have gotten rich inventing a cream that he claimed to cure chilblains, then widespread, particularly among washerwomen. Notable as well was the physician who diagnosed my wife's first pregnancy as a rare and serious disease.

The most popular shopkeeper was Mario Ratti, the barber and greatest gossip of the village, who came to trim our hair – mine always in "media Americana" style. Another frequent visitor was the windmill mechanic Vernetti, whom I phoned every time a strong wind disengaged the brake on our mill's wheel. Then there were the Catalonian hardware man Miserachs, taciturn and unsmiling; Nemesio Vea Murguía, the cheerful and helpful Basque grocer who gave me free wooden boxes for my primitive carpentry projects; the coal seller Peirano, who, contrary to popular belief, was placid despite being red-haired; and the milkwoman popularly known as Fivebums, whose only cow miraculously produced milk for the whole neighborhood.

We only avoided the village's two declared fascists. One was the owner of a fake Italian castle, whose sons belonged to the fascist militia. The other was Alfonso

Corti, the professor of Italian literature who had no dealings with any neighbors, and kept his children under strict vigilance – which did not prevent one of them, a charming and handsome young man, from getting entangled with a local a girl, then taking his own life. I was intimate with René, his brother, with whom we had frequent and lively but rational arguments, both orally and by letter, despite our political divergences.

Sports and Games

I spent my time playing, gardening and reading, outdoors whenever possible, as my father had wished. In leaving me free and in abstaining from ever punishing me, my father was deliberately performing an experiment, as he had warned my mother – whose residual Lutheranism did not accept easily the regime prescribed by her husband. I learned much later that my cousins called me "The Experiment" behind my back.

I spent many hours a day on my bike, and played most of the time with my dogs, who followed me wherever I went. I also had my own gang of friends, with whom I played the usual games, such as cops and robbers. Since nobody wanted to impersonate coppers, we forced the weakest to play this role. We also enacted some of the battles we learned about at school, execution by fire squad being our favorite scene. Again, we forced the weaker among us to play the victim. Most grown-ups do not realize the cruelties children can learn from adult texts and films.

But by far our favorite game was soccer, which we played in the empty lot across the street. It was customary to use the Spanish version of the English standard formulas, such as "Auredi?" for "Are you ready?" Those who despise soccer do not realize what an intelligent and elegant game it is, where competition combines with cooperation, and where honorable behavior is encouraged while cheating is penalized. Besides, this game gives one the chance to improvise, which in turn calls for imagination and initiative, as well as nimbleness and courage. For all these reasons, and because it can be played almost anywhere with anything kickable, soccer is the most popular sport in the world, and the one most likely to elicit heated argument and hooliganism. This is why British snobs hold that soccer is a gentlemen's sport played by hooligans, whereas rugby football is a hooligan's sport played by gentlemen.

In addition to healthy games like soccer, biking, and swimming, I had the stamp collecting vice. True, it elicited my interest in foreign countries and their history, but it became an addiction as strong as smoking. Besides, it slowed down my school work to the point that one of my teachers summoned my mother and asked her to take my stamp collection away – something she did not dare do. On reaching puberty I quit this vice and sold my valuable collection to an executive with an American firm in order to pay for the rent of the house occupied by the workers' school I had founded. My last encounter with philately was the reading of Robert Graves's delightful *Antigua, Penny, Puce*.

Gardening

From a very young age, I was put in charge of much of the gardening, in particular weeding and breaking up the ground with a spade. My father made me work hard, but was a fair employer. For example, he paid me 1 cent per cow's tongue, the insidious weed, as long as its entire root was whole. He taught me how to graft rose bushes but, although I smeared the incisions with mastic and wound them with raffia, they never took. Nothing is harder to teach than know-how. I also tried my hand at carpentry, but no one taught me and I had very few tools. Kites were the only things I knew how to build. I used bamboo canes from our garden, paper and string from Ratti's store, and glue I made with flour and water. My kites were not pretty but they resisted strong winds. I gave them to friends and even to our dentist.

I was allowed to shoot sparrows with a BB gun, and noticed that they stopped chirping as soon as they saw my shining gun. But once, when a wounded sparrow was in agony in my hands, I was so saddened and ashamed that I quit hunting once and for all. By contrast, my playmates killed all the wildlife within sight: lizards, snakes, toads, horned frogs, country mice, guinea pigs, and songbirds. My worse crime in this line was stealing a bunch of newborn country mice, putting them in a pant pocket, and forgetting about them till the next day.

Holidays

We spent holidays either in the hills or at an Atlantic beach. I loved the hills because I could climb them and explore the forests, but I disliked the sea resorts because there was nothing to do but sunbathing and swimming in very cold water.

My first trip abroad was to Paraguay, at age 10 with my father. We took a paddle boat that touched all the ports. The shores of the wide Paraná river were covered in alligators, and between Goya and Asunción we were visited by canoes full of tropical flowers, fruits, and animals such as tiny monkeys and talking *charrúas* (large ravens). In Asunción, the capital city, we boarded what was reported to be the oldest train on the continent, which took us to San Bernardino, a resort on Ypacaraí Lake. This train, of the narrowest gauge, rode at man's walking speed, stopped for cows and goats, and was besieged by fruit vendors who offered delicious mandarins attached to small sticks. On the farms we saw men lying on hammocks while their womenfolk were cultivating the orchard or pounding *mandioca*, a staple root. The hotel was surrounded by gigantic orange-trees, and had a vast and quiet reading room illuminated by kerosene lamps covered by green shades. I revisited this place recently, but now the lake is dead from contamination, and the forest has receded from it.

In the summer of 1932 I took my second memorable trip. We traveled in remote Eastern Patagonia, where my father visited physicians on behalf of the life insurance companies, who wanted to know which doctors were best qualified to check on the state of health of their applicants. Our first stop was Puerto Madryn, then a dismal village and now a tourist destination that attracts people wishing to admire whales and sea lions galore. We donned our swimming suits, but the water was too cold to be enjoyed. While on the beach we were visited by the entire cabinet of the former government, who was confined there in internal exile. I only remember well Ricardo Rojas, a respected university professor and the author of a treatise on Argentinian literature that, according to Jorge Luis Borges, was bulkier than the literature in question. The professor was wearing spats and all despite the mild weather, and posed for my box camera along with his colleagues. My father had mixed feelings: he respected them personally, felt sorry for them, and tried to improve their circumstances. But he had fought their government – a mistake he apologized for when the new government turned to be far worse than the one he had helped depose.

Patagonians

We also visited the fertile valleys of the powerful Río Negro and Chubut rivers that originated in the Andes cordillera. I was particular struck by the Welsh towns like Trelew and Gayman, in one of whose stone and slate houses we had tea with scones served by a lady who spoke only Welsh. (Four decades later, the military secretely shot there a group of guerrilla fighters, among them María del Valle Santucho, my second son's estranged wife, whom I had never met.) We also visited the dismal oil town of Comodoro Rivadavia, whose workers were lodged in small houses built next to one another. Next we went to Río Deseado, San Julián and other windswept ports, that exported lamb meat and wool to Europe. Many years later, at another Patagonian place, I befriended Daniel Candel, a Spaniard expat who had narrowly escaped execution by Franco's hordes, that always started by sending the village teachers to the wall. Candel, a pioneer in children's rights, was teaching primary school in Comodoro, and was exasperated by the permanent Patagonian wind. Our boat reeked of lamb meat, which we had at all meals. At the general store in one of those sad towns I bought for 1 dollar a complete Japanese suit in synthetic cloth for our gardener Alex. Alex, a veteran of the Balkan wars, had once chased me with a large kitchen knife, to punish me for some mischief, but we remained friends.

In those towns I also learned about the legendary feats of the Menéndez-Behety businessmen who owned the shipping company and were said top pay one gold coin per Indian ear. I also learned that in 1920 the first democratically elected national government had ordered the army to massacre the striking workers in the Patagonian meat and wool factories. Back in Buenos Aires, I went to the library of the *La Nación* daily to study these events and wrote an essay, but no one was interested. In 1974 some courageous filmmakers released "La Patagonia rebelde", a film based on Osvaldo Bayer's book that documented and dramatized that episode. Two important characters of that tragedy were the Anarchist activist at the head of the workers union and the company manager who doubled as secretary of the Rural Society, the landowners' association.

The second Perón government censored the film, but it was released after winning an important international prize. Coincidentally, the young Lieutenant Juan D. Perón, who two decades later would became the hero of the working class, led the troops that repressed another two strikes in the early 1920s, at the Vasena metallurgical factory in Buenos Aires, and the Forestal timber mill in northern Santa Fe.

In Río Gallegos we visited some prisoners confined in small isolated homes with nothing to do. One of them told us that his crime had been stealing a chicken. From that town we crossed to Chile, where went to Punta Arenas, the southernmost large town in the world. There we visited a chicken farm owned by a Yugoslav who kept the eggs in large wooden barrels. We also admired the *morros*, isolated stone hills that rise suddenly in the plain, and can also be seen in Australia.

On our way back we boarded the mail plane, which took us from Río Gallegos to Bahía Blanca, an important city some 500 km south of Buenos Aires, and the birthplace of the earliest production cooperatives in the country. That aircraft, built with wood, wire and cloth, had room for four passengers; I was the fifth, seated on my father's lap. Shortly before take-off we caught the pilot having his breakfast: a couple of brandy cups. My father, a physician and teetotaler, was duly alarmed, but our man managed to face the air holes and contrary winds, so strong that at times the plane hardly made 100 km per hour.

Cooking

While picknicking I learned to cook and liked doing it. I had a light, compact yet fairly complete aluminum cooking set for two, where I cooked pasta, as well as a portable grill. I used solid alcohol tablets for the former, and for the grill I gathered dry sticks and, if none were to be had, reasonably dry cow dung. At home, my standard menu was: steak with boiled spinach and mashed potatoes, with fried bananas as desert. Once, when my mother was bed-ridden and we had no servant girl, I filled in the kitchen. My father ate my fare twice in a row without complaining, but the third time he told me that his work at Congress would prevent him from coming home for dinner. An economist would say that he was a perfectionist, whereas I was a satisficer.

I made long bicycle trips or travelled by train in the company of a few dogs, as far as Don Torcuato, a village a few kilometers from the nearest river, the Reconquista. Until an American tyre factory dumped its noxious refuse into the river one could swim in it and catch fish, eel, and large bivalves. The lush adjoining land used to teem with large lizards, wild turkeys, birds of all colors, and huge anthills that floated when the river overflowed. The factory in question literally killed that river in a few months. This was my first environmental experience, and the first time I suffered a consequence of free enterprise.

Josef

While day camping on the Reconquista river I got to know a resident, Josef Mache, a young man, the only son of an elderly Austrian couple who had arrived in the country via Labrador, a Canadian region almost completely covered in ice. Josef's father was a retired locomotive engineer, on whose pension the family lived, and who spent his days working, grumbling, and smoking an evil-smelling pipe. The Mache family lived in a wooden cottage on stakes, and they fished and hunted what they could, mainly eels. Josef had studied at a Viennese technical school, drew meticulous technical diagrams, and painted from memory old picturesque Viennese houses. But he lived so far from town, unemployment was so high, and his ill-tempered father had so intimidated him that he never looked for a job.

After the Reconquista died, I stopped visiting Josef. But one Sunday in 1947 he came by surprise to *El Ombú* while I was chatting with some friends. The talk of the day were the UFOs that, according to reliable sources – such as Jimmy Carter – had been seen in the sky over Roswell, New Mexico. When we asked about them, Josef replied in a matter of fact tone that the report in question was false. He knew, because he had built the only flying saucer there was, which he kept in the courtyard of his house. My friends and I exchanged smiles, and I changed subjects, asking Josef what he had been doing since we had last met. He told us that he had spent some time in the nuthouse, where he had been repeatedly subjected to electroshocks.

Learning at Home

My parents found no time to play with me, except the few times I persuaded my mother to kick a ball so I could practice my goalkeeping skills, or when my father took me for a swim. He and I kissed every time we met or took leave of one another, and while travelling we exchanged long letters, most of which were read only by the political police.

My mother never kissed anyone: ever the nurse, she had grown in the shadow of Pasteur and Koch, regarded kissing as "monkey love," and admired the Chinese for rubbing noses instead of kissing. Of course, she loved me and took good care of me, but never caught me smoking the Egyptian cigarettes that I bought in bulk at 1 cent each, because she never came near my mouth. Yet I did love her sometimes, particularly when she tucked me in bed and I asked her to sing with her sweet voice some popular German songs for me. My favorite was Goethe's about the wild rose in the prairie, which starts thus: *Sah ein Knab ein Röslein stehn,/Röslein auf der Heiden,/War so jung und morgenschön,/Lief er schnell, es nah zu sehn,/Sah's mit vielen Freuden.*

Neither of my parents pampered me. When my father sat me on his lap, it was to talk to me about biology. But he never found the courage to teach me the so-called facts of life. Instead, he talked to me about bees and flowers – a subject that did not interest me as much as the copulation of our pets. The only time my father spanked me was when, perhaps at 4, I climbed a ladder in an effort to grab the reddish full moon that seemed so near. He never even shouted at me.

By contrast, my mother got very irritated when I failed to memorize a multiplication table. As a German, a Lutheran, a stationmaster's daughter, and a nurse, Mariechen did not tolerate slack. She was so punctual that she arrived at the train station at least 15 min before the time indicated in the timetable – which before Perón's time was strictly respected, as the railways had been organized on the British plan. Punctuality, cleanliness, and keeping one's word were perhaps the only virtues my mother succeeded in teaching me. My father taught me gardening, table manners, Argentinian Spanish, and French pronunciation. More importantly, he was my social role model: I wanted to become an honest, concerned, and useful citizen, ready to pay for my beliefs.

First School

My first school served the British colony, which at that time was quite sizable and resided in pretty ghettos of its own. Most of its members had failed to make a living in Britain, and were employed at the local branches of British firms. They earned more than their Argentine subordinates even when they were less competent. They lived in Argentina on sufferance, and kept to themselves like their compatriots in the colonies.

I had only three British acquaintances. One of them, whom I met through my sister-in-law Alicia, worked at the Embassy and generously corrected my earliest English essays. The second, employed at the central post office, returned to England the moment he inherited an earldom with land and cottagers. The third, a Welsh engineer, taught pro bono at the workers' school I had founded. But neither my parents nor I met any Brits socially, and I never heard of a British professor working in an Argentine university, whereas there had been some influential French, German, Italian, and Spanish academics, such as the journalist and historian Paul Groussac, the physicist Richard Gans, the neuroscientist Cristofredo Jacob, and the mathematician Julio Rey Pastor.

The British school in our neighborhood, West Florida, occupied a vast piece of land, with a eucalyptus forest and a swamp. It had a single classroom and two teachers: a tall, stern, and sad-looking man, and a short, red-haired and jolly man who at mid-morning left for the nearest wine-shop followed by a bevy of unleashed dogs, and returned half an hour later on unsteady legs. The two teachers were in charge of about 50 well-behaved pupils who did not mix with the natives. Instruction was imparted in English, a language utterly alien to me. Once in a while, the sober teacher approached my desk, to assign me some homework that I did not understand. The whole thing was a boring farce as far as I was concerned. But at last I was saved from the grim song "Ring around the rosie," dating from the Black Plague, which two decades later my first two sons were forced to sing at Olivos School.

Many of my playmates attended the Catholic school nearby. They did not learn much, but liked their teachers to play soccer with them. Discipline was strict and sometimes cruel. We knew when one of them had misbehaved, because his knees exhibited the marks left by the corn grains he had to kneel on while locked up in a dark room. Incidentally, there were no churches in our village, and the only priest in it went to work in the city.

City Schools

There was no good school in the village, so the more ambitious parents sent their children to schools downtown. We traveled half an hour by train. On the way to the station I passed by the municipal shed that housed the noisy horse-drawn garbage carriages. During elections and national festivities, that locale doubled as headquarters of the Partido Demócrata Nacional, the fraudulent conservative party, which offered its sympathizers free roasted meat, *empanadas* (meat pies), and country music played on accordions – all paid for by the taxpayers. By mid-afternoon some of us went to watch drunken rows, often fought with daggers with beautiful handles wielded by men clad as *gauchos*, with wide belts studded with old silver coins. Such was our introduction to civics.

From Retiro, the train terminal, we walked a few blocks to the nearest state school. Josefa, a very nice chubby girl a bit older than me, held my hand crossing the wide avenue, and sometimes helped me do sums. Once she tried to count my ribs, to prove the story that God had made Eve out of one of Adam's ribs. I was respectful, but did not study much. I preferred playing with the members of my gang, gardening, reading adventure books, and collecting stamps, to relearning the names of Colón's caravels, the sailor who first saw land, and of the two patriots who distributed cockades with the brand-new national colors – sky-blue and white – among the people who had congregated in front of the Buenos Aires city hall to celebrate the birth of our nation in 1810.

First Encounter with Religion

Children were well treated in the state schools, which were compulsory, free, and secular. Most of our teachers were women. Most of them were competent, dedicated, patient, and decently paid, but were not allowed to marry or cohabit, smoke, use lipstick, or paint their fingernails. Their monthly salary of 250 pesos, or 60 dollars, allowed many of them to wear fur coats bought in installments, and even to keep a secret companion. Besides, they taught only one shift, which allowed some of them to enroll in university courses, and others to join voluntary organizations.

Last, but not least, in every neighborhood the teachers, along with the doctor and the pharmacist, were among the leading citizens.

I was punished only twice during my 7 years in elementary school. The first time was when the teacher hit me with the pointer for talking to a classmate. The second time was when the teacher asked the class whether any of us had not been baptized, and I proudly raised my hand. The teacher sent me behind the freestanding blackboard, the place reserved for dunces and reprobates. In both cases, my father went to complain to the school principal. This was a large and sweet woman whom we liked because she gave us a candy every time she reprimanded us. At my father's behest she scolded the guilty teachers, because both corporal punishment and religion had been banished from the state schools four decades earlier. All my contemporaries remember the number – 1420 – of the law in question, because it has been a bone of contention between the Catholic Church and the liberals since 1884.

I happily transferred to the Escuela Argentina Modelo, which at that time was progressive – though not as much as the American school, which taught carpentry and had a sports field. I attended the last 2 years of elementary school guided and stimulated by excellent teachers like Señor Molteni, whom we all liked and respected in spite of his shorter left arm. I flourished at that school, where I was put in charge of the classroom library, was elected senator of our miniature parliament, made some friends, and earned some medals. I looked forward to doing even better in high school. How utterly mistaken I turned out to be!

Chapter 2
Adolescence

High School

It has been said that childhood is shorter in the underdeveloped countries than in the affluent ones, because in the former most children start earlier to fend for themselves: they have no pocket money, and their parents are too busy to spend time on them. They are on their own at around age 12 years. My own childhood ended at the beginning of the autumn of 1932, when I moved from my friendly and protective elementary school to the high school attached to the University of Buenos Aires. This was the old and prestigious Colegio Nacional de Buenos Aires, which occupied a huge building in the French academic style, characterized by fake stonework. It was just one block from the Plaza de Mayo, the main civic square of the republic, born there in 1810.

From the start I viewed my new school more as a correctional facility than as a learning center, because of its emphasis on discipline and tests scores. This regime was enforced by about ten professors, one for each subject, mostly stern and elderly men, as well as a by a whole corps of prefects, sub-prefects, wardens, and monitors who spied and reported on us as if we were potential delinquents. That strict regime did not prevent the cheeky student who sat at the first desk in the front row from masturbating every morning year after year.

Our teachers and guardians instilled in us more fear than respect, and obviously did not realize that we were children going through the most difficult phase of life. Very few of our teachers tried to make learning a pleasant experience: most of them saw their task as that of taming us, rather than stimulating our curiosity and creativity. For example, the flawless recitation of a mediocre poem, or of a page of a dull textbook, was rewarded, whereas questioning, or even asking teachers for clarification, was out of the question. That was standard procedure the world over, except in the U.S.A. and in a few experimental schools in Germany and the U.S.S.R. where memorizing was not compulsory while dialogue was encouraged.

M. Bunge, *Between Two Worlds*, Springer Biographies,
DOI 10.1007/978-3-319-29251-9_2

Near Expulsion

Used as I was to the freedom I had enjoyed at home, as well as to the gentleness of my elementary teachers, I rebelled against the jail-like regime of the high school. For instance, when the calligraphy professor started his course by asserting that his was the most important of the subjects we were to study, I laughed up my sleeve and drew a cartoon that portrayed him as a dressed-up chimpanzee. I placed this cartoon on the front page of my *Magazine against the Professors*. This title might have pleased Sextus Empiricus, but he was no longer around to defend me.

My magazine had an ephemeral existence, for the teacher confiscated it promptly and demanded my immediate expulsion from school. The chief inquisitor, Señor Amoroso [sic], and his minions, agreed that this troublemaker had to be expelled. The principal reduced the penalty to a 14-day suspension, that is, exclusion from classes. This was serious, because a single missing day would have sufficed to expel me without appeal.

Even I was surprised at my irreverence, because I had behaved well in my elementary schools. But I was put off by the solemnity and self-importance of most of my professors. I was also repelled by the disciplinary regime, which I regarded as medieval – as I told the principal when he asked me why I felt unhappy at school. Dr Juan Nielsen – a tall, thin, and dry individual – seemed amused by my insolent reply because he smiled at it – perhaps for the first and last time in that school term.

Given that schools run by mediocre martinets and that measure success only by test scores only breed able yes-men (Golan 2015), who could have expected that only a few decades after my time mine would become co-educational, offer lessons in tango dancing, or that its student body would be mobilized by left-wing militants four decades later? Perhaps deviance is the norm in a culture of amateurs.

Our Teachers

I was a mediocre student because I was neither motivated nor fond of most of my teachers. The exception was Osmán Moyano, our French literature teacher. He was a competent, dedicated, and sensitive man who stimulated and respected us. Having heard about my socialist leanings, he suggested that I read the iconoclastic Villers de l'Isle Adam. As a prize he gave me a copy of the *Petit Larousse Illustré*, which I have used until recently. I remembered Mr Moyano affectionately every time I enjoyed reading one of my favorite French classics: Molière, Diderot, Balzac, Maupassant, Anatole France, Romain Rolland, Marguerite Yourcenar, and Le Clézio.

Juan Batana, the engineer who taught us physics, was competent and dedicated, but his lessons were flat: he had not heard that, as Einstein and Infeld put it, physics is "an adventure in thought." By contrast, his colleague Enrique Butty – regarded as a great scientist because he had claimed to having proved Einstein wrong – was lively and friendly. However, he only came when his many jobs allowed it: he was director of National Public Sanitation, professor of elasticity theory, and dean of the Faculty of Engineering. Fortunately, Jorge Cordero Funes, a talented and generous classmate, tutored some of us once a week at his home.

Most of our professors were not interested in teaching, and some of them were frankly incompetent or even grotesque in their behavior. For example the aesthetics professor, who wrote *sainetes* (vulgar comedies), was always in a bad mood and decorated the floor with huge gobs of phlegm. He claimed to teach us philosophy of art just because he repeated over and over the formula "The sublime, the beautiful, the pretty, and the ugly." He never told us that beauty might be in the eye of the beholder, let alone that aesthetics does not contain any universal principles.

The art professor, an uneducated man, made us copy plastic casts, such as one of Voltaire's famous head, and his own blueprint of a miserable dwelling. His highest praise was "Perfet, no defet." The botany professor taught us only some systematics and nothing about plant physiology, let alone evolution, but warned us several times that the decorative plant that florists called *asparagus* is no such thing, but *Asparagus sprengeri*. The chemistry professor did not explain the synthesis of water out of hydrogen and oxygen, but told us repeatedly that water is "the universal solvent". The lugubrious Señor López was expected to teach the elements of Spanish grammar, but he only succeeded in making us yawn. And the mathematics professor, an engineer who doubled as a Conservative alderman, was stern and demanding but incapable of motivating the learning of algorithms, whose point escaped us. He and his subject terrified me.

Other Strange Teachers

The physical education professor came to the sports field clad in fedora, jacket, vest, and spats. His paunch prevented him from bending, which is why he asked me to go to the front of the class and demonstrate some of the calisthenics exercises that my private gymnastics instructor had taught me. This was Edmundo Blum, a competent, educated, and charming German expat, strong and nimble, but not a muscle builder.

Señor Blum and his family were good friends of ours. His wife Elsie made it clear that, unlike her husband, she was not Jewish. I was close to their two sons, Edgardo and Gerardo. The latter ended up as a colonel in the American air force, whereas Edgardo managed to graduate from medical school. But he was a passionate philatelist, so he neglected everything else, and died young from blood infection. In geography we had to memorize the names of mountains, lakes and rivers – even dry ones, as I discovered when traveling. We got neither geology nor human geography: our professor, the editor of an important daily who got the job because he had traveled as far as Moscow, taught us about territories, not nations. This was in tune with the idealist split between the natural and the "cultural" (social) sciences.

In history we had to memorize the histories of the Spanish, French and British dynasties. Only political and military history seemed to matter. Thus, we were told that the Magna Carta had been the earliest European democratic document, rather than a constitution that enshrined the subordination of the king to the landed aristocracy. We were never told about the war crimes committed from the siege of Jericho on. Nor were we told how many people gathered in the assembly that had proclaimed the independence of our nation: was it 100, 1,000, 10,000?

In Latin we were required to read Cicero's impassioned Catiline Orations. We were not told why Cicero had delivered them: we had to accept that Senator Catiline was wicked, and when one of us misbehaved, the teacher quoted to us Cicero's famous imprecation *Quo usque tandem abutere, Catilina, patientia nostra*? How much more we would have enjoyed Lucretius' immortal poem *De rerum natura*, Marcus Aurelius's *Meditations*, or even Plautus's popular comedies! The lively teacher of history of the Americas told us about the mythical Prester John, the search for whose kingdom may have motivated some of the explorers, but did not tell us how many Amerindians each Spanish "settler" exploited. Nor did he tell us how many Europeans came to "populate" the Americas during the century following "discovery." (According to Fernand Braudel, there were only about 1,000 men per year.) And it was only much later that I learned that the Spaniards sent more executioners than teachers to the New World, and that the monarchs who had financed Colón's expeditions spent all the gold and silver exacted from the Indians on the mercenary armies they employed to loot what are now Belgium and the Netherlands.

In short, the history we were taught was "dry as dust" and superficial, even though the French school of the *Annales*, of "histoire totale," had started more than a decade earlier. To this day, the vast majority of Argentine historians are either chroniclers or biographers. This may explain the success of *Todo es historia* [*Everything is history*], the semi-popular monthly sold in kiosks since 1966.

Classmates

We were a mixed crowd. The first in every subject, Oscar Andrieu, had neither preferences nor friends, and never expressed any emotions – except for the time he shed copious tears for having got an 8 instead of his customary 10. He ended up as a translator of classical Greek. Héctor Genoud, the second in rank, was good-natured and an excellent classmate, always ready to lend a hand. He urged me to join his club, where we played water polo, and I injured my back in overrating my ability at high trampoline jumping. I also made friends with Martín Noël, a member of the wealthy family who owned the factory that manufactured popular evil-tasting chocolate. He was unpretentious and irreverent like me, and became an expert in Argentine literature. I was also close to Jorge Pena, the son of a Socialist alderman, and the only one of us who had heard the admirable slogan of the 1789 French Revolution: *Liberté, égalité, fraternité*. I lost touch with all of them when I was forced to leave that school. But I had friends in other circles, and soon formed my own gang. Besides, I shared most of my parents's friends.

At 17 years I took up rowing, sitting first on a small carriage moving on rails, and later in light cigar-shaped boats that rowed much faster. My favorite excursion was to one of the many islands on the widest branch of the powerful Paraná river. There I would explore the thicket, and would be bitten by mosquitoes with white legs, as befit skillful surgeons who start by injecting an anesthetic. I would also read novels, poetry, history of science, physics, and philosophy under the curious eyes of wild furry otters. Sometimes I would take along new friends interested in subjects quite distant from my favorites. Once, when fast asleep under a full moon, my friend and I felt wet. The river had risen 1 m in a few minutes. We gathered our belongings, jumped into the boat that we had tethered to a tree and, half asleep, drifted back to my club. Our breakfast consisted of tinned condensed milk and plums bought earlier from an itinerant fruit seller. To secure a permanent rowing companion, I submitted my membership application, along with that of my close friend David Jacovkis, to a less fancy rowing club. We were both promptly blackballed. Jews belonged in a rowing club of their own – which, of course, was just as selective and exclusive as the others.

By contrast, the YMCA was open and secular – and welcoming of Jews. My Argentinian sons and I have fond memories of it, particularly of our summer camps in its hilly and wooded Sierra de la Ventana estate, which had a river as well as clouds of voracious gnats. The camp leaders, such as Señor Vignes, organized our activities and assigned us responsibilities but otherwise left us alone. Thus, the Y was a leadership school in addition to a sports club.

Many years later, when I married Marta, who had a safe Italian family name, we were admitted into the *Amérîca* rowing club. In good weather, but in all seasons, we rowed to a quiet island, where I would help her with mathematics, and we would eat a box full of delicious sandwiches prepared by my mother. She doted on Marta, knit beautiful sweaters and gloves for her, and cooked us delicious dinners every Friday night.

Girls

Back to my early youth. The Bivort, Corti, and Miserachs families had very beautiful daughters who could be admired only from afar. The same held for the delicious Anglo-Argentinian schoolgirl who travelled standing, exhibiting her white blouse, in the train carriage that took us downtown. By contrast, I exchanged visits with Mary, the most attractive and daring of the girls in our neighborhood, who tried in vain to teach me to dance. Her father was in jail ostensively for embezzlement, but actually for having been a member of the party unseated by the 1930 military coup. Mary's mother, who welcomed me, worked at the post office where I deposited my savings, which I made by running to school instead of taking the streetcar or the bus. Mary was eventually seduced by a friend of mine, with whom I exchanged fisticuffs over her.

I also had a secret love affair with another pretty girl, with whom I played squash in the street. I jilted her the moment I saw Teba Bronstein, my first serious, albeit unrequited, love. She was charming and very interested in music, literature, and politics. Besides, she treated me, 4 years her junior and still in high school, as if I were grown-up. She was both moved and amused by my long written list of reasons for loving her. We attended together many concerts, lectures, and parties – serious affairs where young people conversed and made music, but abstained from alcohol and hanky-panky. Teba's senior admirer, an advanced medical student, must have taken me as a serious rival, for he once invited me to dine, to discuss our respective territories. We agreed to let Teba choose. I was besotted with Teba for a couple of years, until I made very different friends. The other great loves of my life were Julia, with whom I spent ten good years, and Marta, with whom I have been married for 56 years. I may be called a *serial monogamist.*

My Mother's Most Interesting Friends

My mother was serious and reliable, usually tense, often worried, and sometimes depressed. She relaxed, and occasionally even laughed, only in the company of her dearest friends, most of whom were old colleagues or former patients. I loved them too, and enjoyed their company, which was always brightened by cakes made by my mother or bought at the famous Steinhauser bakery in town. Once Tití, our affectionate monkey, feasted on cakes, and quenched his thirst putting his head in the milk jug but needed my help to get out of it.

My favorite among Mariechen's friends was sweet Ellie, who lived with Herr Lüpnitz, a typesetter and feuilletonist with the democratic German daily. When I asked him what would happen in the next episode, he amazed me by saying that he did not know it beforehand: he would learn it as he wrote. Did the same happen with his more eminent colleagues, Balzac, Dickens, and Dumas *père*?

Paula, another close friend, was a very competent and serious nurse with advanced ideas. Once she laughed at me because I was wearing a patriotic cockade. I was shocked to hear her say that the flag was just a piece of cloth. But a few years later, when the Nazis rose to power, Paula dropped her internationalism, and we never saw her again. Finally, there was Marta Bereny, a pretty and affectionate Hungarian married to an architect who drove her about in his car – an event that invariably caused a stir in the neighborhood.

My mother had a dramatic streak and often felt very sad. I phoned all three friends every time she suffered a severe depression attack, but they always told me that it would soon pass. I tried to console her, but her hospital etiquette forbade her to kiss, hug, or hold hands, so I confined myself to soothing her and leading her to sleep. I blamed my father for neglecting her, but poking at ashes won't start a fire.

El Ombú's *Habitués*

Every Sunday we were visited by people who wished to talk to my father, from neighbors asking for advice to members of his party, journalists, political exiles, or just curious individuals. My favorite was the young and charming economist Raúl Prebisch, who accepted to be my lay godfather, and was later nominated for a Nobel prize. Other close friends were the amiable Roberto Giusti, a prominent literary critic, with whose daughters I had played, and who sat comfortably on his favorite armchair, caressing his Chianti bottle. Every time my mother served fish, Alfredo Bianchi, Giusti's partner in the direction of the influential *Nosotros* magazine, asked her whether it was fresh. He was also known to shed a *furtiva lagrima* when contemplating a beautiful sunset. The poet Juan Burghi, aka Juancho, made a living buying and selling scrap iron, and was duly impressed when my father read a poem he had just written to show him what non-trivial poetry looked like. The prestigious and fearless journalist José P. Barreiro married a girl of the wealthy Cantoni family of wine growers and politicians. Less frequent but equally welcome friends were the fat and friendly Uruguayan Socialist leader Emilio Frugoni, who, even after 10 years, had not forgotten the day the communists took over the newspaper he had founded; and the dermatologist Adolfo Muschietti, whose doctoral dissertation, on prostitution and poverty, had been supervised by my father. He always presented my mother with a box of her favorite bonbons, which she consumed much faster than I.

All these friends treated my mother and me affectionately, and the three littérateurs, as well as my father, patiently read my literary productions, which they subjected to constructive criticisms. And they made fun of my father's tendency to take charge of everything. Juancho said that Augusto would even instruct and correct his own funeral director's operations.

In addition to the above-mentioned friends, there was always some other occasional visitor, like Luis Reissig, expert in Anatole France, now unjustly forgotten. In 1930 Reissig organized the Colegio Libre de Estudios Superiores, the Argentine version of the Collège de France, founded four centuries earlier with the same goal: to study and teach new ideas that the universities ignored or neglected. The idea of the CLES had been Aníbal Ponce's. Ponce, who made a living as a clinical psychologist, was a high-level amateur who succeeded José Ingenieros as the head of his *Revista de Filosofía*, a non-partisan intellectual review. Ponce surprised everyone when he gave a learned lecture in praise of the *Communist Manifesto* of Marx and Engels. He later founded *Dialéctica*, a short-lived review of Marxist scholarship that reprinted Marx's deplorable encyclopedia potboiler on Simón Bolívar, aka El Libertador, whom he treated with Eurocentric arrogance. *Dialéctica* did not publish anything original. Marxism had long ceased to inspire original research.

I was a frequent visitor at the Colegio, where I followed the course on dialectical materialism taught by the distinguished clinician Emilio Troise, as well as that of Rodolfo Mondolfo – just exiled from Italy – on Renaissance philosophy, and a few others. And in the *Revista del CLES* I read Juan Valencia's lectures on genetics, as well as Vicente Fatone's about Indian philosophers. In the same journal my father published in 1933 his well-documented study on Argentine oil and the international oil trusts. The universities would touch neither of these subjects. The Colegio's lectures on genetics had such an impact that my friend Goyo Aráoz, a lawyer, spoke of nothing else.

Once we were visited at *El Ombú* by the professional excentric Omar Viñole, a veterinarian and prolific writer who expressed his contempt for the contemporary Argentine politics by writing incomprehensible books and taking his cow on a leash for a walk in downtown Buenos Aires, which earned him the nickname The Cow's Man. Another occasional visitor was Elsa Jerusalem Widakovich, Einstein's wealthy hostess when he came in 1925 to give a few lectures. She had become famous for her novel *The Golden Scarab*, about a Berlin madam. It was said that this book was autobiographical, which I could not believe because she seemed to me the ugliest woman I had ever encountered. In any event, Einstein expressed his disgust and contempt for the wealthy and otiose ladies and gentlemen who fêted him in Buenos Aires.

One of the strangest unexpected visitors was Otto Strasser, the leader of the Black Front, the left wing of the Nazi movement, murdered in the 1934 purge. (Incredible as it may seem, Hitler belonged to the centrist wing of his movement, whereas Himmler led the right wing.) My father did not waste time on him.

Another surprising visit was that of the wealthy rancher and Conservative boss Antonio Santamarina and his right-hand man Alberto Barceló, the owner of the brothels, gambling dens, and votes in Avellaneda, the industrial town next to Buenos Aires. They had come to offer a parliamentary seat to my father, whose congressional period was about to end. Of course, he declined the offer, but did not foresee Barceló's revenge – a police visit to our home. (It was said that once, referring to a son-in-law he disapproved of, Barceló remarked to his henchmen: "Look at that young man. He looks so wasted, that I fear for his life." The next morning the man in question was discovered mysteriously stabbed to death.)

The Sunday Lunches

Nearly every Sunday we had about a dozen guests for lunch. (Never 13: that was my father's only remaing superstition.) Our guests were strictly divided into political associates and close friends. The politicians came only once a summer to eat a delicious lamb roasted slowly in a fire lit in the early morning by an expert garbed in a splendid gaucho outfit. These lunches, for which I prepared a fruit punch that I did not drink, took place in the cool shade of the *ombú*. These guests, all of them parliamentarians or cabinet ministers, were only interested in the political issues and intrigues of the day. My father, basically an intellectual, was a bit out of his element in those gatherings, though always the gracious and highly respected host.

I was much more interested in the regular Sunday lunches with my parents's closest friends. During those affairs many subjects were discussed, particularly books, such as Leonhard Frank's moving antiwar novel *Man Is Good*, which my father had translated, and Emil Ludwig's thick yet popular *Napoleon*. Other often-discussed authors were Anatole France (ingenious), Sinclair Lewis (realistic), Marcel Proust (boring), Romain Rolland (courageous), George Bernard Shaw (socialist), and Stefan Zweig (tragic). D. H. Lawrence was discussed only when his *Lady Chatterley's Lover* (erotic) was banned. I don't recall anyone in that circle mentioning Thomas Mann, Luigi Pirandello, Eça de Queirós, or even Leo Tolstoy. The Spanish and Latin-American writers were ignored or bashed, except Rubén Darío, a great innovator for some, but kitschy in my view. And everyone would have been shocked if told that Roger Martin du Gard (such a bore!) and François Mauriac (so frivolous!) would eventually be awarded Nobel Prizes, or that Jorge Luis Borges (who?) was repeatedly nominated for it.

A book that provoked spirited discussions was Julien Benda's *La trahison des clercs*, (1927), the earliest and harshest denunciation of the prominent French and German intellectuals – among them the physicist Max Planck, the philosopher Henri Bergson, and the sociologist Max Weber – who had supported their respective governments during the Great War. Benda rightly accused them of betraying the intellectual's commitment to truth. Nobody foresaw the silence of the vast majority of the German (and Argentine!) men of culture when confronted with the coming fascisms.

But of course the most popular cultural novelty of the time was the film. The children filled the local theatre, popularly known as "The fleabag", where we cheered loudly at the deeds of the cowboy Tom Mix and the German shepherd Rin Tin Tin, as well as at the doings of Charlie Chaplin, Buster Keaton, and Harold Lloyd. In my father's circle, *Intolerance*, *Doctor Caligaris's Cabinet*, and the films of the German expressionists, such as Fritz Lang's *Metropolis*, were often discussed.

I had a recurrent nightmare after watching the scene in *Metropolis* where a worker is crushed by a gigantic gearwheel. And I watched several times *Emil and the detectives*, with the evil-looking Peter Lorre. This popular film was based on a novel by Erich Kästner, who had dared call his fatherland "The land where cannon blossom" – a parody of Goethe's name for Italy, "The land where lemon blossom."

We were also occasionally visited by foreign politicians or political exiles, such as the Peruvian Víctor Haya de la Torre, founder of the anti-imperialist movement APRA. I befriended the Mexican journalist Fernando Robles, a friendly, lonely and pious ex *cristero*, or soldier in the bloody and long war that the Catholic Church had waged against the constitutional government (1926–1929). He was shocked when I said that the Mexican landowners, starting with the biggest of all, the Church, treated the peasants like feudal serfs. He saw feudalism as a sequence of colorful tournaments among chivalrous knights. José Vasconcelos, who had been the education minister of the government born from the 1910 popular revolution, came once. I took him for a long walk through the garden and, to my surprise, he expounded on the Nazi doctrine that the telluric, the soil, dominated the life of the peoples. Years later I learnt that such irrationalism was part of the reaction against the oppressive regime of Porfirio Díaz, which called itself 'scientific' just because it had adopted the political philosophy of the later Auguste Comte – a parody of scientism.

I was a frequent participant in those discussions. My father never held me back, unlike his own, who had forbade his eight children from speaking at the table. Once, the young Augusto was severely reprimanded for blurting out "Watch it, Dad!" When, at the end of the meal, his father asked him what he had wanted to say, he replied: "There was a worm on the salad leaf that you were about to eat."

Those entertaining and instructive Sunday lunches, my week's dessert, ceased abruptly in mid-1936, when all of our guests lost their appetite and their thirst overnight. Let us see why.

The Spanish "Civil" War

On July 18th, 1936, the Spanish general Francisco Franco and his staff, who had sworn loyalty to the democratically elected Republican government, revolted, starting a horrendous "civil" war that lasted nearly 3 years. The German and Italian governments participated decisively in this conflict, as did the antifascist International Brigades composed of foreign volunteers. Among them were the 2,800 Americans who composed the Lincoln Brigade – of whom President Reagan later said that they had been "on the wrong side." The war cost half a million dead, an unknown number of "red" prisoners executed by the "nationals" after their victory, and about a million exiles. That war divided not just Spain but also the rest of the West into two sides. Nearly all the intellectuals and artists supported the Republican cause. Franco's only outstanding sympathizers were the excentric Salvador Dalí, one of my favorites painters, and the poets Ezra Pound and T.S. Eliot.

Most of the prestigious Argentine writers, starting with Borges and his circle, kept a prudent silence. The few intellectuals and artists who dared pronounce themselves in favor of the Republicans formed the association AIAPE (Asociación de Artistas, Intelectuales, Periodistas y Escritores), which met at a large basement near the Congress palace, which I visited often.

Franco's uprising emboldened the Argentinian fascists, in particular Manuel A. Fresco, the fascist and clerical governor of the Buenos Aires province, where we resided. The evening of the very day that Franco started his war, a squad of provincial police entered our home and arrested my father, me, and more than a hundred guests, who had come to raise funds to assist jailed political and trade-union activists. The gathering had been peaceful and non-political, without speeches and with nothing to eat or drink. My father, oblivious to the noise, kept working at his desk. His guests, mostly unknown to him, did nothing but engage in conversation and visit the garden. But it is true that about half of them were guilty of having Jewish names. The people under arrest were taken by bus to the jail in La Plata, the provincial capital, about 100 km away. On arrival we were offered wooden benches and a nauseating stew, which I did not touch. All, except me, were set free 24 h later.

Although I was a minor, I had insisted on staying with my father. The next morning, Uncle Jorge came for me and took me to the train station, where he treated me to the most delicious turkey sandwiches I ever tasted. When we arrived in Buenos Aires, he took me to the apartment of my aunt Delfina and her husband Manuel Gálvez, the famous novelist – neither of whom came to greet me. Uncle Manolo, who in his youth had been a leftist and a dear friend of my father's, had become a clerical and a fascist sympathizer. Like a plague victim held in quarantine, I was confined for several hours to a bare room without reading materials. Eventually Uncle Jorge took me to his architect's office, where I stayed alone and worrying about my parents. At sundown Jorge took me to the train station where my father and his jail companions finally arrived, looking triumphant because the authorities had found no incriminating documents on them. *Crítica*, the popular evening paper with circulation of a million, carried a picture of the caravan on its front-page.

We were welcomed by my mother, our cheerful and motherly housekeeper Kathy, and her sullen husband Jakob. Our beds were smeared with the mud left by the policemen who had slept on them with their boots on, after drinking all the fine liqueur they found. Several shelves in my father's library were emptied: the police took away the Baedeker travel books and Edgar Wallace's thrillers because they were bound in red, and this is what they had been ordered to do: to find and take away all the red publications. But they also took away my beautiful Corona typewriter, my alarm clock, and my valuable coin collection, which included some ancient and medieval ítems, as well as a doubloon that an English corsair had stolen from a Spanish caravel, and where the British mint had overprinted with English words. They also confiscated all the letters I had received, and separated those written in Esperanto, on which they pinned a piece of paper saying "Cypher?"

The Friends' Reaction

Our old friends kept silent and never returned to our table. They did not even phone. This disloyalty crushed my father. But it should not have surprised him, because the provincial government headed by Manuel A. Fresco had intimidated nearly everyone. Fresco, who in his office exhibited signed photographs of Hitler and Mussolini, had invented the *voto cantado*, or oral ballot, and put into practice the educational reform designed by the fascist and Thomist philosopher Jordán B. (né Giordano Bruno) Genta, which included a reduction in the duration of primary school. Fresco's police also kept the tradition of torturing with an electric prod initiated in 1930 by Matías Sánchez Sorondo, the minister of the interior of the continent's first fascist government, headed by General José F. Uriburu. That minister too kept photos signed by Hitler and Mussolini, as Ramón Columba (1947), the senate's chief stenographer, attested in his memoirs. Thus, this "enhanced interrogation" technique, as it is called in the USA, came straight from the slaughterhouse.

The vanished friends were soon replaced by the faithful friends of my father's youth: Vivaldi, Bach, Haydn, Mozart, Beethoven, Schubert, and Brahms came uninvited with greater frequency, and gradually new friends joined us to our delight: Bruckner, Chausson, Delius, Fauré, Mahler, Milhaud, Prokofiev, Sibelius, and a few others. We did not care for jazz, which seemed frivolous to us, but I liked some of the old tangos. With my father we played at solving such cases as: Is it old Haydn's or young Mozart's? Brahms's or Beethoven's? These games diverted us from the political news, which was seldom good during that period of the world-wide ascent of fascism.

My father was a Stoic, and kept a copy of Marcus Aurelius's *Meditations* on his bedside table. He was uninterested in gossip, and never discussed personal problems, whether his own or other people's. Every time I started to criticize one of our friends he cut me short, saying "One must take people as they come." But he was always ready to help friends in need, and complained that our close friend Duncan Haymes never told us anything about his own travails and troubles in his regular letters from wartime Washington.

One night, when my father seemed to be in a particularly somber mood, and we were about to play records of his favorites, Beethoven's last quartets, I summoned my courage and asked him what his greatest ambition was. He replied that his aim was to attain peace of mind. I was puzzled but did not dare pressing the matter any further.

I never found out what was troubling him at the time, but suspect that it was to do with the fruits of his 20-year long parliamentary career. His learning and integrity had earned him the respect of his colleagues in all parties, but no one's friendship. And his pet project, of national health insurance – far more advanced than the one Senator Ted Kennedy proposed half a century later – lied buried in the Record of the National Deputies' Congress.

I regarded *Papi* as my very best friend despite our frequent tiffs over trivial problems. His clamming up about his own troubles was due not only to the Bunge stiff upper lip, but perhaps also because he did not wish to unload his burdens on my young shoulders. This must have helped me try and solve my various problems on my own. Thank you for being my role model, *Papi*, and also for helping me develop a similar harsh outer shell, which proved handy in that environment, which was anything but nurturing.

High School

However, back to my high school studies. I had arrived at 1936 in jolts. At the end of the year my school let me "free" because of the low grades that I had gotten in most subjects. I did not feel offended or sad, because I realized that I had been just as bad as the school. Love, politics, and my literary essays had distracted me excessively. The only F I got was in Spanish Literature for failing to memorize Becquer's rhymes angered me, for I thought that my numerous poems, short stories and essays were worth far more than memorizing the verses of a third rank poet, now deservedly forgotten. (I learned only the other day that Borges called "imperceptible" that professor.) My parents did not reprimand me, but trusted that I would find a way to graduate from high school on my own, which I did in 1937, as will be seen later.

One of my closest friends was René Corti, son of the professor of Italian literature mentioned earlier. René was a fascist like his father, and a member of the Legión Cívica Argentina. This was the fascist militia dedicated to beating up antifascists and Jews. (This militia, founded by a certain Patricio Kelly, eventually joined the Peronist movement, and was outlawed after Perón's downfall.) Once René taught me how to manufacture a stick with a leaden head, but I doubt he ever used it. We were always on friendly terms, and argued rationally, both face to face and by mail, on a host of subjects. René and I befriended Lorenzo, the newspaper vendor at the Retiro train station, who kept us up to date with the news and corrected our Italian pronunciation. Our conversations with him were so interesting that René and I missed some trains, so as not to cut them short.

Another dear friend was Luis Bertolino, the level-crossing guard who loved "classical" (i.e., Baroque and Romantic) music. He worked by night and lived humbly with a colleague in a wooden cottage next to the train level crossing in their charge. They could not afford eating anything but fruit, bread, and pasta with tomato sauce. Luis spoke low Spanish, but had good manners and always wore a clean shirt and a tie under his immaculate blue blazer. We first met at the train station, where I had accompanied my father as usual. When the train left, Luis introduced himself and asked me whether it was true that we had large collection of classical music records. I responded affirmatively and invited him to come home and listen to some records.

Thereafter, Luis visited us frequently and played records while I did my homework. When he met my father, both conversed learnedly about music and interpreters. We never found out how his passion for classical music had arisen, but the fact is that, at that time, one could attend for a dollar or less some of the best concerts in the Colón theatre, still world-famous for its excellent acoustics. When Italian operas were offered, the theatre filled with Italians, whereas Wagner's operas attracted not only Germans but also half of the music-loving community. I found opera ridiculous, but enjoyed some arias of Verdi's and even Wagner's.

Luis introduced me to the claque's boss, who allowed me to enjoy the best music for 50 cents – albeit standing in the gods, known as chicken coop. Once, the boss expelleded me for failing to applaud a pianist, perhaps Alfred Cortot, or Arthur Schnabel, whose performance I did not admire.

One summer Luis and I traveled third class (wooden benches) to Bariloche, at the feet of the Andean cordillera, in upper Patagonia, and then by truck to the marvelous Mascardi Lake. We camped in my tent on its shore, and befriended Benito Vereertbrugge, popularly known as "Brujo" (Wizard), the owner of the only hotel in the region and grandson of a legendary physician. Doctor Vereertbrugge had come via Canada from his native Belgium at the beginning of the twentieth century. He rode on horseback to see his patients, scattered throughout a 100-km- radius circle. He exchanged mounts once in a while and charged in kind, particularly horses, for the settlers seldom saw money: theirs was a barter economy, where trust replaces notarized documents and collaterals.

Patagonian Lakes

When I turned 16 years, my parents presented me with an organized tour of the Andean lakes and forests in and around the Patagonian town of Bariloche. My travel companions were a couple of haughty but uneducated landowners. I fell in love with Mascardi Lake, across the Tronador peak, to which I returned several times. In that trip I met veterans of the atrocious southern African Boer war, as well as Scots who had traveled there just to fish for salmon and trout. Two decades later, the only time I tried fishing at a rapid mountain river, I caught without bait 25 trout in so many minutes. They were eaten by the chemistry students camping on the shore of El Bolsón lake.

I also met a friendly young American woman whom I told, with my usual tact, that she resembled Beethoven. There was also a Uruguayan businessman who mailed me several books by Count Hermann Keyserling, much fêted in Argentina, who seemed to me to be a charlatan. The daughter of my new friend, only 16 years old, was serious and beautiful, and looked down on me because her high school taught infinitesimal calculus instead of Latin.

A few months later I took the overnight boat to Montevideo carrying only 50 pesos and a useless and annoying umbrella. I slept in cheap boarding houses, ate meals for 25 cents, talked to restaurant waiters about the latest feats of the ruling dictatorship, and learned that Uruguay had set up the first welfare state on the continent. Gilberto Bellini, who had painted my father's portrait, introduced me to several ladies who invited me to their tables in terraces where they danced for money with handsome young men. I was delighted with the unassuming Uruguayans, as well as with their grapes and watermelons, that tasted better than ours.

Politics

My real adolescence began in 1935, when I exchanged short pants for long ones, and started my fourth school year. That year, my brain was flooded by hormones, and my neurons were radically rewired. These brain changes led me to make several fascinating discoveries: the self, romantic love, literature, classical music, and Marxism.

At that time, communism was rapidly spreading throughout the world, fanned by the persistent economic depression and the economic success of the Soviet Union – the only country without unemployment. There were three additional factors: The crimes of Stalinism were unknown or disbelieved; Marxists claimed to be on the side of science; and the communists were everywhere the main enemies of fascism. For example, in 1935 the Brazilian leftists, led by the legendary lieutenant Luis Carlos Prestes, led a military attack on the fascist *Estado Novo* [New State] led by Getúlio Vagas and inspired by the journalist Plinio Salgado and the clever legal philosopher Miguel Reale, whom I met at congresses two decades later. By contrast, in Argentina, ultraconservative since the 1930 military coup, the communists constituted a minute group that could not even elect an alderman.

The Argentine communists followed the orders of the Communist International (Comintern), though sometimes with remarkable delays. For example, in 1935 a large wall in our neighborhood appeared painted with the slogan "¡All the power to the Soviets!," which the Petrograd communists had proclaimed in 1920. Another communist slogan of that time was directed against the socialists, whom they called "Social fascists," the idiotic and suicidal insult that the German communists had used on the eve of Hitler's ascent to power in 1933. However, the blindness of the local communists was no worse than that of Winston Churchill, who in 1939 praised Mussolini for being a wall against communism, nor that of Pope Pius XI, who, besides colluding with Mussolini, called General Franco a "Knight of Christendom." The fascist's heinous crimes favored the spread of communism and made many people, my father and myself included, discount the news about Stalin's crimes as sheer bourgeois propaganda. Last, but not least, the Marxists promised to solve all the social ills without having to study them – which was not exactly what Marx and Engels had done.

I joined the communist movement in 1935 but was never a member of any cell, because I had no workplace and did not even participate in the student movement. Once in a while a comrade would contact me and ask me to solicit donations for such ghostly organizations as the Association Against Argentina's Participation in the Olympic Games, to be held in Berlin in 1936. I visited various possible donors, none of whom gave me a cent for this ghostly cause. I was increasingly baffled by a number of episodes. I suffered my first shock when Manuel Sadosky, my occasional mathematics tutor and a fervent communist, criticized my essay on the repression of the Patagonian workers in 1920: he told me that we should not criticize the government of the time because it had been progressive. To him, as to all sectarians, all things political were either black or white: none was gray.

I got my second shock on reading the program of the Comintern: I was struck by the difference between its lofty ideals and the miserable reality I was acquainted with. A third shock was the rejection, by the communist monthly, of a book review where I claimed that, because Rousseau had preferred feeling to reason, and had trashed science, he did not belong to the Enlightenment. Then came the party's attempt to control the workers' school that I had founded without consulting the party – but this will be dealt with in the next chapter.

In 1939, the year that the Second World War started, communists the world over were ordered to study Stalin's *Short History of the Bolshevik Party*, and to fight Trotskysm above all. I found that book boring and of no interest at all outside the USSR, so I did not read it. As for Trotskytes, I had met only two of them, who seemed to be totally harmless: a bookish friend of Dr Emilio Troise, my father's friend, and Liborio Justo, the son of the president of the Republic. This obscure engineer got his celebrity minute when he shouted "Down with the dictatorship!" from the Congress's gallery. I could not believe that these two solitaries could threaten anybody, particularly when fascism was advancing everywhere, particularly at home. Thus, little by little, I lost my faith in communism, until I was finally expelled from the party in 1947. I had a single loss: the desertion of a former student and friend who eventually became a professor at the Imperial College of Science and Technology, and whom I recovered only recently.

The Local Communist Intellectuals

During the 1930s and 1940s I got to know a handful of communist intellectuals, none of whom did research of any kind, or even read international journals. My closest friend among them, Manuel Sadosky, whom I mentioned before, studied mathematics and eventually became a well respected if mathematically unproductive university professor, the vice-dean of the Faculty of Science of the University of Buenos Aires and, half a century later, Argentina's first Secretary of Science and Technology. I used to consult him and his wife, Cora Ratto, on mathematical and personal issues. Manuel opined on everything and dispensed advice on anything. He was a great believer in authority. He sided publically with the charlatan Lysenko against genetics; he advised me to read the great Leibniz's *Monadology*, which I found absurd and backward; and he adopted John Bernal's *Social Function of Science* (1939) as his lifetime bible on science and technology policy. The eminent crystallographer had confused both domains to the point of supporting the planning of scientific research, thus earning the well-deserved criticism of Michael Polanyi.

My father was a friend of Emilio Troise, a well-respected clinician and Marxist scholar who, unlike his fellow Marxists, did not quote Marx and Engels at every step. But Troise had nothing to say about the philosophical presuppositions and implications of current science, which was the center of my own philosophical attention.

I also befriended the brilliant journalist Rodolfo Puiggrós, who had started studies in economics and written a couple of books on the economic interpretation of Argentine history. I appointed him professor of Argentine history at the workers' school I had set up. I cut him when, all of a sudden, he converted to Peronism and launched a Peronist magazine. During Perón's third period, Puiggrós was made rector of the Buenos Aires University, which he renamed Universidad Nacional y Popular. He devoted the rest of his life to fighting the military dictatorship as part of a motley group that included Eva Perón's confessor and a number of self-made intellectuals like himself. Some of them fled the country for their life and others, like Rodolfo Walsh, were murdered by the dictatorship. Had I not left the country at the beginning of 1963, I would have been murdered either by the dictatorship or by the *montonero* guerrilla.

Finally, there was Paulino González Alberdi, the Party's economist, who prophesied an imminent economic crisis just at the time when the nation's industry started to grow exponentially. He was deeply offended when, in the company of some fellow prophets, I asked him in a loud voice: "When is your crisis due, Paulino?" Regrettably, Marx and Engels too had engaged in prophesying: every time the

London Stock Exchange dropped brutally, they toasted the coming revolution (Hunt 2009).

In short, the Argentine communist intellectuals were a mediocre coterie. None of them studied seriously the society they lived in or published in scholarly magazines: they were too busy reading old books or the boring *informes* (reports) written by professional activists who followed the Comintern slogans hatched in far-away Moscow. How different were their French, German, British, and even American comrades, who could boast of including some eminent scientists like J. B. S. Haldane, John D. Bernal, and Edward Thomson, as well as the Nobel laureates Fréderic Joliot-Curie and Robert Fogel!

Ethical Intermezzo: The Committed Intellectual

My father, who worked along with some communists at the Federation of Societies to Aid the Allies, had a similar experience: he complained of their authoritarianism and lack of culture. He only respected Cora Ratto, Sadosky's wife and the head de the Junta de la Victoria, the feminine wing of the aid movement. Once both were summoned together by Chief Inspector Kussel, head of the political police and a famous torturer. Cora recalled that, on leaving the infamous police station, mi father, a physician who boasted of his "clinical eye", told her: "Cora, this vermin won't last more than 6 months." Actually the monster lasted only half that time. But my father died shortly after him from a massive stroke, probably caused by police harassment, which had worsened his chronic high blood pressure.

When I asked my father why he did not join the Communist Party, he answered: "Because I want to retain my freedom of thought." (Antonio Gramsci would have said that my father refused to become an "organic intellectual.") He told me this in 1935, a couple of years before a number of "old Boksheviks" were executed for "deviationism" – among them Nikolai Bukharin and Jan Sten, who a decade earlier had tutored Stalin in dialectical materialism, a subject that, to his credit, had found hard to learn (Medvedev 1969).

The surrender of free inquiry is not the only price to be paid for joining a sect. The distinguished British historian Eric Hobsbawm (2002), communist to the last, mentions the conflict between party loyalty and personal loyalty, as in the duty to inform about "deviations" incurred by personal friends. But there is much more: the good sectarian is also an irrationalist, since he must allow passive acceptance of dogma to trump rational debate and free original inquiry. For example, the faithful communist must swallow the mysteries of dialectics along with the confusion of socialization with statization, and must suppress all exploration of peaceful ways to building socialism.

Besides, despite their praise of science, loyal communists must either ignore all unfavorable evidence or disguise it as confirmation – as when a brutal government repression is "interpreted" as a sign of its weakness. Last, but not least, the Marxist-Leninist philosophy involves a radical utilitarian ethics, which is an oxymoron, for it amounts to jettisoning all moral scruples. In short, the dogmatic Marxist is just as inflexible and hostile to research and innovation as the pious religious believer, and just as ready to pillory the dissenter as a religious inquisitor. This attitude explains in part the current worldwide crisis of Marxism, as well as the search for non-Marxist socialisms (see, e.g., Bunge 2009; Bunge and Gabetta 2013). For better or for worse, although I had become interested in politics at age seven, I never saw myself as a political activist, and I avoided politics from the moment I chose an academic career. But before doing so I had to finish high school.

High School: Change of Gear

My parents did not reprimand me for my Spanish Literature failure or for other disappointing results, but trusted that I would find a way to graduate from high school on my own. In fact, I transferred to the Colegio Nacional Sarmiento, where I took exams as a "free student". I got hold of the standard textbooks and made up a timetable that I followed consistently, most of the time in our garden. I finished the year 1936 having to complete only trigonometry, a subject in which I was deservedly flunked twice. These failures were fortunate because they forced me to adopt the proverb "If you can't beat it, join it." In this task I was initially helped by two tutors who quickly became close friends: Manuel Sadosky and David Jacovkis. Manuel was a math student and a born teacher, but could not devote as much time to me as I needed, so I hired David, who was then finishing high school.

David was disciplined and fearless. He would tackle serenely any of the problems that baffled me even when he was only a jump ahead of me. When he met a formula he did not understand, he resorted to the English trick of lighting a cigarette to gain time. At the same time, I also took private English lessons from beautiful Señorita Rubinstein, which helped me to read a pile of books in English.

But the decisive factor was tackling by myself Isaac Todhunter's dusty *Plane Trigonometry* (1859), which I discovered in a second hand bookstore. I studied it conscientiously and solved all the exercises it contained. When I finished it, I passed the examination easily and got my baccalaureate certificate. More importantly, I stayed in love with math for life. So much so that my next project was to study by myself Silvanus P. Thompson's *Calculus Made Easy* (1910), recommended to me by my father's economist friend Ramón Lequerica. That bestseller had been written in the intuitive style of Leibniz and Newton, which prevailed until the Epsilon-Delta Revolution effected by Bolzano, Cauchy and Weierstrass in the mid-nineteenth century. This manual, much maligned by modern mathematicians, taught me to differentiate and integrate some elementary functions. I was exhilarated by calculus.

Mastering Thompson's calculus book gave me confidence and widened my horizon. But it also hindered my access to rigorous mathematical reasoning, which does not involve infinitesimals, or "very small quantities approaching 0," that the philosopher George Berkeley had derisively called "ghosts of defunct quantities." The revolution operated in mid-nineteenth century by Bolzano, Cauchy and Weirstrass eludes infinitesimals, and is thus immune to Berkeley's sharp criticisms, as well as free from the paradoxes that made Marx waste much time trying to master the calculus (Struik 1948).

That profound change translated intuitive expressions of the form "$f(x)$ approaches a when x approaches b", that misleadingly suggest motion, into timeless expressions of the form "For every number ε, there exists a δ such that $|f(x)-a|<\varepsilon$ if $|x-b|<\delta$." The conceptual change was radical, but the calculation methods did not change, and scientists and engineers kept thinking in terms of infinitesimals.

Despite this banning of time and change from pure mathematics, Friedrich Engels and other Marxists believed that mathematics, like the other sciences, mirrors the restless universe – which is why there is no adequate Marxist philosophy of mathematics. I swallowed Engels's dialectical interpretation of mathematics until being disabused by my mail friend José Luis Massera, a Uruguayan mathematician famous for the theorem named after him. He was also one of the earliest to push for the updating of Marxism, and a major victim of the American-supported military dictatorship evoked by Costa-Gavras in his film *State of Siege*. However, let us go back to my adolescence.

The first day of 1937, at age 18 years, I made a New Year's resolution: to study serious intellectual matters. But which ones? I was equally attracted by physics, psychology, and philosophy. Physics, in particular astrophysics and cosmology, dazzled me thanks to the popular books by the eminent British astrophysicists Arthur Eddington and James Jeans.

Psychology had intrigued me since, at 16, I read some of Freud's books, which sold for a few cents at subway kiosks. Through common friends I befriended the psychoanalyst and graphologist Federico Aberastury, whose sister Aminda had recently imported Freudian psychobabble into the country. Federico, charming and persuasive, exchanged a couple of letters with me, and invited me to attend a consultation with a young couple whom he greatly embarrassed. Shortly thereafter he was interned in a psychiatric clinic because he had attempted to fly from his fifth-floor balcony.

I fell in love with philosophy when I read Bertrand Russell's *Problems of Philosophy* (1912). This book persuaded me that psychoanalysis was sheer fantasy. I also read, in no particular order, as is usual with amateurs, many books in the history of philosophy. I was duly impressed by the pre-Socratics, and later on by Spinoza and the philosophers of the French Enlightenment. My father's library had a good edition of Voltaire's complete works, which amused me but did not teach me about the philosophy-science connection.

Encounter with Dialectical Materialism

In my father's library I also found a collection of essays about dialectical materialism. I was immediately attracted to this philosophy because it seemed to explain it all. I was particularly intrigued by dialectics. When I asked my father what this was, he replied: "Master Justo [the neurosurgeon who had founded the Socialist party as well as the first cooperative in the country] thought it is just *hocus-pocus*."

This reply did not satisfy me, so I started a search that was to take far too many years. In 1937 I attended a course on dialectical materialism taught by Emilio Troise. Unlike Lenin, who had criticized Joseph Dietzgen for holding that "thinking is material," Troise – in keeping with the Hippocratic tradition – expounded and exemplified the materialist view that the mental is cerebral. He would have been surprised and shocked to learn that the philosophy of mind official in the Soviet Union was psychoneural dualism.

That year of 1937, so critical for me, I read more than at any other time in my life. In particular, I read some of Hegel in Molitor's French translation, as well *From Hegel to Marx* by the brilliant American pragmatist Sidney Hook (1936), whom Bertrand Russell later called "the CIA philosopher." And, having no filters, I also read piles of rubbish. But luckily I lacked Borges's uncanny ability to discover and read some of the most useless documents ever written – those about the brutalities and betrayals of tenth century Icelanders, and the arcane theologies conceived of by mad scribes.

But I did read lots on the secular theology of the day: Marxism. In particular, I was most impressed by *Materialism and Empirio-Criticism* (Lenin 1909), in which Lenin had criticized the idealist interpretations of the new physics. As I learned much later from an American Jesuit philosopher, Lenin had been right for the wrong reason, namely, for taking Engels's word – a mark of scholasticism. It took me a decade to understand that dialectical materialism contained only two unpolished nuggets, epistemological realism and ontological materialism, under a mound of confused Hegelian verbiage. In time, I also learned that it is not enough to criticize: one should also attempt to invent viable alternatives.

Back to the problem of career choice. My vacillation between the three disciplines that attracted me – philosophy, psychology and physics – did not last long. I soon found out that psychology did not exist in my country as a career, and I discarded professional philosophy after attending a few lectures by Coriolano Alberini and others at the Faculty of Philosophy and Letters. They followed the neo-Hegelian Gentile and the intuitionist Bergson, whose main target was science, whereas I professed scientism before quite knowing what science was. So, after a week I fled from that Faculty to that of Science.

To Physics Through Chemistry

Finally I decided to study physics at university and philosophy at home. Such a combination of profession with vocation is quite common in underdeveloped nations, where biology enthusiasts study medicine, and youngsters attracted to the social sciences enroll in law schools. In those countries higher culture is largely in the brains of dilettanti. The main trouble with this split is of course that the amateurs study what they like, not what they need, so that their production is full of holes and lacking in rigor. They have access only to popularized works, do not sit for exams, do not work in labs, and do not participate in professional communities, so they are neither controlled nor stimulated by experts. That is how nearly all philosophers of science learn about science: from afar. I wished to do better, so I studied and conducted research for 14 years to earn a PhD in physics while studying philosophy on my own and doing odd jobs to pay the bills. My first physics paper appeared in 1944 at age 25 years (Bunge 1944a, b), and my latest 70 years later (Bunge 2015a). And I philosophized for two decades, and tucked a couple of books in English and a dozen papers in good journals under my belt before regarding myself as a professional philosopher.

In 1937 I could not enrol as a physics student because I still had to pass my trigonometry exam. Besides, my father warned me that I would be unable to make a living as a physicist – which was true before the first nuclear bombs were exploded. So, I chose the closest field, chemistry, which my friend David Jacovkis had just began to study. I bought a white lab coat, a copy of Joseph Mellor's *Modern Inorganic Chemistry*, and did the lab work of general and inorganic chemistry. The course was taught by Alfredo Chiodín, a professor incapable of smiling but an excellent expositor respected by everyone.

I crashed the lab with David's help, who made room for me on his bench. The lab director – a certain Angel Bombelli, who during Perón's years acted as the faculty's dean and stole all the movable instruments – expelled me every time he noticed me. During his brief inspection I hid in the balcony, smoked a cigarette, and returned to work when the coast was clear. I was exhilarated when repeating some classical experiments, like Joule's on the equivalence of heat and mechanical work, using the same clever contrivance. I was also moved when watching the traces left by charged particles in a Wilson cloud chamber. At the end of the year I persuaded my father that chemistry bored me, whereas I was still in love with physics and philosophy. I lost no time in registering as a physics student to attain my goal: to join philosophy with science.

New Friends

During that transition year I befriended many students across the fields of physics, chemistry, biology, engineering, and architecture, all lodged in the same old building at the back of my old high school. We used to take coffee breaks at a cafe nearby, and when my father was traveling as a medical consultant, I invited up to ten of my friends to *El ombú*, to chat and listen to music. We also attended plays together at the amateurs' Teatro del Pueblo, rowed on the Paraná, and had delicious meals for 50 cents – roughly an American dime – at *The Hunter's Horn*. None of us drank alcohol, but some smoked. (I had started smoking at age 5, incited by a gardener, and was able to quit only at age 33.)

Among the new friends were Julia Molina y Vedia, my future wife – a recently graduated architect – the engineering students Isaías Segal and Jorge Ruberti, the architecture students Alberto Le Pera and Simón Ungar, the biology student Delia Ingenieros, and Estrella Mazzolli, the only physics student in Buenos Aires. We were all leftists but we seldom discussed politics, and I was the only one interested in Marxist philosophy. David quipped that I even ate with dialectical materialist cutlery.

Delia, perhaps the most interesting of the lot, was tall and bony, did not use make-up, and boasted about her modernity. Her father, the polymath and well-known writer José Ingenieros, was the earliest biological psychologist in the country and, although generally regarded as a positivist, was actually a scientistic materialist. Delia studied biology but never graduated. She went around carrying *La crise du transformisme* (1909), one of the many books by the popular embryologist and materialist philosopher Félix Le Dantec, which she never finished reading. She was friendly and witty, and liked masculine company but abhorred matrimony. She had invented the formula "Function + Variety = Enjoyment." Eventually she became a stage magician, and later a member or Borges's inner circle. With Estrella, who was beautiful and charming, though somewhat slow, I studied a book on calculus in the library. Later on she married Enrique Mathov, a pioneer allergy specialist, who became a very close friend of mine, and taught biology to his two children as well as my own.

I noticed Julia reading Descartes's *Discourse on Method* on a bench in the courtyard of the faculty's old building. This, and the fact that she asked for donations to the Spanish Republicans despite being a fervent Catholic, were enough for me to be attracted to her. Her svelte figure, dressed in Paris fashions, may also have been a factor.

Combining Callings

At the end of 1937 I got my high school certificate at the same time as the estranged classmates of my first high school. I registered as a physics student at the Faculty of Physico-Mathematical Sciences in the La Plata National University, located in the planned city of La Plata, graced by beautiful parks and wide, leafy avenues. I thought it was better than its Buenos Aires counterpart, 80 km away, because it had research laboratories, where the few mathematics, physics and astronomy students studied alongside hundreds of engineering students. Most of my classmates were local or had come from provincial towns, and were free from the arrogance of us *porteños*, or inhabitants of the cosmopolitan Buenos Aires port area.

The young La Plata University was perhaps the most advanced in Latin America, because it assigned priority to the basic sciences. Indeed, instead of being a factory for producing lawyers, physicians and bookish engineers, it rested on three scientific pillars: the Natural Science Museum, home of the extraordinary collection of fossils dug up and studied by the Ameghino brothers; the Physics Institute founded by the German physicist Emil Bose, who on his premature death was replaced by the experimental physicist Richard Gans; and the Astronomical Observatory inspired by Benjamin Gould, the American astronomer hired by President Sarmiento to install the National Observatory, the earliest in the Southern Hemisphere.

I became a regular *platense* student, but went on visiting the University of Buenos Aires, where I met old friends and attended the odd lecture by the mathematician Julio Rey Pastor and the physicist Teófilo Isnardi, as well as all the presentations by foreign visiting scientists. I also befriended Alberto González Domínguez, an analyst interested in physics and engineering, open to new ideas and ever ready to help with some mathematical problem or other and offer us a cup of coffee. He introduced us to cybernetics, game theory, and the theory of distributions.

Getting Rid of Freud and Ready to Start University

During my last free summer of 1937–1938, I took delight in studying a beautiful physics textbook on the Paraná Delta, where I often went rowing. During that time I also wrote a small book against psychoanalysis titled *Marx* vs. *Freud*, which I did not attempt to publish. It was little more than a detailed critical review of Reuben Osborn's *Freud and Marx: A Dialectical Study*. This book, which made quite a stir, had just been published by the Left Book Club, to which my father subscribed, and whose monthly releases he and I devoured. Osborn agreed with Erich Fromm and Wilhelm Reich, whose attempt to combine Marx with Freud was continued by the Frankfurt School, Herbert Marcuse, and some French postmoderns starting with Louis Althusser and his protégé Michel Foucault.

I remember only two of my objections to the fashionable creed of psychoanalysis. One was that Freud ignored the biological psychology suggested by Pavlov's experiments, and that my future colleague Donald Hebb was to revive two decades later in his epochal *Organization of Behavior* (Hebb 1949). My second objection was that the Oedipal account of social rebellion (hatred of the father or the father-figure) was not only false but also reactionary, for presenting social strife as caused by a personal issue rather than as a conflict of economic interests. Years later I added that psychoanalyis is not an ongoing scientific endeavor but a sect of fantasists who had not even learned that mind and sexual drive occur in the brain (Bunge 1978a, 1979a, d, j, 1980a, e, 1981a, d, 1985b, c, e, 1987a, c, 1990b, 1991a, 1999c, 2010a).

This essay got lost along with two novels and *Espartaco*, a long drama in free verse. Fearing a police search, I collected the lot in a bound volume, which I confided to my good friend Enrique Mathov. Years later, driven by the same fear, Enrique got rid of that adolescent leftover. Good riddance! Adolescence is to be enjoyed and suffered, not published. Mine finished in March of 1938, when, high with expectations, I entered the Faculty of my choice, where two decades later I was to return as a professor.

Chapter 3
University *Et Alia*

Beginning University

To reach La Plata, the provincial capital where my university was located, I had to travel about 100 km, crossing Buenos Aires city from North to South. During my first two university years I woke up at 5:30, and 30 min later boarded the train to Retiro, where I took the subway to the Constitución station, where another train took me to La Plata, where I arrived at 8:00. (Two decades later, the same train took twice or thrice as long.) The Faculty was only three blocks away from the station, and we walked there along with some of our professors.

I knew that Spanish was hardly used in the international scientific community, not for lack of expressive power but because the Hispanic world, ruled by the cross and the sword, had remained outside science. This is why, at the kiosk of the Florida station, I bought, in addition to the tabloid *El Mundo,* the English, French, democratic German, and fascist Italian newspapers. I skimmed them all quickly between Florida and Retiro, to learn the increasingly alarming news of the day, and enrich my knowledge of the foreign languages I had studied, as well as to learn some Italian. I needed all five languages because my textbooks were written in them. But much of my reading was guesswork, for I was too much in a hurry to consult dictionaries.

That early train to La Plata was full of professors and students at my university. Since my fellow *porteños* had the bad habit of speaking loudly, I could not help overhearing conversations among my travel companions. Thus I once heard the Socialist Congressman Américo (later nicknamed Norteamérico) Ghioldi flattering Professor Ricardo Levene, who claimed to teach sociology, and refused to comment on the *Communist Manifesto* of 1848 in the presence of ladies. Another time, two well-dressed ladies were praising their respective physicians. One of them was enthusiastic about Dr. Emilio Troise, but regretted that he consorted with Jews. (In fact, Troise presided over the Committee Against Racism and Anti-Semitism.) Two decades later, when travelling to the same university to teach quantum physics,

M. Bunge, *Between Two Worlds*, Springer Biographies,
DOI 10.1007/978-3-319-29251-9_3

I often enjoyed the company of interesting colleagues, such as the philosopher Risieri Frondizi and the Egyptologist Abraham Rosenwasser.

While still a student, one of my most interesting travel companions was the Law professor and socialist politician Carlos Sánchez Viamonte, whom I befriended. Another was the Jesuit priest Juan A. Bussolini, an astronomy student. He was the only of us who did not worry about his professional future: he boasted that, as soon as he graduated, he would be appointed director of the San Miguel Institute of Cosmic Physics. In fact, he got this job, as well as President Perón's personal protection, without ever graduating, so he had no need to keep learning. His familiarity with the terrestrial powers compensated for his ignorance of celestial bodies.

However, I spent most of my travel time reading the physics and philosophy books I carried in my bulging briefcase. In good weather, I also read during lunch breaks in a park planted with lovely gingko trees, the oldest known ones and the only ones that benefit from air pollution. It took me several years to learn that many of the philosophy books I had read at the time, particularly Hegel's, were at best useless. I still fail to understand why I respected that most inexact stuff while studying rigorous mathematics and physics.

In 1938 there were less than ten physics students. We recognized ourselves as such only gradually, because we shared the first 2 years with the mathematics, astronomy, and engineering students. During those first 2 years I befriended a handful of serious and kind engineering students, and studied several subjects with them. We met at their modest homes, where their mothers offered us cookies and large cups of *mate* tea, the national beverage. (*Mate*, the dried leaf of a gigantic tropical tree, is a rich source of caffeine, and is popularly believed to have many medical virtues.) We parted company in third year, when the physics students enrolled in courses deemed to be irrelevant to engineering. I lost touch with all but one of them, who liked to do geodesy jobs in remote places, where he camped in his tent until old age. Since then it has seemed to me that, by and large, engineers are rational, practical, competent, honest, reliable, tranquil, and lacking in political passion as well as in a sense of humor.

The only totally new subject to me was projective geometry, which at that time was assigned an exaggerated importance, though much less so in La Plata than in Buenos Aires, where it was the fiefdom of two brothers who were said to share a lover. It is a beautiful and well-organized theory, with theorems and meta-theorems, but stagnant and disconnected from the rest of mathematics and of no use to physicists.

It was then that I committed the only fraud in my career: my engineering friends made my projective geometry drawings, and later on also those for descriptive geometry, in exchange for the resolution of calculus problems. (I did not even have a drawing board.) Much later we learned that our calculus professor, a certain J. C. Vignaux, was doing something much worse: he published under his own name some theorems proved by my friend Mischa Cotlar, who could not get a university degree because he had not even a high school certificate – until the University of Chicago overlooked this gap and awarded him the doctorate he richly deserved.

Eventually Mischa's former employer joined the Peronist party, and was appointed director of the Electronics Institute, a field to which he was a total stranger. But his academic career finished in 1955, when the students barred his entrance to the Faculty of Sciences in Buenos Aires. Yet he was not a fool: he delivered clear lectures, and, in a milieu with higher scientific and moral standards, he might have been a productive mathematician. As the Spanish philosopher Ortega y Gasset famously said, "I am I and my circumstances."

My Mathematics Teachers

Although none of our mathematics teachers published in refereed international journals, they all knew their subject and were intelligible. The latter was partly due to the fact that they tended to omit proofs, and partly to their use of intuitive means to introduce new concepts. For example, the sine function was defined as the space-time trajectory of a linear oscillator. The Nazis would have approved, for they claimed that the Aryan soul is intuitive, whereas the Jewish one is abstract.

For better or for worse, there were no accessible textbooks of higher mathematics in Spanish, so the more ambitious students used texts in foreign languages. The American book industry had yet to invent the voluminous text brimming with trivial exercises, and occupying up to 1000 pages, which can be sold at high prices, sometimes to the benefit of the departmental boss who chose the textbooks for the most popular courses. So far as I know, nobody has reported the sighting of any student who completed the reading of such heavy tomes. Their Japanese counterparts cover roughly the same material in a quarter of space.

The Teaching of Physics

The two general physics courses were excellent. They were taught by Dr. Ramón Loyarte (1888–1944), a spectroscopist as well as a Conservative congressman. He had studied in Germany, had published a few papers on ionization potentials in local journals, and continued the great didactic tradition initiated by his predecessors, the German physicists Emil Bose and Richard Gans, who in turn imitated the famous "Pohl circus" in Göttingen. Our cohort did not benefit from Gans's lectures: he had gone back to Germany in the 1920s, and stayed there even after the ascent of Nazism to power because he was declared a "valuable Jew," and even made director of the once-prestigious Kaiser Wilhelm Society. Two decades later, when he returned to La Plata, I asked Gans with patriotic pride whether he had noted any changes since his previous stay, and he replied: "Sure, trains take now twice as long as before."

Loyarte taught all but one of the advanced courses. He used the blackboard and followed his own textbook, which was clear if outdated. He did not use the vector and tensor calculi, so he wrote thrice as many formulas as needed – to the detriment of clarity. But, in contrast with his colleagues in the other Latin American universities, Loyarte illustrated some important points with "demonstrations," that is, measurements or experiments visible from the farthest desks. They were set up by assistants trained in Germany – which, until 1933, had been the world leader in all the natural sciences.

Some of those demos were spectacular. For instance, a thick electric cable connected to a row of batteries was suspended from the opposite walls of the main classroom, and the lights were extinguished. When the current started to flow, the cable became increasingly red, and at the same time it sagged because of dilation due to the heat. In this way, two different effects of an electric current were shown at once. A third, the magnetic field created by the current, would have been shown too if a large magnetic compass had been hung in the vicinity of the cable.

Laboratory practices were another asset of my alma mater. They constituted a separate course, also 2 years long and under a different professor, Dr. Héctor Isnardi, aka Pucho (Cigarette Butt). He was also a spectroscopist as well the owner of a small local bank, and the father-in-law of two incompetent assistants. Pucho Isnardi was usually seen in the midst of a cloud of cigarette smoke, busy measuring the spectral lines of cesium. Nobody knew what he did with those data, for he never published. Although we regarded him as a scientist, actually he was no better than a stamp collector. But at least he was amiable and a good lab manager.

We performed the elementary lab exercises in groups of three or four, so they were fun as well as instructive. They started 1 year after passing the theoretical exam, for we were expected to understand what we were doing. This phase shift had a tacit philosophical rationale: whereas ordinary knowledge may start with observation, the way empiricists demand, scientific knowledge starts from background knowledge and involves hypotheses, which guide both experimental design and data interpretation. In other words, experimental scientists have got to know why they do what they are doing, and how their instruments work. For example, mechanics explains the shift of the scale's needle, and electrodynamics that of the galvanometer's. These needles or pointers are so many observable *indicators* of imperceptible things or processes, in our case the gravitational field and the electric current respectively. Empiricist philosophy, in particular the positivism inherent in the standard account of the quantum theory, ignores indicators, and therefore fails to explain measurement, even though it claims to stick to it (Bunge 2012a).

For that reason, Kohlrausch's classical textbook of physical measurement always started with theory. I was lucky to find a copy of it in a second hand bookstore. Incidentally, all the students of my generation spent many hours a week hunting for treasures hidden in such stores, particularly during the war years.

Theoretical Physics Courses

We took five courses in theoretical physics: analytical mechanics, thermodynamics, electrodynamics, quantum mechanics, and geophysics. Loyarte taught the first three, Teófilo Isnardi the fourth, and Simón Gershánik the fifth.

Analytical mechanics is fascinating because of its generality and abstractness, as it may be interpreted not only in mechanical terms but also in thermodynamic, electrodynamic, quantum, and even economic terms. Hence, it should called 'Hamilton-Lagrange formalism' instead. This semantic neutrality of the theory in question generated a confusion that gave rise to a philosophical dispute that involved Maxwell and Lord Kelvin. Kelvin, like others, tried to reduce the whole of physics to mechanics. And, since Maxwell's theory is about fields, not bodies, Kelvin refused to admit it as long as Maxwell did not succeed in fitting it into the Hamiltonian formalism that had been accidentally born in analytical mechanics (see Bunge 1957b). This question led me, several decades later, to analyze the semantic concept of interpretation into its two components: reference or denotation, and sense or connotation (Bunge 1974b).

Loyarte taught Maxwell's six basic equations without using the vector calculus, needed to grasp them globally and interpret them correctly, as well as to save muscle and chalk, since every one of Maxwell's two triplets can be condensed into a single vector formula. When I realized this, I took a break from physics and studied Richard Gans's excellent textbook on vectors, as well as an engineering book on electric motors, that used the tensor calculus. This calculus allows one to identify the invariants (constants) under certain coordinate transformations, that is, the features of things that do not depend upon the choice of a reference frame, which is merely a matter of convenience.

The Faraday-Maxwell theory of the electromagnetic field fascinated me, as it had captured young Einstein's imagination. I remember distinctly the moment I understood that theory – the morning of March 27th, 1941 – because the epiphany happened at the waiting room of the clinic where Julia gave birth to our first child, Carlos Federico, aka Cantarito. That was also the subject of my first university lecture, which I gave at the Universidad Nacional del Litoral at the invitation of the historian of science Aldo Mieli and the mathematician José Babini, who wished to check whether I was a suitable candidate to teach history of science.

While preparing that lecture, which was subsequently published (Bunge 1943), I discovered that the person who had introduced that theory into the country was the Belgian engineer Germán Ave Lallemant, who had also been a pioneer of Socialism in the country (Tarcus 2007). That booklet, as well as the one devoted to Newton, were published by the Universidad Obrera Argentina, and both were reviewed favorably in *Nature*. However, back to my course work.

Learning Quantum Mechanics

The last course I took with Loyarte was the one on quantum mechanics. He explained Heisenberg's matrix formulation as clearly as possible, including its philosophical motivation, namely to use only measurable quantities – an ideal that, as Heisenberg himself realized two decades later, his theory had failed to meet. But the mathematical formalism of that theory is so awkward that it was used only in a couple of cases. Thus, Heisenberg had not forged the tool required to tackle the problems that had motivated his search for a new theory: the calculation of atomic spectra (quantum statics) and of the effective cross sections of inter-particle collisions (quantum dynamics and quantum chemistry). But he had laid the first bricks – at age 25!

Those problems were solved by Schrödinger's "wave mechanics", which at the time was taught by Teófilo Isnardi (1890–1966) in Buenos Aires. Isnardi had learned it by himself – no mean feat to perform in that scientific desert. Moreover, he subjected it to a remarkably deep philosophical examination (Isnardi 1927), which unfortunately I read only three decades after his death. The same professor also taught thermodynamics and statistical mechanics in La Plata. I studied the former with the book of Enrico Fermi, the great nuclear physicist, and statistical mechanics with that of Landau and Lifshitz. A few years earlier, the amazing Fermi had visited Buenos Aires sponsored by his government, lectured wearing a black shirt, and saluted the public with his right arm stretched out. That did not help him: in 1938 he was dismissed from his Roman chair because his wife Laura was Jewish, and left for the USA. His collaborator Franco Rasetti declined Fermi's invitation to work with him on the Manhattan Project, on moral grounds, and came to Canada, where he designed and directed the physics lab at Laval University. Rasetti's book on nuclear physics, which I studied, was one of the earliest on the subject.

Fermi's excellent book on thermodynamics led me to study *Dimensional Analysis*, by Percy W. Bridgman (Bridgman 1922). Thanks to this book I learned to perform the dimensional analysis of any formula, but it did not teach me what the dimension of a physical magnitude is, a problem I solved much later (Bunge 1971). Finding conceptual holes and trying to fill them has always been a hobby of mine.

However, back to Bridgman. Bridgman fathered high pressure physics – for which he earned a Nobel prize – as well as operationism, a positivist doctrine that I later examined several times (e.g., Bunge 1958a). In 1960 I met the great man at Harvard and argued with him after a lecture of mine. A few months later he took his own life on learning that he had cancer. His assistant Gerald Holton told us that his boss proceeded with the same care he showed in his experiments: he covered his garage with a linoleum sheet, shortened his shotgun, and shot himself.

Landau and Lifshitz were two Russian theoretical physicists, important not only for their scientific and didactic work, but also because they had dared defend the relativistic and quantum theories. These had been attacked by powerful Stalinist ideologues in the name of dialectical materialism. It took me several decades to realize that Marxism, when ossified and made official, had proved a serious hindrance to the development of all the sciences in the U.S.S.R, particularly the social ones. What else could be expected from the uncritical adoption of dialectics, a doctrine that is at best false, and at worst confused and esoteric, as I argued much later at an international symposium (Bunge 1975c).

Teófilo Isnardi, my examiner in thermodynamics and statistical mechanics, shared the patience and tolerance of nearly all our professors. The only exception was the geophysics professor, Dr. Simón Gershánik, who seemed to us unclear, pedantic, and distant. What is worse, he taught his seismology classes at the distant observatory, from 8:00 p.m. to 10:00 p.m. During the examination he scolded me for having forgotten a certain formula. I answered indignantly that I knew how to deduce it, which was more important than memorizing it. He gave me only a six, far below my usual ten.

I disliked this professor's classes but learned to love his subject when I studied an American textbook on seismology. Then I learned that the problem of finding the epicenter of a quake by analyzing the waves registered in a seismogram is hard for being *inverse*: it goes from a given effect to its unknown cause(s). In any event, that was the last subject I had to be examined on, when I had already started to do research. However, because at that time the La Plata University conferred only doctorates, I had yet 7 years to go before getting a university degree. So, between 1947 and 1952 I was an underpaid instructor.

Other Interesting Teachers

I interacted often with Rafael Grinfeld, an assistant professor, and Ernesto Sabato, who held the uncommon job of assistant doctor. Once in a while we saw Enrique Loedel Palumbo, an adjunct professor, our expert in Einstein's theory of gravitation, but found him haughty and unsociable. During the October 17th 1945 events that had started with Perón's arrest and his return thanks to Evita's courage, Loedel and Grinfeld interrupted a meeting of the physical society in Buenos Aires, urging us to leave and join the large crowd that was gathering in the nearby Plaza de Mayo. Nobody followed them, because they were unknown to most, and because Enrique Gaviola, the president of our society, reminded us that we had formed it to protect physics from politics. I agreed wholeheartedly with his view that politics does not belong in labs or classrooms.

Rafael Grinfeld, the second to the experimental physics professor, and the Institute director when I was a professor, was accessible and chatty – too much so in my view. He had published a few papers, one of them on the light-emission mechanism of fireflies. And, jointly with Alberto Sagastume Berra, our most distinguished mathematics professor, Grinfeld had written a clear book on Bohr's atomic theory, published three decades earlier.

Grinfeld, an armchair anarchist, was fired in 1931 by the fascist government of the day, and forced to migrate. That made him a member of the first emigration of Argentine scientists. We all knew that story and respected him until my classmate Horacio Brodsky asked him and Héctor Isnardi, who were sitting together, how to calculate the field of a magnet of an arbitrary shape, such as a croissant. Both teachers put on a solemn face and answered: "Ah! that requires general relativity!" This answer was absurd, for the said theory concerns gravity, not magnetism. Horacio and I were sadly disappointed. But I confess that I still don't know the correct answer.

Ernesto Sabato

Sabato, soon to become a famous novelist, held the post of assistant doctor, charged with guiding the first steps of budding researchers. He was popular with us because he was accessible, friendly and generous, and always ready to expatiate on a variety of subjects: literature, art, the tango, politics, and occasionally even physics – except magnetism, which he professed to hate. The Sabato who was an enemy of science, believer in the occult, in a permanent state of indignation, and who used poetry books as shooting targets, emerged much later.

I befriended Sabato as I was finishing my first year, when he saved me from being flunked for failing to answer an elementary question that Professor Loyarte had asked me. In the summer of 1939 he asked me whether I was interested in working on cosmic rays, which were intriguing because no one knew how or where they originated: we only knew that they were a motley of electrically charged extraterrestrial particles. I answered him immediately with alacrity. Sabato replied saying that he was only waiting for some pieces of apparatus that the Institute had bought in France. But that equipment never arrived, so the project aborted.

In 1941, Sabato assigned my buddy Jorge Bertomeu and me the projects we were to complete in our fourth year. They consisted in finding the characteristic curve of a triode, and in analyzing the spectrum of the blood of a crime victim, to try and identify the poison he had been given. The first task was one in electronics, the second in spectroscopy. We had a hard time performing both tasks. We had to surmount three obstacles: bibliographic scarcity, our supervisor's scientific deficiency, and his insistence on dragging us to his home when he got tired, to talk about what interested him at the moment – never a physics problem. The first problem was that electronics had not yet reached our university, although it had been born three decades earlier. The only ones with some knowledge of the subject were short-wave radio amateurs, who had some know-how but no theoretical knowledge at all. Of course there were plenty of electronics books abroad, but they were not available to us because there were no courses on the subject.

Sabato suggested that we learn from popular magazines for radio buffs. We got from them some intuitive knowledge about circuits with electronic valves, but they were scientifically contemptible, for they did not tell their readers that what moves electrons across valves is an electromagnetic field: they pretended that the interactions among currents were mediated by particles. Thus, they unknowingly drew cartoons of Feynman diagrams.

Something similar happened with spectroscopy. Despite being the Institute's specialty, and the fact that all the doctoral dissertations had been about it, none of the La Plata experts had bothered to write a manual on the subject. Maybe they too had only know-how. I was lucky to find Gerhard Herzberg's recently published book on the subject. Herzberg had migrated from his native Germany to Canada, where he performed the work that earned him a Nobel Prize three decades after Jorge and I were learning on our own how to use a spectrograph and the associated instruments.

All of Sabato's physics was what he had learned in La Plata: he never studied any of the articles in the main physics reviews, even though they were available in our Institute's library. He did not do research and consequently did not publish in research journals, not even the local ones, where Loyarte and a few others had published some modest papers.

Presumably, Sabato had not advanced any further because no one demanded more from him; because the project he had been assigned as a postdoctoral fellow in Madame Curie's laboratory was a routine one; and because he was seduced by literature and politics. Only very talented and focused individuals with a strong scientific vocation, like Bernardo Houssay and his close associates, as well as Richard Gans, Enrique Gaviola, Félix Cernuschi, Eduardo De Robertis, Braulio Moyano, Cristofredo Jakob and a few others scientists kept producing high level work, even without any demand, stimulation, or peer recognition. These were absent in Argentina at that time: anyone with a doctorate could get a job that allowed him to survive. In particular, Sabato, his patient wife Matilde, and their quiet son Jorgito subsisted on Ernesto's modest salary.

In sum, Ernesto failed in science for lack of guidance and discipline, and left the field before really getting to know it intimately – through doing original research. Besides, he was basically an artist: he wrote and drew with great ease, and preferred the company of Parisian bohemians to that of scientists. The latter did not seek Sabato's company because he had done nothing to deserve being addressed as *cher collègue*. He had been an outsider in the Paris scientific world, and could not be an insider in La Plata, where at the time there was no inside.

Sabato had also been passionately interested in politics. Indeed, he made his first trip to Europe representing the Argentine communists, but someone in Brussels told him the truth about Soviet domestic politics, and converted him to Trotskysm. In short, when I first met him, Sabato had lost his youthful interest in both science and political activism, though not in political chat, and was trying to become a professional writer, a goal he soon attained. So, despite his good will and friendly attitude to us physics students, Ernesto could not perform well his task of giving direction to budding scientists – something that only experienced researchers can do.

My Closest Fellow Students

We physics students were few and interacted strongly and well. Initially I had frequent long talks about physics with Marco Antonio Poggio, who drove us in his uncle's black Citroën to the shore of Río Santiago, where we chatted on physics under a tree. Marco was indignant when, sharing with our teacher Dr. Grinfeld his repugnance over the way the Nazis were treating the German Jews, he compared it with the discrimination against Blacks in the USA. Grinfeld replied angrily: "You cannot equate Jews with Negroes!".

Later on Marco joined the circle of Fidel Alsina Fuertes. Fidel, several years our senior, had an engineering degree, held clear and firm if mostly wrong ideas, and was charismatics. He also had a strong spatial intuition, which helped him in some areas while obstructing his access to abstract matters. A few years later, along with Fidel, Marco and a few others, we formed AEF (Agrupación de Estudiantes de Física). There were about 30 of us, among them my friends E. Jorge Bertomeu and Jacobo Meyer Goldschvarz. Our task was obvious: since our teachers did not read journals, conduct research, or discuss physics with colleagues, we would do all that. Unexpectedly, we succeeded. AEF's main activity was the seminar that met Saturdays in Marco's large drawing room with a blackboard. His charming wife and her friends served us delicious *empanadas* and non-alcoholic beverages. At the seminar we expounded on papers recently published in the best physics journals, and reported on our own research. Among the latter were a report by one of us who was building a micro-cyclotron in his garage; a mechanical model of an organism imagined by Fidel; and my own failed attempt to build a theory of the electron that, unlike Dirac's, would comply with the positivist demand *Thou shalt employ only measurable quantities!*

Other Classmates

Horacio Brodsky had entered university at age 14. He could easily be spotted due to his large blond mane, which earned him the nickname The Mane Kid. He was shy and affectionate, and sought my advice on everything. He boarded the same train to La Plata as I, carrying a bunch of bananas and some comic magazines – the same that my two children read. Although he looked and acted as a preadolescent, Horacio's brain was remarkably adept at higher mathematics. He understood quickly and proved ingeniously any theorem one threw at him. We studied together functions of a complex variable in the apartment where I lived with Julia and our two children. Our first-born, while still a baby, once peed all over Horacio when he bent down to look at him.

In the laboratory Horacio and I worked on the same team. Taking advantage of his young age and timidity, we exploited him shamelessly: we assigned him the most tedious tasks, such as revealing photographic plates in the dark room. Horacio worked conscientiously and with few complaints. No one doubted that he would develop into a brilliant scientist. Regrettably his father, a pharmacist, ordered him to transfer to the pharmacy faculty, so as to eventually succeed him in the family business. Horacio obeyed and thus lost not only a brilliant career but also his beautiful mane and, with it, his nickname.

Another close and dear classmate was Jorge Bertomeu. He had all the virtues of his Catalonian ancestors: he was serious, hardworking, responsible, and taciturn. With him I studied analytical mechanics and electrical measurements, a subject on the electrical engineering curriculum. Jorge and I studied it because we wished to understand how our electrical and electronic measurement instruments worked. Since we found no textbook on the subject, we used the excellent lecture notes that Juan Sabato, Ernesto's brother, had written for his course in the Faculty of

Engineering. I used to arrive in Bertomeu's modest home a little after 8:00 a.m., and a couple of hours later his mother would serve us large cups of white *mate*.

An even closer classmate and friend was Jacobo Meyer Goldschvarz, whom we called Cobo, a nickname that stuck. He was very different from the rest of us, for he had been a naval cadet and had manual skills. Cobo loved baroque music and admired his sister Julieta, the only Argentine clavichord virtuosa and a student of the famous Wanda Landowska. Their father, an *haute couture* tailor, had traveled from Russia to Zürich during the First World War to meet Lenin, one of the few socialists who had opposed the war. Nobody knows what gave him the idea that his son could make a career in the navy of a nation that was even more antisemitic than conservative. The very idea of an Admiral Goldschvarz was just as ludicrous of that of a Bishop Goldschvarz.

Cobo and I visited each other frequently, read some of the same books, and for a while were loved by the same woman, Liliana, a pretty and elegant Italian physicist. Once, when he and his wife Coca were driven to *El ombú* by another couple of friends to help celebrate one of my birthdays, they were hit by a drunken playboy. All four friends were taken to hospital. Coca's gall bladder had exploded, and Cobo suffered a concussion that wiped out his recent memories and left him mentally incapacitated for a year. This accident did not alter our encounters, but it persuaded him to quit his studies.

Cobo, who had earned a BSc, made a living building measuring instruments for the industry, and after Perón's fall was appointed technical secretary of the Physics Department at Buenos Aires University. He and our friend Waldemar Kowalevsky were the only ones in our circle who knew how to use their hands, and could not care less for the latest sophisticated if ephemeral fashions in theoretical physics. Failed at Buenos Aires, Cobo triumphed at the prestigious Delft Polytechnic in the Netherlands, where he could materialize his love for low-temperature physics and buy the car of his dreams: an MG that he crashed the same day he got it.

In Delft, Cobo learned Dutch, got his PhD, set up his own lab, and supervised 50 doctoral dissertations. Upon retiring, he returned to Buenos Aires, where he was ignored once again. Shortly thereafter he died in hospital from an ill-diagnosed disease. We met only twice during all those years: once in Delft, where he drove me to admire Vermeer's tender art, and later in Geneva. In Buenos Aires I had taught Adriana, his only daughter, and advised her to switch from philosophy to sociology. She took my advice and was happy in her new field, but we never met again.

Other good classmates, but not as close, were Abelardo Tejo, who had a strong physical intuition and taught the explosives course at the naval school despite being an anarchist; Gustavo Rothe, an amiable giant so absent-minded that he once rested his elbow on a Bunsen burner; Antero M. Bueno, a Peruvian professor who, having published a book on atomic physics, realized that it was time to learn something about it; and Cecilio Weinstein, Grinfeld's only PhD student, who was also enrolled in the Faculty of Philosophy, where he must have been a model student.

Students of Other Disciplines

I got to know several students at different faculties at the university restaurant, where one could eat lunch for 45 cents, the equivalent of a dime. The menu varied only slightly, and always started with a greasy soup covered in cabbage aphids. One of the habitués was an indigent provincial, who poured on it all the salad oil and the grated cheese that remained. At the beginning, just to keep up with the others, I also drank a glass of wine for five extra cents, but it made me dizzy and sleepy, so I did not repeat the experience.

At those tables for four we always had interesting discussions about a large variety of topics. Television, cell phones and the like were still in the future, so that face-to-face conversation was still practiced and enjoyed. Such direct communication, which we are losing, gives and demands far more than contacts via electronic media, the abuse of which is weakening the sensory and affective bonds that used to generate and maintain cohesive social systems. What we have gained in information and speed, we have lost in warmth, solidarity, and cohesion.

Two of my friends outside physics, whom I saw again many years later, were the mathematician Eduardo Zarantonello and the biochemist Samuel Kaplan, whom I called Sammy. Eduardo, who later taught in the USA, finally settled in beautiful Mendoza, at the feet of the majestic Andes cordillera. He had a clear mind and explained to me Hamilton's potent and beautiful equations in a few minutes while we stopped in the Institute's park in midwinter. With Sammy we talked about science and politics, and exchanged Jewish jokes. One of his favorites was this: One individual tells another that Jews and bicycles are to be blamed for everything that is wrong in the world. Invariably, the listener asks, why bikes?

Integration in La Plata Society

I was not as well integrated as most of my classmates in the local society because I led two distinct parallel lives in addition to the one in La Plata: my family life about 100 km away, and my extracurricular activities midway, in Buenos Aires. I slept so little that I often dozed off in theoretical classes. I never participated in the student movement because it seemed to me to be more political than academic, and what I expected from university was knowledge, not power – which anyway did not interest me. All the student parties, even the ones inspired in the 1918 university reform, only wanted to gain delegates to the faculties' governing boards. With few exceptions, such as the Buenos Aires science students in the 1950s – some of whom I had the privilege to teach two decades later – the student leaders did not support scientific research. As I write, the Catholic students of the national university of Córdoba are organizing a ceremony of "lecture notes blessing." Apparently, they have not heard about Gutenberg's achievement, which rendered lecture notes obsolete – only a means for bad teachers to hold on to ill-gained chairs.

It is true that the 1918 Córdoba Manifesto mentioned science and the need for the university to keep up with it. But in fact the famous university reform was political rather than academic, and harmful in the long run because it politicized the Argentine university and set a bad example for the rest of Latin America. So much so, that just 1 year later, when the physiology chair at Buenos Aires University had to be filled, the student representative voted against the only candidate who had engaged in scientific research – Bernardo Houssay, the first Argentinian to earn a Nobel Prize in science.

Before entering university, I had attended some meetings and marches organized by the "reformist" student leaders. One of them, Ernesto Giúdice, a medical student, was also a communist and a reporter for *Crítica*, the popular evening daily. When I asked him when he expected to graduate, he told me that he was delaying taking his last examination in order to keep up his student leadership standing. I was shocked, for the Argentine university was free, and presumably taxpayers paid for students to study, not to train as politicians.

In short, I never participated in university politics. However, because I had an academic reform agenda, I accepted to run as a councillor, and was elected for 1 year, at the Faculty of Philosophy in Buenos Aires in 1958, and at my alma mater a year later. In the former, our democratic slate defeated the conservative one headed by Jorge Luis Borges, the exquisite miniaturist who held the grossest political opinions (see Peicovich 2007).

In the Faculty of Philosophy, the student representative, Eliseo Verón, a leader of the "reformist" movement, joined my learned conservative colleague León Dujovne in opposing every one of my proposals for academic reform, from keeping the faculty open on Saturdays to adding teaching assistants to all the chairs, to gathering the anthropologists into a new department, to allowing Gregorio Klimovsky, an excellent logic teacher, to teach an alternative course to the one taught by Eugenio Pucciarelli, a remnant of the Peronist university who had not learned any modern logic. While Dujovne was our best expert in Spinoza and his period, Verón went to France, where he became an adept of Claude Lévi Strauss's armchair sociology, and became known for his publications on the social as linguistic, and on social studies as a part of semiotics. (Social issues can wait, whereas academic careers cannot.) On returning to Argentina, Verón became the guru of Eduardo Duhalde, a prominent right-wing Peronist leader and statesman. However, let us go back to my freshman year.

Universidad Obrera Argentina

On starting university I realized that, since my compatriots were paying for my studies, I had a duty to repay them. After gathering some information on what used to be called popular education, I decided to found the Universidad Obrera Argentina (UOA). I wanted this school to teach both vocational and humanistic studies to adult workers, because there were no schools where they could learn anything other than some household and office skills, and union activists had nowhere to undertake social studies.

I drew up a charter, rented an office downtown, bought a desk and a few chairs, had some letterhead paper printed, and invited my friends to join me. None of the engineering or science students replied. Only Julia, the commerce student Elsa Karothy, and two poets – the anarchist Emma Barrandeguy and the communist Gerardo Pisarello – volunteered. We paid visits to the secretaries of several unions, and offered them courses in Spanish language – the only subject we could offer at the time. Only the metallurgical and the wood unions were interested, and they got our two poets, who volunteered enthusiastically.

This modest activity alerted three very different institutions: the Social Order division of the Federal Police, the Order of the Calatrava Cross, and the Construction Union. The police summoned me to tell me that I had to keep them informed about all of our public activities. The Calatrava fascists, whose den was next to us, attempted to invade our office, and made such a racket, that the building manager gave us notice. Ironically, four decades later, along with the Prince of Asturias prize I was given a pin with the strange and beautiful Calatrava cross.

My meeting with Rubens Iscaro, a leader of the powerful Construction Union, then dominated by communists, was discouraging. He told me that, since they were about to launch their own workers' school, they saw mine as a nuisance. I assured him that we would not offer any courses in construction, but on the other hand we would teach social sciences, which were not included in their curriculum. (I suppose that he believed that Marxism had rendered social studies redundant.) He was not satisfied, and I never saw him again.

The following year we moved to a larger room, also downtown. There I installed a blackboard and a lot of chairs. We offered a number of public lectures, as well as a course on Diesel engines, which were becoming increasingly popular. Our most popular lecturer was the law professor and socialist parliamentarian Carlos Sánchez Viamonte, aka Carloncho, who also accepted to figure as our director. The Diesel engines course was taught by a charming Welsh engineer familiar with the noble British tradition of taking science and technology to the masses. The plainclothes policeman assigned to report on us soon tired of attending his lectures, and invited me to the nearby café for a chat. There he proposed to me a gentlemen's agreement: he would stay in the café, and at the end of every lecture I would inform him about "what had been discussed and resolved." So we did. For example, I would inform him in all seriousness: "We compared Diesel engines with internal combustion engines, and decided that the former are more efficient than the latter." He took note, and we were both satisfied. I miss the time when gentlemen's agreements were kept without any fuss.

Marriage

At the beginning of 1940 the UOA moved to a house across from Parliament. Before moving we had to clean and vacate the former locale. My mother came to help me, holding a broom. During a pause she told me without preambles, as was her wont:

"Papi and I think you should marry Julia." This news surprised me, because marriage was not on my horizon, and because my father did not have a good opinion of Julia's intelligence. Besides, she was 7 years older than me. Still, I agreed, and that same day proposed to Julia, who consented. We agreed to get married in Uruguay, so as to be able to divorce if necessary. (Eventually Julia reneged on this pact.)

A few weeks later the four of us crossed the Río de la Plata accompanied by our wedding witnesses, Manuel Sadosky and Cora Ratto, and boarded the train to Maldonado, near Punta del Este. There Julia and I were married at the public registry by a stuttering officer wearing a colorful band. The first decade of our marriage was happy, and we had two sons, Carlos, aka Cantarito, and Mario Jr., aka Bambi. Eventually, Cantarito married his classmate Annick Vivier, and became a reputable computational physicist. After trying Venezuela and Brazil, they settled in México with their four children. Bambi became a calculus teacher and married a charming physician, with whom he had two children.

At the beginning of our marriage, Julia talked to me about her long study trip to France and Italy, taught me art history, and incorporated me into her big extravagant family. I became close friends with her brother and colleague Mario, and their sister Alicia – a helpful and lovely if empty-headed woman married to Roberto Ledesma, my favorite Argentinian poet. She was also a close friend of Borges, and kept us informed about his household.

Julia soon lost her interest in architecture, hardly worked at her profession, and stopped learning. Consequently we drifted apart and eventually divorced. She had been the star pupil of *Monsieur*, the old-fashioned French director of the School of Architecture in Buenos Aires, and predisposed me against modernist architecture. Much later my son Eric, a creative architect, taught Marta and me to appreciate modern architecture to the point that, while in Geneva, we lived at 1, rue Le Corbusier, and now occupy a glass-and-steel apartment in one of the dark towers designed by Mies van der Rohe, of Bauhus fame.

First Children

My Argentinian children Cantarito and Bambi survived the Peronist schools, though with some incidents. For instance, neither of them cared for religious instruction, which had been reintroduced by Perón. On noticing this, Miss Porota, their second-grade teacher at Olivos School, asked them:

- *If you are not Christians, which gods do you worship?*
- None. We are atheists.
- *What's that?*
- Atheists do not believe in any gods.
- *What a stupid race!*

When Bambi finished fifth grade, I suggested that he quit school and study at home, to avoid rereading Evita's pathetic *My life's reason*, said to have been produced by a ghost-writer. Bambi did very well and, on passing his last elementary school exam, I asked him to name his graduation present. He chose an afternoon at the amusement park, on whose trains and wheels he had a great time while I felt sick. I suppose that developmental psychologists have found out why children feel so attracted to peril, that they love the Ferris wheel and can make fearless soldiers.

Cantarito was always a model student: smart, quick, and disciplined. He was also very sociable, to the point that, when I asked him to count his intimate friends, he took his address book and counted 75. Bambi was remarkable for his generosity and readiness to help slow children and stray dogs. He too was curious and hard-working, but impatient. He was happiest at the work-shop, and his dexterous fingers looked to me like screw drivers. I therefore suggested that he enroll in a vocational school rather than in a bookish high school. He accepted enthusiastically, and we went to enroll him at the reputable Otto Krause Industrial School. He was rejected on the basis of a report signed, surprisingly, by a priest who declared that "children of good families should not work in manual professions."

That happened in 1955, when Perón's honeymoon with the Catholic Church had ended, but the Church still exerted a strong influence on public education. So, Bambi attended the standard high school and ended up studying mathematics, which he taught to the beginning students of the Faculty of Sciences of the Buenos Aires University. He was a dedicated and popular teacher; and, when his salary became inadequate, he put computers together, with the programs that his customers needed. So, vocation won in the end. But the law terminated his teaching career at age 65, whereas I was able to keep teaching till 90.

Two years after that disgraceful ecclesiastical interference, I decided to divorce Julia, and asked our children to choose the parent whom they wished to live with. Cantarito chose me, and lived with Marta and me until he married Annick, and both went to the USA for their doctoral studies. Bambi chose Julia and we became estranged until Marta, my new wife, sought him out and persuaded him to see me.

Marta and I keep in close touch with our two Canadian children and my two Argentinian ones, who are scattered over four countries, and have given me nine grandchildren and nine great-grandchildren. I ought to apologize for having contributed to everpopulation, but thinking of my offspring makes me happy and often proud.

Universidad Obrera on Congress Square

Shortly before getting married, I had rented a whole house overlooking Congress Square. The UOA occupied an office and two classrooms. I sublet an office to the auctioneers Cajigal and Velón, a room to a Lithuanian woman, and the garret to an individual who read *Pampero*, the Nazi evening daily. Our tenants, none of whom seemed busy, paid only for the first month, but caused no trouble. I filled one of the large rooms with drawing boards, and the other with chairs. An engineer taught a course on industrial design, and I taught one on the physics of everyday life. I also recruited a governing body, that gave three men the chance to talk at great length on matters they knew little about: the delegate of the state employees union, a socialist who knew and respected our charter much better than me, and a communist who regretted that we did not teach his hobby – the theatre.

A socialist friend of the UOA noted that we were ill situated, and suggested that we move to a workers' quarter, where our potential students resided. I found a suitable house near the Constitución train hub, owned by the boss of the Cinzano bottling plant. His office, where I went to sign the location contract, was decorated with a large portrait of Mussolini. He disliked my declining the glass of vermouth that was expected to guarantee our contract even more strongly than my signature. After all, who was I but a student of a nonstandard subject?

I spent the summer of 1941 outfitting the new locale, designing and ordering cheap but sturdy desks and chairs, as well as metallic plates with the names of the classrooms, among them Michael Faraday and James Watt. The school occupied the entire house except for a small apartment at the back, which was occupied by Arsenio Florentino Sesto, the electrician recommended by the state employees union, whom I hired as caretaker in lieu of rent for his lodging. His wife Teresa, an Italian seamstress born in Tunis, kept the house clean, while he became increasingly involved in the school's administration.

Arsenio – short, strong, skillful, honest, and as ugly as intelligent – had read a lot in his youth, and had a healthy common sense. He soon became my second in command. Together with my secretary Adolfo Morinigo, a nephew of the Paraguayan dictator, we formed the triumvirate that took care of the day-to-day management problems. Adolfo too was a well-read electrician but, unlike Arsenio, quite shy. Both Adolfo and Arsenio soon became my trusted friends and my line to the working class.

Three years later, when the police closed down the UOA, evicting Arsenio and his wife, I offered them one of the small houses that my father had left me, payable in monthly installments of five pesos – roughly one dollar. Since his home was across from mine in West Florida, Arsenio and Teresa attended all my parties. When Teresa gave birth to their only child, I sent her to the obstetrician who had been a classmate of my father's, and he returned my cheque for his work. The medical profession has been accused of commercialism. This is slander, particularly in the Third World, where doctors treat many patients without expecting remuneration.

Arsenio died a few years later from an undiagnosed illness. To avoid the complicated and expensive transaction required to transport his corpse from the province to the nation's capital, it was placed on the passenger seat of a car, with a cigarette butt in his mouth. When the car crossed the checkpoint, the guard was told that the passenger was sleeping it off. The next day a colleague of his and mutual friend told me that Arsenio had lately converted to Peronism and, given his education, had been assigned to read aloud the daily dose of Evita's book *La razón de mi vida*, the Peronist bible. I was surprised, because Arsenio had never told me about this. Maybe it was a ruse to keep his job at the Ministry of Public Work, or maybe he saw in early Peronism merits that we, the "contras," refused to acknowledge. Perhaps this is why our opposition was ineffective and even perverse, to the point that, when the Peronist movement was outlawed by a military dictatorship, the famous writers

Borges and Sabato went to the Casa Rosada to have lunch with the sinister and bibulos president, General Jorge Videla. Borges declared then that "the President would be mad if he called elections," and Sabato called him "*una bellísima persona*" – an excellent person.

In any case, an objective study of Peronism is yet to be conducted. It will be a hard task to perform, because that movement defies the classical political categories imported from Europe, and because it is still very much alive, though even more complex and fragmented than half a century ago. This should not surprise the admirers and detractors of Theodore Roosevelt or Winston Churchill who only see one side of these shrewd political polygons.

What the UOA Taught

The UOA offered elementary courses in mechanical, electrical, and chemical engineering, as well as a 2-year course in the humanities and social sciences for union activists. This course included Spanish language (taught by Señorita Konstantinovsky, whose enthusiasm compensated for her inability to master the letter R), world history (taught by Julia), Argentinian geography and history (Rodolfo Puiggrós), economics (Ricardo Olivari), labor law (Arturo Frondizi), and history of the Argentine labor movement (Juan Atilio Bramuglia).

A few years later, Puiggrós jumped from Marxism to Peronism, and Bramuglia from socialism to Peronism. (One of General Perón's abilities was to attract able individuals from all traditional parties, offering them important positions that they would never have attained in their original parties.) Overnight, the obscure Bramuglia became the country's brilliant minister of foreign affairs and champion of the "third position," a precursor of the movement of non-aligned nations that Nehru, Sukarno, Tito, Nasser, and Nkrumah hatched 6 years later to the chagrin of the US government and its main allies, the three remaining colonial powers. Arturo Frondizi – brother of my friend Risieri, the philosopher – who was a serious student of the Argentinian economy, remained in the Unión Cívica Radical party, which had been in power between 1916 and 1930. He was elected president in 1958, and retained this post until 1962, when the CIA deposed him with the help of the usual suspects – the military and the Church.

Frondizi's political career was unusual. It started in 1938, when I asked him to speak at a dinner organized by the union of gastronomic workers in support of the Spanish Republic, which was on its last legs. My father, whom I had consulted on this matter, suggested that I try his lawyer, a certain Arturo Frondizi, who shared his own interest in the oil industry and belonged to the left wing of the "radical" (liberal) party. Frondizi was taken aback by my invitation, and told me that he had never spoken in public, to which I replied: "Well, this is your chance to start." He accepted, and his rousing speech won resounding applause from the 1000 or so diners in the huge dining hall.

This event launched Frondizi's public career – the only political success in my life. His presidency was harshly criticized for involving compromises with all the political groups – and thus earned the dissatisfaction of all. But who, except for the Mexican president Lázaro Cárdenas, succeeded in Latin American politics during the last century? However, back to the UOA's teaching staff.

One of our teachers was an engineer called Picasso. He was an amiable Italian expat who missed his favorite dish, the *bagna cauda,* whose recipe he would recite in detail and with poetic unction. Another teacher was the electrician Eduardo Lapacó, who never tired of drawing on the blackboard what he called "Ohm's medallion." He was also a communist and a member of our governing board, at whose meetings he opposed every one of my proposals but suggested no constructive alternatives.

Lapacó's campaign culminated with a meeting convened by my good friend Manuel Sadosky, who, after praising my commitment to Marxism, accused me of behaving in a dictatorial manner. He was just applying the traditional communist's tactic of fighting whatever they could not control. I paid no heed to the accusation, and kept managing the school according to my own lights. My horsefly flew away shortly thereafter, and I took over his course, emphasizing alternating current circuits, which he had ignored. This episode illustrates a general law: because NGOs are self-governing organizations of public service, they do not like authoritarian politicians.

In 1941 I rented a second site for the UOA: a large four-level seigneural mansion near the other seat. We set up chemical and electrical engineering laboratories in the basement, a mechanical workshop on the first floor, and classrooms on the two upper levels. Near the entrance there was a place for our volunteer bureaucrats, as well as an office where the teachers could meet, and write on a large book the subject they would teach each time.

One of our teachers, a charming engineer whom I will call Denpirasi, held court every day. He was gregarious and had good friends at City Hall, where he was employed. He phoned a colleague every time we got a letter ordering us to vacate the labs we had set up in the basement, in clear violation of a municipal law. Denpirasi's connections worked every time.

From that time on, I concentrated on our teaching work, which involved about 60 teachers and 1000 students. Every evening, in an office with a few chairs, I met with staff members to address their concerns, and with the student representatives (one per course) to discuss their grievances and suggestions. To discourage people who came just to chat, I hung a poster with the legend "If you have nothing to do, do it elsewhere."

We offered courses in a dozen skilled occupations, as well as in humanistic studies. All the teachers in technological subjects were engineers, vocational schools' graduates, scientists, or students like myself. Every teacher was paid 40 pesos per month per course, and was expected to write three pages of lecture notes for every class hour, which were mimeographed and distributed at the start of each lecture. Our students came to class three times a week in one of two shifts: from 6:00 to 8:00 or from 8:00 to 10:00 p.m. Since they arrived straight from their workplaces, they were tired, and so could not keep up their attention during lectures. Consequently I replaced the traditional lecture with self-study in groups of four. The teacher was available for consultations. To implement this didactic change, I replaced the classical classroom with small desks and chairs for four persons each. The students read the lecture notes, discussed them among themselves, and asked the teacher for help only when none of the four could resolve a difficulty.

In addition, the students in technical courses did some lab or workshop work. The chemical practices were supervised by Naum Mittelmann, a research chemist; the electrical engineering ones by the university professor Carlos Bianchi; and work at the lathe and drill by a couple of articulate foremen. My friend Isaías Segal, an engineer who worked at the factory of the Beretta handguns, taught theoretical mechanics. A few years later he felt honored to share a jail cell with the legendary Alfredo Palacios, the first socialist parliamentarian on the continent.

We also held small economics and philosophy seminars. But, since there was no research in either field, nothing original came out of those meetings. We also had a library of 5000 donated volumes, with my friend Friedman in charge, who catalogued them and took note of the loans. A large table and chairs for readers occupied the center of the library.

The UOA was self-supporting. We charged 6 pesos per course per month – a little more than the daily wage of an unqualified industrial worker. The students in the humanities courses held scholarships paid for by their unions. We also got a few small donations, such as books and the demijohns of chlorhydric and sulphuric acids brought by a student who owned a small chemical factory.

Public Impact of the UOA

The UOA kept a low public profile but it was a member of the Argentine Federation of Popular Education Societies, which counted about 400 schools and libraries scattered around the country. I was elected secretary general of this society, and organized its first and last congress, which was held at the UOA. There were about 50 delegates, every one of whom read a paper. Mine was included in my book *Temas de educación popular,* which was published in 1943 by the bookstore El Ateneo. In that book I gathered several essays of mine on adult education, a subject that had been overlooked by nearly all the experts in education science.

The leaders of that Federation were a mixed lot: from retired schoolteachers to the Lieutenant Colonel (retired) José A. Vila Melo, whom I befriended. He was a friendly and courteous provincial, an anticlerical who presided over a secular Boy Scouts society dedicated to cultivating the so-called national traditions. In the winter of 1943 we met at a café across from Plaza de Mayo, where he told me in confidence that the man destined to save our country from the decadence and corruption into which it had sunk, was about to return from Italy, where he had been sent as our military attaché.

I was then one of the few whom the appearance and meteoric ascent of Colonel Juan D. Perón a few months later did not surprise. What did amaze me was that a retired army officer, Vila Melo, was well informed about the plans and activities of a secret society, the GOU (Group of United Officers), which quietly put Perón at the helm of a brand new ministry set up for him: the Labor Secretariat, from which he contacted many union activists and leaders of non-governmental organizations. The president of our federation, an old retired teacher, let me read the friendly letter Perón had written him out of the blue.

Shortly thereafter, as soon as the UOA gave out the year's certificates, a police squad searched, ransacked, and closed down the school's two houses. It was said that the order had come from Perón's office. One by one, many other NGOs, among them the Colegio Libre de Estudios Superiores, suffered the same fate. But of course the main victims were the trade unions, some of which were assaulted by newly minted bullies faithful to The Leader. Authoritarian regimes destroy the civil society (sometimes wrongly called 'social capital') because they stand between the individual and the state. As Mussolini had written: under such regimes "Everything in the state, nothing outside the state, and no one against the state."

The policemen who invaded the two sites of the UOA destroyed in minutes what had been functioning with increasing success for 6 years. And they stole everything they could, from the petty-cash box to the machines, measuring instruments, and books. The latter, marked with a large UOA seal, surfaced shortly thereafter in second-hand bookstores. I used the balance of our money, deposited in the Bank of Boston, to pay for the rent in accordance with the rental contracts that I had signed. The only document the police left behind was the large teachers' attendance book. Not a single copy of the leaflets, posters, and other printed material was left. Hence, anyone wishing to write the UOA's history would have a hard time. Presumably, the same holds for all the other Argentinian NGOs of that period. The most radical historical revisionism is the one that involves the destruction of the relevant documents.

When my father died, on August 1st, 1943 I resigned my job at the UOA. My vacancy was filled by Germán López, a lawyer whom I met again in 1985, when President Raúl Alfonsín invited me to visit him at the presidential residence in Olivos. The school's governing body organized a farewell ceremony and presented me with a beautiful gold medal bearing the UOA badge. A pickpocket stole it the next day while I was riding a bus.

Father's Death

At this time, 1943, my father suffered a massive brain stroke. He had always endured high blood pressure, which at the time was untreatable. His condition was likely worsened by the police harassment he had been subjected to. In fact, he was under surveillance, had to comply with police summonses, and was jailed just because he presided over the federation of associations that raised funds for the allies that fought the Axis powers. The government that issued from the military coup of June 4th was openly sympathetic with the Axis, and it was the one that dismissed Bernardo Houssay from his chair and institute.

My mother phoned me immediately on my father's arrest and, although it was already night, I went to see the Socialist senator Alfredo L. Palacios to ask for his help. As young men, he and my father had been active in the Socialist party, but parted company two decades later. Palacios, respected by everyone for his integrity, reacted immediately. His driver took us to the central site of the federal police. Unshaved and haggard, my father hugged and kissed me as usual, and was released a few days later.

As the repression worsened, I suggested to my father that we hide in the provinces until the police forgot about us. He went to Córdoba, and I to Santa Fe. The news of his death came over the phone while I was having dinner with my friend, the mathematician José Babini, and his family. Earlier that same day, Professor Babini had been sacked by the new rector appointed by the military government. This was the Thomist philosopher and fascist activist Jordán B. Genta, who three decades later became the ideologist of the air force until murdered by a guerrilla group.

My father's wake and burial, in a rainy and cold day, were attended only by his brothers Jorge, the architect, and Eduardo, the rancher, his secretary and friend Guillermo Cavazza, some notable neighbors, former TB patients whom my father had treated without charge, my old gymnastics instructor, a few coworkers in the anti-fascist organization, my philosopher friends Hernán Rodríguez and Isidoro Flaumbaum, and many plainclothes policemen. None of the Sunday guests of yore dared put in an appearance nor even send condolences. The newspapers were just as cautious, but emphasized a fact that I did not know: that my father had won the gold medal of his 1900 medical class. Fear is the political equivalent of curare.

My father's death radically changed life in *El ombú*. I was now the head of a family composed of six people, but was jobless and without the possibility of obtaining regular employment, for I had no professional qualifications, the right connections, or the "good behavior" certificate that the police gave only to individuals with correct political opinions.

We made do, first squandering my father's small legacy, then trying my hand at several trades: physics tutor, translator, insurance agent, transporter, and manager of a family construction company. I was swindled five times, until at last I succeeded with the company I controlled. These adventures taught me that free enterprise benefits only those in control.

Mother in Jail

We vacated *El ombú*, hoping to sell it along with the surrounding land – which proved impossible due to the recession. My mother moved to an apartment in Belgrano, her favorite neighborhood in town, where she made some new friends.

She was invited to join a clandestine group that opposed the military dictatorship, but one of the conspirators, a retired colonel, denounced them to the political police, and all of them were jailed. My mother was sent to the Good Shepherd Asylum, run by nuns and reserved for the prostitutes who did not pay their dues to the local chief of police.

I went to seek the help of various notables, among them Carlos Saavedra Lamas, who had been awarded the Nobel Prize for brokering the peace treaty between Paraguay and Bolivia. Another was Monsignor Miguel de Andrea, the only antifascist Argentinian prelate and a follower of Don Luigi Sturzo, who had to leave Italy when the pope surrendered to Mussolini. They all excused themselves: they did not know anyone in the new government. My mother was freed a month later, but was not intimidated. Three decades later, during the dictatorship of Isabelita, Perón's widow, Mariechen was shot at while reading in her armchair after returning from the open-air fair, where she had been bad-mouthing that sinister government. The bullet grazed her and left a hole in the windowpane. But Mariechen, a nonagenarian, went on saying what she thought. She felt at home in her neighborhood, where she had lived for over half a century, and had made many good friends, who kept an eye on her. When she went to the Black Forest to visit Marta and me, she felt homesick after a few days of sightseeing. In her case, soil trumped blood, thus falsifying the Nazi myth of blood and soil.

As 1943 was drawing to its close, I faced the future with as much trepidation as expectation: although I was a born optimist, I also realized that, at age 24 and with a family to support, I had no regular employment and no prospects of getting one. I was also aware that my passion for philosophy both sustained and slowed down the scientific apprenticeship that I had just began. Yet, on the whole I felt happy and looked forward to completing my dual apprenceticeship – more on which in the next two chapters.

Chapter 4
Scientific Apprenticeship

From Student to Budding Scientist

My research apprenticeship began unexpectedly in the fall of 1943, when Ernesto Sabato introduced me to a Dr Guido Beck at a modest boarding house in a Jewish quarter of Buenos Aires. Beck was a cheerful, friendly, and unassuming middle-aged man, who smoked almost non-stop – his only vice. He did not volunteer to tell us about his checkered life. It took me several years to learn, bit by bit, why and how he jumped from one place to another ever since he earned his doctorate at Vienna University in 1925. He told me that one decade later he left for Odessa because the Austrian clerical fascists had overrun his beloved Vienna, but did not tell me why he left that place, which he had liked. Anyway, eventually he landed in Paris, which was still one of the world's scientific centers, but soon to be occupied by German soldiers and dumbed down by postmodernism and psychoanalysis.

We may never learn how Beck got from Paris to Lyon fleeing the German invaders, nor how he escaped from a Vichy camp and landed in fascist Portugal. Despite all his misfortunes, Beck was forward-looking, and far more interested in theoretical physics and in how to help it flourish in the remote corner where he had landed almost by chance – and only shortly before the pro-Nazi military coup that dismissed Professor Bernardo A. Houssay, the greatest scientist the country had produced.

When Sabato left, Beck invited me to sit next to him, and started writing formulas, commenting on them in a mixture of Portuguese, French, German, and English. It took me a while to learn that those vaguely familiar formulas were about the peculiar forces that hold the constituents of atomic nuclei together. I was just as frightened as excited. Beck was never known for his clarity, and anyway the subject was still so immature that someone had quipped that 'nuclear' was a misprint of 'unclear'.

Beck had just arrived from Lisbon to work at the National Astronomical Observatory in Córdoba, at the invitation of its director, Dr. Enrique Gaviola.

M. Bunge, *Between Two Worlds*, Springer Biographies,
DOI 10.1007/978-3-319-29251-9_4

The latter was our first observational astrophysicist, who had earned his doctorate in Berlin, and had learned his superb optical craftsmanship at Johns Hopkins under the fabled Robert Wood. (When I first met American physicists, nearly two decades later, they only wanted to know about Gaviola.) Beck would be my supervisor until 1951, an abnormally long period anywhere, especially in La Plata, where doctoral theses seldom took more than a year.

Guido Beck

Beck had been born in 1903 in the Bohemian town of Reichenberg (now Liberec), an outpost of the Austro-Hungarian Empire, known as Kakania, for being both *königlich* (royal) and *kaiserlich* (imperial). He attended high school in Switzerland during the First World War, and did some research under the supervision of Professor Hans Thirring. Since then Beck had held temporary positions in various places: in Leipzig (1929–1932) as Werner Heisenberg's assistant; in Odessa (1935–1937), where he supervised his first doctoral students; in Kansas City as a visiting professor; in Copenhagen as Niels Bohr's research fellow; in Cambridge as Lord Rutherford's guest; and in Paris (1937–1940) invited by Louis de Broglie, the only prince to earn a Nobel Prize. While in Leipzig, Beck wrote the chapter on relativity for the prestigious *Handbuch der Physik*, and in Kansas City the chapter on relativistic quantum mechanics in the famous Frank-von Mises treatise, successor to the Riemann-Weber (de Abren 1989; Nussenzveig 1995).

The German invasion of France pushed Beck to Lyon, where the Vichy government interned him in a concentration camp for scientists. The clarion woke them every morning calling the inmates "*À la recherche*!" In 1941 Beck escaped, and somehow landed in Coimbra, the main Portuguese center of higher learning, where he got a professorship thanks to De Broglie's recommendation. He adapted quickly, learned Portuguese to the point of reciting from memory entire cantos of Camões's long and pompous national epic poem, and pursued his deep study of Dirac's theory of the electron. His only doctoral student, Rodrigues Martins, had to move to Mozambique. Once, at a café, Beck was overheard making fun of Oliveira Salazar's fascist dictatorship. That joke landed him for several months in the famous Caldas da Rainha jail. The generous de Broglie intervened once again and wrote to Gaviola, who got him a job at the Córdoba observatory. When he arrived in Buenos Aires, Beck wore a black patch on an eye, which was diseased due to the bad diet he had suffered in jail.

Beck had no news from his mother and sister, who had remained in Vienna. At the end of the war, he learned that they had been murdered in a Nazi concentration camp. During one of his peregrinations to Bohr's quantum temple in Copenhagen, he had married a Danish girl who was forced to leave him when the Germans occupied Denmark under penalty of being interned in a camp for having married a Jew. By contrast, Kurt Sitte, a former doctoral student of Beck's whom I befriended much later, refused to divorce his Jewish wife and was interned in another camp, where he survived by writing doctoral theses for camp guards. When the war ended, Kurt rejoined his wife, and Beck's wife returned to him.

Beck worked in Córdoba between 1943 and 1951, the year he left for Brazil only to return to Argentina a year later to succeed his former pupil José Antonio Balseiro. The indefatigable Balseiro, who had died recently, had set up a physical research center later named after him. This was situated in Bariloche, at the feet of the majestic Andean cordillera, and some 1,000 km from politically restive Buenos Aires. This Center had the full support of the national government, in particular the armed forces, which disliked and feared city universities. In Bariloche one had to do science full time, which had always been Beck's aspiration.

Beck was happy in Bariloche, not only because he could concentrate on his research, but also because he could keep a horse, that he mounted wearing gaucho-style pantaloons and a *poncho*. However, in his interview with John Heilbron, Beck complained that his Latin American students lacked something characteristic of his Ukrainian students, and which my father too had noticed when he visited the USSR at about the same time: enthusiasm (Heilbron 1967). Could this be because the Soviet youths were building a new society, whereas their Latin American counterparts lived in fear of the next military coup?

Still, I believe that Beck underrated the young Latin American physicists's enthusiasm: his own best students, Balseiro and Nussenzveig, were surely as enthusiastic as they were capable, and so were Daniel Amati, Carlos Bollini, Juan José Giambiagi, Andrés Kálnay, Juan Roederer, and Miguel Angel Virasoro, all of whom distinguished themselves – and all of whom had to exile themselves. Nor did I lack in enthusiasm when working on projects of my own rather than on Beck's. After all, I have been publishing in physics, albeit intermittently, for six decades after Beck unleashed me in 1951.

In Córdoba, where Gaviola had given him *carte blanche*, Beck recovered his health. (The two men were on excellent terms but kept addressing each other as 'Doctor'.) He also adapted well to his new milieu, which was advanced by comparison with the Portuguese. Beck was even accepted as a member of the traditional Jockey Club, where he used to dine. Did his oligarchic fellow members know that he was Jewish and a socialist sympathizer? During the first 2 years of our association, I visited Beck every month, but never stayed more than a few days because I missed my family – something he neither understood nor forgave. The first time he took me to the observatory, he grabbed a medium-size telescope, and aimed it, without vacillating, at Jupiter and then at Saturn. That was the first time he had handled a telescope.

When summer arrived, Beck took me by bus to the Astrophysical Observatory in Alta Gracia, which Gaviola and Julio Platzeck, his disciple, had set up with their own hands, starting with the reflection telescope, which they had fashioned by hand and from scratch. In the village Beck and I rented a couple of horses and took a long tour of the hills. That was Beck's first horseback experience. Apparently he suffered no ill effects, whereas I could hardly sleep that night on the sleeper that took me back home. (Beck had got me a berth in the J carriage, which of course we interpreted as "Jewish.") Seeing that I enjoyed galloping, Beck christened me "Gauchito," while I called him "Chief." And he called Balseiro "El Niño" ("The Child") from the moment he learned that this is what the servants in Balseiro's home called him. Damián Canals Frau, another of his students, got the nickname "El Pibe." Beck treated the three of us as his children – the children of a solicitous but strict European. Beck was friendly, talkative, and fond of telling good jokes, many of them ethnic, which he embroidered with *gusto*. Regrettably for me, a family man, he had the Bohemian habit of working at night. He would check my calculations when I started to yawn.

Working Under Beck

The first problem Beck assigned me was to find out whether the specific nuclear forces conjectured at that time – one of which was named after the mysterious Majorana – were invariant under rotations and other transformations. I had misunderstood the statement of the problem, and so I performed a pile of unnecessary calculations on extra-large sheets of paper, and multiplied matrices by pairing rows with columns the way my teacher did, instead of using the more abstract and elegant method of using only the algebraic equations defining the matrices in question.

It took me several years to realize that, although Beck hit on interesting problems, some of them were too ambitious and that, lacking physical intuition, he centered on calculation problems and employed "brute force" to attack them. As I once said at a meeting of our physical society, we had been trained to obey the Duce's slogan, "Believe, obey, fight." In his last letter to me, Beck supported my criticisms of this view, which he had practised over decades, and went even further: he came to believe that the entire quantum theory was obscure by comparison with special relativity.

The results of that first research experience were a rather long paper in the local journal (Bunge 1944a) and a short one published in *Physical Review* (Bunge 1944b) – which was like peeking at heaven. Several other short papers followed, one of which appeared in *Nature*, another high shrine of science (Bunge 1945). Seventy years later, I experienced a similar elation when giving a talk on the so-called Aharonov-Bohm effect (Bunge 2014c) to Anton Zeilinger and his quantum optics group. I also discussed problems in the foundations of quantum theory with Zeigler, famous for his ingenious experiments on entanglement at large distances. In my view, these corroborate the hypothesis "Once a system, always a system," which supports systemism and refutes the long-held belief that there is no history at the microphysical level.

The set of my early papers lacks unity. One of the reasons is that Beck, following the recent European (or rather German) fashion, dragged his doctoral students around the field before locking them up in what American students used to call "concentration camp," that is, the doctoral dissertation. Such hyper-specialization may just produce many people conversant with a single arcane theme, such as string theory, and therefore likely to lose interest in the field, as well as being incapable of teaching elementary courses.

Beck's second motivation for postponing the start of my thesis was that he feared that, as was usual in my country, passing the last test would qualify me for a professorship, and he thought that I was not yet ready for that. He should not have worried on this count, for I lacked the political requisite, namely, membership in the Peronist party – a condition that most of my colleagues fulfilled without qualms, claiming that it was just like getting a vaccination certificate.

Gaviola, whom I befriended later, told me that Beck filled holes instead of constructing buildings – perhaps because he never felt rooted anywhere. Others believed that he was inherently restless – forgetting that he had been a serial refugee. The truth is that Beck pioneered the study of beta decay, as well as the layer model of the atomic nucleus. The physicists who continued working on these projects – namely Wolgang Pauli, Enrico Fermi, and Maria Goeppert-Mayer – were rewarded with Nobel Prizes.

Who is to know whether Beck too would have been invited to Stockholm had he settled in North America instead of South America? Speculation aside, Beck's most important accomplishment was to supervise more than 50 creditable doctoral dissertations in half a dozen countries.

During my monthly visits to my supervisor, I stayed at the guest room of his chalet in the observatory's large garden, where Gaviola cultivated his dahlias. I ate always the same unhealthy meals in the kitchen of one of the employees. I hardly interacted with the astronomers, who were rather withdrawn. Gaviola held that they had chosen to work in astronomy because their job forced them to work by night, without having to interact with normal daytime people. Incidentally, Enrique, who read widely in several sciences and in philosophy, was an amateur psychologist. Once he sent me a paper on the various kinds of intelligence, asking me to pass it on to a McGill psychologist. My colleague looked at it and thought that it was serious but outdated. Still, it was far more respectable than the psychoanalytic fables then in vogue in Argentina.

The senior computer in the observatory was no exception to the rule that observatory workers are withdrawn. I saw him excited only once, when he banged on my door to show me the errors he had found in the highly respected Jahnke and Emde table of functions – which for him was almost like meeting God at a house of ill repute. The man's only known hobby was to take off his lab coat at the end of his work day, and recalculate by hand values of important functions in hopes of finding mistakes in the standard tables. He was a natural Popperian.

My Doctoral Thesis

Beck proposed to me a dissertation problem that he rightly assumed would interest me for its philosophical flavor, namely, to find out whether Dirac's quantum-relativistic theory of the hydrogen atom yielded kinematic results similar to Bohr's semi-classical model, such as the electron's velocity. He also wanted me to use the complicated state (or wave) functions he himself had recently introduced. With hindsight, I realize that this was a strategic mistake, because new results come from new problems or new assumptions, not from making cosmetic alterations in the solutions.

In any event, I did not obtain the results we had hoped for. To begin with, Beck's state functions, which involved matrices, complicated things without clarifying them. I believe he had fallen for the Viennese saying "Why make it simply if it also can be done in a complicated fashion?" Second, Beck erred both in adopting the standard velocity operator and in the goal of the calculations – the spatial averages of the said operator not only in the case of the denumerable spectrum (the successor to Bohr's orbits) but also in that of the continuous one. The first mistake was hardly avoidable at that time, when we tended to believe Dirac's word as scriptural even though his velocity operator, namely $c\alpha$, is a 4×4 matrix whose eigen values are + infinity and – infinity, which would require an infinite energy for an electron to move. Later on I proposed more reasonable velocity operators (Bunge 1955a, 2003a).

Our second mistake was to use "wave" functions corresponding to stationary (or steady) states to tackle a kinematic problem. It ought to have been obvious that an electron in such a state does not move at all – just as Bohm's 1952 theory predicts. And this was precisely the main result of my thesis (Bunge 1960d). In other words, I proved that, contrary to what my thesis' title promises, there was no *kinematics* of the relativistic electron (or positron, or proton). In general, quantum mechanics does not compute trajectories nor consequently, velocities. Hence it is not a *mechanical* theory. The reason is simply that it does not contain a particle coordinate: its spatial coordinates x, φ and the like only identify points in ordinary space. However, it took me nearly two decades to understand this point (Bunge 1967a).

Jailed

The year 1951 was crucial for both the Argentine government and me: it was the first time that an important sector of the working class, namely the railroad employees, went on strike, and the year I spent time in jail for more than one night. A police squad came at midnight in late February shortly after a party celebrating a family anniversary, and proceeded to search my papers, neatly ordered in a hanging-files cabinet.

Fortunately, the detective disregarded the files containing sheets covered with mathematical symbols: he was not one of those learned postmodern idiots who hold that "science is politics by other means." The only document that caught his attention, and which he confiscated, was the file titled *Fundación*. He suspected that it dealt with the Eva Perón Foundation, a target of the opposition. Actually my file contained my proposal for a private foundation in support of writers and intellectuals, which I had written at the request of my friend Rogelio Galarce for his client, the bibliophile and Jorge M. Furt, owner of the old and famous ranch "Los Talas". (All Furt gave me for my effort was a copy of *Numerorum mysteria*, printed in 1599, by Petrus Bungus, a late Renaissance Italian scholar with a Pythagorean passion.)

To go to the police station I dressed up roughly as I am dressed as I write: white shirt, gray flannel trousers, and a blue Harris tweed jacket. I was taken to the West Florida police station, where I surrendered documents, money, wristwatch, pens, belt, shoelaces, cigarettes, and matches. I spent the rest of the first night sitting on the floor of a leaking zinc shed, under a heavy rain, wondering what would happen next. I felt greatly relieved when a drunk homeless person was thrown in.

I spent the rest of the week incommunicado in a bathroom. I was allowed to take baths and read, sitting on the floor, the books that Julia brought to the station. One of them was Francis Bacon's *New Atlantis*. I failed to realize that she had sent me a cyphered message by underlining certain words in it. I got no information; I did not even know what I was being accused of. But I saw clearly the savage beatings some of the prisoners got. Justice worked fast and cheaply under Peronism.

After a week I was transferred to the basement of the Central Police Department of La Plata, the provincial capital. The first thing I saw was a row of doorless toilets – like the ones in YMCA hostels – two of which were occupied by young identical twins. Everyone else surrounded me, asking me for cigarettes and matches, and wishing to know for the cause of my detention. I suppose they did not believe me when I said that I did not know it.

We were about 120 men, half of them common criminals and the rest "political," a category that included union activists and a couple of individuals – a physician and myself – who were neither. Among the former there were murderers, burglars, pickpockets, and a swindler. The latter and a few others, like the man who had sold a streetcar, were proud of their exploits, and exhibited their professional CVs: lots of newspaper cuttings about their accomplishments.

The most interesting of those delinquents was Dedos Brujos (Wizard Fingers), the nickname he had earned as a master pickpocket. He was the only friendly and cooperative one of the lot. Before becoming a professional thief, he had earned good wages working as an untangler in cotton mills. Once he stole something, was caught, and incarcerated for a while, after which he quickly got another job. But the police followed him, and informed his new employer about my friend's criminal record, as a consequence of which he was sacked. This went on until Wizard Fingers understood that his career as an honest worker was over, and that to support his family he had to steal. May this anecdote help criminologists understand that penal codes are worthless in societies that breed unwilling delinquents.

Most of the so-called "politicals" among us were union activists. One of these was a young Peronist who had been active in the railway workers's union, and refused to believe that Perón had betrayed the working class. I shared with him and a medical doctor the blanket and the fruits that my friends Raúl Fernández and Rogelio Galarce had conveyed to me. But our Peronist comrade was beyond consolation: he had lost not only his freedom but also his faith.

Another interesting inmate was Próspero Malvestiti (Prosperous Ill-dressed), the secretary of the textile union, who was the first to speak at the seminar that I proposed. He asked whether destiny existed, which led us to further questions, for instance about free will and the power of individuals. Our seminar, which met every evening, was very interesting, and lasted the entire week I spent there. It was the only thing missed when I was freed. I doubt that the one led by the existentialist professor at the Faculty of Philosophy in Buenos Aires engaged its participants with the same degree of interest.

Another memorable episode was the "report" that a communist leader foisted on us. Actually it was not an actual report, for none of us had any fresh news. But lack of data did not prevent him from performing a "Marxist analysis of the situation." He had all he needed: the principles of Marxism-Leninism, and the thesis of Victorio Codovilla, the Comintern's representative and head of the Argentine communists, that Peronism was but the local version of Nazism – which of course precluded any scientific study of that perverse but original mass movement. The most remarkable statement of our political analyst was that our incarceration only showed the government's weakness. I proposed that, since a strong government would not keep us locked up, we should support the authorities instead of fighting them. My joke did not go over well. O, the mysteries of dialectics!

It was only when I was handcuffed to a provincial communist notary, who had avoided me, and we were taken to the main penitentiary in Buenos Aires city, that we found out, from the judge's mouth, why we had been jailed: we were accused of inciting the railway workers strike, the first since Perón had risen to power 6 years earlier. The General and Evita were indignant at such a show of ingratitude. The accusation was doubly absurd in my case, because in those days I traveled by car, and the only railway worker I knew was my good neighbor Rosales, aka Negro, a faithful Peronist whose family were friends of mine. At all events, I was set free, though not declared innocent.

Those 2 weeks were very instructive. I learned that (a) jails are delinquency schools, where veteran criminals instruct novices and plan new crimes; (b) with few exceptions, like Wizard Fingers, common criminals are sullen, uncommunicative, and selfish; (c) lack of writing implements forces one to make mental calculations, as well as to evaluate what one has been doing; and (d) life is more precarious than I had thought before.

The last two teachings led me to resume my thesis work with new intuitions and renewed energy. It also helped that my supervisor had moved to Río de Janeiro, where he was better treated, had good friends, and got a new exceptional pupil, Moish Nussenzveig, perhaps his best Latin American student. In any event, a lack of maps is a powerful incentive for exploring new lands. I finished my dissertation in November of 1952, got a nice affidavit from Beck, and defended it in December. My former students and friends Daniel Amati and Tito Sirlin, who had worked briefly under Richard Gans, had got their degrees a few days earlier. They never talked about their unpublished theses.

Thesis Defense

My thesis defense went well despite its bad start. Indeed, I arrived at the Physics Institute an hour later than scheduled, just as my examiners were preparing to leave. This delay happened because the day before, December 21st, I had sprained an ankle while playing soccer with my son Bambi and his friends, who were celebrating his ninth birthday. Because of my limp, I missed the first train, arrived late for the second, and so on.

Fortunately Antonio Rodríguez, the chair of the examination team, asked his colleagues for patience and treated me cordially, even though I was neither a Catholic nor a Peronist. (After all, physics theses were not plentiful.) At the end of my presentation, he asked me what I thought about David Bohm's revision of quantum mechanics, published a few months earlier in *Physical Review*. I was able to discuss that theory in some detail because I had studied it, and had engaged in frequent mail exchanges with Bohm. My examiners were satisfied, and I went back home waiting for the official notification that I had earned a doctorate in physico-mathematical sciences.

The notification I expected never arrived. Instead, what came was a letter from the secretary of the faculty, a probability specialist, asking me to see him. When I arrived in his office he received me civilly, asked me to sit down, made sure that the door kept open, and told me that he would give me my diploma there and then, for the authorities did not want me to attend the graduation ceremony. I suppose they did not wish to be seen shaking the hand of someone who only the year before had enjoyed the hospitality of the local police. But it is more likely that they got wind of the plan of some friends of mine to applaud my appearance for longer than usual. One cannot be too careful in matters political or parapolitical.

After handing me my diploma, the secretary asked my opinion about the two most popular interpretations of probability: as long-term frequency and as degree of certainty. I got carried away because I found the subject fascinating, and the year before had published a paper (Bunge 1951a) defending a third view: probability measures real possibility. The man seemed interested but in a hurry to say goodbye, for he did not even offer me the traditional cup of coffee. Nobody dared publish my thesis. It was published only in 1960 by the Universidad Nacional de Tucumán (Bunge 1960d), thanks to Augusto Battig, a good physics professor and a regular participant in the meetings of our physical society.

Parallel Jobs

While apprenticed to Beck (1943–1951), I had several other occupations, two of them intellectually challenging: my private philosophy studies, and a teaching assistantship to my favorite professor, Dr Teófilo Isnardi, in the Faculty of Sciences of the University of Buenos Aires. I had contacted him by mail while still in high school 1937, asking for his opinion on my alleged generalization of special relativity to three dimensions. Don Teófilo corrected me with this usual patience and courtesy, and we met 5 years later, when I took his thermodynamics course.

Around 1944 I joined the Núcleo de Física, the physics students' seminar started by my friend José F. Westerkamp, aka Pipo, along with Julio Iribarne, Estrella Mazzolli, Cecilia Mossin Kotin, Waldemar Kowalewsky, Máximo Valentinuzzi, and a few others. Don Teófilo attended some of our meetings, and participated every time problems of interpretation of the mathematical formalism of quantum mechanics arose. Isnardi was a keen critic of this theory, whereas we students swallowed it whole, and thought that he was a scientific conservative. A few years later I reached a similar position to Isnardi's: the theory is basically true, but the philosophy attached to it – the Copenhagen interpretation – was wrong because it was subjectivistic, and thus contrary to the aim of science, which is to describe and understand the real world (Isnardi 1927; Bunge 1955b, d, 1959b).

In 1942 I had been an honorary teaching assistant in experimental physics, but that was routine work. Without my asking, my friend the mathematician Alberto González Domínguez, whom I used to consult often on mathematical problems, got me the teaching assistantship I wanted. One afternoon, while we were driving in a taxi together with Isnardi, Alberto asked him all of a sudden: "Don Teófilo, why don't you appoint Mario as your T.A.?" Taken aback, Don Teófilo only managed to mumble that he solved problems on the blackboard. The subject was not touched again that day, but a few days later Don Teófilo called me and gave me carte blanche: I could teach whatever I wanted. The succeeding year I got the official appointment, and started to earn a salary that paid just for my subscriptions to the two top physics journals. I enjoyed this job tremendously, as it gave me the chance to study new subjects and teach them to students who, for the most part, were bright and motivated.

Soon thereafter my salary suffered a 10 % discount –my "voluntary" contribution to Evita's charitable foundation, which gave away domestic goods in exchange for votes. (In a remote northwestern village, I saw soldiers unloading sewing machines from an army truck, and taking them to humble houses.) When I complained about the discount, I was told to ask in writing, every month, for its cessation. That went on until the end of 1952, when my appointment was terminated. I suppose that those 7×12 letters of mine are still in my bulky police record. However, those 7 years as a teaching assistant helped some students and were crucial to my training as a university professor.

The first year I had only two students, whom I met in the commons of the Physics Institute to solve problems in special relativity, such as the free fall of a body. The number of students rose every year, until I had more than 30. We dealt with quantum mechanics, classical electrodynamics, antennas, waveguides, Laplace transformations, and various kinds of delta "functions" – which at the time provoked the ire of mathematicians, but they were in the process of being polished by Laurent Schwartz. I handled these useful monsters with the nonchalance typical of physicists: the deltas worked, and that was all that mattered to us.

I am sure I learned much more than my students. In particular, my course on antennas taught me that the task of the radio or TV engineer – namely to design antennas that radiating waves of the desired shape in the desired direction – is far more complicated than that of the physicist, who studies the radiation emitted by a given antenna. He tackles an inverse problem, which requires some guesswork, whereas direct problems can often be solved with the help of known algorithms.

Interesting Students

About half of my students of that time eventually distinguished themselves, and I befriended some of them. I remember in particular Daniel Amati, whom I had tutored in high school physics, and who ended up working at CERN in Geneva. He came often to Florida in the company of his chum Tito Sirlin, to discuss Einstein's latest gravitation theory. At the end of 1952, when it was rumored that I was about to be sacked because of my refusal to join the Peronist party, they and two mathematics teachers pleaded with me, arguing that joining the party was just a formality, and that I would surely be replaced with someone incompetent and submissive.

Another outstanding student, Juan Carlos Mallmann, started his own university in Bariloche, which hosted some social scientists, and invited me to join it. The charming Eduardo Ortiz began as a mathematician and ended up as a professor of the history of science at the Imperial College of Science and Technology. He visited me often to discuss philosophical issues, until the Communist Party demanded that he cut me out. I also taught Horacio Bosch, Juan José Giambiagi, Mario Gutiérrez Burzaco, Juan Roederer, and Carlos Abeledo. They were all very bright, strongly motivated, and achieved distinction.

I also tried to teach Mabel Molinelli Wells, a pretty, lively, popular chatterbox and the only declared fascist, who was the center of attention wherever she went. Eventually she caught the great mathematician Alberto Calderón, a timid and affable provincial engineer who had been discovered and stimulated by Alberto González Domínguez, and who ended up as a full professor of mathematics at the University of Chicago. He became famous for his work on a hard inverse problem, which may be restated as the Fireman's Problem: given the air temperatures in the periphery of a house on fire, calculate the temperature at its center. This problem is still being worked on, seven decades after Calderón met it while working at a Buenos Aires factory. Many years later he widowed and married his colleague Alexandra Ionescu-Tulcea, who had taught Marta at Penn.

Beck's Argentine Students

Beck had several Argentine students, the best known of whom was José Antonio Balseiro. El Niño, as Beck called him, was the ideal pupil: ambitious, intelligent, hard-working, and docile, on top of which he came from a well-to-do family, had already earned his doctorate with Gans; he was also single, right-minded, and without philosophical or political worries, or cultural interests outside physics, mathematics, and religion.

Balseiro had studied in La Plata together with his buddy and co-religionist Antonio Rodríguez. They were Loyarte's last students, who died of cancer at the time of their graduation. Their theses, never published – hence non-existent according to Gaviola's standards – were modest laboratory projects. Although the Balseiro family had an ample and comfortable home in Córdoba city, he shared with Beck a modest chalet in the observatory's garden. There he stayed, day and night, within call of our teacher, until he married and left for Manchester. He went there on a modest fellowship to study under Leo Rosenfeld, whom Beck privately called "Bohr's hitman." Shortly thereafter I criticized Rosenfeld's quantum orthodoxy, which was even narrower than his erstwhile master's (Bunge 1955b.)

Balseiro's stay at Manchester was unenviable. The cradle of the Industrial Revolution was cold and gray, Rosenfeld seemed to be running like a detached caboose, and he was a Marxist while his postdoc was a conservative uninterested in either physical interpretations or philosophical presuppositions. This formalist stance, which Balseiro shared with Beck, drove Gaviola up the wall: he demanded that every physical formula be read "operationally," that is, in terms of laboratory operations. This is why we were all surprised when Balseiro was appointed professor of experimental physics when Héctor Isnardi retired. We learned only later about his strong link with the Catholic hierarchy of La Plata, headed by Monsignor Octavio N. Derisi, an able neo-Thomist philosopher and Catholic militant, complicit with all the authoritarian regimes (López-Dávalos and Badino 2000).

Argentine Physical Society

Beck and Gaviola formed a peerless team: between them they upgraded Argentine physics overnight. They did it through AFA (Asociación Física Argentina), which they set up along with about 20 students and a single teacher, the associate professor Ernesto Galloni, an engineer who did X-ray crystallography (see Westerkamp 1975). None of our full professors attended the AFA's foundational meeting: they were too busy teaching at various places. Now, all of a sudden, we had a leading team: Gaviola and Beck. I am the proud owner of the bronze plate that some years ago AFA presented to the few survivors: *La AFA a su fundador*. I got a similar plate on the occasion of its 100th scientific meeting in 2015.

We first met at a La Plata milk bar, where we elected the governing board, with Gaviola as president and Galloni as secretary. Gaviola and Galloni constituted an ideal team, as the former stood for principles while the latter was an effective organizer. Both men governed the AFA until 1974, when it self-destructed following the 1969 anti-science campaign led by Rolando García and Oscar Varsavsky. García, who had been an effective dean of the Faculty of Science during its 1955–1966 golden period, joined the Peronist movement and wrote about "national and popular science". His friend Varsavsky, a former colleague of mine, was a failed scientist who, like García, preached a combination of Peronist demagoguery and the post-modernist obscurantism imported from Paris following the May, 1968 student revolt (see Garzón Valdés 2000). Up until then, AFA had practised the progressive and apolitical science policy sketched by Houssay and Gaviola. AFA's suicide may be unique in the history of science.

AFA did more than the conventional learned society: in addition to convening periodic meetings at all the university towns, it controlled the quality of scientific production, and offered crash courses to high school physics teachers. Our first meetings were attended by 20–30 people, who presented between 10 and 20 supposedly original papers, plus the occasional long report on a vast subject, like Gaviola's on Smith's manual for assembling nuclear bombs. Nowadays the attendance has risen to nearly 1,000.

Once in a while we were visited by Brazilian physicists, such as Mário Schemberg, whom I met again in his home turf, and José Leite Lópes, whom I befriended years later in France. The most spectacular of all the Brazilian visitors was Joaquim da Costa Ribero, an original experimentalist enthusiast of carnauba wax, who filled blackboards that his black servant erased. Inevitably, as in all scientific meetings, there were a few lousy presentations. One of them was by a chemist, supposedly an expert in surface tension, who claimed to have disproved the existence of transuranics. Later on, the same specialist proposed dividing every physics department into separate divisions, each specialized in measuring a single physical magnitude, the way governmental weights and measures bureaus are organized. (Presumably, he saw himself as the Director of Surface Tension Measurements.) While employed at the industrial lab of my friend Andrea Levialdi, the same individual invented a shoe polish enriched with chlorophyll. And once, while holding a position of power, he insulted my friend Pipo Westerkamp over a political disagreement. Pipo, the most peaceful of men, punched him in the face, as a consequence of which one of his fingers got infected.

Gaviola, who had had the best scientific education of us all – in Berlin, under the likes of Einstein, Born, and Noether– and was anything but shy, commented on nearly every presentation, and kept us up to date about his pet project on Eta Carinae. This is the stellar system so vast, changing, and bright that the ancient Chinese had called it Heaven's Shrine. Beck, far less loquatious, started every one of our early meetings writing on the blackboard a list of topical but still poorly understood subjects, such as superconductivity, superfluidity, and above all nuclear forces that he hoped would tempt new recruits.

It was then that I learned that, in science, *in the beginning was the problem* – to put in scriptural terms. That is, research does not start by observing, as the empiricists believe, nor by exploding a myth, as Popper believed, but by discovering a hole in the body of extant knowledge and attempting to fill it. Thus, unwittingly, Beck was my first teacher in the philosophy of science in the making.

Four Frauds

Our young professional association controlled the quality of academic work in our field, supporting the physicists who demanded that every academic post be filled by public competition, as well as by exposing plagiarisms, diploma theft, and other crimes common in underdeveloped countries. Gaviola was the most relentless and ferocious of inquisitors. He was, in fact, a master of scientific integrity (see Bernaola 2001).

The first case of fraud Gaviola denounced was Balseiro's, who apparently had stolen from Antonio Valeiras, an obscure mathematics professor, his original paper on anthoidal bodies (or functions). Valeiras reported the theft to Gaviola, who accused Balseiro at a scientific meeting held in the observatory. I remember distinctly this episode because I was sitting next to the defendant. Gaviola exhibited the two publications in question, and demanded from the culprit a confession and an apology to Valeiras. Balseiro did not deny the charge, but refused to apologize. He made a grimace of disgust, and gave no indication of remorse. I suppose he was used to discussing such issues in the confessional. (Will a few Hail Maries suffice, Reverend Father?) Besides, Balseiro must have guessed that Beck, his supervisor, would protect him for fear of losing his star pupil. But Gaviola was a bulldog in ethical matters, and he insisted until Balseiro apologized to Valeiras. The latter, who held a minor academic job and was one of the few high school teachers who refused to pay his due to Evita's Foundation, accepted the apology.

Balseiro's Brilliant Career

I do not remember Balseiro's contributions to physics other than the creation of the research center that has borne his name since his death from leukemia at age 43. Since its foundation in 1955, the Instituto Balseiro, dependent on the Atomic Energy Commission and located in Bariloche, has trained a number of full-time physics students, several of whom distinguished themselves in the USA, such as Juan Maldacena, famous for the conjecture that bears his name and belongs to a string-theoretic version of quantum gravity, which some scientists regard as a piece of science fiction.

From the birth of the Balseiro Institute, some of us have wondered whether it is legitimate to spend taxpayers's money on a specialized institute located in a remote place, disconnected from the main cultural centers of the country and, at least in principle, subject to military authority through the National Atomic Energy Commission. A detailed answer to this question requires an investigation that falls outside the frame of these memoirs.

However, there is a question that may be examined without additional empirical data, namely that of the reach of the sectoral (or sectional) approach compared with the systemic one. The sectoral approach assumes that the object of study can be isolated from the rest of the universe, as in the cases of clipping fingernails and learning an algorithm or computational rule. In the case under consideration, the sectoral viewpoint presupposes that physics may be detached from the rest of science, and the latter from the society that hosts it. But of course both presuppositions are false: physics has always gained from a close relation with mathematics, chemistry, and engineering, and in recent years computation as well.

The systemic view of the matter at hand is that science is a system, not a heap of disciplines, so that every one of its constituents flourishes or wilts along with others. It suffices to remember that computational physics would not exist without digital computers, which were invented and built in the light of electronics, solid-state physics (based on quantum physics), and the abstract theory of computation. (See more on science policies in Bunge 1980b).

In addition, there is the problem of the uneducated scientific expert who, for lack of a broad perspective, may be a good problem-solver but can be neither a bold innovator nor an inspiring teacher. When the eminent nuclear physicist Emilio Segré told me that he was about to visit La Plata, I suggested that he not miss its Natural Sciences Museum, which has a unique collection of fossils. He answered me in astonishment: "But I am not a paleontologist!" By contrast Ernest Rutherford, the founder of the science that Segré cultivated, tried once to engage the Bishop of Cambridge in a conversation, and asked him whether he had seen the latest play by George Bernhard Shaw. The bishop excused himself: "Regrettably, my duties are so demanding, that they leave me no time for the theatre." Rutherford: "I understand, Your Grace, and how about literature? Have you read J.B. Priestley‘s latest novel?" Bishop: "Unfortunately, I have no time left for that either, my Lord." Rutherford: "I understand, your Grace. Ours are dogs's lives." Obviously, Rutherford was a very well trained dog, which explains in part the strong impression he made on all his visitors, Beck included. Back to Argentina in the 1950s.

Since the smart crook Ronald Richter sold President Perón the ambitious Huemul Project, the nuclear physicists and their hangers-on suddenly got so many resources that they were soon wasted, at the same time that the other sciences languished. Anyone with some link to physics, however weak, was offered a secure, undemanding and well remunerated job at the Atomic Energy Commission, whereas Houssay and Leloir – soon to be awarded a Nobel Prize each –had to resort to private benefactors, such as the Campomar and Sauberan foundations, to maintain their productive research teams, and many able individuals had to emigrate or take nonscientific jobs. When the Huemul Project started, cement became scarce in the construction industry, where I worked at that time. One consequence was of course its steep price increase, another was costly delays, and a third was that many bricklayers mixed cement with brick powder, which resulted in weaker concrete structures, which accounts for the collapse of several buildings at that time.

The Pap and Klein Cases

Another case of scientific fraud in which AFA got involved was that of the Hungarian historian of science Desiderio Papp. This individual had written several popular books on the history of science, one of them on Henri Poincaré, whose original contributions are so complex, that only a team of experts could grasp. Papp had been admitted to AFA for claiming to hold a doctorate in astrophysics from Vienna University. He also attended some of our meetings, but when Gaviola chatted with him, he immediately suspected that there had been no such doctorate. Soon after the war's end, he wrote to Vienna University, whose Registrar assured him that Papp was not among their graduates. This explained why his books were so superficial: they only listed the names of some scientific achievements, and showed no evidence of first-hand scholarship. In view of all this, AFA expelled Papp.

However, this contretemps did not affect Papp's popularity, because Aldo Mieli, the only scholar among us trained as a historian of science, died in 1950. I attended his funeral in the company of our common friend the mathematician José Babini, whom I drove to *El Ombú*, where I played for him a record of Rousseu's *Le devin du village*. After listening to a Beethoven record, Babini opined that he was boring –which of course vexed me.

This episode led me to imagining how different the Century of Lights would have been if Rousseau, that complex genius, had concentrated his talents on music – and how different Mieli's, Babini's and my own lives would have been, had Mieli not been sacked from his Rome University chair because of this triple crime (anti-fascism, Jewishness, and gay activism), which had forced him to accept the invitation of the Universidad Nacional del Litoral, where he and Babini had been trying to lure me as Mieli's successor until both of them were sacked by an Argentine fascist official.

Guillermo Klein had studied physics in Germany, and presented at AFA meetings several papers on quantum mechanics in curved spaces, which looked to be a promising way of joining quanta with gravitation. Nobody knew anything else about him until Gaviola stepped in. With his keen nose and his international connections, Gaviola found out that Klein had lifted his presentations from papers that Erwin Schrödinger – the inventor of the "wave" mechanics that everyone had been using since 1926 – had published during the war in the obscure *Proceedings of the Royal Irish Academy*. Consequently we expelled Klein. I was sorry for him, for he seemed to understand the stolen formulas and was a modest and quiet man who might have been redeemed.

The Richter Case

The worst physics fraud, but one that did not involve AFA, was that of Ronald Richter, the German Czech who sold Perón a plan for building an original nuclear bomb, whose possession would have allowed him to dominate the Southern Cone. But in 1952, after spending a fortune without any tangible results, the government consulted a panel of physicists and engineers, among them Balseiro, Báncora, Gamba, and Gans. Although none of them had worked in nuclear physics – let alone in fusion, which is what Richter claimed to be capable of achieving – the fraud was so obvious that they produced a crushing report (Mariscotti 1985).

Shortly thereafter, Gans told me that when Richter showed them a large instrument panel, he, Gans, went behind it and found that the instruments were not connected to anything: they were only for show. This multi-million fraud might not have occurred if the government had consulted AFA. In fact, long before that report was issued, Gaviola found that, although Richter had indeed graduated from the German university in Prague, his thesis was unknown because it had not been published. Gaviola decreed: "Unpublished theses are nonexistent."

The National Atomic Energy Commission (CNEA), created by Perón by a 1950 decree with the purpose of "supporting and using" Richter's work, never published a full report on the Richter affair from his hiring in 1949 till his fall in 1952. It suffices to peruse its luxuriously produced journal, sold in kiosks, to realize that CNEA lacked quality filters. Among other pearls, one will find an "experimental curve" about cosmic rays that was a segment of a straight line because only two values of a certain function had been measured. If only a third value had been measured, a bump might have been seen. The Commission employed a few serious researchers, but it left no memorable scientific legacy. Its initial objective was not to give researchers a chance to work, but to gain their political silence and to fool the laity, which it accomplished. Only ten or so physicists and chemists stayed out of the CNEA, refusing to work under military authority and join the ruling party (see Ortiz and Rubinstein 2009). I was one of them.

Uselessness of My Doctorate

My doctoral diploma did me no good, because it was not accompanied by the Peronist party card, without which I could not even get a job as a dogcatcher. I realized this a few days after my thesis defense: Balseiro, who had succeeded Isnardi as the director of the Physics Institute at the Buenos Aires university, did not renew my teaching assistantship. Thus, he let me go instead of promoting me, for I did not even seek the required warrant signed by a Peronist leader. The same happened to my friends and colleagues Pipo Westerkamp, Julio Iribarne, and a couple of others.

When I lost my only job I complained to Balseiro. He behaved the way *gallegos* (Galicians) do according to their enemies: when you meet them at a landing in the middle of a stairway, you don't know whether they are going up or down. In other words, Balseiro treated the matter as something that did not concern him and, anyway, about which he could do nothing.

The students too were victims of the government's hardening: from 1953 on they were required to pass the National Doctrine test. I saved them this indignity by signing their practices booklets before that requirement was implemented. Such deceits happen under all authoritarian regimes and, as the Polish dissident philosopher Leszek Kolakowsky noted, that is how minor dissidents survive.

In cases of need a good Catholic will say that God will provide. Well, this time He did. In fact, at the beginning of 1953, David Bohm (1917–1992) unexpectedly invited me to join him at the Theoretical Physics Institute in São Paulo University through a postdoctoral fellowship paid for by the Conselho Nacional de Pesquisas Físicas do Brasil. He invited me to continue the recent discussion we had by mail on his reformulation of quantum mechanics (Bohm 1952). He also wished to discuss the hypothesis, in my article on chance (Bunge 1951a), which he had liked, that causality may emerge from the randomness prevailing on the next level of organization, and conversely. Several decades later Ilya Prigogine wrote some bestsellers on this subject.

I boarded a ship bound for Santos, São Paulo's busy port. I spent most of the time on deck to escape the bad smell reserved for third class passengers. David and his postdoc Ralph Schiller, as well his wife Berenice, were waiting for me at the dock. A few weeks later we were joined by another two unorthodox postdocs: the New Yorker George Yewick – physicist, amateur painter and political aspirant, and Jean-Pierre Vigier, a Parisian mathematician protégé of Louis de Broglie, and a communist decorated for his courage in the anti-Nazi resistance. The enthusiasm and optimism of both George and Jean-Pierre contrasted with Ralph's depression. He had just finished his dissertation under Peter Bergmann – Einstein's co-worker and future friend of mine – and was not relishing the prospect of returning to the USA, then in the midst of the McCarthy's hysteria, from which Bohm had recently fled.

I stayed at four different bed-and-breakfasts in the course of 6 months. I left one of them because its owner, an English spinster, kept howling for having lost her youth; in another, the toilet bowl was teeming with myriads of squirming intestinal worms; and so on. But the city amazed me for its modernity and drive, in stark contrast with Buenos Aires's decay and lethargy. São Paulo is in modern Brazil what Chicago had been in the USA around 1900.

I went daily in *bonde* (streetcar) to my office at the Institute of Theoretical Physics of the University. (You'll be amazed at the origin of the word *bonde*.) This institute occupied ample premisses, had a rich library, and a research tradition that went back to 1934, when it was started from scratch by the Ukranian Gleb Wataghin, the imaginative theoretical and experimental physicist who had formed the first generation of Brazilian physicists – thus unwittingly dashing the fashionable thesis that in all cases scientific communities precede individual scientists. Every day, at mid-morning, Donha Terezinha served us a delicious *cafezinho*. But there was no dining hall, so that at noon I left for lunch in the cheerful company of a couple of colleagues, secretaries, or librarians. There I made three good, enthusiastic and talkative friends: Abramzinho Zimmerman, George Leal Ferreyra, and Shigueo Watanabe, with whom we set up a junior seminar. I also had dealings with some of David's colleagues, particularly Mário Schemberg – an imaginative physicist, art critic, nationalist, a communist militant, and believer in telepathy. He was just back from Brussels, where he had enjoyed the hospitality of Ilya Prigogine, whom I befriended decades later.

Schemberg told me that he was not interested in particles, and that Gibbs's formulation of statistical mechanics, unlike Boltzmann's, was materialist – but he did not bother to explain. (I happen to believe just the opposite.) Mário was brilliant, very sure of himself, and popular with students. But as soon as he arrived, he and

David engaged in an ugly fight over the Institute's directorship. In the seminar Bohm put in evidence his unstable temper as well as his prodigious imagination.

The Dutch Touch

One of the most memorable foreign visitors was Hendrik Casimir, famous for discovering the force that the vacuum exerts on a charged particle, and that bears his name. He was displeased by my remark that his employer, the Philips radio company, had closed down its lab in Buenos Aires as soon as the war was over. Casimir felt offended, and reminded me that Philips was a Dutch firm with no obligations to the countries that hosted its branches.

Another interesting visitor was Leo Rosenfeld, Bohr's erstwhile assistant. He subjected Bohm's heterodoxy to a harsh criticism, and finished by saying: "The standard theory supplies all we need. Why ask for more?" As an example of the Berkeley-Bohr philosophy, that the universe is observer-dependent, he mentioned the steam engine, which according to him would not work were it not under the constant survey of engineer and coal man. Rosenfeld had not realized that the laws of thermodynamics do not involve the observer, that the stars had been "obeying" those laws long before any observers emerged, and that locomotives can be automated.

Bohmian Mechanics and Its Critics

I spent the first month of my stay telling Bohm the objections that his recent papers (Bohm 1952) had raised. His papers had caused a commotion in the quantum community (see George 1953; Freire 1999), and they are still being discussed. David answered my questions quickly and clearly. He got me fully converted after a month. I compared it in detail with its rivals (Bunge 1956b), and taught it as dessert in my quantum mechanics courses both in Argentina and in the USA (Bunge 1956c).

But, as I polished my philosophy, I became increasingly critical of Bohm's views. Firstly, he had proposed more than a "reinterpretation" of the mathematical formalism of quantum mechanics: his was a new theory, since it contained two "hidden" (scatter-free) variables, the classical position and a newly defined momentum, that allowed him to define precise trajectories plus a new unobservable force, that would explain the quantum fluctuations.

Secondly, Bohm had not succeeded in eliminating randomness, for he had not even attempted to derive probabilities from nonprobabilistic assumptions. So, it was wrong to call his theory 'causal'. Just like the standard theory, Bohm's had both causal and probabilistic features.

Thirdly, like Einstein – who had exerted a decisive influence on Bohm when they met in Princeton at the older man's request – he had confused the concepts of causality and realism with what I called *classicism*, that is, the description of physical reality *par figures et mouvements*, in the Cartesian style (Bunge 1979f).

Fourth, as Alfred Landé pointed out to me when we met in Venice, Bohm's theory did not yield new testable results, and it did not suggest any new experiments – nor could it, since its distinctive variables were unmeasurable.

Finally, both sides of the controversy mixed up three mutually independent (non-interdefinable) philosophical categories: realism, causality, and classicism. It took me several years to notice these confusions (Bunge 1979f). And two decades later, when the Bell inequalities were experimentally refuted, some physicists worsened the prevailing conceptual chaos by introducing the oxymoron 'local realism'.

Despite Bohmism's weaknesses, the quantum "establishment" was alarmed. Indeed, several of its heavy weights, such as Born, Pauli, and Rosenfeld, responded aggressively (see George 1953; Freire 1999). But none of them, not even the sharp Pauli, noticed the confusions I pointed out above: their main concern was that Bohm had been the first to challenge John von Neumann's alleged proof of the impossibility of "hidden variables" theories. Parallel: intuitionist logic, the earliest example of a consistent logical theory without the excluded-middle law.

Pauli's Objection

Wolfgang Pauli was famous for his incisive criticisms. His complaint, that Bohm's theory did not include a measurement theory, greatly worried David. But this charge was unfair, since the orthodox theory did not have one either pace von Neumann's claim that he had produced a general theory of measurement. I realized this point years later, when I noted that such theory is impossible since there can be no universal measuring instrument, if only because there cannot be a universal indicator. I also emphasized the crucial role of the indicators of imperceptible things and properties in the design and interpretation of measurements and experiments (Bunge 1967a, 2010a). Pauli's unfair criticism had an unexpected effect on Bohm and his postdocs Vigier and Yewick: they started thinking of a theory containing only fractionary numbers, since these are the ones that measurements supply. (I had tried the same approach a few years earlier, but my friend Manuel Sadosky disabused me, pointing out that the set of fractions has zero measure: the rational numbers, of the form m/n, are rare hills emerging from a sea of real numbers, the vast majority of which are irrational.)

To this end, Bohm and his associates looked frantically into the mathematical literature, until they found a few arcane papers that contained the "integral part of" operator. But this attempt to meet an empiricist's objection by building a theory palatable to them failed. Bohm got discouraged, and tried next a topological approach, that did not prosper either.

The Bohm Cult

Reading Hegel in English translation must have added to Bohm's confusion, and anyway turned him into a holist. When I asked him why he wasted his time reading that garbage, he replied that Hegel inspired him. Two decades later Bohm met the Dalai Lama and went on a lecture tour sponsored by the Hare Krishna sect. When he came to Montreal as part of that tour, I refused to have him invited by my university. Nowadays his holistic writings belong to the New Age canon. There is even a firm that offers business people to cultivate customer loyalty on the basis of Bohm's insights into "the wholistic universe." That once-brilliant scientist may be regarded as one of the many victims of McCarthyism, which was in turn a Cold War episode.

In conclusion, Bohm's alternative to the Copenhagen orthodoxy failed. However, it had an important unintended consequence: the disproof of von Neumann's alleged proof of the impossibility of hidden variables theory encouraged John Bell (1964) to try again, and his attempt led to important experiments. Unlike Bohm, Bell did not build a theory, but only a set of formulas, in particular the famous Bell inequalities involving hidden variables. The greatest merit of these formulas was that they are experimentally testable consequences of standard quantum mechanics.

Two decades later, Alain Aspect put those inequalities to the experimental test, and found them false. Further experiments, such as Anton Zeilinger's, disproved also the locality assumption: they showed that long-distance correlations, or "entanglements," are real (see Schlosshauer 2007). I put it thus: Once a system, always a system. But I confess that entanglement still bothers me, and I hope that eventually a mechanism for it, perhaps ripples in the quantm vacuum, will be found.

In short, Bohm's brave project failed: nature does not dress up classically. Worse, his project backfired, for Aspect's negative result was misinterpreted as a refutation of realism – as if nature did not exist while not under observation. This conclusion was false and even silly, if only because modern physics is a very late newcomer to the real world. But Einstein, De Broglie and Bohm were ultimately to be blamed for that false conclusion, for they had conflated realism with classicism (Bunge 1979f). Thus, philosophical sloppiness had an unexpected result: the discredit of the most important philosophical presupposition of the exploration of reality – that the observed preexists its observers. I was unable to assist Bohm at that time, because my philosophy was still embryonic.

Distractions in São Paulo

In São Paulo I had some distractions but few amusements, for I missed my home. On Sundays I visited the huge town, trying to escape the loud howlings from the gigantic soccer stadiums every time a goal was either made or frustrated. This is how I saw giant sloths on trees in city parks, and discovered that the African magico-religious beliefs and practices were still alive; that the spiritualist temples by far outnumbered the Christian ones; and that not even the poorest families were willing to sell their cats – which I was trying to buy for Berenice's experiments.

Sometimes I boarded a train that took me to the edge of a large eucalyptus forest spotted with fern trees and 2-m high red anthills. (Imagine the huge caves underneath!) At other times I visited rural villages whose shops offered only Japanese goods, from stonehard dried fish to comics. To overcome fatigue, I snacked on delicious Brazilian sweets made from milk, coconut, or *abobora*, a kind of pumpkin, and drank *guaraná*, a delicious beverage that was later discovered to be toxic.

Some Brazilian cinemas showed funny movies centered on the *cangaceiros*, legendary bandits that robbed rich ranches. My favorite was *Mulher rendeira*, where one can hear the sticky song, still popular, that goes like this:

Olé, mulher rendeira
Olé, mulher rendá
Tu me ensina a fazer renda
Eu te ensino a namorá.

Once, our gang went to a Japanese cinema that featured movies about the Japanese occupation of China, that only showed the hardships that the Emperor's silent and heroic servants went through: their Chinese victims did not appear. We were charged half the usual price because, so the ticket vendor assured us, we would understand only half the stuff shown. On another occasion we went by bus to Rio de Janeiro, to witness Richard Feynman's show. While lecturing, he moved around with a monkey's nimbleness, and drummed on the table with his hands like a Carnival musician: obviously, he was enjoying himself. Feynman filled blackboards with the famous diagrams that bear his name, and were intended to give intuitive images of the interactions between electrons and photons, thus belying his own claim that "no one understands the quantum theory." But Feynman achieved intuitability by dispensing with the field concept and, in particular, by picturing photons as particles – which was plain cheating. But it helped calculation, which was his main goal (Bunge 1955e). Feynman's lectures were the most spectacular I have ever witnessed, and they made him the most successful student attractor of his generation. However, it is also possible that his emphasis on computation and diagramming hid the conceptual problems of the theory, which he glossed over.

On the evening of the great dinner in Feynman's honor, a large black car brought the director of the Manhattan Project, J. Robert Oppenheimer, surrounded by bodyguards. Bohm – who had been a member of his inner circle until Oppenheimer denounced him to clear his own name with the FBI – approached him and asked for his help in returning to his country. Oppenheimer did nothing of the sort, just as he had done nothing to get an American job for the great Max Born, one of the founders of quantum mechanics and his former teacher and collaborator: Born got stuck in dark and cold Edinburgh. The banquet for Feynman filled an enormous hall. At the end, some words were spoken. My friend Erasmo Ferreyra, emboldened by wine, got up and asked in a loud voice: "Professor Feynman, what do you think of Senator McCarthy?" Feynman, usually loquacious, kept silent. The recent execution of the Rosenbergs had apparently intimidated the sons of The Land of the Brave.

Would-Be Migrants

Julia sold the few shares we owned, and came to join me a couple of months before the termination of my fellowship. Since we had no jobs in our country, we made a half-hearted attempt to settle in Brazil. But I was yet to make my mark in physics, and Julia had no connections anywhere. A Paulista friend got us an introduction to Lúcio Costa, who a few years later, jointly with the great Oscar Niemeyer, would design Brasilia, the amazing new national capital. Costa, an eminent member of Le Corbusier's school, received us politely and looked over Julia's designs, which did not impress him. After all, she had been a star pupil of the legendary Monsieur, the neoclassical tyrant of the Buenos Aires School of Architecture. We gave up our hope of settling in Brazil, which was decades ahead of Argentina in all fields. They did not need us.

It was only 25 years later that a number of Argentines, who had studied during the golden 1956–1966 decade but felt uncomfortable or even threatened under the 1976–1983 military dictatorship, found suitable jobs in Brazilian universities. Unlike their Argentine colleagues, the Brazilian military appreciated schools – as long as they did not harbor "subversives." Like the Prussian king Frederick the Great, they understood that a powerful army needs well-educated if docile brains.

Bohm's Migrations

Bohm migrated from Brasil to Israel, then to Bristol University, and finally to Birkbeck College in London. Here he was a colleague of the physicist John D. Bernal and the historian Eric Hobsbawm, both eminent scientists, and yet both unreconstructed Stalinists. Bohm was eventually received by the three-centuries old Royal Society of London.

Before leaving Brazil, David came to visit me, and stayed 2 weeks at my mother's house, next to *El Ombú*. We travelled by train to visit Manuel and Cora Sadosky at the chalet they had rented in Miramar, a resort on the Atlantic. There, while I was trying to sleep next door, I overheard the three of them peeling me alive for shortcomings I was not aware of. This experience led me to wonder about the strength of the friendship-loyalty connection.

In 1958 I visited David and his wife in Bristol. At the physics seminar I witnessed a spirited exchange he had with a colleague whom David exasperated by using the term 'reality'. David attempted, unsuccessfully, to explain what he meant by it. None of his colleagues defended him. Definitely, philosophical realism was unpopular among the keenest students of reality. Worse, neither the realists nor their adversaries bothered to offer arguments: they seemed to think that evidence and rational discussion only matter in marginal issues. No wonder that, as I write, the best physics jobs go to believers in the "participant" (observer-centered) universe, many-worlds cosmology, and above all to believers in string theory – that weird and barren IOU.

It took me two decades to understand that the matter of the objectivity of a given theory could not be settled without the help of a theory of reference or denotation. My own theory (Bunge 1974a) allows one to uncover the reference class of any mathematically well-defined predicate. To accomplish this task, it suffices to analyze a predicate as a mathematical function, as Frege had proposed, and identify its domain. For example, "metabolizes" applies only to living beings, whereas "interacts" applies to pairs of concrete entities. Other predicates may pose problems – which is normal in science.

In 1961, when Marta and I visited Bohm in London, he seemed demoralized, but soon recovered his usual euphoria. This happened because he became instantly popular by exploiting Schrödinger's observation on the inseparability or "entanglement" of the constituents of a quantum system. He built on it an entire holistic, idealistic, and irrationalist world view, according to which – just as the guru Ramana Maharashi had said – "thought is the ultimate source of all evil" (Bohm 1980).

Back Home

While in Brazil I rethought some physics but produced none: I failed to Bohmize relativistic quantum mechanics, and had nothing to say about plasma physics, a subject David had suggested. During the last weeks as Bohm's postoc I only studied Bertrand Russell's splendid monograph on Leibniz (1900). This was the first serious philosophical work I read after skimming Boole's *Laws of Thought*, which the year before had given me a glimpse of modern logic. Russell showed me how to rethink tough problems, which had intrigued a great brain, in the light of what had been learned since. What a difference with the eloquent but purely destructive pamphlets of Marx's against Proudhon, Engels's against Dühring, and Lenin's against the Russian followers of Mach's! Russell had sought truth instead of ridiculing political rivals. As a postmodernist might say, for Marxists philosophy is politics by other means.

My daily life did not change much when back from Brazil, because I had no job and no prospects of getting one. So, I resumed my pre-Brazilian ways: private lessons, articles for encyclopedias, poorly paid translations, and administering the minute family building enterprise. Still, I kept doing some physics and some philosophy, greatly stimulated by our private seminars. Two of my best papers date back to that period: my attempt to refute Berkeley's subjectivism, which was generally regarded as irrefutable (Bunge 1954), and my paper on a new image of the electron (Bunge

1955a). In the latter I put as affiliation "Servicio Técnico-Científico", a ghostly firm invented by my friends José F. Westerkamp and Julio Iribarne, which never got any paying clients. My paper is the only extant witness to that adventure.

However, my main occupation was to research my book on causality, which I submitted for publication in 1956 but was published only 3 years later (Bunge 1959a). I worked at it hard and enthusiastically, but kept it to myself, because nobody seemed to be interested in the subject, that was generally believed to have been killed twice: first by David Hume, and two centuries later by the quantum theory.

Private Seminars

From the moment my friends and I were sacked from our modest university assistantships, we tried to revive the physics seminar that we had kept active between 1945 and the end of 1952. We met every Monday at the downtown apartment of Pipo and Angelita Westerkamp, who had installed a blackboard. We invited all the local physicists, including those working at the National Atomic Energy Commission, a few of whom dared to come.

The attendance grew from 6 to 12. David Bohm gave a talk, and so did my former student Tito Sirlin, an enthusiastic convert to "our [Feynman's] intuitive interpretation" of quantum electrodynamics, and who later did well at New York University. Two of the most assiduous participants were Héctor Rubinstein and Samuel Schiminovich, who had recently earned their physics baccalaureates. A few years later, when I returned to the University, I got Schiminovich the teaching assistantship for my course in Buenos Aires, and helped him get a fellowship to do his doctoral work under my friend Ralph Schiller – but he failed because he was a know-all. By contrast, Héctor was successful because he knew he always had a lot to learn. He got his PhD at Columbia University, worked for a while at CERN, ended up as a professor at Uppsala University, and fulfilled the dream of every Argentine man: he married a Swedish girl of a traditional family.

At our unofficial seminars we presented some work of our own, and summarized and discussed papers recently published in high-impact journals. We discussed freely, something that was no longer possible to do at the university, where there was no physics seminar any longer. Our more cautious colleagues did not attend our meetings, but somehow they got to know what we did, and I believe they admired us for it. The evidence for this conjecture is the success of our slate in the election held in 1955 to renew the governing board of AFA, our physical society.

The Atomic Establishment

When the Atomic Energy Commission (CNEA) was set up in 1950, our physics community split into two: the great majority rushed to get employment in it, whether or not they knew anything about nuclear physics, and a tiny minority criticized it, though not all for the same reason. My friend Enrique Gaviola and I objected to CNEA being a dependency of the Navy, and because its ultimate goal was the manufacture of a weapon of mass destruction; we also objected to its chief admission requisite, namely membership in the Peronist Party; finally, we disapproved of the inequitable allotment of the national science budget: the CNEA got the lion's share – an inordinately voracious and lazy feline. Gaviola and I wanted the CNEA to go under civilian control, that its budget be adjusted to its real needs, and that the positions be filled by open competition. A few years later we told all this to my friend Arturo Frondizi on the eve of the election that led him to the Presidency of the Republic. But obviously he had to face far more important and urgent issues, many of them emergent from his wish to please everyone.

In any case, shortly before the said AFA election, I proposed Gaviola, Westerkamp, and a few like-minded colleagues, to present our own slate of candidates. They agreed, I made up the list, had it printed, and mailed it to all the AFA members. Gaviola headed our slate, followed by Westerkamp as Secretary, I as Publications Secretary, and a few other undesirables. To everyone's amazement, we won by a large margin. Obviously, many CNEA employees had voted for us, and we all wished to safeguard AFA despite our differences.

When the result of the AFA election became known, Gaviola phoned me, and asked: "What are your orders, Chief?" I answered him: "You are the chief. Let's get to work." AFA kept sailing under its new authorities, which had only to follow our tradition.

What Makes Scientists Tick?

Here is where the question "What makes scientists tick?" emerges unexpectedly. Aristotle believed it was sheer curiosity; Robert Merton added peer recognition; the constructivist-relativists claim that it is love of power (or wish to increase one's "social capital"); and the cynics vote for vanity and the wish to climb. I believe that it is all of the above plus something else. In 1955 I hurried to finish and submit my paper "A picture of the electron" (Bunge 1955a) to show the AFA fellow members that the dissidents, though not employed as physicists, were still bona fide physicists. But of course that paper, one of my best, did not start like that.

The chief motivation of that paper was philosophical. Indeed, a decade earlier, while working on a failed theory of the electron without unmeasurable variables – as demanded by operationism (Bridgman 1927) – a new position coordinate "appeared," which intrigued me because it consisted of two very different terms: the ordinary purely geometric coordinate, and one that obviously "belonged" to the electron, since it was contained in the latter's energy operator. This intriguing coordinate, $X^{\mu} = x^{\mu} + (\Lambda/2)\, i\gamma^{\mu}$, where $\Lambda = h/mc$, was at the center of the paper in question. Years later I found out that Richard Feynman and Herbert C. Corben too had mentioned that operator, as a consequence of which it was called the *Feynman-Bunge-Corben* operator, although neither of my co-owners did anything with it, whereas I had included it in six new constants of the motion.

This is one more example of what the sociologist of science Robert K. Merton called "Matthew effect": the famous get the credit for discoveries of unknowns. In short, a false philosophical principle, which I would harshly criticize later on, led me to make my first independent scientific contribution. In turn, that paper was the starting point for other papers that my student Andrés Kálnay and I used for further work (Bunge 1958a, b; Bunge and Kálnay 1969, Bunge 2003a, b, c, d, e).

Perón's Downfall

Shortly thereafter, the so-called Revolución Libertadora happened, which undid a number of Peronist monstrosities but also made some grave political mistakes, such as excluding the Peronists from the political arena and executing some military loyal to the former regime.

The new regime gave academic leaders the chance to clean up the universities, which had been seriously downgraded under Peronism. Public competitions for all university chairs, most of which had been occupied by incompetent and servile individuals, were announced. This gave a chance of employment to hundreds of dissident intellectuals, who had being doing odd jobs for a decade.

Suddenly our private seminars became redundant. This was a bittersweet experience, for we had become used to meeting frequently, and now our new responsibilities kept us so busy, that we seldom met. We went overnight from underemployment to overemployment. For example, I was appointed to teach physics at Buenos Aires and La Plata. On top of this unexpected academic upgrading, I finally got proper identity documents, and slept without fear of being taken away by the political police.

Chapter 5
Philosophical Apprenticeship

My First Philosophy Textbook

My first philosophy manual, which I studied in 1936, was the standard book of logic and philosophy of science then used in the fifth and final high school year. It was well organized and clear but outdated. It was divided into two parts, the first of which dealt with classical syllogism theory, the oldest yet most difficult part of logic, now included in predicate logic.

Almost two decades later, when my first-born took his logic examination, he wrote the square of opposition using the modern symbols, which baffled his examiners. They told him frankly: "We don't understand your symbols, but we'll give you the benefit of the doubt, and pass you." And in 1957, when I joined the Faculty of Philosophy in Buenos Aires, the professor of logic had to look up the truth tables from a little card, he had never published an original paper, and repeated all the nonsense he had learned from Dilthey and Husserl. Yet the only Logic Center in the country bears his name.

The above-mentioned high-school textbook expounded logic just as it had been taught half a millennium earlier: *Barbara*, *Celarent*, *Darii*, etc. Undoubtedly, it helped in reasoning correctly, but it could have been modern instead of medieval. A contemporary logic manual would consist of the propositional calculus and some predicate logic. But, sure enough, that antiquated but clear book was soon replaced with one co-authored by the said colleague, who attempted to teach phenomenology, that obscure and obscurantist ancestor of existentialism. This is what happens in the absence of well-functioning bullshit detectors.

The second half of our textbook dealt with science, and it could have been written in France half a century earlier. It exaggerated the importance of the classification of the sciences, but did not even mention the formal/factual (or a priori/empirical) dichotomy that Leibniz had proposed two centuries earlier. Nor did it mention the existence of biosciences, such as geography, demography, and psychology, whose mere existence knocks down the wall between the natural and the

M. Bunge, *Between Two Worlds*, Springer Biographies,
DOI 10.1007/978-3-319-29251-9_5

social sciences that had been erected by the German idealists, from Kant to Dilthey. Yet, my antiquated textbook kindled my curiosity for the philosophy of science.

Other Books That Nurtured My Love of Philosophy

In Chap. 2 I mentioned reading Russell's *Problems of Philosophy* (Russell 1912) along with some classics of dialectical materialism. Later on I also read Harald Høffding's *Brief History of Modern Philosophy* (Høffding 1912), Émile Meyerson's *Identité et réalité* (Meyerson 1908), and a short book by Hans Reichenbach. I also read a pile of wordy and ephemeral English, French, and Italian books of no relevance to my project – from Giordano Bruno to Benedetto Croce. Dilettante that I was, I had no guidance, and had yet to construct my own filter. Thus, while I detected the obscurantism of Schopenhauer, Nietzsche, Bergson, and Husserl, I failed to detect Hegel's.

Later on I consulted Spinoza's *Ethics* whenever I encountered a metaphysical problem, as well as André Lalande's rich if outdated *Vocabulaire* in three volumes (1937) every time I encountered an unfamiliar term. And the Russian Marxist Georgi Plekhanov introduced me to the radical wing of the French Enlightenment: the materialists Thiry d'Holbach, Denis Diderot, and Claude Helvétius, so much deeper and influential than Ludwig Feuerbach, Marx's favorite. But I wasted too much time trying to decipher Hegel's riddles, and it took me several years to realize that he had started the Counter-Enlightenment and had invented the trick of passing off absurdity as depth. But at least from the start I was more interested in problems than in authors. Witness my choice of subjects for my philosophical papers and books: natural law, causality, chance, process, dialectics, matter, space, time, system, emergence, level, mind, meaning, truth, intuition, problem, hypothesis, theory, explanation, simplicity, exactness, science, pseudoscience, scientism, value, justice, social structure, socialism, and other problematic concepts whose elucidation required combining my two intellectual passions.

Science or Philosophy?

I never had to opt between science and philosophy. Both attracted me equally from the moment of my intellectual awakening at about 16, as I discussed in Chap. 2. True, I once thought I had found the demarcation line between them: whereas science is the study of reality, philosophy equals metascience, or the study of the study of reality (Bunge 1944c). But eventually I realized that science in the making involves a number of philosophical presuppositions – such as that the world preexists us and can be explored – so that the two domains have a partial overlap (e.g., Bunge 1967b, 1983b). In other words, whereas *scientia lata* can be described without mentioning philosophy, *scientia ferenda* cannot, so that scientific investigators cannot help but philosophize.

From 1936 on, when I started to complete high school on my own, I read much philosophy, mostly bad, as well as some semipopular physics books, in particular those by Arthur Eddington and James Jeans, both of them distinguished scientists and eloquent writers. They wished to "sell" philosophies that seemed wrong to me: Kant's subjective idealism, and Plato's objective realism respectively. Indeed, Eddington had stated that we discover what is already in our minds, whereas Jeans held that the universe is a mathematical construction. Moreover, both claimed that those are results of contemporary science. Anyone could see that, if Eddington were right, anyone could understand physics without studying it; and that, if Jeans were right, pencil and paper would suffice to discover reality. But disproving the claim that physics is idealistic requires knowing a lot of physics, and I was far from meeting this requisite. This is what motivated me to start studying physics at the university, as described in Chap. 3, and I have kept doing so ever since 1938, though slowly and sporadically. For example, two summers ago I discovered that the so-called Aharonov-Bohm effect is not a physical fact but a misunderstanding that can be cleared through a semantic analysis (Bunge 2015a).

Interlude: Of Amateurs and Professionals

While studying physics at the university, I kept reading philosophy on the side: both my library and my briefcase contained publications in both fields in roughly equal parts. I read them at home, on trains, subways and streetcars and, in good weather, in the garden or in public parks as well. But, whereas I studied the physics (and mathematics) I needed, I used to read only the philosophy I wanted – as befits a dilettante.

My case was far from unique: all the so-called developing countries suffer from a plethora of amateurs and a deficit of specialists. One of my old Argentine friends, who has not even finished high school, has supported himself teaching logic at a private university, has won an important poetry prize, and talks unabashedly about the toughest philosophic-scientific problems, as a consequence of which he is regarded as an important philosopher. In the developed countries, by contrast, most philosophers devote their whole lives to a single subject or author. For example, a former colleague at McGill knew everything about David Hume, except that his skepticism about imperceptibles led him to rejecting the best science of his time, with the result that his epistemology was dogmatic and obscurantist.

Authentic philosophers are eminently generalists: the expression 'specialized philosopher' is an oxymoron, because all important issues come in bundles (see Bunge 2014b). Highly specialized philosophers seldom tackle bundles of problems, and seldom take a look at what is going on in other fields. For example, philosophers of mathematics may know all about the paradoxes in set theory that caused so much worry a century ago, but they are unlikely to know that category theory replaced set theory as the foundation of mathematics half a century ago. And most philosophers of mind do not usually read journals in the sciences of mind, so that they think as dualists even if they call themselves materialists. A consequence of such slanted interests is that such individuals cannot be useful discussants about issues that mix philosophy with science, such as the nature of mathematical objects or of biological species, the independent existence of photons and ideas, or the competence of standard economic theory to help design effective economic policies.

From the start I sought philosophical ideas capable of clarifying the philosophical issues that science seemed to generate. I could not foresee that acquiring such ability would take me more than two decades of weeding, planting, fertilizing, and pruning. Nor did I foresee that my intellectual gardening would be just as enjoyable as a discovery expedition.

First Philosophical Paper and Seminar

My earliest philosophical paper bore the pretentious title "Introduction to the study of the great thinkers", and appeared in two successive issues of the monthly *Conferencias* (Bunge 1939). It was the text of the two lectures I gave to full audiences at AIAPE in September, 1939, a few days before turning 20. (For AIAPE see Chap. 2.) I guess that paper was a sort of distillate of my disorderly philosophical readings. I hope it fed at least some moths, by contrast to the plain clothes policeman who, in the report published in his division's rag, wrote that he left the lecture hall whistling the tango "Mostery," which had just been released, and whose idiotic lyrics started thus: *Everything in life is mostery/ And no one knows why.*"

In 1942, while encountering quantum mechanics, I decided to tackle the problem of causality, for I intuitively doubted the popular claim that that theory, being essentially probabilistic, had rendered causality obsolete. I read here and there, wrote a lot, and invited my friends to discuss the problem in a seminar to be held at the Universidad Obrera Argentina. Only two persons were interested: the German expatriate Isidoro Flaumbaum, who was studying philosophy at the university, and his silent friend Roberto Salama, an education student at the same Faculty.

Our discussions were stimulating but produced no edible fruits: the subject by far exceeded our competence. Besides, there was hardly any literature on it, because only the Thomists were seriously interested in causality, but they could not handle it competently because of their disconnection from modern science. However, the problem continued to interest me, and I tackled it more seriously a decade later, after interacting closely with David Bohm, who claimed to have produced a causal interpretation of quantum mechanics, as we saw in the previous chapter.

Earliest Philosophical Problems

It took me 5 years to find my earliest genuine philosophical problems. They were three, and were what later were called "demarcation problems": How does philosophy differ from science, what distinguishes the latter from pseudoscience, and whether philosophers can fruitfully discourse on nature without using natural science? (As we learned from his autobiography, the young Karl Popper too had tackled some of these problems.)

In particular, I wished to find out why Eddington's and Jeans's philosophies of physics were wrong if popular, why telepathy and Oedipus's complex were fantasies, and why the Romantic philosophy of nature had failed to replace the science of nature. I found no answer to these questions in the meager philosophical literature within my reach. In particular, the French and British Marxists I had read were not interested in contemporary problems: their forte was the past, which has the great advantage that it cannot bite back. Ditto the neo-Kantians and the neo-Hegelians, as well as Dilthey, Bergson, Croce, and Husserl: none of them had written anything useful to my project. Moreover, all of them took it for granted that philosophy and science were mutually disjoined – a thesis that I intuitively rejected.

True, the logical positivists were keenly interested in science, but at the time their works were unavailable in Buenos Aires. And anyway, they had not criticized physical idealism, psychoanalysis, or parapsychology. Worse, as I was to discover much later, the neo-positivists were largely responsible for the subjectivism inherent in quantum orthodoxy, and some of them were sympathetic to psychoanalysis, while others warned that future experiments might confirm parapsychology. In short, none of the extant philosophies of science helped me answer my earliest philosophical questions. So, I had to try and answer them by myself.

Minerva

The philosophy student Isidoro Flaumbaum came often for lunch, and we had long conversations on contemporary German philosophy, which he knew far better than I. He was also my liaison with the Faculty of Philosophy. I suppose that it was after one of those conversations that I had the idea of organizing a sort of rationalist common front to fight irrationalism, in particular existentialism. This pseudo-philosophy had started to rule in the Latin American schools of humanities: it rode on the fascist wave, and hid behind the phenomenological veil, which looked respectable because Husserl dealt only with the self, and had never pronounced himself about any life–and–death issues of his time.

Other friends of mine, in particular Hernán Rodríguez, found the idea of a philosophical common front absurd: they were used to the vulgar idea that philosophy was condemned to be split into closed rival schools. This idea seemed natural in countries with no tradition of philosophical seminars, societies, congresses, or other fora of rational debate. We had not even the debating societies that had played such important intellectual and political roles in Britain.

Since our philosophers were individualistic, the organ of the rational front I envisaged had to be a periodical. This is how *Minerva* was born: as the "continental journal of philosophy." Though high-sounding, this announcement happens to be true: *Minerva* was the only philosophy journal to circulate throughout the continent. All the others were local, and the Thomist or Marxist among them were partisan as well.

Before launching the journal I consulted often with Risieri Frondizi, who was then teaching at the young University of Tucumán, after having written his Harvard dissertation on the gestalt (or emergent whole) concept, supervised by Alfred N. Whitehead (see Gracia 1980). I invited submissions from many philosophers, some of them recommended by Francisco Romero, then the senior Argentine philosopher (see Speroni 2001).

I got a few submissions, for which I paid 100 U.S. dollars, which at the time was the salary of a full professor. One of the personalities I tried to contact was Richard von Mises, the mathematician and former outstanding member of the Vienna Circle. But I addressed my letter to his brother Ludwig, the economist and conservative ideologue, who was angered by my confusion of the two of them.

What *Minerva* Published

Minerva was born in the Fall of 1944, and died a year later for lack of both submissions and funds. It contained articles, short notes, and book reviews. Here is a list of the articles contained in *Minerva*'s two volumes:

Rodolfo Mondolfo, "La filosofía de Giordano Bruno"
Mario Bunge, "¿Qué es la epistemología?"
Simón N. Neuschloss, "El irracionalismo en la física contemporánea"
Isidoro Flaumbaum, "Meister Eckart y Martin Heidegger"
Hernán Rodríguez, "Conflicto de vida y muerte de Antonio Machado"
Risieri Frondizi, "Panorama de la filosofía latinoamericana contemporánea"
Hans A. Lindemann, "El 'Círculo de Viena' y la filosofía científica"
Luis Farré, "Los valores en Platón"
Alfred Stern, "El significado de la fenomenología"
Mario Bunge, "Auge y fracaso de la filosofía de la naturaleza"
Elisabeth Goguel de Labrousse, "La degradación de la moral estoica"
Víctor Litter, "La etnología científica y la escuela histórico-cultural"
Hernán Rodríguez, "La formación filosófica de [Mariano] Moreno"
C. Anglés B., "Introducción a una crítica de la fenomenología"
Maximilian Beck, "¿Decaerá nuestra civilización como lo predijo Spengler?"
José Juan Bruera", "La lógica, el Derecho y la escuela de Viena"
Werner Bock, "Nietzsche y el círculo de Stefan George"

Mario Bunge, "Nietzsche y la ciencia"
Louis O. Katsoff, "La lógica del hecho"
Rodolfo Mondolfo, "Determinismo contra voluntarismo en la filosofía de Nietzsche"
Raúl A. Piérola, "Nietzsche y Sócrates"
Julio Rey Pastor, "La filosofía ficcionista"
Francisco Romero, "Indicaciones sobre el problema de la concepción del mundo"
Alfred Stern, "Max Scheler, filósofo de la guerra total y del Estado totalitario."

In retrospect, I believe that the best papers were those by Bruera, Mondolfo, Rey Pastor, and Stern. Bruera's was one of the earliest indictments of the political servilism preached by legal positivism; Mondolfo's papers were at a level that Italian philosophy has not regained; Rey Pastor's was the most scholarly evaluation of Hans Vaihinger' fictionism; and Stern's denunciation of the complicity of Scheler's first philosophy with the German establishment was courageous, given phenomenology's prestige at the time.

Minerva's Reception

Minerva's reception was between lukewarm and hostile. It is likely to have had more lay than professional readers, had no visible academic impact, and is now very hard to find. The Thomists liked my review of Octavio N. Derisi's doctoral dissertation, but later attacked *Minerva*, asserting that every issue was worse than the preceding one. Stern's paper angered Francisco Romero. He and his younger brother José Luis had strong democratic convictions and paid the price for them, but they admired above all Wilhelm Dilthey's antiscientism, failing to realize that it had helped pave the way for Nazism, in promoting intuition over reason and experience. Moreover, José Luis did not see the incompatibility between his academic field, social historiography, and Dilthey's preference for *Verstehen*, in the sense of sympathetic understanding. (There is more on Francisco Romero in Bunge 2001b, and on Dilthey's *Verstehen* in Bunge 1996a).

I heard two reactions from the Marxist camp. I got praise from the Cuban Carlos Rafael Rodríguez, the only Marxist to accompany Fidel Castro and Che Guevara in Sierra Maestra. We met four decades later at his office when he was the minister of economy. When I told him that I found Cuban philosophers backward, he replied: "That's because they follow Soviet philosophy, which has remained stagnant for decades."

By contrast, my countryman and friend Rodolfo Puiggrós told me that he and his comrades thought that my paper on the philosophy of science, in the first issue of *Minerva*, revealed that I was a follower of Ernst Mach, hence an enemy of Marxism. The same paper earned me the esteem of Paco Miró Quesada, aka Paco, the chief Peruvian philosopher and a very interesting and warm person. We first met in 1956, and four decades later we had an intensive exchange of letters. Paco was curious about all scientific novelties, and a great admirer of rationality, but never got entirely rid of the phenomenology he had absorbed in his youth. Paco had no intellectual filters: he got as enthusiastic over chaos theory and Marta's work in category theory as over paraconsistent logic, Newton da Costa's monster, which admits contradiction because, according to Hegel's dialectics, it is the source of novelty.

How Minerva Affected My Life

The publication of *Minerva* hurt me in various ways. First, I lost all the money I had invested in it, for the bookstores and kiosks that had sold it did not pay me back for selling the 1,000 copies of each of the six published issues. Second, the Argentine philosophical community followed the national custom of rubbishing whoever did something new. In particular, Monsignor Derisi did not invite me to take part in the first National Congress of Philosophy (1949), which enjoyed strong official support, and where President Perón gave a memorable speech against what he called the "insectification of man." (Years later, while we were meeting in Chile, Derisi boasted that every one of the participants had been given a luxurious vicuña *poncho*.) And Francisco Romero's faithful did not invite me to the inaugural meeting of the Argentine Philosophical Society (1956) that he presided over nominally. However, we had friendly relations before and after that episode, and he led me to reading Eça de Queirós's novels.

The third price I paid for launching *Minerva* was the steepest of all: Guido Beck, my thesis supervisor and friend, expelled me from his flock. He wrote me an aggressive and cruel letter, and denounced me at an AFA meeting for violating the full-time condition. He even denounced the modest application of Boolean algebra in my first paper in that journal. Fortunately Rey Pastor, the most respected mathematician among us, wrote him that my formula was correct.

Left without a master, I did what any abandoned pet does: I fended for myself. To begin with, I changed subject: I tried to build a theory to explain the puzzling Cherenkov radiation, emitted by electrons moving faster than light in a dielectric. I sent my formulas to Beck, which he liked because the exponent two-third occurred in one of them – which is like buying a house because it contains a pretty flowerpot. He took the trouble to spot my mistakes and readmitted me to his flock, for I had shown him that my affair with physics was not just a flirtation. In short, I resumed work on my thesis under Beck's guidance, but in the course of it several mathematical problems emerged that took us both by surprise. The worst of them were certain divergent series and singular integrals. Fortunately, my friends Rey Pastor and González Domínguez, who delighted in mathematical pathologies, taught me tricks to tame, evade, hide, or disguise such beasts. But before using them to the full there was another interruption.

Political Intermezzo

Towards the end of 1945, a national election was called to restore the political democracy that had been suspended by the fascist coup in 1943. Two large antagonistic blocks were built: the new Peronist party, and a coalition of the traditional parties. The latter called itself Unión Democrática, even though it included two parties that did not believe in democracy, the Conservative and the Communist ones. What kept this conglomerate together was the diagnosis of Peronism as the local variety of fascism. This thesis was put forth by Victorio Codovilla, the leader of the Communist Party, a very small outfit but one capable of filling a public park on short notice.

Ironically, the US State Department shared Codovilla's diagnosis, although it was not based on even a minimally objective study of Peronism. Such a study would have shown that, although Peronism did have a fascist component, this did not explain its phenomenal success. Peronisn was a new and strong mass movement that, though supported by the army and the Church, answered some legitimate grievances by the poor. In addition, instead of jailing the trade union leaders, it coopted and corrupted them, as well as using elections to its own advantage instead of forbidding them or even buying votes the way American politicians do.

Peronism was doing what none of the fascist parties in other countries had done: it gave women the right to vote, modernized the labor code, seduced leaders of all the rival parties, created a ministry of public health, maintained an independent foreign policy, and held free elections, at the same time that it set up a clientelist state, put gangsters at the head of most labor unions, converted the schools into centers of political indoctrination, destroyed civil society, muted the press, and marginalized its political opponents – in sum, it Peronized society with the support of about 80 % of it.

Neither of the self-styled political experts at that time, whether in the country or abroad, in universities or in parties, understood this new political movement because, instead of investigating it, they tried to fit it into the classical European categories. I, for one, was an enthusiastic activist in the Unión Democrática, and even got a praiseful letter from Spruille Braden, the American Ambassador, who interfered in the political process as much as he wished. Unwittingly, the ambassador was a major contributor to manufacturing Perón's reputation as the savior of national independence. However, the diplomat is not to be blamed for it: he was just following the long American tradition of meddling in other countries against their own long-term interests.

A group of neighbors asked me to lead the local Unión Democrática committee, which I did during the summer of 1945. This activity caused some fanatic Peronists to disfigure the two entrance pillars to *El Ombú* with threats written in ominous red paint. Consequently I walked around carrying a pistol I borrowed from Michele, our gardener. But it was just for show, since I did not learn to use it. My activity was confined to going every evening to our committee, which occupied a garage, and visiting some local politicians. One of them was no less than Roberto Uzal, the local conservative boss. Yes, Max Weber was right: politicians are sometimes forced to make dealings with the devil. Even the saintly Gandhi and Mandela did it.

In the committee we organized public meetings and the affixing of posters, and we discussed the little news and many rumors of the day. Although my speaking at a meeting was announced, I did not even attend it. I was also included on the communist slate, although I never belonged to any cell, and my faith had weakened so much that 2 years later I was formally expelled from the party.

My participation in the Unión Democrática earned me some good apolitical friends, in particular the lawyers and accountants around Rogelio Galarce. This lawyer, to whom I had been introduced by Ernesto Malaccorto – protégé of my lay godfather Raúl Prebisch and a habitual guest of the Sunday lunches in *El Ombú* – who headed an important law firm. Rogelio and I helped each other in good times and bad. Contrary to my expectation, Beck did not scold me for my political escapade, perhaps because he and his family had been victims of fascism, and perhaps also because his fancy Jockey Club friends told him that Peronism was the worst plague. In any event, after the elections I was free to work on physics all day long.

Centered in Physics with Occasional Forays in Philosophy

Between 1945 and 1953 I concentrated on physics, but never forgot philosophy. For example, in 1951 my criticism of phenomenology and my first writing on chance appeared – the former plagued by misprints and the latter censored by the editor of the journal where it appeared, who made me treat Henri Poincaré without the respect I felt for him. I also kept interacting with other philosophy lovers, at first occasionally and later regularly, in what I called the Buenos Aires Circle. But before this circle emerged, I met with several other people to discuss philosophical problems.

The first of those conversations was convened by the friendly and fat Carlos Prélat, a chemist who taught at the Faculty of Science. He was a pioneer in the philosophy of chemistry, for his book (Prélat 1947) appeared half a century before its subject was recognized as a legitimate branch of philosophy. Prélat, who had studied long before quantum chemistry arrived in Argentina, held that his science was irreducible to physics. This assertion was enough to kindle a spirited if friendly discussion among us. We shared a good friend, the printer and publisher Domingo Landolfi, who offered us his conveniently located office to meet with a few other philosophy amateurs. There I met a few times with Prélat and a couple of others, in particular the bright and charming professional dilettante Tomás Moro Simpson (1964), who two decades later published a book in analytic philosophy which exerted an unexpected positive influence during the dark age of the last military dictatorship (1976–1983).

Prélat opened our first meeting asserting that quantum mechanics had destroyed materialism, by showing that particles are just wave packets, which in turn are nothing but symbols. He wondered why I, who was working on quantics, kept calling myself a materialist. I answered as best I could, but Prélat's question worried me, and I kept working on it for several years. Obviously, it required either a reinterpretation of quantum mechanics, a redefinition of "matter", or both.

I believe I performed both tasks much later, in several articles and books, particularly *Foundations of Physics* (Bunge 1967a), *Scientific Materialism* (Bunge 1981d), and *Matter and Mind* (Bunge 2010a). There I proposed a reinterpretation of quantum mechanics that kept its standard mathematical formalism but avoided the concept of observation, which belongs in experimental physics, and I equated the predicate "is material" with "is capable of changing."

A few years later, Prélat was sacked from his university job for refusing to join the ruling party. He returned to his native town, Concepción del Uruguay, one of the most beautiful and cultivated Argentine cities. As soon as Perón fell, I wrote Prélat, urging him to return to Buenos Aires. He answered saying that his chemistry was not up to date, but accepted the invitation of the newly created university in Bahía Blanca. A year later, when I visited him there, I found him happier and fatter than ever. Ten years later, when the military replaced the constitutional regime, Prélat had to flee to Brazil, where he reconstructed his life once more, all the while teaching and writing chemistry textbooks. Argentina does not eat its children, but gives them away to its neighbors.

Other Occasional Philosophical Meetings

Around 1950, my fellow teaching assistant Oscar Varsavsky invited me to participate in the colloquium that he, the logician Gregorio Klimovsky, and the chemist Heberto Puente were convening to discuss one of the central problems of philosophy since Kant: the distinction between analytic and synthetic propositions, which was then a hot topic. All three conveners were philosophical amateurs and enthusiastic adherents of logical positivism – the only philosophical movement that took science seriously. Varsavsky and Klimovsky were bright and hardworking, and both would flourish a decade later as enthusiastic teachers. All three took themselves very seriously, and worked hard to support themselves: Varsavsky by writing one Western per month, and Klimovsky by teaching elementary mathematics and praising psychoanalysis at a private school.

I accepted Varsavsky's invitation but, when I started to write my presentation, I realized my ignorance of the subject. I had not even read Wittgenstein's *Tractatus logico-philosophicus*, back then regarded as the bible of logical positivism. Since I did not find it in the library of the Faculty of Philosophy, I resorted to Hans A. Lindemann, who had attended the last meetings of the Vienna Circle.

Lindemann was an Austrian businessman who had planned to get rich enough to take early retirement and devote the rest of his life to philosophy, which he did in Buenos Aires. He lent me Wittgenstein's book for 24 h, during which I devoured it. It disappointed me: it seemed no more than a haphazard collection of disconnected aphorisms of questionable value. But the early Wittgenstein had the undeniable merit of reviving interest in ontological problems, which had been discarded by the empiricists. In particular, Wittgenstein asked what are the ultimate constituents of the world? Astonishingly he answered without taking modern physics into account. His theory of knowledge, akin to the Marxist theory of "reflection," discarded invention, while his assertion that mathematics is a collection of tautologies (logical truths) seemed to me to be false, since mathematics involves extralogical concepts, such as those of function, limit, and even the humble relation ≤.

A couple of years later, while trying to stay awake reading Wittgenstein's *Logical Investigations*, I suffered an even worse disappointment: I learned that Wittgenstein denied the very existence of philosophical problems. This was perhaps because he confined himself to inventing and analyzing questions such as "What is knowing the color red?," which is trivial unless one looks at what the sciences of color perception have to say about it. Wittgenstein's answer, that "knowing red consists in knowing English," is just silly, as only a witch doctor would confuse a fact with its name. In short, I did not make a presentation at that meeting when I realized that, although I disagreed with much of what was said in it, I was unable to propose alternatives. I tackled the analyticity question a decade later (Bunge 1961a), and argued that the word 'analytic' is polysemic, that is, it designates different concepts.

What Is an Electron?

The large classroom at the Faculty of Science was full with teachers and students eager to learn the latest philosophy. I learned something amazing. When someone mentioned electrons, Varsavsky asked Puente: "What is an electron, Heberto?" Puente answered in his usual solemn tone: "The electron is an incomplete symbol." Everyone seemed satisfied with this answer, although I doubt that anyone knew what an incomplete symbol is. I still don't know, but I understood that the answer in question involved the confusion, typical of magic-religious thought, between things and the symbols representing them.

Puente's position was consistent with his rejection of the atomic theory, which led him to flunking any students who dared mention the word 'atom'. But Varsavsky's silence surprised me, because his recent doctoral dissertation (unpublished) was about quantum mechanics. And this theory happens to deal with electrons and the like, which have properties, such as mass, charge, and spin, that symbols lack. Of course, one can perform computations without worrying about the meaning of the symbols occurring in them – but that's not science.

In any event, the ignorance and arrogance of the local positivists did not affect my appreciation of the Vienna Circle, with whose members one could debate rationally. I was particularly fond of Philipp Frank, Hans Reichenbach, and Richard von Mises, for they wrote clearly and tackled problems in contemporary physics, though rarely with my approval. But I shared Einstein's and Planck's vehement criticism of Mach's anti-atomism. I attacked the latter in a paper that I submitted to the journal published by the Chemistry Student Center, which rejected it because the communists who ruled it at the time had damned me. So, I gave my paper to a visiting Peruvian chemist, who had my paper published in his country (Bunge 1951c).

A decade later, in two seminars for physics students, one in Buenos Aires and the other in La Plata, we examined in detail Frank's *Foundations of Physics* (Frank 1946). I also persuaded my fiancée, Marta, to translate Rudolf Carnap's *Foundations of Logic and Mathematics* (Carnap 1939), which I still regard highly. (To review her translation, we met clandestinely at a desolate and cold milk bar. But that happened a decade later). Because of this, but mainly because of the vulgar confusion between scientism and positivism, I am often regarded as a positivist despite my many criticisms of positivism. Labeling and libeling are easier than analyzing and arguing.

The Buenos Aires Circle

Once, around 1950, I proposed to my close friends the mathematician Manuel Sadosky, the allergist Enrique Mathov, and the philosophy student and IBM employee Hernán Rodríguez, to meet regularly to discuss philosophy and comment on papers published in five different journals, which we paid for jointly. Sometimes the chemist Pipo Westerkamp and the medical student Hersch M. Gerschenfeld, aka Coco, joined us. The clinician Emilio Troise (the only acquaintance of ours who had published a philosophical book), the psychiatrist Gregorio Berman, and a few others came once each. We never got a professional philosopher. Our meetings stimulated us greatly.

Once, while we were discussing Pavlov's tirades against psychologists ignorant about his work, Gerschenfeld brought a fresh brain that he planted on the table, described its visible parts, and the functions then attributed to them on the basis of electrophysiological experiments. Coco ended up as a professor of neuroscience at the École Normale Supérieure in Paris, and wrote about his useless political activities in Buenos Aires (Gerschenfeld 2009).

One cold winter Sunday we met in a garage to discuss Sadosky's paean to computers. He had swallowed the hype then in vogue, that computers can think. I adopted a critical attitude, inspired by my reading realistic computer experts in the *Proceedings of the Institute of Radio Engineers* (Bunge 1956a). Sadosky admitted his error, but kept his enthusiasm for computers, although he never learned to use them. In this he was more of a coach than a trainer. A decade later he set up the Computation Centre at the Buenos Aires University (see Jacovkis 2013).

Mathov taught me more than any other members of our circle, because he did not allow me to make any groundless claims. He also attempted to initiate me in immunology: he assigned me the task of injecting cow milk in geraniums, and looking for differences with the plants in the control group. But geraniums have wooden stems, which not even a dentist's syringe could pierce, so the experiment did not even get going. Enrique talked like a street boy, but was smart, frank and generous. In the view of all of us, he was a diamond in the rough. Eventually he became the leader of the Argentine allergy community. Once Enrique told me in all seriousness that I would never be recognized as a philosopher, because my writings were intelligible. Sadosky attended all the meetings of our circle, most of which were held at his apartment. With him, Enrique and Hernán, we organized a parallel school for our children to overcome the limitations and distortions of the Peronist educational system. Manuel taught them mathematics, Enrique human biology, Hernán history, and I physics. Three of our five children ended up as scientists, and two as physicians.

Discovering Modern Logic

I discovered modern logic only around 1950, but did not study it seriously until 1959, when I got the Hilbert and Ackermann textbook. It was only in 1964 that I felt confident enough to teach undergraduate logic, first at Temple University in Philadelphia, and later at McGill. The logic textbooks I found most useful were those of Copi, Hilbert and Ackermann, Suppes, Beth, Stoll, and later Tarski's monumental volume on semantics and metamathematics. The study of logic did more than adding a bit of knowledge: it changed my way of thinking. I understood the value of exact concepts and valid arguments, which were absent from philosophers such as Hegel and his successors. From then on, part of my work has consisted in what I call *exactifying* key concepts and assumptions. For example, my main technical objection to Milton Friedman's monetarism is that it contains far too many formulas of the form "variable y depends on variable x," which is just a programmatic hypothesis: one that challenges one to look for the particular function f such that $y = f(x)$ fits the data.

But of course not all concepts are exactifiable. The oldest examples of inherently inexact concepts are those of theology. And those characteristic of dialectics, such as "dialectical opposite," "contradiction," and "sublation" (*Aufhebung*) are no clearer than "transubstantiation" and "Holy Trinity." At the other extreme of the exactness spectrum we find trivialities that sound like deep insights if they were first stated by famous authors. One such sacred cow is Wittgensgtein's closing statement in his *Tractatus*: "Whereof one cannot speak, thereof one must be silent." Here is another in the same class: "Levitation is forbidden because it is impossible."

However, my respect for logic never went as far as worshipping every word written by great logicians such as Frege, Russell, Tarski, or Quine. For example, I corrected Frege's construal of a predicate as a function from individuals to truth values (Bunge 1974c, 1975a); I simplified Russell's theory of definite descriptions (Bunge 1974d); I criticized Tarski's conflation of formal and factual truth (Bunge 1974d); and I refuted Quine's thesis about the ontological commitment of the so-called existential quantifier (Bunge 2012a). Nor did my respect for logic degenerate into formalism, the approach that cares more for form than for content, as is the case with the philosophers of science who confine themselves to ferreting out the logical structure of theories but have no semantics capable of discovering the entities such theories are expected to describe. However, let us now come down from the clouds, as Aristophanes would say.

How I (Nearly) Made Ends Meet

I tried several transient jobs during Perón's leaden decade (1945–1955). Manuel Sadosky found me the task of giving private physics lessons to the brothers Amati: Daniel, who was to distinguish himself in physics, and Paolo, who became a zoologist. Manuel also got me my first philosophy teaching experience: a private philosophical primer for a group of about 15 very bright and curious teenagers who paid me well. I chose reading and commenting on *Theaetetos*, Plato's dialogue on knowledge – a model of rational debate. I read that masterpiece for the first time, along with my students. They were just as delighted as I, and they posed many tough and interesting questions, as usually happens with children when released from dogmatic cages. Since then I have tried to follow the Socratic method in my philosophy classes. In fact, these have consisted of question-and-answer sessions on a preassigned subject and the accompanying minimal list of references.

When I completed my postdoc fellowship with Bohm in São Paulo, at the beginning of the spring of 1953, I started to work on two long-term projects that demanded more and more of my time, and were to keep me busy between 1954 and 1970: Causality, and the Foundations and Philosophy of Physics. When I visited Quine in 1966, we talked about long-term projects and short-term essays, or long-distance runners and sprinters. We agreed in preferring the former to the latter. His own dual project, that of expanding the uses of logic and naturalizing it, was obviously of the first kind, whereas his colleague Hilary Putnam, who flittered among different subjects without going deeply into any, seemed to him as hopeless as he was brilliant.

Back in Buenos Aires, with great projects but no income, my good friend and former business partner Raúl Fernández encouraged me to apply to the Ernesto Santamarina Foundation for a fellowship that would allow me to concentrate on my scholarly work for a year. (Raúl assured me that Ernesto Santamarina, brother of Antonio, the conservative leader who two decades earlier had attempted to win over my father, was a progressive rancher, who employed a teacher and a barber for his *gauchos*). I won the scholarship, and started right away working on the foundations of quantum mechanics. My friend and former student Horacio Bosch, who worked at the Atomic Energy Commission, got me photocopies of all the papers I needed and that had been published in journals that were available only at his workplace – which I regarded as enemy territory.

Encounters with the Great Houssay

My only obligation as an Ernesto Santamarina fellow was to report monthly on the progress of my research to the supervisor that the Foundation had assigned me. This was no less than Bernardo A. Houssay (1887–1971), a Nobel laureate in physiology and medicine, whom I visited at the Institute of Physiology and Experimental Medicine. This was a private research center supported by the Sauberan Foundation, where he had been working since he had been sacked from his chair at the Faculty of Medicine by the military dictatorship that had seized power in June, 1943.

All scientists had heard some of what Hussay had been doing to his laboratory dogs and frogs. Some of us had also read some of his calls for full time jobs, the relation between teaching and research, the need to fill labs with curious brains, the role of basic research in national development, and other conditions of fruitful scientific work (see Barrios-Medina and Paladini 1989). Houssay was one of my heroes. I fully agreed with his science policy, and disagreed with the self-styled reformist student leaders who fought him: I thought them ignorant and obscurantist.

Dr Houssay was a genius and a teacher of teachers, he had a vast scientific culture, and a long experience in advising fellowship holders, on top of which he had earned a reputation for severity, tactlessness, and bad temper. (It was even rumored that his children addressed him as Doctor.) For these reasons, I prepared my interviews with Houssay with as much trepidation as care. I used to arrive much earlier to leaf through science journals, attend the coffee breaks, and chat with my cousin Julia Uranga Bunge, who worked there as a lab assistant before devoting her life to Opus Dei. My first interview with Houssay did not go well at all. He started by criticizing my father for not having devoted himself to research after having won the gold medal of the 1900 cohort. Next he asked me whether I had been a good high school student, and made his characteristic disgusted grimace, wiggling his rabbit-like mustache, when I told him the truth. (He had been promoted from grade 3 to high school in the course of a few months, and had finished at age 13.) The master also asked me why I was not working at the Atomic Energy Commission. My answer did not satisfy him, for he knew a chemist, Santos Mayo, who was doing good work there. Finally, Houssay asked me about my research project, and asked me a couple of relevant and perspicacious question about it. My answers must have satisfied him, for his attitude changed, and he saw me off with a cordial handshake. He even made an effort to smile. The 11 subsequent interviews went very well. We never met again, but almost half a century later the Buenos Aires University granted me the honor of occupying the Bernardo Houssay Chair for a week.

A Productive Year

That year 1954 was the first in my life, although I was already 34 years old, that I had an adequate monthly income, which allowed me to work full-time on what I liked the most. I worked simultaneously on my fellowship's theme and my long-term philosophical project: the determination categories, from spontaneity and efficient causation to chance and final causation. My preliminary research led me to visit frequently the National Library and the very poor and outdated libraries of the faculties of science and philosophy. In the first of them I read parts of Pierre Duhem's 10-volume treatise *Le système du monde*. Duhem was the famous physicist, positivist philosopher and militant Catholic who held that modern mechanics had been invented by some obscure Paduan scholars rather than Galileo. In that same library I also bumped into Ettore Bignone's curious and learned monograph *L'Aristotele perduto e la formazione filosofica di Epicuro* (Bignone 1936), which helped me to understand ancient science. (Unexpectedly, a decade later this same book would help me settle in Canada, as will be seen in Chap. 8.) I used both Duhem's and Bignone's works to write a paper on self-determination. I never published it, but I gave it to David Bohm while he was staying in my mother's house. He included it in one of his books (Bohm 1957) almost verbatim and without attribution.

The Argentine libraries had not been enriched since about 1900, and they had been especially neglected under the regime whose most vociferous supporters shouted *Espadrilles yes, books no*! But at least there had been no public book-burnings except for the one presided over by the prelate Raúl Sánchez Abelenda at the Faculty of Philosophy in 1966, and the far worse one 10 years later. According to Carlos Alchourrón, a witness of the former purifying auto da fe, my books had been consigned to the flames in the company of those of Freud and Piaget. In the Faculty of Science I found the complete works of Galileo and other classics, and at that of Philosophy I read the two tomes of Descartes's that were never cited by Angloamerican scholars: *Traité du monde* and *Traité de l'homme*. Because they were close to materialism and deism, its author had not dared publish them, but they were published after his death and had strongly influenced the members of the radical wing of the French Enlightenment, particularly Diderot, Holbach, Helvétius, and La Mettrie.

Why were those Cartesian works not translated into English until the 1970s, and why are they usually ignored in the history of philosophy courses, whereas they never forget Descartes's *Traité de l'âme*, which any idealist can accept? I tend to think that this eclipse of the materialist phase of the "masked philosopher" resulted from a combination of ignorance and ideological censorship. This is quite different from the caution that all the cryptomaterialists philosophers, such as Averroes, Galileo, Spinoza, and Locke had to observe. Even now, under liberal regimes, it is more convenient to pass for a naturalist than for a materialist, for agnostic rather than atheist, to teach Husserl rather than Russell, and, above all, to comment on the well known rather than to question it.

My research projects resulted in my book *Causality* (Bunge 1959a) and a handful of papers published in international journals: *American Journal of Physics*, *British Journal for the Philosophy of Science*, and *Methodos*. I collected most of those articles in *Metascientific Queries* (Bunge 1959c), which was published at the same time as *Causality*. Popper's comment on my *Queries* was characteristic: he just told me that he too had once used the term *metascientific*. But he also disclosed to me that he had reviewed favorably my paper against Rosenfeld in the *British Journal for the Philosophy of Science*.

Bolivia and Chile

At the beginning of 1955, a former student of mine invited my friend Pipo Westerkamp and me to attend the course on modern physics organized by the Laboratory of Cosmic Physics in Chacaltaya, Bolivia, financed by UNESCO. When we landed in La Paz, at 4,000 m above the sea level, we were informed that, due to the absence of two famous physicists, we had been promoted from participants to professors. Pipo and I engaged enthusiastically in our unexpected occupations despite our severe *apunamiento* (headache caused by oxygen deficiency). Among other lectures, I gave one on Bohr's complementarity principle in a theatre filled to the brim (Bunge 1955b). Since the media had described me as a member of the Atomic Commission, I demanded a rectification.

I offered the Laboratory my long essay on the age of the universe, which they published quickly as a small book (Bunge 1955c). In that essay I examined the cosmological theories of the day, which I had seen in *Astrophysical Journal*, and concluded that no astronomical data confirmed the creationist myth. The Big Bang did occur, but it was only the start of a new phase in the history of the world. An absolute beginning would have violated all the conservation laws.

The organizers of the course took us by bus, first to Chacaltaya, then to the Lake Titicaca and the Puno village, in the tropical jungle. Chacaltaya, situated at 5,220 m above the sea level, was where a bevy of new particles, such as the pion, had been detected by exposing photographic plates to cosmic ray showers. Their analysis involved the laws of energy and of linear momentum conservation, which the creationists conveniently forget. For example, a fork that seems to appear from nowhere on a photographic plate is interpreted as the disintegration product of a neutral particle at the fork's vertex. The discovery of black holes is similar. Along the way we saw shoeless Indians sawing ice blocks destined for fridges in the city. They lived on location in stone caves, and their daily wage was equivalent to the breakfast we had at our hotel. No one told us the life expectancy of those workers, but from a trip to the sulphur mines in Argentina, at the same altitude, that I had visited earlier, I knew that they were employed for only 7 years, after which they were sacked for not being profitable any longer. In La Paz I also saw a man carrying a piano on his back.

When we visited President Víctor Paz Estenssoro, he showed us the holes left by bullets shot in past coups. I was then told that his deputy, Hernán Siles Suazo, employed the former Argentine communist Carlos Dujovne. In Buenos Aires Carlos had set up the Problemas publishing house and library, which had entrusted me with some translations, in particular the *Selected Correspondence* of Marx and Engels, which paid just for my son Bambi's tonsils surgery. When I visited him at his cold office on top of the presidential palace, Carlos – the brother of León Dujovne, a future colleague of mine – told me that, when he realized that his party was going nowhere, he offered his services to Siles, whose government was progressive and so popular that it completed its term. The next year Siles was elected president, and again finished his term – an exception in a country that had suffered more than 100 coups since its birth. Carlos finally felt useful after decades of peddling unrealistic slogans manufactured far away by ideologues who knew nothing about local conditions.

Chile

In La Paz I made some new friends, in particular the Chileans Alberto Cordero, aka Mouton, and Sergio Aburto. A few months later they invited me to go for a month to Santiago to give a talk on physics, as well as a philosophy course, at the prestigious Instituto Pedagógico in Santiago. My impressions of Santiago were mixed. Near the center was the San Cristóbal Hill, a pleasant park rising about 300 m above its surroundings.

Downtown looked prosperous and elegant by day, but by night it was overrun by many feral greyhounds than ran silently in packs and scavenged. Everybody was amused by the ingenuity and sense of humor of the *rotos* (broken), or members of the underclass, but no one seemed to do anything about them: they provided local color. And I was told, as if it were natural, that the police would immerse the legs of reincident delinquents into fresh cement and push them down the harbor.

I gave my philosophical audience the first draft of my *Causality* book. The learned Professor Félix Schwartzmann made many interesting remarks, and drew my attention to the French mathematician and philosopher Jean Ullmo, whom I visited in Paris a few years later, and invited to write a contribution to the Festschrift for Popper that I was preparing. On inspecting the bookshelves of the university library, I noticed Popper's *Open Society*, published in 1945 but unknown in my country. It impressed me immediately for his attacks on Plato as a reactionary, and on Hegel as both reactionary and obscurantist. Back home, I wrote to Popper, and we quickly became penpals, for we shared rationality and realism. Our friendship lasted until, two decades later, I criticized his three-worlds fantasy. Karl exalted criticism as long as it was not directed at him. Worse, as he said in 1969 at a meeting in his honor, he did not believe in constructive criticism – which shows that he was unfamiliar with the way scientific communities work.

Between lectures I had several memorable experiences in Santiago. On hearing that I was in town, my lay godfather Raúl Prebisch – the director of the UN's CEPAL (Comisión Ecnómica para América Latina) – whom I had not seen in two decades, invited me to dine at his home in the company of his first wife, Adelita Moll, whom he had known and courted at *El Ombú*. The Prebishes told me about the miserable *callampa* neighborhoods (shanty towns), as well as about the rumor that Perón's government was about to invite the Standard Oil Company to initiate oil prospecting in our country. "Oil tends to corrupt políticians", said Raúl.

Clearly, the Prebisches were better informed than I – or Milton Friedman, or Frederick Hayek – about the so-called developing world. Incidentally, two decades later those two economists and "libertarian" ideologues sang the praises of General Pinochet's murderous dictatorship, whereas Prebisch (1981) criticized them both in his sharp and funny "Dialogue on Friedman and Hayek."

Another unexpected encounter was the one I had with Teba, my first love, whom I had not met in a quarter of a century. When she read in the newspaper that I was in Santiago, she invited me for dinner with her companion, an old reader of *Minerva* and a widower beyond consolation. I felt no emotions on seeing her after so many years, as if the neural circuits that had made me feel so sad long ago had been completely dismantled. Teba was working as a free-lance translator, an occasionally interesting but ill-paid sort of work in that part of the world, where all intellectuals know at least one foreign language. By contrast, translating can be profitable in North America, where most people are monoglots. My late friend Henry Mins, an amateur philosopher, made a good living by translating technical stuff from 33 different languages and running a translations firm in New York. In the 1970s his firm paid fair fees to a few former philosophy students of mine who would have starved without that income.

Out of the blue, the legendary poet Nicanor Parra invited me to his ultramodern home, which he shared with his young Scandinavian wife. We discussed classical mechanics, which he taught to make a living. He claimed that Newton had discovered his laws by induction. I told him that this was impossible, because the data of planetary astronomy did not contain the key concepts of force, mass, or acceleration. But he knew better.

In one of my lectures I met a certain Marcelo Segal, a charming if crookish dilettante who stuck to me and did me three signaled favors. One of them was to introduce me to a group of anthropologists who taught me a lot about Easter Island, whose gigantic statues are still a matter of controversy. One of the anthropologists introduced me to Benjamín Subercaseaux, a famous amateur anthropologist and essayist who boasted of having a grandmother so dark that just looking at her evoked dragged chains.

Segal's second favor was to organize a public debate with a certain Luisi, a well-known blind Marxist philosopher. The question to be debated was the nature of philosophical problems. Luisi was of the opinion that all such problems are of the form "Who said what, where and when?" I held that every philosophical issue had three aspects: logical, ontological, and epistemological. Paul Weiss, the editor of the *Review of Metaphysics*, to whom I submitted a short paper on this topic, told me that it contained nothing new to the readership of his journal. I welcomed this reply as a sign that I was able to write intelligibly.

Segal's third favor was to take me to Valparaíso, known as "the jewel of the Pacific" for being situated on the border of the most beautiful bay in the world. Eduardo Rey López, my one-eyed gardner, disagreed: he voted for the Alhucema bay, in Morocco, which he had visited as a young sailor. I spent hours contemplating the bay. Every few minutes I also saw the streetcar bearing a sign with a Calvinist flavor: "From Pleasures to Cemetery."

1955

Shortly after my return from Chile, Argentinians experienced something that many of us wished but few expected: a military coup that ousted Perón and his thugs. The so-called Liberation Revolution made many errors and committed some crimes, but it gave the opponents the chance to clean up the educational system. In particular, my friends and I were finally able to get identity documents, and to participate in public competitions for all the university chairs. I would finally be able to earn a living teaching subjects I loved and for which I believed myself to be qualified.

But the change of government had also some unexpected negative effects. One of them was the sudden cessation of our two private seminars, and the concomitant weakening of the bonds among us. In particular, Sadosky lost his academic interests and plunged into university politics, where he thrived. He returned to the Faculty of Science and was soon appointed its vice dean. He confided to me that he and the Dean, the crafty Rolando García, made up an ideal team: Manuel supplied initiatives while Rolando, who had good connections and bureaucratic abilities on top of an unobjectionable family name (as common and harmless as Smith), put them into practice. However, that would happen a couple of years later.

The new government appointed the respected historian José Luis Romero as the interim rector of the University of Buenos Aires, who in turn appointed the mathematician and historian of science José Babini as the interim dean of the Faculty of Science. He summoned the physicists connected with the faculty, to pick those willing to teach the various courses during the 1956 academic year, while the public competition to fill the chairs followed its course. When the turn of quantum mechanics came, I was chosen unanimously for being the only one in our group who had published on the subject. A few days later something similar happened in La Plata. So, overnight I was promoted from nobody to interim professor of theoretical physics at my two universities. Those were my first regular jobs.

Andrés Kálnay

In both courses I was blessed with some exceptional students, such as Andrés Kálnay. We had become friends the year before, when he came to consult with me wearing his recruit's uniform. Andrés always asked interesting and tough questions. After taking my course, I had him appointed teaching assistant in my La Plata chair, and later on we wrote together some papers that got published in international journals (Bunge and Kálnay 1969, 1983a, b). As soon as Andrés got his doctorate, Gaviola took him to his brand-new Mathematics, Physics and Astronomy Institute in Córdoba. Andrés taught there all the branches of theoretical physics, and formed a bevy of students who continued his work. It took me several years to wean him from the standard semi-subjectivistic interpretation of quantum mechanics, which he had learned when studying the orthodox and famous Cohen-Tanoudji textbook.

At the end of 1966, while I was arriving in democratic Canada, Andrés was escaping the dictatorship headed by General Onganía. He crossed the Andean cordillera without identity documents. From Chile he passed to Lima, where he got employed at the National University of Technology. He felt happy there because he could do his research, and even train some students. One of them, Holger Valqui, took the crash course on the foundations of physics that I taught in México City during the summer of 1968.

Andrés approved of the social reforms that the government of General Velasco Alvarado seemed, at least on paper, to favor the poor. (It was said that the Peruvian land reform, just like the Mexican one half a century earlier, gave the military officers the best tracts of land.) But the student body opposed the government because it originated in a military coup, and they boycotted the pro-government teachers. So, Andrés had to emigrate once more, this time to Caracas, the Venezuelan capital. What first strikes the foreign visitor to Caracas is the contrast between the white haves, who live quite well in the bottom of the valley, and the dark have-nots, who subsist on the slopes of the mountain, and are currently the core of the proChávez movement.

Andrés was immediately hired by IVIC (Instituto Venezolano de Investigaciones Científicas), which was still flourishing under the direction of my friend Marcel Roche, physiologist, humanist, and cellist. IVIC was the ideal workplace for any investigator, if only because it demanded full-time dedication, attracted the best graduates of the local university, employed dozens of exiles from other Latin American universities, and had a library that subscribed to 1,500 scientific journals. Andrés stayed at IVIC till the last. During those years we visited one another several times. We met in Caracas, Montreal, Aarhus, Freiburg, and Uppsala. He travelled with two suitcases, one of them filled with pills – a sign of his hypochondria. He died alone from many diseases, some of them real.

Vlady, Fatone, and García

The most outstanding student in my first La Plata course was Vladimiro Ern, aka Vlady. He belonged to a Mennonite family that had fled from Central Europe to Russia, then to Paraguay, and finally to Argentina. Vlady had an engineering degree and was married to Haydée, an aeronautical engineer. He was curious, hardworking, and cheerful. At the end of every lecture we had to walk past the office of the Institute's new director, Rafael Grinfeld. We walked on tiptoe to avoid being caught by him, for he liked to talk at length, but we rarely escaped his detector.

When the philosopher Vicente Fatone was appointed rector of the newly created Universidad Nacional del Sur, in Bahía Blanca, he invited me to discuss the possibility of setting up a physics institute. I took Vlady along, and we talked to Fatone and Rolando García, his adviser, friend and protégé. During a long bus trip to the muddy shore teeming with large crabs, I exchanged some confidences with Rolando. I told him that my worst vice was envy, and he confessed that he had only one and a half publications: his dissertation, and a paper written together with a colleague.

I was soon to learn that Rolando had decided to drop research and seek positions of academic power and prestige. He fulfilled this ambition and, in the process, showed that he was a crafty operator who did not shy away from intrigue and adulation. However, it took me a couple of years to realize this. When I did, I exposed him, but nobody wanted to believe it, for it is advantageous to be on good terms with the powerful. Even now, after Rolando joined the Peronist ranks, denounced basic science, and demanded a "national and popular science," in his country he passes for a great scientist with admirable integrity.

When back from Bahía Blanca, I sketched a plan for the physics institute and added a list of candidates My proposal was to gather half a dozen physicists working in a single area, so they could interact and prosper in Bahía Blanca's culturally poor milieu. Fatone never replied. When we met again, he told me that the Ministry of Finance had opposed the project. But I heard that García – who saw me as a rival in philosophy for having attended a course of Carnap's – had advised Fatone against my project.

When I resigned my La Plata chair I lost contact with Vlady, but we met again during my first visit to Yale University. He was working there under the supervision of Lars Onsager, famous for both his work, which was soon to earn him a Nobel Prize, and for the impenetrability of his lectures. Vlady called him "Lars the Obscure." But maybe Vlady was the one at fault: he had trained as an engineer, and engineers are bent on changing things rather than studying them.

Teaching Quantum Mechanics

When planning my first quantum mechanics courses I made the then fashionable mistake of emphasizing the continuity of that theory with classical physics: I spent too much time discussing the optical-mechanical analogy that Goldstein had treated so well in his beautiful if misleading book (Goldstein 1950). But once we arrived at the Schrödinger equation we moved quickly, and succeeded in covering the scattering of particles.

In my courses I discussed Bohm's theory at the end. My students liked it so much, that they asked me why I had left it for the end, for thanks to it they had started to understand quanta. I now believe that it is mistaken to emphasize the said quantum-classical similarity, because it obscures the novel features of quanta. It is best to adopt a shock approach, and talk about *quantons* instead of particles and waves, for in quantum mechanics these are just analogies (Bunge 1967d).

All of my quantum mechanics students showed interest in the historical and philosophical comments I interpolated between formulas. This is not surprising: most young science students are curious and have little respect for the interdisciplinary frontiers invented by myopic specialists. When José Babini offered a course on the history of science at the Faculty of Science, he attracted many students – among them my former student Eduardo Ortiz and my wife Marta, a beginning mathematics student. Perhaps I too would have attracted many good students if I had offered philosophy of science courses at the same faculty. But the chair in this subject was in the Faculty of Philosophy – a theme for the coming chapter.

Chapter 6
First Jobs

The 1956–1962 period was the first stretch of my adult life without political, pecuniary, or emotional anxieties. I was suddenly admitted into the international philosophical community without having to abandon physics, made a living teaching both disciplines, started my second long-term philosophical project, put an end to a 17 year-old marriage that had been unravelling, married my last great love, and left for good my beloved country just in time. In short, during those 6 years I renewed myself in the light of reason and the warmth of young love.

Accepted by Philosophers

I had lived at the margins of the philosophical community, though not by choice. This marginality ended abruptly in the winter of 1956, when I received an invitation to participate in the Inter American Congress of Philosophy, to be held in Santiago de Chile and Viña del Mar. This invitation came as a big surprise, in the midst of my physics courses in Buenos Aires and in La Plata. I never learned whose idea it had been, though I suspect it was one of my students in the course I had taught the year before in Santiago.

In any event, I packed at the last minute and went to the international airport, where a small aircraft was waiting. There I met my old friend Risieri Frondizi and three individuals new to me: the Thomist philosopher and Catholic activist Monsignor Octavio Derisi, the legal philosopher Carlos Cossio, and a young man who looked like a rugby player, whom Cossio introduced as a professor. As I was about to board the plane, a photographer asked me courteously to step aside. The next day, the main newspaper published a photo of the Argentine delegation, with me missing: I was eliminated, as in the Stalinist histories of the Bolshevik movement. Two decades later, the same newspaper was eager to interview me. This is the great advantage of media executions over the other sort: one may survive them.

M. Bunge, *Between Two Worlds*, Springer Biographies,
DOI 10.1007/978-3-319-29251-9_6

When flying over the Andean cordillera, Cossio informed us that we were above the Comechingones hills, and Risieri told me: “This is the last time I believe what Cossio says.” At the Santiago airport he immediately identified as a philosopher the individual with a fanatic’s expression who came to greet us. He turned out to be a Catholic philosopher well connected with the owners of *fundos*, or ranches, one of which offered us an opulent lunch that lasted several hours.

The congress was well attended by philosophers from all the Americas and Europe. Like every meeting of its kind, it was very uneven. There were logicians, historians of philosophy, philosophers, ideologues, and mystics. Miguel Reale, who had been an ideologist of the fascist Estado Novo headed by Getúlio Vargas, proposed a tribute to Giovanni Papini. Everyone stood up except me: no one else knew that Papini had been a minor fascist writer.

I made two presentations, one critical and the other constructive. In the former I criticized the paper by the Belgian phenomenologist Herman van Breda, a friendly Franciscan in charge of the Husserl Archive in Leuven. I guess I drew from my paper on that barren doctrine (Bunge 1951b, Bunge 2007b). Van Breda did not attend that congress, but we met much later several times, and once he invited me to give a talk at the Catholic University in Leuven. He liked to offer advice. His recipe for attaining a ripe old age was sexual abstinence, but he only lived 63 years.

In my second presentation I argued that the properties of an object, whether concrete or abstract, come in bundles or systems rather than in mutual isolation, as Hume believed. More precisely, I defended the thesis that, for any property of an object, there is at least one other property of the same object, related to the former by some law or invariant relation. This was my debut in systemic ontology. Like many another metaphysical conjecture, it poses the epistemological, in particular methodological, problem of finding the trait(s) related by a law to another feature. That is, it has heuristic power.

During the breaks and excursions I made several new friends: Willard van Orman Quine and Henry Margenau, whom I had read, and others new to me, like Roderick Chisholm, Jorge Millas, Paco Miró-Quesada, and Gerold Stahl. Millas arranged for me to lecture in Concepción, where I stayed at the home of a geneticist who kept chickens from Easter Island. I struck a lifelong friendship with Miró-Quesada, with whom I met again at other congresses, as well as in Perú. Gerold Stahl, a German logic professor who had been forced to fight in the war, introduced his subject in Chile, but had to emigrate when Henry Kissenger installed General Pinochet in the La Moneda palace. I also befriended Euryalo Cannabrava, an expansive and boastful Brazilian, who claimed to be the only philosopher in his country, and boasted to have hunted wild meek buffaloes at the mouth of the Amazon.

Quine

The famous Harvard logician Willard van Orman Quine, aka Van, attended all the lectures and, unlike most of his fellow Americans, was curious about the tongue of the land. We made friends chatting about physics and after he listened to my talk about systems of properties. We discovered each other as members of the small fraternity of materialists – whom he, a radical reductionist, preferred to call 'physicalists' – and friends of the philosophy that I called *enlightened*, that is, clear, debatable, and pro-scientific. Later in his memoirs Quine wrote: "The star of the philosophical congress was Mario Bunge, an energetic and articulate young Argentinian of broad background and broad, if headstrong, intellectual concerns. He seemed to feel that the burden of bringing South America up to a northern scientific and intellectual level rested on his shoulders" (Quine 1985, p. 266).

One evening I gathered my courage and gave Quine the original of my *Causality* manuscript, which had been rejected by half a dozen publishers. The next morning Quine told me that he liked my book, and urged me to submit it to Harvard University Press, which I did on my return home. Harvard asked two physicist-philosophers, Henry Margenau and Victor Lenzen, to referee it, which they did in detail and favorably. The Press accepted my book, and suggested that I visit them to sign the contract for it, which I did the following year. The book was copy-edited by a former physicist, a very slow worker, and appeared in 1959. Surprisingly for a book on an unfashionable subject by an unknown author from a forgotten nation, *Causality* enjoyed at once *succès d'estime* and *succès de librairie. Scientific American* devoted a full page to it, and it was translated into German, Hungarian, Italian, Japanese, Polish, Russian, and even Spanish. Its ultimate accolade was Dover's paperback reissue (Bunge 1979e). The French translator of this book did not find a publisher. At that time, Paris had long ceased to be the City of Light, and its faculties of philosophy had become the homes of the most obscurantist and reactionary of all schools: of Nietzsche, phenomenology, existentalism, Lévi-Strauss's structuralism, Lacan's psychoanalysis, Althusser's version of Marxism, hermeneutics, and general semiotics.

Faculty of Philosophy and Letters

In 1956, the philosophy of science chair at the Buenos Aires University was opened to public competition. At the time its occupant was the author of a self-help manual on how to write doctoral dissertations – something he himself had not done. He taught not only his specialty, but also a seminar in demonography, for he was a learned esotericist. This individual was the disciple of a physician who had taken a course in projective geometry, whose vocabulary he used in his extravagant articles in the philosophy of science. The same physician had also written a book on man, in which he attacked Francisco Romero's idealistic philosophical anthropology. He did little more than remind people that we are basically animals. Unsurprisingly, neither this physician nor his disciple had ever published anything in a refereed journal.

I regarded myself as competent to teach, or rather to learn to teach, philosophy of science because I was a scientist and had published some papers in international philosophy journals. But when I saw the list of my nine competitors, I realized that my c.v. might be insufficient or even a drawback: that connections were at least as important as background – and I had none, or so I thought. Besides, the subject enjoyed no prestige whatsoever in Argentina, to the point that few cared who taught it.

Ten individuals entered the competition. Nine of them had never published anything in international philosophical journals. Two of them, Gregorio Klimovsky and Rolando García, had not even published book reviews. Rolando's wife, a charming and chubby Swede, confided in me that her husband was brimming with brilliant ideas but suffered from writer's block, and that he would be very pleased if I listened to him, took notes, and wrote them down for him. I suppose that he might have consented to adding my name as a coauthor. A few decades later García published in Spanish a couple of popular books on some fashionable subjects.

Raymundo Pardo was the only one of the ten postulants who had published a book – on what he called "the constituents of reason," that was at least remotely related to the philosophy of science. He had done his entire academic career under Peronism – or rather perched on it – and circulated denunciations of his colleagues, accusing them of incompetence and missing lectures, with details that only a private eye could have obtained. But at the time that did not advance his academic career.

I was extremely lucky with my panel of judges, for none of its members belonged to any clique, and all three were old acquaintances: the legal philosopher José Juan Bruera, the mathematician Beppo Levi, and the historian of philosophy Rodolfo Mondolfo. Bruera had written a brilliant criticism of legal positivism for *Minerva*, and had visited me in *El Ombú*. I had met Levi at several scientific meetings, we had exchanged reprints, and he had replied *in extenso* to a mathematical question of mine. Besides, I was a close friend of his son-in-law Ricardo Resta. As for Mondolfo, I had been in touch with him since his arrival on the eve of the war, and had published two papers of his in *Minerva*. Besides, we had exchanged reprints and I had visited him in Córdoba, where he had to teach the Greek language instead of Greek philosophy.

In sum, my jury was partial to me, so that its unanimous vote was to be expected, though certainly not by me, for I learned its composition only recently. I was far luckier than Bernardo A. Houssay when, in 1919, he had competed for the physiology chair at the same university. Argentina' first Nobel Prize in science was then judged by a panel that included two student representatives – a product of the university reformation carried out the year before. Both student delegates voted against the scientist. The result was a tie that the dean, the research physician Dr Julio Méndez, broke in favor of Houssay. Houssay worked in that chair until sacked in 1943 by the fascist dictatorship following the military coup that had occurred shortly before. At that time, Houssay's laboratory included about 100 biomedical researchers, among them the endocrinologist and McGill graduate Christiane Dosne Pasqualini (2012), who had settled in the country.

The dignitary who handed me the professorial diploma was an old medicine professor who had been a classmate of my father's, whom he remembered fondly. The traditional daily *La Nación*, which had published laudatory notes on all the newly appointed professors, ignored me. But a few decades later the same newspaper invited me to become a regular contributor to it.

First Steps in the Faculty of Philosophy's Corridors

On receiving my first philosophy appointment I proceeded to prepare my inaugural lecture, titled "Philosophizing scientifically and doing science philosophically" (Bunge 1957a). I wrote it in solitude, in one of my favorite places: the Uruguayan beach of Maldonado, whose pine-tree forest almost reached the seashore. Since I was on a very restricted budget, I lodged in the attic of the only hotel, which was so hot that I got a pruritus that the local doctor treated with a high table-salt dose.

My next task was to prepare my course. Since none of the textbooks known to me was satisfactory, I resorted to the maieutic method that Plato had attributed to Socrates. My lectures consisted essentially in questions that I distributed beforehand to the students, and an abstract of the research that they had prompted. I wrote each question on a 6 × 8 card. I had adopted this procedure a few years earlier for my own work, so I did not start from scratch. Eventually I filled several hundreds of such cards, classed them by subject, and placed them in boxes. When a box filled up, it was time to write an article or a book chapter.

The boxes complemented my hanging-files cabinet, containing sketches of papers, some of them aborted, as well as some letters. Some of this material is now in the *Fondo Bunge*, at the Buenos Aires Facultad de Ciencias. At *El Ombú* my study contained those filing cabinets, two bookshelves, a typewriter, and two tables, one for physics and the other for philosophy. After I left *El Ombú*, my study contracted or was confined to a corner of the bedroom. Some universities assigned me offices for myself and my students and coworkers, but only my home offered me the privacy I needed to concentrate. Regrettably, I had not inherited my father's ability to concentrate and work in the midst of a noisy party.

When I went to take charge of my philosophy of science chair, I was met at the door by the janitor, a tall, dark-skinned, and friendly provincial, who put me at ease, saying: "You must be the new professor. I'm Flores. Welcome, professor. Come in, I'll introduce you to the faculty's secretary." Flores turned out to be as helpful as friendly, by contrast with his boss, Costantini, the chief porter, a disagreeable individual who put on airs, and was said to have helped the Peronist police in spotting troublemakers. Peregrino, his second, was a friendly and helpful fellow; we used to exchange jokes, something that was frowned upon in that stiff environment.

The secretary turned out to be a friendly young man. He talked to me enthusiastically about his doctoral dissertation, which he was about to defend: it was about accents in the work of Lope de Vega, the great playwright. He did not know that Lope had been at the head of a writing workshop that produced over 1,000 comedies, and that before falling into the printer's hands, the manuscript had to pass through those of the copy-editor, the theatre director, the type-setter, and the printer. I was appalled. To what level of triviality and incompetence had I fallen? I had yet to meet my junior colleagues and the professor who claimed to teach logic.

The large classroom was full of friends and curious people in addition to students. I did not recognize any colleagues. I was wearing for the first time the suit that a humble Paraguayan tailor had cut for this occasion. Since nobody introduced me, I felt as though I was intruding in foreign territory. I read my lecture with the expected monotony and, contrary to my habit, made no use of the blackboard. The clapping at the end woke up a few people. There were neither questions nor objections, for that was not the Faculty of Science routine.

The year before, when Francisco Romero – the dogan of Argentine philosophers – had given his lecture on science at the inaugural meeting of the Argentine Philosophical Association, a high school teacher stood up and dared to say that Romero's definition of 'science' did not fit his own discipline, geography. Romero remained silent but his face was getting so red that the lady who presided over the occasion stood up and said: "This is a meeting of friends. One does not argue here. I declare this session closed."

A few days later, Rolando García – the meteorologist and amateur philosopher – boasted that he had sat across from Romero at the dinner celebrating the birth of the new society – as if to corroborate that it was a society of friends, and that I was not one of them. On another occasion, Romero confided to me: "Discussions only serve to confirm one's own convictions." I suppose that, if Socrates had visited Argentina, he would have been given a generous dose of hemlock right after his first dialogue.

First Philosophy of Science Course

When I started to teach my course, the large classroom was full, but I guess that half of those present had been attracted by the rumor that I was a strange fellow. It also transpired that half of the students enrolled in the course were absent because the Student Center was boycotting the new professors that were replacing the ones who had been appointed by the old regime. I was one of the professors impugned by the self-styled reformist leaders. I turned a deaf ear, and 2 months later welcomed the failed strikers as if nothing had happened.

I noticed the contrast between the pseudo-reformists at the Faculty of Philosophy, who played politics, and the genuine reformists at the Faculty of Science, who had participated decisively in the academic cleansing process. As for the strike ringleaders, a pipesmoking fellow and his girlfriend, we met again 15 years later in Paris, after a lecture of mine sponsored by the physicist Lévy-Leblond. They were now at the feet of Louis Althusser, the structuralist neoMarxist who a few years later would strangle his wife, and all but confessed that he had only a second-hand knowledge of Marx.

Back to my first course. I started each class by stating a problem, describing some background, and inviting questions and objections. Before the bell rang, I dictated the handful of questions that the students were expected to discuss on 6×9 cards. The response to this heterodox procedure was amazingly good: the next time nearly all the students handed in their cards, which I read and corrected on the train and at home, and commented on at the next class without revealing the names of the responders. Most of the students welcomed the change from passive listeners to active participants – as María Montessori and John Dewey had anticipated.

I thought that these micro-tests rendered the traditional oral examinations unnecessary. But I gave them because they were compulsory. However, I reduced them to a formality, asking my students beforehand to deliver a short oral presentation on any of their favourite subjects. Thus, I took the sting out of the final examination, which is always stressful and never productive. This new regime favored only the good students. The others waited in the wings for better times. These came when I took my first sabbatical leave (1960–1961), and then when I left the country at the beginning of 1963. In both cases my replacement, Gregorio Klimovsky – who loved to lecture although his ignorance of both science and methodology was so vast, that he spoke in favor of psychoanalysis – restored the old teaching regime and its concomitants: boredom, passivity, and anxiety.

To write the cards, the worst students stole verbatim from easily traceable sources, such as Ferrater Mora's superb *Dicccionario de filosofía*, and some of them sunk in the postmodern swamp. But, as was to be expected, the distribution was bell-shaped: most cards were acceptable and a few were very good, and some of their authors distinguished themselves later in life. Yet by far the most visible of that group was Mario Pantaleo, aka Father Mario, a bad scholar but good person. His ecclesiastic superiors had sent him to assist the dying in public hospitals, where he was assigned tiny bedrooms, so he could be on call around the clock. He hated this job and bitterly complained to me about it, and I advised him to organize a union to fight the hierarchy. In any event, that is how Father Mario's hands got a reputation for effecting miraculous healings. Many more flocked to him than to any one else in our group, and his many fans in Buenos Aires, Rome, and even San José de Costa Rica, expect his promotion to sainthood any day.

The very best essays, those combining precise information with relevant philosophical reflections, were those signed Marta Cavallo. After a couple of months I asked her to identify herself. From the back of the classroom, Marta timidly raised her hand, and informed us that she was auditing my course, and that she was enrolled in the teachers' college – a disgraceful American invention. Thus began an affair that has lasted till today, more than half a century later.

1958

The year 1958 was one of great changes for me. My son Carlos and I left *El Ombú* and moved to an apartment-hotel downtown. I got a fulltime position in the philosophy faculty, and devoted Thursdays to physics. My salary was now roughly the sum of what I had been paid for teaching physics in Buenos Aires and La Plata, and philosophy in Buenos Aires: around 170 dollars per month.

I may have been the first philosophy professor to hold a single chair. It was customary to accumulate as many chairs as possible, for original research was rare, and one had to count on political calamities. One of my rivals in the public competition

I won listed three full-time jobs on his c.v. – which was as nothing compared with the seven such *chambas* (jobs) held by a Mexican colleague in as many places. His only problem was to collect his seven salaries on a single payday twice a month. That year some students and I translated about 50 papers that I found useful for the study of our discipline. I got them printed and collected in the series *Cuadernos de Epistemología.* None of my successors used them. It is hard to find traces of my passage through the University of Buenos Aires, except for the memories kept by my few surviving students. Working there proved to be little more than sowing in the desert.

American Interlude

My first visit to the United States was to attend the 5th Inter American Congress of Philosophy in Washington, D.C. Since I arrived a few minutes after its closure, I confined myself to quick visits to Washington, New York, Boston, and New Haven. I had been late because the American consulate initially refused to give me a visa, and the Pan American plane developed a fault that grounded it for 2 days in Rio de Janeiro. My visa had been denied because I answered truthfully the questions about my political past, but it was finally granted thanks to the intervention of the congress organizers. The consular officer in charge of my file told me that I should have lied, to which I retorted that philosophers are expected to tell the truth. However, my journey proved to be useful, for I made new friends and signed a book contract. In Washington I visited the Lincoln memorial and befriended a junior Argentine diplomat, who showed me around town and told me how many colleagues of his got piles of goods for free. Shortly before returning to their country they bought things in installments, and after paying the first of these, forgot about the rest and left the USA for good.

In New York I stayed at a YMCA hostel, and visited the translation workshop of my postal friend Henry Mins, who made a good living translating from 33 languages. I put him in touch with a few students of mine, whom he paid well in hard times. Henry invited me to stay a couple of days at his country house, and his wife drove me to Princeton, where I talked to a couple of scientists and peeked at ENIAC, then the largest computer.

In New York City I also met with Irving Louis Horowitz, who held a doctorate in philosophy, and persuaded him that what really interested him was sociology. He spent the next year as a visiting professor of sociology at my Faculty of Philosophy. His widow, Mary Curtis, told me recently that that job saved Irving's intellectual life. Irving quickly made friends with my friends Gino Germany, the founder of the Department of Sociology, and Hernán Rodríguez, who acted as Irving's interpreter. Irving also captivated the most beautiful girl in the faculty, who obviously saw beyond his harelip and corresponding speech impediment.

During that 1 year Horowitz helped Germani – a follower of Talcott Parsons – enlarge his horizons by importing such authors as the deep Robert Merton and the brilliant C. Wright Mills – whose explosive book *The Power Elite* had tacitly answered Franklin D. Roosevelt's question when he faced the powerful groups that opposed his progressive reforms: Who owns this country? When back in the USA, Horowitz made a brilliant and fast academic career at Rutgers University, and published a number of insightful papers on topical subjects. In one of them he made a pessimistic but correct assessment of Argentine politics, which angered me at the time. He realized that President Frondizi's attempt to please everyone was actually multiplying the number of his enemies, for everyone felt betrayed. Besides, as I learned only recently, the Argentine president won the enmity of the CIA when he voted against expelling Cuba from the Organization of American States – and in those times one did not snub the Big Boss impunely. At the time I failed to see all this because I refused to look beyond my academic work. I re-awoke to politics only three decades later, when the Quebec separatists threatened to break up Canada. However, back to my first visit to the USA.

In New Haven I looked around Yale University – where my daughter would study neurobiology four decades later – and tapped in vain on the doors of the few professors I knew by name, but were obviously away on holidays. I was enthralled to attend an open door performance of Katherine Hepburn, one of my favorite film stars.

The next day I went to Harvard, where I signed the contract for my *Causality*, which the Press took 2 years to produce. Its director took me for lunch, and was annoyed when, following the Argentine tradition, I insisted on paying my part. I had never heard of expense accounts. A few years later, when a Prentice-Hall representative invited Marta and me for dinner, he and his friend had lobster and wine. But of course that was peanuts compared with the expense accounts of the 4,500 lobbyists registered in Washington with the aim of bribing members of Congress and civil servants. Yet, a moral philosopher won acclaim when he stated, with a straight face, that business is a school of morality.

University Administration

I had decided never to occupy an administrative post. But the year 1958 was exceptional, for the university had decided to devote the year to its own reconstruction. Consequently I accepted the invitation to run as a member of the governing board of the philosophy faculty, and was elected for that year.

I proposed about a dozen changes, all of which were passed. Among them were having problem-solving sessions in all subjects, giving Gregorio Klimovsky *venia docendi* to teach a twentieth-century logic course, gathering the anthropologists into a department, and opening the building on Saturdays. All these proposals were passed despite the systematic opposition of three fellow councilors: my esteemed colleague León Dujovne – who accused me of turning *his* faculty on its head; the graduate's representative Mercedes Bergadá – a translator of Thomas Aquinas; and Eliseo Verón, the "reformist" student delegate, who is now a successful postmodern author. By contrast, the dean, Marcos Morínigo – a gentle and respected Paraguayan linguist – lent me his full support, and saw to it that the reforms were put into practice. He supported them because he had taught at an American university.

That year I struck up relations with some colleagues in other fields, in particular the anthropologists, the art historian Jorge Romero Brest, and Manuel Solari, an education expert. But I also had tense relations with the logic professor Eugenio Pucciarelli – nicknamed Corky because he floated on all political waters – as well as with Juan Mantovani, who had won notoriety fighting the "positivist" (proscience) pedagogy introduced at the start of the century by my uncle Carlos Octavio.

I never got near my only world-famous colleague, Jorge Luis Borges, for he went around surrounded by half a dozen tutelary virgins. Besides, we belonged to rival political bands, he the conservative and I the progressive. In 1958 my faction won the election that sat me at the Faculty's Governing Body during that year. This brief but time consuming foray into university administration persuaded me once and for all that this function is incompatible with research. I kept my resolution to avoid such entanglements during my 45 years-long professorial tenure in Canada. I was powerless and sometimes treated meanly, but kept reasonably happy learning, writing, teaching, advising, and working for learned societies.

First Visit to Europe

That same year of 1958, I fulfilled two old dreams: visiting Europe, and attending a world congress of philosophy. In order to participate in the 12th International Congress of Philosophy to be held in Venice, I asked our brand new National Research Council for a travel grant. My file fell into the hands of its hard-working if wily vice-president, the former meteorologist and failed philosopher Rolando García. When I asked about the status of my request, I was informed that, regrettably, Dr García had lost my file. I complained to the council's president, the eminent Bernardo Houssay, and got the subsidy just in time.

Gregorio Klimovsky, García's buddy, was furious – not with him but with me. I had broken the code of García's academic mafia, which also included Oscar Varsavsky – who a few years later would become the head of the anti-science movement – as well as my old friend Manuel Sadosky, who three decades later would become the first Minister of Science and Technology, a post he held with more honesty than imagination or efficacy. Though not an original thinker, Manuel was a superb teacher: knowledgeable, clear, enthusiastic, and stimulating (see Bunge et al. 2004). His promotion to top science-policy maker was just an example of the Peter Principle.

During the long journey to Rome in a propeller aircraft, I interacted with two eminent biomedical researchers: Eduardo Braun Menéndez and Eduardo De Robertis. Braun, Houssay's right-hand man, had liked my booklet *What is science*?, which had been published by the engineering students. He was one of the handful of Catholic scientists, but he opposed the official recognition of Catholic universities because no research was done in them.

I had earlier befriended De Robertis, the discoverer of the synaptic vessel in the neuron, when he invited me to talk to his histology students. I had first seen him much earlier in La Plata, where he had been the only user of the electron microscope. Once I suggested to him to implant a battery of microelectrodes in the primary visual cortex of a human subject and connecting them to a TV screen, to visualize the subject's dreams. Eduardo did not answer. Brain imaging would take a few decades to arrive.

During a long stopover in Dakar, due to a mechanical problem with our aircraft, we admired the beauty and elegance of the Senegalese, and discussed the unpredictability of machines and humans. While flying over the Sahara desert, we could see clearly a caravan of about 50 camels loaded to the hilt. Over Madrid we were struck by a violent windstorm. The aircraft was shaken so badly that Braun's wife started frantically rolling her rosary, while her husband kept calm. He died a year later in an aviation accident.

We arrived in Rome past midnight, and I slept until car honks woke me up. After changing some lowly Argentine pesos in the lofty Banca di Santo Spirito, I went sightseeing. I started at the Coliseo, still colossal after being deprived of thousands of blocs by Romans who used them to repair their homes. When discovering a colleague, my guide made the horns sign. Clearly, paganism was alive and well in the capital of Catholicism. At the Coliseum I met a cousin who did not hide his displeasure at seeing me. Apparently he told his sons that I intended to kill good Christians with the pikes that the guards were bearing for show.

Congress in Venice

I had read a lot about Venice, but was not prepared to admire so much beauty built over centuries with water, brick, tile, and fantasy. I stayed at the enormous Santa María della Salute convent, built in the baroque style on the Canal Grande. There were few showers but the nuns were helpful and friendly. And we enjoyed the protection of the ferocious condottiero Colleoni, whom Verrochio had immortalized in his equestrian bronze statue five centuries earlier, but looked ready to charge.

The congress was presided over by Felice Battaglia, an unfortunate name for a philosopher, even at the height of the Cold War. I heard and met many philosophers whom I had only read before, and many more who were totally new to me. Freddy Ayer, happy over his recent notoriety and paternity, talked with his usual loud voice and haughtiness. Karl Popper, with whom I had been corresponding, gave signs of pleasure when we met at a vaporetto station. He did not seem annoyed when I criticized his adoption of von Mises's probability theory in his *Logic of Scientific Discovery*, which I would soon review (Bunge 1959a, b, c).

Emile Bréhier and Hermann Randall Jr, the famous historians of philosophy, seemed arrogant and choleric to me. The Spaniard Julián Marías, then in his

Wittgensteinian phase, bored us with his talk on the various meanings of the verb "to see," among them as "to understand." The Cuban delegates met at a café to criticize the organization of the congress, but did not bother to attend the meetings. The Western and Eastern Germans did nothing but attack one another. John Somerville and his fellow American Marxists kept to themselves and made no significant contribution: they had already learned all they needed. Seeing that most speakers abused the time allotted them, the chairman of a huge session asked a very tall and strong Catholic priest in black to drag away the culprits, a task that he performed with gusto to the amusement of the audience.

The Soviet delegates invited Ayer, Popper, me and a few others to a private meeting where Bonifati M. Kedrov, a friendly fellow credited with 1,000 publications, expounded on what he claimed to be the kernel of dialectical materialism. In his eagerness to pass it off as a respectable philosophy, he avoided its pitfalls, did not mention a single successful application, and did not engage with any of its critics, such as the learned Gustav Wetter, SJ, who was attending the congress. No one expressed their disappointment, but Popper kept murmuring "Murderers" under his breath.

I befriended Melvin, a much younger man, a war veteran and an expert in Charles Peirce, arguably the greatest of American philosophers, whose technical innovations cannot be understood without a solid knowledge of mathematical logic – a blind spot of Marxist philosophers. Melvin's open-mindedness was limited, though. When I told him that both the epistemology and the action theory of Marx and Engels were even more pragmatist than Peirce's own, he took refuge in history and informed me about the Kantian root of the term 'pragmatism'.

Still, Melvin and I kept on friendly terms for a number of years, and he arranged for the Russian translation of my *Causality* (1962). This was followed by the Russian translations of my *Intuition and Science* (1967) and *Philosophy of Physics* (1974). But of course Marxists disapproved of my frontal attacks on both dialectics and mind-body dualism, and I guess that they have not yet taken notice of my democratic and cooperative version of socialism, despite the sensational failure of authoritarian and statist socialism in the ex-USSR and elsewhere.

My congress paper, on the concept of a level of organization – physical, biological, social, etc. – was well received by several participants. It was originally intended to lead off a whole book on the various concepts of a level. I discarded this project a year later, when I realized that I had been misled by linguistic philosophy. But I used it as raw material for a paper published in the *Review of Metaphysics* (Bunge 1960a). My former student David Blitz, who wrote his dissertation on the said concept (Blitz 1992), kept the original of that typescript, but I hope he loses it.

One of the congress meetings was held on the delightful San Giorgio Maggiore island, owned by the aristocratic Cini family. During a break I roamed in the beautiful park and sat on a secluded bench to write to my beloved Marta. I mailed the letter to a common friend, because her family kept her under strict surveillance to prevent her from communicating with me.

Years later, while in Rome, I interacted with Marcello, Prince Cini, a highly respected physics professor who wore jeans and voted red, and the first to publish a realistic theory of measurement. Another congress session was held at the once-famous Padua University, where Galileo had taught under the protection of the *Serenissima* republic, which barred entrance to the Inquisition.

In the beautiful university library I found a copy of Bernard Bolzano's multi-volume *Wissenschaftslehre* (1837), which my colleague Francisco Romero had recommended to me. My admiration for this mathematical and philosophical genius has only increased with the years. And the pleasure that library gave me mitigates my contempt for the university hospital, where the physician who failed to diagnose my painful stomach ulcer examined me with a wooden stethoscope instead of performing an endoscopy.

New Acquaintances

During the many breaks between meetings, I made many interesting acquaintances. One of them was the physicist Alfred Landé, who was just as angry with Bohm as with Bohr. Other memorable men were the Armenian Lebanese who tackled the problem of time with no tool other than ordinary language; the Finnish logician who had invented a negation-less logic on the assumptions that logic deals with facts, and that there are no negative facts; Oskar Becker, the German phenomenologist and modal logician who had been Jürgen Habermas's teacher and a fervent Nazi supporter; and Mme. Hirsch, a crystallographer who used her craft to unveil and analyze the deeper layers of old bronze sculptures.

We were invited to a show in La Fenice, that jewel of an opera theatre built in 1792 – just 5 years before the Republic of Venice's demise. We paid for that privilege by having to visit the Murano crystal works, where exquisite craftmen made ingenious if kitschy, useless and pricey adornments.

Once the congress was over, I went to take a peak at Florence, an open museum and the cradle of the Renaissance. The first thing I did on entering the *città delle colline* was to walk up to Fiesole, one of the earliest Etruscan settlements, but today only one of the five hills from whose tops one gets a panoramic view of Dante's city of artistic wonders. My lunch, at a *trattoría* on the river Arno, was pasta with tomato sauce, which cost me 60 Lire, or 1 dollar – all I could afford.

My next stop was Naples. I just wanted to see some of Pompeii, whose frescoes left me as marveled as it had left millions before. I doubt that today's *nouveau riches* would commission anything similar to those combinations of delicacy and voluptuousness. After feeding the voracious Neapolitan mosquitoes, and resting three days in Ischia – the Capri of the poor – I flew to Paris.

Touring Paris I confirmed what my colleague Jorge Romero Brest had told me: that it was a harmonious whole composed mostly of ugly parts – none uglier than the Sacré-Coeur basilica, which spoils the highest summit of the city. But entering the Louvre I felt transfixed by the Winged Victory of Samothrace at the top of the staircase. When in the bowels of Notre Dame I felt deeply moved, though I like to think that my emotion was artistic, not religious. I revisit it every time I return to what used to be the City of Light.

I was overwhelmed by the Garnier bookstore in the Quartier Latin, the largest I had ever seen. After rummaging for a long while through its many bookshelves I asked a salesman to use the *salle des bains*, and he asked me why I wanted to take a bath, which is not something Frenchmen did lightly. Of course, I should have asked for a *toilette* or, more delicately, for *le petit coin*. By the way, I was mildly shocked at the liberal use that Parisians of all social classes made of the word *merde* (shit). And I was seriously shocked when, strolling on an evening in the Champs Elysées, I had to fend off two prostitutes, each of whom had grabbed an arm of mine and attempted to drag me in opposite directions.

At the Institut Henri Poincaré I met the exuberant Jean-Pierre Vigier, whom I had befriended when we were both guests of David Bohm's, but we found nothing to talk about, for he was more interested in politics than in philosophy. Doctor Hirsch, the crystallographer I had met in Venice, invited me to have dinner in the company of her granddaughter and the Dominican friar Dominique Dubarle. It would have been hard to set up a more dissonant quartet. She made up for this faux pas by taking me the next day to *l'Oubliette*, the underground bar where trouble-makers used to be locked up for life. There I listened to Rameau's moving song about the rebellious bandit Mandrin, that begins thus:

Du haut de ma potence, From the height of my scaffold
Je regardais la France. I looked down at France.

Karl Popper

From Paris I flew to London, where I attended a pair of lectures by Karl Popper at the LSE (London School of Economics). They were lively but disorganized, and yet he recorded them on one of the large and heavy recorders of the time, that his assistant Joseph Agassi lugged for him. Popper's courses must have been stimulating, for he was clear and enthusiastic. But I doubt that they were of much use to his students, all of whom were enrolled in social science, for he had no systematic philosophy of social science, and had swallowed all the dogmas of standard economic theory, but had not read Keynes, much less the socialists who had built the school where he lectured.

At that time Popper was still fighting the logical positivists, so their ideas is what he lectured on. In his first lecture he asked the students to observe and note carefully everything they would perceive during the next 5 min, while he kept silent. The young men looked around puzzled. Popper asked them what general conclusions they had reached. Silence. So, Popper exclaimed triumphantly: "As you can see, inductivism is false." But he had not defined 'inductivism', or even 'induction.' And he never explained why induction cannot deliver any high-level laws like Newton's. My own explanation is this: empirical data do not contain any of the non-empirical concepts, such as those of mass, atom, and evolution, characteristic of those laws (Bunge 2012e).

Popper's deputy at the LSE, John O. Wisdom, was an amiable fellow whom I had met briefly in Venice, where he went around in the company of his lovely girlfriend. He invited me to lunch, and told me that he was a practicing psychoanalyst as well as a Popperian. He could afford this incongruence because his knowledge of science, like Popper's, was bookish – and perhaps also because he met his boss's paramount requirement, namely unconditional loyalty.

I also met Popper's assistant, the lively and friendly Israeli Joseph Agassi. He had so much curiosity, enthusiasm, and speed, that he found no time to study anything in detail and depth. I also met his interesting wife, the sociologist Judith Buber, the daughter of Margarete Buber, a writer and political activist who had been a victim of both Hitler and Stalin. Years later Marta and I befriended the Agassis, corresponded with them, and met with them in several countries, including their own. I believe Joseph is Popper's best pupil, and one of the few who neither flattered nor betrayed him.

Marta

Back in Buenos Aires I resumed my course work and wrote a paper on integral spin particles. I did with Kemmer's theory what I had done previously with Dirac's, and found a less "trembling" position coordinate, the corresponding velocity, and constants of the motion parallel to those I had found for spin one-half particles. But this time, hurried by my emotional problem, I had no patience to submit my work to an international journal, and sent it to the national version of *Science* (Bunge 1958b).

Marta, 19 years my junior, was serious, intelligent, charming, courageous, beautiful, and elegant. She used to study together with three classmates a couple of years older than her. All four audited my philosophy course, and were enrolled in the teacher's college, which under Peronism had decayed less than the university. I started accompanying her to the Retiro railway hub, where we sat under the English Tower, distinguished by a copy of Big Ben and a reminder of the great power that Britain had exerted in our country. At first we took the first trains available, but we soon started missing trains. After a couple of months we confessed our mutual love. From then on we also met in parks, museums, and milk bars.

When Marta's parents discovered our secret, they forbade Marta to meet me, and sent her two younger sisters to spy on her. But we kept communicating through friends. One early evening, wearing my best clothes, gloves, and a fedora, I went to ask her parents, Ricardo and María-Teresa Cavallo, for Marta's hand, who, at 20, was still a minor. Marta, looking lovely as ever, was silent and subdued but serene, and stood next to the crackling logs in the fire-place, that did not compensate for the icy psychological temperature.

The Cavallo spouses drily refused my request, arguing that they were a Catholic family, whereas I was divorced and held political views they disapproved of.

But they could not object to our age disparity, for there was the same difference – 19 years – between Ricardo and María-Teresa. A few weeks later, my friend Enrique Gaviola went to vouch for my character, but was sent away. I guess they did not know or care that he was the doyen of Argentine physicists.

So, Marta agreed that the only solution left to us was to elope, which we did on December 30, 1958, after boarding a plane that took us to Córdoba city. There I hired a taxi that drove us to the hotel I had booked in a tourist resort 36 km away. At about the same time my son Carlos, aka Cantarito, delivered to Marta's parents the letter wherein she notified them of what she was about to do. Ours was a risky adventure, because she was still under age. But her parents feared scandal, so they resigned themselves. After all, Ricardo Cavallo was a respected civil engineer and the CEO of the national railway line.

Marta and I spent our honeymoon at a hotel on the large Carlos Paz reservoir, where we swam daily, and from which we took long walks in the surrounding hills. One day we met a club friend of Marta's, but he kept our secret. Everything would have been perfect had Hegel not meddled. In fact, to get her degree Marta had yet to pass an examination on Hegel's enigmatic *Phenomenology of Mind*. But she bravely attacked the large volume I had lent her, and a couple of months later she passed the last hurdle.

Marta's parents forgave us and we met with them at the office of a lawyer who married us by proxy in México. Nearly two decades later we discovered that the date on our marriage certificate was a national Mexican holiday, so that the document was forged. (Even professional crooks are amateurs in our country!) Just in case, nearly three decades later we got legally married in Buenos Aires, when Eric was 20 and Silvia 14. Our witnesses were my second son Bambi, a mathematics professor, and his lovely wife Lilia, a medical doctor.

Newly Married

When back from our honeymoon, my friends welcomed us affectionately, but Marta's cut her off. She enrolled as a mathematics student in the Faculty of Sciences and took the six introductory courses, which that year were taught by newly appointed full professors, all of whom had done original research. Our original plan was that eventually she would use mathematics in psychology or social science. But she fell in love with mathematics, where she has flourished for half a century. Today she is a professor emerita of mathematics at McGill University, but keeps attending meetings and publishing papers impenetrable to me.

At the beginning Marta, my son Cantarito, and I dwelled in an apartment on the 20th floor of a modern building belonging to the El Hogar Obrero, the oldest cooperative in the country. As befits an Argentine middle-class family, we had a domestic employee who cleaned the apartment, did the laundry, and cooked dinner for us. We took the subway to go to our respective faculties for our course work, but studied mostly at home, as there were no offices in the university. On Friday evenings we visited my mother, who had loved Marta from the moment she saw her, and knitted beautiful sweaters for both of us. The next morning we traveled to our rowing club in Tigre, at the mouth of the Paraná delta, where we got a rowboat that in a couple of hours of vigorous rowing took us to a wild island. There I coached Marta in math while huge mosquitoes with white gloves lunched on us, and curious wild otters spied on us. In sum, that year of 1959 Marta switched tracks while I got a few years younger.

The Buenos Aires Philosophy Congress

That same year the 6th Inter-American Congress of Philosophy met in Buenos Aires. It was well organized by the university rector, my old friend the philosopher Risieri Frondizi, and his team. I took an active part in several discussions and in guiding some participants around town. I read two papers: one on the (modest) role of induction in science (Bunge 1960a), and another on the various concepts of a level (Bunge 1960b).

During breaks, and at a party in our apartment, I made several new friends, among them the Americans Charles Hawthorne, Paul Weiss, Betty Flower, and Peter Caws, all of whom we met again in the USA. Weiss, who had edited the complete works of Charles Peirce jointly with Hawthorne, wanted to see with his own eyes the quarter where, as I had told him, Jews lived together with Arabs – as they had done in Moorish Spain, and still do in Montreal. Incidentally, Alfred Stern, who had written for *Minerva*, was staying at the Jousten Hotel, a Nazi meeting place. I had him and his wife moved to a cleaner place.

There were also some eccentrics, like the Peruvian psychiatrist Honorio Delgado, a fascist existentialist; the smart and enterprising Brazilian legal philosopher Miguel Reale, who wore fascism on his sleeve; the unsmiling Spanish-American Eduardo Nicol, who had compressed phenomenalism into the formula "Being is in sight," and forced his pretty wife to wear flat shoes, so she would not look taller than him; the boastful Euryalo Cannabrava, of whom I never heard again, and a countryman of his who mentioned many fashionable theories without analyzing any of them.

Another anomaly, whom I took for a long walk in the Rose Garden, was the Belgian Charles De Koninck. He had fled to Canada at the end of the war for having collaborated with the German occupiers, and was now teaching at Laval University in Quebec City. He told me that Aristotle's physics, far from having been just an important early phase of science, was the true physics – but of course he did not bother to offer any evidence. His son was to inherit the chairmanship of his Department.

Betty Flower, as elegant, confused, generous, and affectionate as ever, took to Marta and myself, and later had me invited as a visiting professor at her university. This was the University of Pennsylvania, aka Penn, in Philadelphia, aka Philly, founded by Benjamin Franklin, the first American scientist.

To the USA

In August of 1960 we flew to Philly via Los Angeles and San Francisco. Our travel agent Verónica Kleiber – daughter of the famous conductor and wife of my old friend, the physicist Andrea Levialdi – organized our journey, throwing in a week in México City, which was then still livable, with a population of four million people, one-fifth its present population. There we visited the old Archaeology Museum, which was then just an antiques deposit. We also visited the famous muralist David Alfaro Siqueiros, not my favorite, who had visited *El Ombú* in the 1930s. He received us with a big pistol in hand, for fear of an assault. Finally, a very amiable philosopher, whom I had never heard of, invited us to dinner at a restaurant patronized by bullfighters, with trumpeters that blew so hard, that they made conversation impossible – which of course was a blessing.

The next day we flew to Los Angeles, where we enjoyed the warm hospitality of Alfred Stern and his wife. She was a Puerto Rican poet who spoke an Americanized Spanish he corrected relentlessly. (For example, she translated *lane* as *lana*, the Spanish for wool, and he corrected angrily: *¡carril!*). Stern, whose *Philosophie des valeurs* (1936) I had reviewed favorably in *Minerva*, had been forced to emigrate from Vienna to México, and was now teaching at the famous California Institute of Technology, aka Caltech. In Los Angeles he had had trouble renting a house for not having a "Caucasian" profile. But he had no trouble making new friends, such as the great theoretical chemist and nuclear peace activist Linus Pauling, a double Nobel laureate.

Through Stern's good offices, the University of Southern California invited me to give my first two lectures in English. The Sterns drove us to Caltech, Hollywood, and other places, among them our first supermarket. There we saw the largest and most tasteless plums ever – our first experience with American giantism.

First International Congress of Philosophy of Science et al.

Our next stop was Palo Alto, near San Francisco, where Stanford University is located, and where the First International Congress of Logic, Methodology, and Philosophy of Science was about to meet. We could not imagine that four decades later our daughter Silvia would obtain there her doctorate in cognitive neuroscience – particularly since at that time Stanford psychology was dominated by the brainless neobehaviorism of William K. Estes, R. Duncan Luce, and Patrick Suppes.

Had Silvia attended Stanford when we stayed there, she would have been forced to study the so-called measurement theory, a conceptual misfit based on the confusions of the German *Maß* (measure) with *Messung* (measurement) originally due to Ernest Nagel, who had been Suppes's thesis supervisor (see Bunge 1973d). They had mixed a concept with an empirical procedure.

We stayed at a students' dormitory, and the next day ate our first American breakfast, abundant as befits the start of a workday. This time, our work consisted in trying to understand some of the many presentations given mostly by Americans. Marta and I had no problem with written English, but spoken English was quite foreign to us. For instance, I pronounced *laughter* by analogy with *daughter*. I had yet to learn that English pronunciation had been forged in markets and workplaces, not in academies. But Americans, unlike Europeans, are tolerant in phonetic matters, perhaps because they themselves are not sure how to pronounce *almanac*, *cylinder*, *nuclear*, *terrorist*, and other words.

I read my paper, "The complexity of simplicity," chaired a meeting, and at the banquet was seated at the same table with the congress's organizer, Patrick Suppes, and the great mathematician Alfred Tarski, and was asked to deliver a short speech as well.

The talks I best remember were Popper's on verisimilitude as improbability, Tarski's on Suppes's failed attempt to infer models from data, and Stanislas Ulam's on the superhuman abilities of computers. When Ulam finished his panegyric, I stood up and timidly asked him whether it might be possible to design a program capable of posing new problems. He remained silent for a long while, at the end of which he admitted that he had never thought about that – which to me sounded as if a biologist had never thought of embryos.

I had long conversations with a number of philosophers, who, with the sole exception of the arrogant Donald Davidson, were patient with this unknown from a remote and negligible corner of the world. I had the most memorable talks with Arend Heyting, L.E.J. Brouwer's successor, who concurred with my statement that mathematical intuitionism – a foundational strategy – has little to do with philosophical intuitionism; with Pat Suppes, whom I admired until I read his paper where he attempted to legitimize with three-valued logics expressions like "that is 2 long," which is just ill-formed for it omits the length unit; with Popper, on whether machines would ever come up with original ideas – a possibility he did not discard; with Georg Henrik von Wright, perhaps the only gentleman in the philosophical community, whom I tried to persuade that inductive logic ought to be subjected to empirical tests; with Józef Bocheński OP, who told me that he had entered a seminary because the priesthood was the only means a Polish peasant had of acquiring an education; and with Gerold Stahl, my old friend from Santiago de Chile, with whom I had planned writing a paper in Heidegger's style but never succeeded – an experience that taught us that philosophical forgery, like banknote forgery, is a highly skilled occupation.

In sum, the Stanford Congress was of a high level, taught me a lot, and gave me the chance of making some interesting acquaintances. Nobody seemed to regret the absence of Thomists, Kantians, Hegelians, Marxists, phenomenologists, or existentialists. But I regretted the preeminence of logical positivists, who attempted to fill their ontological vacuum with logic, or to find use for non-standard logics. However, I had no name for my own philosophy, because it was still *in statu nascendi*, and it would take me more than two decades to develop it into a system. My stay at Penn was a decisive early phase of that development.

Penn

At the end of the congress we took the night flight to Philadelphia, where we would remain for the next 9 months. Our arrival must have been similar to that of a Patagonian Indian just arrived in Buenos Aires at the end of the preceding century. This was the first time we belonged in a world-famous bicentennial university, with a wide and beautiful campus, renowned professors, well-prepared students, weekly seminars in all the important subjects, rich libraries, a unique museum of Sumerian antiquities, a welcoming faculty club, a decent bookstore, and plenty of interesting lectures and movies announced in the daily bulletin that we found every day in our mail boxes. All that in the third city of the country, which had not entirely forgotten that it had been the cradle of the nation and the keeper of the noble Quaker traditions of tolerance, equality, and pacifism.

Marta was admitted in the mathematics graduate school and was offered to teach an elementary course. She enrolled in several courses as well as in the pro-seminar. This consisted in solving problems that required more imagination than information – just what she needed and wanted. Her imperfect English phonetics caused some misunderstandings, as when she asked her students to take a *shit* (meaning *sheet*) of paper, or when she wondered about the meaning of *armoyiuls* (R-modules). But she overcame all the obstacles with her characteristic courage and tenacity.

Marta also made a number of friends of various backgrounds, among them the number-theory professor Emil Grosswald. He was a serial expat (Romania – France – Cuba – Puerto Rico – USA) who had polished crystals in Cuba before getting an American visa. Emil did not believe that women could produce good mathematics. He could not have foreseen that Marta would achieve distinction in a new field. But he should have foreseen that his colleague and compatriot Alexandra Ionescu-Tulcea, who was then teaching Marta, would climb much higher than himself.

Marta suffered only one disappointment at Penn: that was when the provost summoned and upbraided her for flunking a student. When she replied that the student in question had not learned the subject, the dignitary responded: "That is immaterial. What matters is that his father is a very important donor to the university." Unfortunately, that was not an isolated incident: most American universities give "gentlemen's (or ladies's) Cs" to bad students with moneyed connections.

My Courses and Stray Lectures

I was assigned the undergraduate philosophy of science course and a graduate seminar. In the latter we discussed, among other topics, the concepts of approximate truth and of real (as different from conceptual) possibility. That is when a student informed me that modal logic was the philosophical fashion of the day. I soon realized that it is foolish to handle the various concepts of possibility – conceptual, physical, technological, moral, legal, and perhaps more – with a single theory. For example, conceptual possibility equals logical consistency, whereas physical possibility is short for compatibility with the known natural laws. This discovery allowed me later on to dispose quickly of the many-worlds philosophies of Saul Kripke, David Lewis, Hilary Putnam, and others. The sole merit of these *jeux d' esprit* is that, as David Hilbert once said of other mathematical curiosities, they help get promotions.

At Penn I also gave a couple of lectures about intuition and intuitionism sponsored by the departments of philosophy, mathematics, and physics that drew a large attendance and provoked several questions. An unexpected bonus was that a member of the audience was a representative of Prentice-Hall, who offered me a contract for a book on the subject, which they eventually published (Bunge 1962c). This book also appeared in Russian and Spanish.

The original motivation for writing that little book had just been to stop a bad student of mine in Buenos Aires, who used to pester me in class quoting the claims that Wilhelm Dilthey, Henri Bergson, G. E. Moore, and the Nazi ideologists had made about the superiority of intuition over reason. While I criticized philosophical intuitionism, I respected mathematical intuitionism and admitted the multiplicity of concepts of intuition and their roles in scientific research: though absent from the finished building, intellectual intuition, in the sense of insight, is helpful in the scaffolding.

That same academic year I also gave lectures at Harvard, Yale, Pennsylvania State College, Swarthmore College, Wayne, Kansas, and a Catholic convent. I was often asked interesting questions and offered stimulating objections – another new experience. Van Quine, Philip Frank – of Unified Science fame – Percy W. Bridgman – the father of operationism – and other celebrities attended my Harvard talk and raised interesting issues.

Neither Marta nor I foresaw that our son Eric, who was to come 7 years later, would graduate from Harvard, teach there, and meet his charming wife and partner Mimi Hoang, also a graduate of the Harvard Graduate School of Design. In fact we never foresaw any of the major events in our lives, except for our children's births. The reason for that unpredictability is, of course, that the isolated and rational individual exists only in rational-choice theories: real people are immersed in tightly knit social systems that nobody fully controls.

Nelson Goodman

The most interesting and also the most abrasive of my Penn colleagues was Nelson Goodman, the author *The Structure of Appearance* (Goodman 1951), which I had read in Buenos Aires. When he took me for lunch at the Faculty Club, I told him that I admired this book for its clarity but not for its content, which was phenomenalist: like Hume and Kant, Goodman regarded the universe as a collection of disconnected phenomena or appearances. This commonsensical worldview seemed to me to be anthropocentric and therefore inconsistent with the modern scientific worldview born around 1600. Goodman replied that content did not matter: that what mattered most was form or logical structure – which is why he admired above all Rudolf Carnap's *Der logische Aufbau der Welt* (*The Logical Construction of the World*), of 1928, which I had not yet read. That was the subject of his advanced course, which I attended only a few times because it did not interest me. Goodman was a dull lecturer: he did not pose substantive problems, did not engage with philosophers other than Berkeley and Carnap, and never even mentioned scientific controversies relevant to ontology. Besides, he did not dialogue with his students, and none of his few doctoral students achieved anything. But Goodman enjoyed an extraordinary reputation in the American philosophical community. For example, William Craig, the famous logician, told me that he regarded Goodman as one of the most profound philosophers.

Other Colleagues

The weekly departmental seminar was lively. It was attended by philosophers like Goodman, ordinary-language philosophers like Howard Zinn and Zeno Vendler, and linguists like Henry Hiż. There were often passionate arguments, particularly between the Aristotelian Hiż and the positivist Goodman. Hiż, a friendly Polish linguist, had tried his hand at mathematical linguistics, and tackled the problem of the grammatical structure of questions – though not of the semantic and epistemological features of problems.

Glenn Morrow, a respected but somewhat distant and frosty colleague, was soon to become the greatest living Plato expert. He held that, contrary to received opinion, Plato's ideal city was not in Sparta but in Crete. I understand that this is now the ruling view, but it does not seem to excite anyone at a time when Plato is no longer either revered or reviled. Morrow and his wife had us for Thanksgiving dinner, and played records of touristy Greek music. He was offended when I told him that it sounded Oriental to me. It still does after half a century.

William Fontaine, one of the very few Black philosophers in the USA, confided to me that he disliked the vehemence of some of the discussants. He was humble and taciturn, and communism worried him far more than racial segregation. A follower of Martin Luther King Jr's would have said that Fontaine was an Uncle Tom. But the March on Washington was a decade ahead.

Mathematical Psychology

I was delighted to discover mathematical psychology, a subject whose very existence was unknown in my homeland. From the beginning I interacted with the three Penn mathematical psychologists, in particular Eugene Galanter, the only one to build a utility curve out of experimental data. He found it to be a sigmoid, like for most other goods: initially increasing, then increasing with decreasing speed, until reaching a plateau. Obviously, Gene had studied only normal people, not tycoons or tyrants, who wish to own more, the more they possess.

I also befriended his friendly colleague R. Duncan Luce, a mathematician stuck in stimulus- response psychology. He also believed in the existence of a single psychological law and a single theory of "measurement." He, Galanter, and others were then in the process of launching the Psychonomics Society, to help shift attention from myth and anecdote to law, and from clinical to experimental psychology. But the great breakthrough would occur only three or four decades later, and it would be the work of cognitive neuroscientists, not mathematical psychologists.

Marta and I also befriended Fritz Stahl, Eugenio Chang-Rodríguez, and Robert McNaughton. Fritz was a Dutchman specialized in Indian logic; Eugenio, a Chinese Peruvian, achieved distinction in the study of Latin American literature, and exhibited his ear-to-ear smile even when recounting horrible events; and Bob was in transit from philosophy to theoretical computer science. We met again Fritz and his beautiful Indian wife and children, in Amsterdam and Corfú; I met Eugenio by chance in Helsinki, and Bob at an airport.

Interesting Students

I had several interesting students. One of them was Chaim (né Hermann) Potok, a rabbi who had officiated as a chaplain in the Korean war. He was anything but a Talmudist: he was interested in important philosophical problems, but his ambition was to write novels. I warned him that it was becoming increasingly difficult to get fiction published at a time when the small publishing firms were being acquired by big businesses alien to culture, and were looking only for bestsellers. Fortunately, Potok disregarded my warning. His first novel, *The Chosen*, was an instant success. One of its main characters, apparently, is modelled after me.

Two other outstanding students, who became close friends of us, were the inseparable James Hullett, aka Jay, and Bob Schwartz. Both taught us the ropes around Penn. Jay taught me how to trace philosophical publications in a journal that has become extinct, and he photocopied and mailed to me articles that were unavailable in Argentina. After getting his PhD under Goodman, he taught for a few years at Boston University, and then became director of the Hackett publishing firm, which has done the good service of publishing core philosophical texts at very reasonable prices.

Bob Schwartz's father was a successful plumber who wanted his son to succeed him as the head of his firm. But Bob was far more interested in philosophy than in plumbing. So, I got him to invite me for supper at his home, and I persuaded his father to allow Bob to follow his calling. Bob too wrote his dissertation under Goodman, who had become the day's Berkeley. Once Bob came to McGill accompanying Goodman. He gave a talk, which started by promising that he would make a star there and then – just with words. Like all constructivists, he mistook things for their symbols. I have regretted since then having meddled with his career choice.

I also have a clear recollection of Charlie Parker, my worst Penn student. He was nice and timid, but studied to little profit. He deserved to be flunked, but I felt sorry for him. I engaged him in conversation and, to make him feel comfortable, I started by asking how he got his shoes to shine so beautifully. He confided his secret in me: spit and polish. Next I asked him what he had for breakfast at his fraternity. He replied: a cup of coffee and a doughnut. So, malnutrition was a factor in his low scholastic efficiency.

Goodman's Lie and Hook's Bait

Nelson Goodman and I were invited to participate in a symposium on simplicity at the annual meeting of the AAAS (American Association for the Advancement of Science), to be held in New York City at the end of 1960. At one point in my presentation, I held that Goodman had adopted the old idea that simplicity is the sign of truth (*simplex sygillum veri*). When Goodman indignantly denied this charge, I opened my briefcase to extract and exhibit the reprint of the recent paper where he had made that statement, and that he had mailed to me (Goodman 1958). He kept silent but treated me frostily from then on, and refused to renew my appointment for 1 year.

A few days later we went to Yale University – where three decades later our daughter would get her BSc in neurobiology – to attend the annual meeting of the Eastern Division of the American Philosophical Association. At that meeting Julius Kraft, *Ratio*'s editor, died from a heart attack when the moderator, a Nietzsche scholar, refused to allow him a 2 min extension. A Frenchman might say that this was a clear case of *le mort saisit le vif.*

We spent Christmas Eve at Paul Weiss's apartment on campus, along with other participants in the APA meeting. Shortly before midnight Sydney Hook, with unsteady step and holding a wine glass, approached us and, without preliminaries, offered me the editorship of the Spanish-language edition of the journal of the Freedom of Culture Congress, over which he presided. He assured me that financing would not be a problem. I declined his invitation because, although I too was disenchanted with communism, I refused to become a professional Cold Warrior like himself.

Shortly thereafter it became known that Hook had not only cooperated with McCarthy's House Un-American Activities but that, along with Sir Isaiah Berlin and other British and American philosophers, he had been paid to report on left-leaning colleagues (Stonor Saunders 1999). In his memoirs, Bertrand Russell called Hook "the CIA philosopher." But it would be short-sighted to accuse Hook of corruption: the communists themselves, particularly those in power, did commit some of the crimes Hook accused them of, starting with their travesty of socialism.

A few months later I had an even stranger experience. George, a physicist whom I had befriended in Brazil, invited us for dinner at his New York apartment. There he introduced us to a high official of the dying American Communist Party. This man asked me, as if I were the Delphi oracle, what the party ought to do in view of the severe crisis it was going through. I told him that they ought to close shop in view of the worldwide failure of communism. The man seemed bewildered, for I was probably questioning his very raison d'être. A decade later, when George visited us in Montreal, he was running for Congress on the Democratic ticket.

Approximate Truth

During the Spring of 1961 I built a mathematical theory of partial or approximate truths such as "Plato was a Tibetan philosopher", "The Earth is spherical," and "$\pi = 3$." The great majority of logicians and mathematicians reject the very idea of a half-truth: they believe that logic contains the principle of bivalence, according to which every proposition is either completely true or fully false. I claim that this statement is false: that logic is about formal validity, not truth. Besides, scientists, technologists and approximation theorists use approximate truths all the time.

I thought that only elementary logic teachers, who introduce the logical functors via truth tables rather than via axioms, would fail to be persuaded by the above. Yet even our friend Paul Bernays, who had been Hilbert's second, was among the deniers. By contrast, Popper got interested, and in short order constructed his own theory of what he called 'verisimilitude'. In my opinion he failed because he construed verisimilitude in probabilistic terms, and probability quantitates possibility, not partial truth. In any event, my first trial (Bunge 1963b) was not the last: I proposed others (e.g., Bunge 2010a), and am still looking sporadically for closer approximations.

Europe with Marta

At the end of the academic year we flew to Europe by the cheapest airline, the Irish one. I was flanked by two gigantic priests who overflowed their seats. In Dublin, lugging our suitcases and dazed with sleep, we were refused admission to the first hotel we tried because Marta's passport was in her maiden name. Once we were finally welcomed at another place, we walked on streets blackened by priests and nuns, and visited sad churches decorated with the semitransparent banners of Irish regiments that had been decimated by German guns during the First World War. A bus took us to the green countryside, and we had a Chinese dinner – the only edible and affordable food in Dublin at the time.

The next day we flew to London, and took the first bus to Oxford, from which we went to look at the still enigmatic Stonehenge temple-observatory-bastion. Later we admired the Canterbury gothic cathedral, the peer of Notre Dame and Chartres. When making our ritual offer for the upkeep of that jewel, we picked up a religious pamphlet by the famous quantum chemist Charles Coulson – proof that scientists can be religious as long as they abstain from studying problems that require the explicit rejection of religious dogmas. I blame Professor Coulson not for his philosophical naivety but for having blocked Margaret Thatcher's academic career. Had he helped her get a PhD, she would have worked silently in a chemical or pharmaceutical laboratory instead of turning the British social clock back by several decades.

Our Canterbury innkeeper got furious when he found out that we had taken baths on a day other than Saturday. In Oxford we visited the fabled Bodleian Library, where we admired the electromechanical arm that picked the desired book and placed it on the conveyor belt that took the book to the reception desk. When it started raining cats and dogs, we took refuge in a dirty tearoom patronized by hookers.

The high point of our voyage to England was the home of Karl and Hennie Popper in Penn, Buckinghamshire. They had kept the furniture that Karl had made in Vienna as a cabinet-maker apprentice. His father, like my paternal grandfather, believed that everyone ought to be able to earn a living with his hands. They also kept a grand piano that Karl beat with great enthusiasm. In fact, he was a very well rounded amateur.

Karl's artistic preferences had not changed in decades. He would not listen to any music other than Mozart's, and when I recommended to him some recently published novels, he replied that re-reading Jane Austen was good enough for him. Of course her novels are delightful, but closer to the sentimental feuilleton than to the great art of Cervantes, Balzac, or Tolstoy.

On the Saturday the Poppers drove us, at high speed, to an inn at Stratford upon Avon, where we chatted before an open fireplace and lunched on river trout. In the afternoon we went to the theatre to see Shakespeare. Marta and I were shocked when, just before the show started, the public stood up and sang *God save the Queen*! The next day came Ernest Gombrich, the famous historian and psychologist of art. A few months earlier he had visited us in our humble Philadelphia apartment, where we served him a TV dinner on a precarious little table. That evening he drew for us the itinerary for the Italian tour we were about to take. Had it not been for him, we might have missed San Gimignano and other jewels.

After talking for a while about academic matters, Popper and Gombrich switched to their native language. First, Karl complained that his name was not on the Honours List in the day's *Times*. Next they commented on current events. They agreed that the Congolese freedom fighters led by Patrice Lumumba and his comrades were *lauter Affen* (just monkeys). This opinion of Popper's was congruent with views that he voiced at other times on other imperialist feats: that the British settlers in New Zealand had not decimated the Maori natives, that "the British had done much for India," and that he could not believe that British armed forces had invaded Buenos Aires twice in the early nineteenth century. His transformation from a young Viennese socialist into a good British subject had been complete (see Hacohen 2000).

France and Italy

From London we went to Paris, where we saw some of the obligatory sights, in particular the Louvre and Notre Dame. We also met Coco Gerschenfeld, who was happy studying the nervous system of the turtle under Professor Fessard at the École Normale Supérieure, and made us addicted *Le Monde*, which was then the best daily in the world.

We had lunch at Jean Ullmo's, the mathematician and philosopher that the Chilean philosopher Félix Schwartzmann had talked to me about. Ullmo had promised me to contribute to the Festschrift for Popper that I had started to compile. The idea occurred to me when, at the Stanford congress, I listened to Karl's many health complaints, which made me believe that he would not live much longer. (In fact, he lived for another 33 years.) The other guests at Ullmo's table were an Indian civil servant and his family, who were touring Europe courtesy of the Indian taxpayers.

We rented a "beetle" and, with the help of a very detailed *Michelin* guide, we visited Chartres, the Loire valley, the beautiful Dordogne – including the astonishing Lascaux cave – Avignon, Marseilles, Arles, Nimes, and Nice, and tried to swim under the Pont du Garde. We arrived in Nice at midnight and ordered the bouillabaisse fish soup, that my father had praised to heaven, but we could not eat it because of the excess of pepper.

Then we drove to Geneva, Lausanne, and the Swiss Alps – before the great tunnels were opened – where we became almost drunk on *fondue*, and finally arrived in northern Italy. We followed the route that Gombrich had drawn for us, which included the amazing Giotto paintings and Donatello sculptures, and the Monte dei Paschi di Siena bank, which had been in business for five centuries because it had escaped the attention of Wall Street raiders and economics professors.

We rested a couple of days In Capri, not to be missed for readers of Axel Munthe's *History of San Michele* and the gardner who once in a while dug up *roba di Timberio*. After swimming in blue waters, stepping on a couple of sea urchins, and admiring the *grotta azurra*, we returned to our gray home.

Return Home

In late 1962, after a year of absence, we returned to Argentina with great enthusiasm, believing that we could use what I had learned abroad to improve things at home. This illusion lasted only 1 year, after which the Army split into two bands, the reds and the blues, who fought one another – which gave us the idea of leaving for good. However, let us not skip that year, for it was an enjoyable one.

Upon our return we stayed for a while at my mother's chalet. When we opened the trunk where we had saved our woolens, a thick cloud of moths flew out the window. A few days later we found an apartment for the three of us: Marta, my first-born son Cantarito, and myself. It was situated near the botanical garden, where

I went to study weather permitting, and we bought it with a mortgage. Since we had no furniture, and were on a very limited budget, we ordered from craftsmen bookshelves, tables, and chairs that I designed. I paid in advance for the double bed from a fancy Italian store, but it never came, so we slept on a mattress placed on the floor until I had saved enough money to buy a second bed. The table I had ordered took long in coming, so meanwhile we used trays placed on our laps. This is how the Catalonian publisher Alejandro Argullós ate when he came to visit us, and that was enough to make friends with him.

Cantarito, always cheerful and used to camping, did not complain. Anyway, a few months later he married his classmate Annick Vivier, and both left on scholarships for the University of Florida in Gainesville to get their doctorates in theoretical chemistry. That was the seat of the Quantum Theory Project, founded by the Uppsala physicist and chemist Per-Olov Löwdin. Their wedding was celebrated at a bar with cold drinks and sandwiches. At the end of 1964 Marta and I drove our "Beetle" from Philadelphia to visit them, I got to attend a couple of stimulating lectures by Löwdin and Slater, and argued with my son on perturbation theory, which he believed to be a part of physics whereas I regarded as a mathematical ancillary. However, let us return to our homeland.

We spent January of 1963 in the beautiful northern city of Salta, my fee for a course on the philosophy of physics for high-school teachers, organized by the Asociación Física Argentina. The students asked many philosophical and pedagogical questions, such as how best to introduce the mass and charge concepts, whether reference frames are the same as coordinate systems, and whether physical dimension, as LT^{-1} in the case of velocity, is the same as size – as Suppes stated in his article in the *Encyclopaedia Britannica.*

Birth of *Scientific Research*

Back from Salta I started to write *Scientific Research* (Bunge1967b), my treatise on methodology of science. I believe this text was the first to tackle systematically all the main concepts in the field, from those of problem and theory to those of empirical generalization and theoretical law, natural law and law statement, indicator, and hypotheses evaluation, as well as the differences between pseudoscience and science, and between basic scientific research and technological design. It was also the first to adopt a stance that was as distant from empiricism (in particular inductivism) as from rationalism (in particular deductivism). However, let us not get ahead of things.

When I resumed my courses, I had two agreeable surprises. One was that the enrolment in my class had doubled. Another was to detect, standing in the back of the classroom, a man with a round face and a wide smile who asked intelligent questions. He turned out to be the eminent biologist Osvaldo Reig, with whom I immediately struck a close friendship that lasted through several migrations of both of us until his death three decades later. He invited me to meet his best students, among them Jorge Rabinovich, to discuss some scientifico-philosophical problems, such as the nature of biospecies, the status of the competitive exclusion principle, and the role of mathematics in biology. We also met in his tiny apartment, which he shared with his wife Estela Santilli and their first baby, to study Joseph H. Woodger's *Biological Principles* (Woodger 1929), which disappointed us.

Woodger, a delightful biologist-philosopher whom I met in London in 1965, believed in psychoanalysis, and had an exaggerated faith in set theory as *the* tool for constructing biological theories. Two years later I invited him and Nicholas Rashevsky, the editor of the *Bulletin of Mathematical Biophysics*, to a colloquium in Amsterdam that produced no sparks and shed no light. I soon learned that the set concept is indispensable to understand that of a species, but that theoretical biology can use the entire mathematical panoply. In any event, contemporary theoretical biology bears no traces of Woodger's work.

Taking the Gospel Across the River

One evening in 1962 Osvaldo Reig and I boarded the boat to Montevideo, to help my dear friend the philosopher Mario H. Otero organize the Grupo Uruguayo de Lógica y Epistemología. We first met the mathematicians – Rafael Laguardia, Juan J. Schäffer, José Luis Massera, and their assistants – with whom we discussed the two main concepts of mathematical truth: those of satisfaction in a model, or example of an abstract theory, and demonstrability or theoremhood. In the evening we met with the professors and advanced students of botany and zoology. We exhorted them to fuse both departments around the more general disciplines of genetics and evolutionary biology. Predictably, the professors defended their turfs, whereas the students cheered our suggestion, and eventually succeeded in having it implemented. This illustrates the idea that philosophy can help science in addition to using it. It is also an example of the principle that such missionary work can be accomplished without grants.

The same year, 1962, the Universidad de la República and the teacher's college in Montevideo invited me to lecture at both places. I accepted because I like Uruguayans – Argentinians without arrogance – and because they could use some modern philosophy of science. I also needed the extra money, as inflation had seriously devalued the Argentine peso. I flew Thursday mornings and returned in the evenings, finding Marta in tears whenever my flight was delayed.

I had few students in Montevideo, but we engaged in lively discussions. During the noon break I had long and interesting conversations with Mario H. Otero, a refugee from phenomenology who loved mathematics. He was a widely read and soft-spoken man who, during the military dictatorship (1973–1985), was tortured and subjected to a simulated execution for the crime of being the Dean of the Faculty of Humanities and Science. When freed, he found asylum in México's Instituto de Investigaciones Filosóficas while his wife, the mycologist Lina Betucci, was hired by the Faculty of Science in the same Universidad Nacional Autónoma de México, aka UNAM, where Marta and I joined them a few years later.

Mario H. was lucky compared with his colleague José Luis Massera (1915–2002), a world-renowned mathematician who spent 9 years in jail for having been a communist congressman. His wife Martha was incarcerated for being married to him. Costa-Gavras's film *State of Siege* paints a vivid picture of that cruel regime and the role of the American Embassy and its agent Dan Mitrione, a doctor in torture.

Research Projects

I have always had a central research project surrounded by one or more satellites projects plus the occasional paper. Between 1956 and 1965, my central project was *Scientific Research*. At that time there was a lively discussion between the upholders of the steady-state cosmological theory, proposed by Hermann Bondi and Fred Hoyle, and their critics. This theory was popular because, contrary to the creationist cosmologies, it postulated that the universe had had no origin and was globally homogeneous.

I did not object to these assumptions, but could not accept that matter kept emerging out of nothing at the same rate that it got lost from view due to the expansion of the universe. Bondi and Hoyle had added this ad hoc hypothesis to reconcile homogeneity with expansion. In my paper "Cosmology and magic," published in *The Monist* (Bunge 1962a), I held their creation *ex nihilo* hypothesis to be false for contradicting all the known conservation principles of physics.

I went even further: I stated that, to deserve being counted as scientific, a hypothesis had to meet the condition I called *external consistency*, or compatibility with the bulk of extant knowledge. Since the Bondi-Hoyle creation hypothesis violated that condition, I decreed that the stationary-state theory of the universe was pseudoscientific, as was any cosmology other than megaphysics.

Shortly thereafter, new astronomical findings falsified conclusively the said theory, a fact that was not noticed by John D. North, the author of the acclaimed *The Measure of the Universe* (North 1965). He defended the moribund theory and attacked my criticism of it. But his derisive treatment of the latter did not prevent him from quietly lifting my thesis that the relativity principle is not a law but a metalaw, or law of laws, so it cannot be empirically tested in a direct fashion (Bunge 1961c). *The British Journal for the Philosophy of Science* rejected the brief note where I pointed out North's fault.

That same paper of mine also angered Popper, for whom falsifiability was the necessary and sufficient condition of scientificity. Maybe this disagreement showed him that I was too independent to be entrusted with his legacy. By contrast, it served me as a test of the conception of science I was forging as I wrote *Scientific Research*, as well as to confirm my suspicion that even the best known philosophies of science – Popper's, Reichenbach's, Frank's, and Carnap's – were simplistic. Whereas

Popper and the logical positivists looked at hypotheses one by one and regarded empirical data as decisive, I looked for *systems* of hypotheses (i.e., theories), and regarded science as a whole, so that I valued external consistency as much as data fitting.

Another project I finished during those days was putting together and proofreading *The Myth of Simplicity* (Bunge 1963b), my fourth English language book – perhaps my least successful.

The Silent Coup Against Presidente Frondizi

In March of 1962, while my friends and I were enjoying the freedoms that we had unexpectedly received from above in 1955, the armed forces, apparently encouraged by the CIA, replaced President Arturo Frondizi with his vice president, José María Guido, an alcoholic who signed whatever papers were placed on his desk. The coup, allegedly performed to protect Western Christian Civilization, was performed silently. Six months later, the army's two factions, the reds and the blues, fought bloody battles, but kept in place the puppet president and his cabinet.

Ideological censorship was only one of the novelties, and it, too, was done silently and with the complicity of the media. I felt it every week when, returning from Montevideo, a customs official rummaged through my briefcase and looked at some of my notes. Another occasion was when SIDE, the army's information service, summoned me to acknowledge receipt of a parcel containing a copy of the Russian translation of my *Causality*. I was not threatened, but read the warning: Leave! None of my colleagues worried: the deposed President had been unpopular, and there was too much to do.

Leaving Home for Good

The serious deterioration of the political situation led Marta and me to decide to leave our country at the first opportunity. I wrote to some friends abroad. Only Quine offered help: he was willing to support me if I applied to the New School of Social Research in New York, that had been established by the founders of the Frankfurt School. But I regarded it as philosophically backward, so I replied to him with an ungracious letter: With friends like you, who needs enemies?

I next entered the competition for the philosophy of science chair at the new University of East Anglia. The selection committee preferred an Oxford alumnus who had failed his PhD and had no publications. As an Oxford don explained to me years later: *but at least he was one of us*. Besides, he had the grace of dying of cirrhosis of the liver soon after his appointment – a gentleman to the last.

But help was on the way. In January of 1963 I got a phone call from the University of Texas at Austin offering me a position to replace for one semester a professor who was on leave for that period. I accepted the offer with alacrity. We asked my former student and good friend Julio Colacilli de Muro to take charge of all our affairs, and exchanged for clothes and luggage the government bonds with which my last salaries had been paid. We flew to Austin, Texas, at the beginning of February 1963. Three years later, the military replaced their puppet president with General Onganía, the mildest of the brutes.

The Peronists were voted back in 1972, and 4 years later another military dictatorship, the worst in the country's history, began misgoverning the country until 1983. The highlights of that period were the "disappearances," the Falklands war, and the massive brain drain. Free elections were then called and won by the centrist party headed by Raúl Alfonsín, a decent man surrounded my mediocres.

During those three decades, every step forward has been followed by a step backwards, so that the country's net speed has been null. For example, science recovered the level it had when we left, but philosophy has fallen to even lower depths. Worst of all, nowadays no one seems to feel the optimism, enthusiasm, and drive that the intellectuals of my generation felt in 1955. The mood is similar to that of the Europeans who, when remembering the 1914–1918 massacres, were getting ready for those of 1939–1945.

Since 1985 Marta and I have returned almost yearly to our beautiful country for short visits. Every time we have had a great time visiting old friends and making new ones, as well as lecturing and sight-seeing. And every time we congratulated ourselves on having settled in a country with boring politics. Better bored than dead.

Chapter 7
Itinerant Professor

Texas

Our long march from Argentina to Canada started at the beginning of February, 1963. We boarded a Canadian Pacific plane in Buenos Aires and, flying over the snow-white Andean cordillera and the bluegreen Caribbean, landing in México City, where we took a plane to Austin, Texas. That stopover was enough to arouse the suspicions of the Austin customs officers, who scrutinized our luggage, including Marta's sewing kit. You never know what foreigners may try to smuggle in!

John Silber, the chairman of the philosophy department, an active member of the movement for the abolition of the death penalty, and later President of Boston University, welcomed us and drove us to a drafty apartment that had just been vacated by a visiting professor who, being British, put up with inconveniences that middle-class Argentinians do not tolerate. Hence we started hunting for a better place, which we found easily. I walked to the university, which moved several drivers passing by to offer me a lift and ask what was wrong with my car. My walking past bushes occupied by squirrels always provoked their angry complaints.

I taught a philosophy of science course and a graduate seminar about the philosophy of mind and related matters. Most of my students were tall Texans with a low education. Many of them drove huge convertibles and wore large Stetsons that they kept on even in class. Most of them were apathetic, and asked few questions, and no ideas excited them. They would have been amazed, or even shocked, only if a Mexican-American or an Afro-American had suddenly appeared among them. How different from my cosmopolitan, refined, and questioning students at the University of Pennsylvania!

My graduate seminar was much more lively. I remember especially two young men, one of whom could not stand me while the other admired me. The former got angry every time I spoke the word *real*: he claimed not to understand the meaning of sentences like "These chairs are real." Consequently, he did not grasp the contrast between hypotheses and reality, which is central to scientific methodology.

M. Bunge, *Between Two Worlds*, Springer Biographies,
DOI 10.1007/978-3-319-29251-9_7

By contrast, his admiring classmate declared that I was "a one-man university," just because I used examples drawn from different disciplines.

Colleagues

I never had more interesting and helpful colleagues. I found myself immediately surrounded by philosophers, biologists, anthropologists, and historians who were active in research, and who looked for me to debate philosophical problems. We met in offices, cafeterias, clubs, and at barbecues in gardens with gigantic oaks and pecan trees. My most frequent interlocutors were the anthropologist Richard N. Adams aka Rick, and the Guatemalan philosopher Héctor-Neri Castañeda. Gina, Rick's charming Guatemalan wife, gave me driving lessons that allowed me to get a driver's license, without which nobody is anybody in Texas. I had to relearn almost *da capo*, because in Argentina, two decades earlier, there had been neither streetlights nor intensive transit.

Rick wanted Marta and me to help him exactify the general notion of power he had read in the English translation of a work of Max Weber's. According to it, power would be the *probability* of altering other people's behavior. I wrote some pages that did not satisfy Rick. By contrast, Marta built a whole algebraic theory that he liked, and for which he thanked her in one of his books. Three decades later, when studying Weber in the German original, I found that the key word was not *Wahrscheinlichkeit* (probability) but *Chance*, which in German, as in French, means opportunity. When I said this at a meeting of Weber experts (Bunge 2007a), most of them were as surprised as the scholars who, 15 centuries earlier, had discovered the mistranslations of Aramaic into Greek that had given rise to dangerous heresies. This episode moved me to propose my own definition of "power" (e.g., Bunge 2009) in terms of the shrinking of the actor's space of possible states.

Castañeda was Rick's polar opposite: while equally competent, curious, hard-working, and willing to argue, he was coarse and aggressive, boastful, and duplicitous. Héctor criticized Max Black for not having had any original ideas, but he himself produced no theories of his own until he came up with his "pre-dia-philosophical theory of guise." When asked at the ninth Inter American Philosophy Congress (Caracas 1977) what he meant by "guise," he was at a loss: he could not even translate 'guise' into Spanish. Héctor came every morning to my office to discuss problems that interested him far more than me, until I asked him to let me keep working on my *Scientific Research*, which was in its last stage.

I was making good use of the large and well-stocked university library, from whose high tower, 3 years later, a former marine would be shooting people at his leisure for 96 min. There I found all I was looking for, including the obscure booklet by A. C. Ewing from which one of my students had lifted an entire page for his term paper. Ewing, whom I met a decade later, was a gentle Cambridge moral philosopher who, as Ayer told me, was eager to meet his Maker to ask Him whether there are any synthetic a priori statements.

I learned mostly from Rick Adams and a couple of evolutionary biologists, who suspected that life had emerged several times on our planet, as a consequence of which Darwin's tree of life might have to be replaced by the bush of life. I also had several unproductive talks with the excitable Ferrucio Landi Rossi, who was trying to combine Marxism with Morris's semiotics and Wittgenstein's ordinary language non-philosophy.

Memorable Texan Experiences

Here is an assorted collection of memorable Texan experiences: seeing trees loaded with caterpillars, large pits full of squirming rattle-snakes, and huge fields covered in bluebells; eating stakes more tender and tasty than in Argentina; attending a packed recital by the brave yet sweet Marian Anderson; reading the announcement of a lecture in praise of capitalism at a trade-union local; visiting showy and expensive dental offices where they did much work on our mouths that had to be redone 2 years later; hearing lots of people who badmouthed President Kennedy with intense hatred a few months before his assassination; learning about the Institute of Latin American Studies, where Mexican historians had to go if they wished to learn about their country; using the well-stocked co-op competently managed by the student union; visiting the executive director of a foundation who could not give me more than 2 min of his valuable time because he was busy setting up a society of foundation executives on the principle that charity begins at home; and being taken to exclusive eating clubs whose entrance doors opened only to individuals with the proper key.

Marta Discovers Corfu

Marta did not like the University of Texas because she was the only woman in a sexist environment, and was not enrolled in any courses, as I had been hired for a single term. Professor R. L. Moore, a famous topologist who kept working after 80, did not allow her to audit his seminar. Like Wittgenstein's, his seminars were open only to his faithful. Like a good Texan, Moore may have wished to privatize mathematics, whereas in normal scientific communities knowledge is shared. And in the many social gatherings organized by my kind colleagues, Marta was directed to the gynecaeum of housewives wearing long gowns.

Seduced by *Prospero's cell* and other novels by Lawrence Durrell, Marta studied Greek with the excellent manual written for American soldiers in case they were sent to invade Greece. She learned it so well that, when disembarking in Piraeus a few months later, she spoke it quite well. Besides, guided by Durrell, she identified the place where we would spend the coming summer months: Palaiokastritsa, on the Northwest cost of Corfu, the ancient Kerkyra, Athens's ally in the Peloponnesian war but a refuge of Spartans when the Ottoman Turks invaded Greece. On our way to Greece we made a stopover in Philadelphia to formalize my contract with Temple University, my next employer, re-enroll Marta at the University of Pennsylvania, and reserve an apartment.

In between these chores, our good friend Jay Hullett threw a party for us, where we met a few acquaintances from our earlier visit, among them Bob Schwartz, the rabbi and future bestseller Chaim Potok, and the rising mathematical star Peter Freyd, who would exert a decisive influence on Marta's career – and who, perched on the best chair in the room, pontificated cleverly on all topics.

We flew from Philly to Milano, where we admired the Duomo and met Ludovico Geymonat, a mathematician and historian of science. He had persuaded his friend, the powerful publisher Giangiacomo Feltrinelli – the "discoverer" of *Doctor Zhivago* and *Il Gattopardo* – to publish an Italian version of my book on intuition and intuitionism. But while having a delicious tea we argued over politics and disagreed about Maoism, which Geymonat admired whereas I criticized for its anti-intellectualism and costly improvisations. We said politely goodbye and I never heard again about my book, a victim of political fanaticism.

The next day we traveled by train to Brindisi, where we boarded a ship to Piraeus, Athens's port. In downtown Athens we registered at a cheap hotel and went to climb the Acropolis, to get a closer look at what I still regard as the most beautiful of all buildings: the Parthenon. Its sight has moved me every time I've looked at it.

We rented a car, drove to Delphi and the immense olive-tree forest below it, passed by the Gulf of Corinth, looked at our first mosque in Ioánnina, and finally arrived in Corfú town, that has kept many Venetian jewels. On the *Listón*, the pedestrian street formerly reserved for the families whose names figured in that Venetian list, we found the Pension Suisse, long since gone. I entrusted to its manager my only valuable possession: the typescript of my *Scientific Research* minus chapter 1, which filled the briefcase I had bought with the bonds that the Argentine government used in lieu of money.

We drove straight to Aghia Triada (Holy Trinity), the cove on the West coast that Marta had chosen in Austin. There, under a giant mulberry tree, we met Herr Alfred Hürter, a friendly and helpful man exhibiting an enormous surgical scar on his torso. He was a retired German consular official trained to work in China but who, when the war started, had been sent to Jordan.

After the war Hürter decided to enjoy life, took early retirement, learned modern Greek, and lived half the year in Palaiokastrita, going around in his smoky motor boat. He was popular among the villagers, who called him Kirios Alfredos, and struck a close friendship with the most interesting of them, Miltiadis Mikhalás, a descendant of the Spartans who had fled the Ottoman invasion of the rest of Greece.

Militiadis was a fisherman and the secretary of the Commune of Lakones, a dismal village of about 700 people, situated on top of a hill, and offering a spectacular view of the coast below, which none of the locals looked at. He was thin, fit, tense, and smart, spoke French and Italian in addition to Byzantine Greek, and was the only newspaper reader in the village. Miltiadis was the natural son of the taciturn

Paradiso woman and an equally silent man who walked past his son every day without greeting him.

Militades had had a very poor and joyless childhood. Sokratis, the village shoemaker, had made him his first pair of shoes for free. He was the man I was told to meet when I asked around whether they had heard about Socrates, Plato, or Aristotle. During the Italian occupation, Miltiadis had run errands for the soldiers, mostly peasants like himself. When the war ended, both he and the shoemaker joined the ELAS communist guerrillas. Miltiadis deserted after the first defeat at the hands of the Anglo-Greek army, whereas his benefactor kept fighting until he was taken prisoner and locked in a prison camp for several years. However, back to the Aghia Triada beach.

Miltiadis

Kirios Alfredos asked a child who happened to pass by to climb to Lakones and ask Miltiadis to come. After a while Miltiadis appeared on his motorized bike, and showed us a small house of his that was not being used. We proposed to him that we rent it for 2 months, on the condition that he instal an outdoor shower. We sealed the deal with a deposit.

On our return from a month-long tour of the country, we saw that Militadis had kept his word: there was now a tiny shower that released a trickle of water. His two young sons, Spiros and Yorgo, filled the tank with the ice-cold water they drew with a bucket from a deep well nearby.

Spiros was placid and pleasant like his mother, and he liked to counsel "Sigá, sigá" (slowly, slowly), whereas his younger brother was highly strung like his father, and proud to announce that he was very busy. Eventually Spiros became a high school teacher and married a pretty city girl who went to the beach just to get tanned and exhibit her jewels. Yorgos married an English tourist, thin and lively like himself, succeeded his father in town hall, and conducted multiple business deals on the side. In short, the Mikhalás children confirmed the popular formula *Genome is destiny*, but I do not know of any other case.

In our tiny apartment, light came from two kerosene lamps, and the kitchen stove was fired with coal. But we hardly used it, for neither Marta nor I knew how to cook. Kiría Nina, Miltiadis's kind wife, taught Marta how to cook a chicken in the local style, but chickens were hard to find. The enterprising Spiros Bourbos brought some edibles from Corfú town, 30 km distant, on his asthmatic motor tricycle. He also brought daily an ice bar for our icebox, but the fierce sunshine reduced it to a quarter of its initial size. For mail we depended on the good will of the bus driver Nikos, one of our beach friends, who collected it at the post office after showing our passports. Actually we expected only one piece of mail: the letter from the American consulate in Athens, notifying us that our visas were ready. But the letter only came at the last minute.

Daily Life in the Enchanted Island

Our daily routine was as follows. We woke up early, breakfasted on local toast and honey, and worked or studied till noon, when we went for a long swim, snorkeled among schools of colored fish, or rowed in a heavy fisherman's boat. On the beach we chatted with Nikos and his friends. We went to town only when we got a lift from Kirios Alfredos.

Back from the beach we showered with very cold water from the well, and lunched on tinned macquerel, and whatever else Bourbos had found for us in town. (Once a farmer invited us to eat all the figs we wanted for 1 dollar per person.) Then we had a long nap, sometimes interrupted by the fisherman Eftigios, who came to sell the last fish of his catch of the day, hoping to get a glimpse of Marta in her undies.

Tanned and fit from swimming and rowing, Marta was then at the peak of her beauty. And wearing the regional blue velvet dress that Kiría Nina forced on her, she may have looked like Nausikaa when – according to a local legend – Odysseus first saw her three millennia earlier when awakening from his shipwreck a few miles down the coast.

Once Eftigios took us in his smoky and coughing boat to the northern island of Mathraki, where some Englishmen went to shoot exhausted migratory birds from camouflaged shooting ranges. The hunters claimed that their easy slaugther was a gentlemanly sport. We spent the night at a fisherman's whitewashed house beautifully decorated with semi-antique tapestries.

Alter a nap we read novels and socialized with our hosts and new friends, particularly Nikos and Maria Tsilimbaris, with whom we struck a very close relation that was passed down to their daughters and their husbands. Nikos descended from Albanese who had worked or fought for the Turks. By contrast, Maria, a shy, sweet, and blonde Venetian beauty, descended from Cretans who had fled along with the Venetians when the Turks occupied their island. (The Versailles Treaty had given Crete and other islands to Italy as a prize for siding with the Allies, while at the same time making ponderous pronouncements about the right of peoples to self-determination.)

Nikos, whose father had been a prisoner of war of the Italian occupying military, had a degree in economics from the Athens university, and managed the local branch of a state bank in Corfú town, whereas Maria had taught home economics at various schools, to which she travelled on donkeyback. Both lived most of the year in the outskirts of Skriperó, a village on the slope of the Pantokrator, the highest mountain in the island. Nikos tended a beautiful garden cum orchard that reminded me of *El Ombú*.

We have known their two daughters since birth, and have followed closely their lives for nearly half a century. Margarita graduated in geology but no one seems interested in exploring the geological wealth of the country. Veatriki has a master's degree in biology and works as a hospital lab technician at the island's hospital.

This happens in all of the underdeveloped countries: they train some scientists but do not use them as such.

When Bourbos opened his restaurant, we often had dinner there. It had only five tables and offered a single menu (steak, fries, tomato salad, and crème caramel) for 1 dollar. There was also a gramophone with a single record left by a French tourist, which every night played repeatedly and at full volume a fashionable song that began thus:

Oh, Monsieur Cannibale,
Je n'veux pas mourir.
Monsieur Cannibale
Laissez-moi partir.

Two decades later, Bourbos owned a medium-sized hotel that was always full of German tourists. He became wealthy but a survivor of several heart bypasses, and missed his former life of physical effort and organizational work, particularly during the long winter months, when his only distraction was to drink in the tavern in the company of friends. Miltiades's old age was better, because he read the official newspaper and watched TV.

Exploring Corfu

Everything was new to us in Corfú: the Greek and Byzantine ruins, the orthodox religious services, the elegant if derelict Venetian houses, the tasty if heavy local dishes, the French arcade, the British cemetery, the elegant Ionian Parliament set up by the Brits, the Reading Room full of useless old books, and a couple of bookstores with English paperbacks that catered to the dwindling British colony.

There were some ruined Byzantine noblemen with a reputation for honesty and had managed to keep their estates in the wooded interior, plus a few exotic foreigners. One of them was the *lordos*, an English businessman who had been knighted, and owned a large villa on the secluded beach of Rovignà, that we visited often because of its clear albeit frigid waters. There we befriended Dimitri, a kind fisherman who had made a living when the sea was full of fish, and now caught just enough fish to feed his cats and the two seagulls who always came to demand their share with their shrill voices.

The *lordos*'s daughter Emma wrote novels for which she had an infallible recipe: she started a book where the author of an English classic had left. For instance, while in Jane Austen's *Pride and Prejudice* Elizabeth and Mr. Darcy are blissfully married forever after, in Lady Emma's follow up Mr. Darcy gets lost at sea, and Elizabeth, forced to support herself and her family, becomes the first woman publisher, etc. (This plot is my own, but I trust it gives you the flavor.) I guess that Lady Emma's cottage industry folded, for the last time we met her she was expecting the new owner, a Russian oligarch who came in his huge yacht, which looked like a battleship, with its long cannon and revolving radar outfit.

In the course of half a century we met with several other interesting characters. Once, when walking up to Lakones, I met a tall gypsy with his gigantic bear walking next to him on his hind legs. We engaged in conversation and he told me he hailed from Romania. When I told him that I came from Argentina, he spat contemptuously and resumed his walk downhill. Another time we saw a red convertible tumbling down to the sea because its drivers, two Canadian young women, had parked it on an incline without putting on the brakes.

Every time I entered the monastery perched on a hilltop to watch a religious service, I had to rent a skirt at the entrance to cover my naked legs. It was a proper tribute to the beautiful liturgical singing of the officiating *papás*, an overfed family man who was not inclined to discuss theology. The village idiot greeted us cheerfully when we passed by the tiny public square in Lakones, one of the many villages founded by Spartans fleeing the Turks. Nearby Spartilla was another such village.

Marta and I, and later on our children too, spent many happy and productive summers on that unique island, coveted by Angevins, Genoese, Venetians, Ottomans, Brits, French, and even Russians. We made several good friends with whom we are still in touch. Those generous people made us feel at home, while the landscape of blue waters and green olive trees and cypresses has been a peaceful balm and a source of inspiration over half a century. There, surrounded by trees and in view of the sea, I planned or wrote dozens of articles and book chapters in the course of half a century. The first of my Greek writings was chapter 1 of *Scientific Research* (since first chapters are best written last), and I am revising this page across the blue sea and in view of the old Venetian fortress and the hills beyond the channel.

Back in the USA

At the end of our first visit to Corfu, after spending some weeks waiting anxiously for the American visas, we returned to Philly, where Marta resumed her mathematics studies while I taught at Temple University, Penn's poor relative. I taught quantum mechanics, philosophy of science, and logic. I had some outstanding students in the logic courses, and most of my physics students worked at the Dupont chemical factory, where they earned more than I.

At Temple I had no memorable colleagues except for a friendly English instructor whom I hired to correct my English pronunciation. He missed Botswna, where he had taught for a while, and we shared our admiration for the early George Orwell. Once Marta delighted him by cooking a chicken Kiría Nina style. But we had most of our meals at student cafeterias or faculty clubs. I spent much time in the deserted library, where by chance I found a book by the great American-Canadian neuroscientist Wilder Penfield, which taught me a lot about the mental functions of the brain. We traveled often to New York for talks, films, and good meals. There I attended the famous meeting of the American Physical Society where Dirac gave a memorable talk, and whose hand can be seen on the cover of an issue of *Physics Today*.

At that time I also attended a number of meetings of the American Philosophical Association, popularly known as "the meat markets," because they were filled with colleagues searching for a job. Some chairmen were literally beleaguered by hungry candidates willing to teach anything. I had only two memorable experiences of this kind. The longest of them involved a drawn-out but interesting lunch with faculty, who recommended my appointment, but their choice was overruled by the university president, who was only interested in my political opinions. The shortest interview was by a chairman who wanted to know only one thing: what I thought about beauty. My popular answer, that beauty was in the eye of the beholder, did not please him, for he got rid of me and turned to the next in line. The right answer was to parrot the opinions of famous philosophers like Aristotle, Kant, or Hegel, none of whom had had any artistic experiences. This is the best mechanism for keep producing mediocre and pliable philosophers unlikely to annoy anyone with unorthodox ideas or teaching methods.

Truesdell and Neoclassical Mechanics

That was also the time when I struck up a lifelong friendship with Clifford Truesdell, a man with as many foibles as talents. He taught me much through his many letters and learned publications, the first of which I had discovered in São Paulo one decade earlier. As the greatest living authority on classical mechanics, he was was incensed by the amateurish forays into this field by Patrick Suppes and his school, who did nothing for science but bring confusion to the philosophical community.

I disagreed with Truesdell on two major points: his global rejection of quantum mechanics and his contention that rational mechanics, and indeed the whole of theoretical physics, was just an application of mathematics. His view of physics was more rationalistic (or apriorist) than rational.

When at home in Baltimore, Clifford and Charlotte Truesdell played at living in Euler's time. I visited them several times. The first time, when he came to the door, I took him for the butler because he was short, stocky, and muscular. When I stayed there as their houseguest, they dressed in period clothes, and we dined in their candle-lit dining room, which he had restored with his own hands. Their bedroom was dominated by a gigantic painting representing Charlotte in the nude holding Clifford's severed head. When I asked him why he refused to listen to music written after the baroque period, Truesdell answered: "What was good enough for Euler is good enough for me."

Truesdell liked being taken for an "idiot" (excentric), but he started the "neoclassical mechanics" movement, which revived interest in classical physics, in particular continuum mechanics, that was then relegated to engineering faculties but in urgent need of updating in view of the fresh findings on jet propulsion and nonclassical liquids such as petroleum.

President Kennedy's Assassination

The most sensational political event of that period was of course President Kennedy's assassination. Like millions of contemporaries, I distinctly remember the circumstances of the moment I heard the news of it: I was riding a taxi from a medical office. That same evening I commented on the event with my physics students. I told them that nothing of the sort would happen in my country, backward as it was. But I was not surprised, because I remembered the visceral hatred that rank-and-file Republican Texans expressed at the sole mention of the name of the only American president since FDR to inspire millions of people worldwide. Even I had been seduced by his Alliance for Progress.

Shortly before the end of my tenure at Temple, I finished my *Scientific Research*. At about the same time I accepted the invitation of the University of Delaware to spend the next academic year as a Distinguished Visiting Professor of Philosophy and Physics. But first we went to Greece, to recover from a hard academic year.

Corfu Revisited

We flew from Philadelphia to Frankfurt, where we bought our first "Beetle," which I drove through Italy to Brindisi, where a ferry took us straight to Corfu. This time we occupied two rooms of the Mikhalás family: a bedroom and a study with two desks. We were soon adopted by Snoopy, a delightful stray bitch, playful and affectionate, who was very clean, slept on our beds, and accompanied us wherever we went. She sat on the prow of the boat and chased all the cars that passed by. We never found out where she did her business.

Yorgo Mikhalakis, María's brother, and the young men he had befriended while doing their military service, were staying in a tent under some tall olive trees in the adjoining lot. Marta and I visited them often to chat and listen to their beautiful popular songs from various islands, such as "Samiotisa." Our friendship was strengthened when we offered them asylum the day a deluge revealed holes in their tent.

We met most of them several times over the years, particularly Yorgos, a helpful, affectionate and cheerful giant, always ready to help, as well as to eat pistachios, feta, and grilled fish. (The fact that the Greeks are the fattest of Europeans suggests either that they do not stick by the famous Mediterranean diet, or that it does not work.) Yorgos had studied electronic engineering both in Athens and in Toronto, and taught this subject at the local technical school until retiring. He was not in love with his specialty, but he was a responsible and popular teacher.

First Visit to Israel

Towards the end of that summer of 1964 we drove to Athens to have our passports stamped by the American consulate, and we shipped our car to Philadelphia. In Piraeus we boarded a ship bound for Haifa, but had to wait 2 days in dry dock for a major repair. In the Haifa train station we waited a long time standing in the sun, and when the train finally arrived, we were rudely elbowed aside by a platoon of healthy army recruits who occupied all the seats. So, we had to travel standing to Jerusalem, to attend the 2nd International Congress of Logic, Methodology, and Philosophy of Science.

After a while we were approached by Azarya Polikarov, a Bulgarian philosopher who spoke the fifteenth century Spanish of his ancestors. He told us that during the war the Bulgarian Jews were protected by the locals because they had successfully integrated into Bulgarian society: they did not segregate themselves in ghettoes.

When arriving at our lodgings in the splendid new campus of the Hebrew University, I was feeling languid – the first symptoms of the "bug" I had caught at the dry dock in Piraeus. I was so sick that I had lost the zest for life. I was promptly interned in the university's excellent hospital, where a Yemenite female physician took good care of me, although she could not medicate me because no one knew what ailed me. But her company cheered me, although she kept complaining about the discrimination she was subjected to by her Ashkenazi colleagues. While at hospital I was visited by Rom Harré, Mary Hesse, and Imre Lakatos. All three were very kind on that occasion, and all three attacked me later when they saw me as a competitor.

When I started to recover, Marta went sightseeing on a bus tour, which proved to clinch her mathematical career. Indeed, that was her first encounter with Bill Lawvere, her future main thesis advisor, and Jim Lambek, her future academic host at McGill. Both Bill and Jim kept in touch with us during the succeeding half a century.

As soon as I recovered I started attending some talks, but my memories of them are hazy, no doubt because of my recent debilitating illness. However, I was impressed by Abraham Robinson's resurrection of infinitesimals, as well as by Tarski's lament over the multiplication of set theories and the consequent fragmentation of the foundations of mathematics. He seemed really distraught over these recent developments, which not only questioned his tidy view of mathematics, but also his own stewardship of the whole mathematical system, that seemed to be unraveling under his watch. At every break he would hold court and expatiate on his nominalist credo, which the Fefermans (Feferman and Feferman 2004) have discussed brilliantly.

That time Tarski scolded Robinson for confusing the mathematical concept of set inclusion with the ontological part-whole relation. But he himself had been guilty of far worse confusions: those between sign and concept, between conceptual and material existence, and between formal (a priori) and factual (a posteriori) truth. Tarski was only trying to practice the nominalism he professed, and consistent nominalists are single-level materialists.

My own presentation concerned the explanatory power of scientific theories. It included the unorthodox notions of partial theoretical truth (or truth in a theory) and partial empirical truth (or truth relative to a data base), both of which may be assigned a numerical value between 0 and 1. My paper was well received, and it moved Michael Polányi, the Anglo-Hungarian chemist, to approach me. We talked about the various kinds of intuition, as well as about planned science – a major source of his anti-communism. Neither of us could foresee that his son John would become a peace activist, earn a Nobel prize, or have a strong impact on my Canadian academic career. When the congress ended, we went to Newark, Delaware, the location of my next employer.

Delaware

The University of Delaware is small, has a beautiful campus, and is located in a dismal village without personality, but surrounded by great mansions of industry captains related to the huge Dupont chemical plant. There were only two places to eat: the lovely Faculty Club and a dreary pizza parlor. The University assigned us a well-equipped house next to the one occupied by an instructor whose 5-year old daughter led a band of children who ran around screaming at the top of their voices. Marta stayed most of the week at a small apartment in Philly, near Penn.

We had recovereded our "beetle," and drove it through the surroundings, sometimes as far as the highly polluted Atlantic shore. On Fridays I picked up Marta at the empty train station, and we ate fish at a good restaurant in the middle of nowhere. I returned her on Mondays. On the weekends we often did what American children do 6 h a day: we watched TV. Our favorites were the *Mary Tyler Moore* and *My Three Sons* series. Thus, we did what millions of commuters do all their lives.

On weekdays I met daily with Bill Reese, the head of the tiny philosophy department, and the other professor, Lucio Chiaraviglio. We soon became friends, and used to swim together at the gym over the noon break. Bill was religious but tolerant and affectionate, and would have liked me to stay on. Lucio, belonging to a well-to-do Italian family forced to emigrate for being liberal, was a singular type. He had grown up in Buenos Aires, and his curiosity about philosophy was piqued by a copy of Kant's first *Critique*, in his dentist's waiting room. No one believed this story. I believed it because I knew that Argentina is a land of curious amateurs.

Lucio had studied at the University of Chicago, where he had obtained three degrees: first in theology, then in mathematics, and finally in philosophy. He fell in love with molecular biology and published several mathematical papers on genetic combinatorics.

Lucio's philosophy was positivist: he had fallen for Carnap's extraordinary opinion that propositions get their meaning from empirical tests such as chemical assays. I held just the opposite thesis, that meaning precedes truth. However, this philosophical divergence was no obstacle to our friendship. We met again in Atlanta, Montreal, and Brighton, UK. I visited Lucio often, and had discussions with him and his charming wife Ida, who had transformed their living room into a greenhouse full of ferns. She had studied polítical science, and her PhD thesis was about the US Supreme Court, at that time presided over by Earl Warren, which had a progressive and activist social agenda. Ida was not interested in current events, such as the Black movement or the Vietnam War, but she would not miss a lead story of *The Atlanta Constitution* daily. I could not become enthusiastic about that aseptic politology which was remote from the real arenas of politics, the streets and the corporations boards. However, the popularity of the works of John Rawls and Ronald Dworkin suggests that this is what most political science students like: long-distance politics.

My duties at Delaware were light: one course in philosophy and another in physics. I divided the latter into two: the two relativities in the first semester, and the foundations of physics in the second. The course on both special and general relativity produced several results: a paper by Bill Sutcliffe, my best student and soon a good friend, the edited volume *Delaware Seminar in the Foundations of Physics*, and three papers of mine. When she moved to Berkeley, my daughter Silvia inherited Bill and his wife Sally, who took her to concerts and theatre performances – common entertainments in my time but rarities in the USA after the 1960s.

Bill's paper originated in a mistake I discovered in Wolfgang Pauli's famous book on relativity. Pauli had held that temperature behaves like length ("contracts" with movement), whereas we showed that it behaves like time (it "expands" with movement). The paper, Bill's first, appeared in *Il Nuovo Cimento*, the European journal of physics. The *Delaware Seminar* (Bunge 1967e) was a collection of previously unpublished papers, among them those of my friend Peter G. Bergmann – a former coworker of Einstein's – Edwin T. Jaynes and Harold Grad, both experts in statistical mechanics. There was also a paper by my friend Clifford Truesdell – the greatest expert in classical mechanics and thermodynamics, as well as the demolisher of the Suppes-Sneed-Stegmüller-Moulines "structuralist" school, which involved the confusion between the scientific and the model-theoretic concepts of a model (Truesdell 1984).

My contribution to that volume dealt with the structure and content of physical theories. In it I presented my thesis that the difference between axiomatized physical theories and their mathematical formalisms is that every primitive (undefined) concept in the former ought to be characterized by two postulates: one specifying its mathematical structure (function, operator, space, etc.), and another mentioning its physical referent (body, field, speed, stress tensor, etc.). I called *semantic postulate* this bridge between concept and physical entity or trait. I illustrated these ideas with a theory invented ad hoc, representing an entity with spin but without mass – perhaps like a neutrino.

I put my thesis to work in two papers. One of them, on physical axiomatics, was published in the high-impact *Reviews of Modern Physics* (Bunge 1967c). I presented the second paper, on my axiomatization of the relativistic theory of gravitation, at the well attended 4th International Conference on Relativistic

Theories of Gravitation, held in London in 1965, that is, in the middle of the revival (1960–1975) of that field, which followed the discovery of Mössbauer's effect and technique.

Relativities Congress

The above-mentioned relativities meeting, held at the Imperial College of Science and Technology, was attended by many stars in the field and environs, among them Peter Bergmann, Vladimir Fock, Leopold Infeld, Pascual Jordan, Roger Penrose, and Leo Rosenfeld. Clive Kilmister and I were the only philosopher-scientists present.

I asked a member of Fock's entourage, who taught at Lumumba University in Moscow, whether he or any of his colleagues were interested in philosophy. He replied: "Very much so, to the point that we hold a regular philosophy of physics seminar. But we don't admit any philosophers in it, for they had been hostile to the relativities and to quantum theory."

In one of the parties we attended, the Labour minister of science and technology told us: "The party [of science and technology] is over." That is, the government was about to cut down on British science and technology, which had remained far behind those of the most advanced nations. No one suspected that the Conservatives, who would rise to power 14 years later, would be even worse: not indifferent but hostile to learning.

My presentation of my axiomatization of Einstein's theory of gravitation was well received except by Jürgen Ehlers. He held that my semantic postulates were redundant, as David Hilbert had proved that physical content can be extracted from mathematical form. He did not bother to quote chapter and verse. But of course no such document existed: Hilbert had declared that physics is more complex than mathematics, since mathematical objects do not refer to real things, which is why they can be assigned multiple interpretations. For example, the exponential function can be taken either in itself or as representing the growth or decay of a population of some kind, depending on the semantic assumption to which it is paired.

The Carnap-Popper Debate

Right after the gravitation congress I attended the big conference in honor of Karl Popper at Bedford College, that Imre Lakatos had organized. Again, there were several heavyweights, among them Tarski, Bergmann, Carnap, Kuhn, Quine, and Suppes.

My talk discussed the differences between shallow and deep scientific theories, or those that merely organize data, and those that go beyond them, for they involve concepts that do not occur in data (Bunge 1968b). I also offered comments on some presentations, notably on Abraham Robinson's, that had given the old infinitesimals, ridiculed by Berkeley, a new lease on life.

That was the conference where two great debates took place: Carnap vs. Popper, and Kuhn vs. Popper. In the former, Popper sent his faithful pupil, David Miller, to represent him. Right at the beginning of his talk, Miller committed an error in elementary probability theory. Carnap was quick to detect it, and in few minutes he tore down the Miller-Popper criticism of inductive logic. Everyone saw this as Popper's defeat, and proof of Carnap's intellectual superiority over him. I saw it as a skirmish between two factions of the same party alien to real science.

In retrospect, I think that Popper lost that debate because he shared Carnap's belief that propositions can be assigned probabilities. And obviously Carnap had given much more thought than Popper to probability theory. But neither of them gave any reasons for treating propositions as if they were random, and none of them drew the shallow/deep distinction I had made in my paper. This distinction shows why it is impossible to generate deep theories out of empirical data – whence the utter uselessness of both inductive logic and Popper's alternative verisimilitude calculus.

Besides, both rivals dealt with single hypotheses rather than with hypothetic-deductive systems, and both believed that a hypothesis stands or falls according as it fits or fails to fit the relevant empirical data: they ignored the condition I call *external consistency*, or compatibility with the bulk of antecedent knowledge. For example, most scientists will reject out of hand any hypothesis violating energy conservation.

My verdict about the Carnap-Popper match is then quite different from the prevailing one: as the physical chemist Margot Bergmann put it to me, neither Carnap nor Popper knew what they were talking about – they were philosophers of second-hand science. Miller put his tail between his legs, and later on stretched Popper's skepticism to a ridiculous extreme. Scientists are moderate or methodological skeptics, not radical or systemic ones. For logical reasons, it is impossible to criticize everything at the same time: one evaluates *A* assuming *B*, and when questioning *B* one assumes *C*, even if only provisionally.

The Kuhn-Popper Match

There was also consensus about the Kuhn vs. Popper match. In my view Popper lost it because he shared Kuhn's false assumption that scientific revolutions are so radical, that they sweep away all past knowledge, while in real science newcomers are required to contain their predecessors as first approximations.

Worse, Popper started his talk by stating that he was not interested in what Kuhn called "normal" science: he too was only interested in breakthroughs. But this concession did not appease Kuhn. Nor did it help Karl to ask Kuhn whether he might call him Tom: Kuhn kept his cold arrogance.

Every serious historian of ideas knows that, although there are breakthroughs once in a while, revolutions are rare. In particular, successive scientific theories have partial overlaps. So much so that the first test a new theory is subjected to is to check whether it reproduces, at least to a first approximation, the true components of its rivals – the Bohr-Einstein “correspondence principle.” And a research grant application that promises a total break with antecedent knowledge does not even reach the refereeing stage. In sum, a claim to total epistemic revolution is a reliable indicator of crankiness.

This remark should suffice to shoot down the Feyerabend-Kuhn thesis about the “incommensurability” (incomparability) of rival theories like relativistic and classical mechanics. In particular, it is false that the corresponding concepts of mass are incomparable, as both refer to bodies; moreover, relativistic mass approaches classical mass as velocity drops (see Bunge 1974a). One concludes that one theory is truer than the other only because it gives comparable results sensitive to empirical data and compatible (or not) with classical electrodynamics, which Einstein had taken for granted.

In addition, the discussants in that debate had only two breakthroughs in mind, those of Copernicus and Einstein, at a time when many others were in progress behind their backs, in particular quantum chemistry, molecular biology, cognitive neuroscience, and the *Annales* historiographic school. In any event, the constructivist-relativistic denial that scientific progress is cumulative is false. Science, like capital, starts from an initial endowment rather than from nothing, and it grows through positive feedback: new findings fuel the epistemic engine. This is why total revolutions are impossible in any field.

In short, Popper lost the second debate for conceding too much to his critic, and for failing to realize that his skepticism, and even more so Lakatos’s, paved the way for the Kuhn-Feyerabend epistemological relativism. This is a form of radical skepticism, which in turn is nothing but epistemological nihilism – as the Australian David Stove (1982) realized at the time. But rational criticism proved ineffectual against the “radical philosophy” front, which joined up all the objectors to rational debate, in particular the Nietzscheans, Kuhnians, academic feminists, phenomenologists, existentialists, and neo-Marxians.

I was disappointed by the Popper conference, for it taught me nothing new. To top it all, a strong shower fell suddenly while I was watching a Shakespeare play in a park, so that I had to dry up my clothes in front of a gas radiator that swallowed one shilling per hour. But the prospect of rejoining Marta on our favorite retreat cheered me up.

To the Black Forest

At the end of the summer, feeling fit and happy, Marta and I drove to Freiburg, in the South-west corner of Western Germany, via Yugoslavia. I had obtained a generous Alexander von Humboldt research fellowship to work on the axiomatic foundation of physics at the Institute of Theoretical Physics in Freiburg.

Its director, the gentle Helmut Hönl, had studied under Arnold Sommerfeld, one of the few non-Nazi physicists who had remained in the country, and co-author of a paper that Werner Heisenberg – the Foundation's president – had used to build his matrix quantum mechanics. Hönl told me that once, during the war and in the middle of the night, he got a phone call from Hitler, perhaps because he taught at the university headed by his protégé, Heidegger the Obscure.

I chose Freiburg because it was the German university town nearest Zürich, where Bill Lawvere, Marta's chief thesis director, would stay during the 1965–1966 academic year. We made good use of that year: Marta finished her dissertation, and I wrote my *Foundations of Physics* (Bunge 1967a). I also organized a successful international symposium on the foundations of physics on behalf of the Académie Internationale de Philosophie des Sciences that was held in Oberwolfach, in the heart of the Black Forest. In addition, I made a lasting connection with the Springer publishing firm, and Marta and I were hired by McGill University, our academic home for half a century since 1966.

We rented a modern, comfortable and quiet apartment on the Schlossberg, the hill overlooking the town, just 400 m from the Black Forest, and with a good view of the Rhine plain. I worked in a small room in a semi-basement, and my office at the Institute was across from the beautiful red stone cathedral, the twin of the cathedral in the sister city of Fribourg, in French Switzerland. Every time Marta proved a theorem, I took her to eat frogs' legs at a fancy restaurant across from the cathedral.

Marta and I greatly enjoyed Freiburg and its unspoilt surroundings. Sometimes we crossed over to Alsace, to buy vegetables unknown to the Germans, still faithful to potatoes and cabbage. Once, by mistake, we took the Death Road on the Schauinsland mountain while Marta was still learning to drive. On seeing a skull-and-bones sign at a corner, she got confused and pressed the accelerator instead of the brake pedal. The car turned over, and we remained upside down hanging from our seat belts. We thought at the same time and calmly: "So, that's it." But it was not it. The windshield was splintered, and the car roof flattened on our heads, causing me a neck luxation that hurt and made a cracking sound for decades. Marta suffered not even a scratch. At my request, a group of young men came and turned the car over. The car was repaired, and Marta hired a more demanding driving instructor. Don't tell me there is no such thing as good luck.

Freiburg Friends

Our best Freiburg friend was Walter Felscher, a young mathematician specialized, like Marta, in category theory. He was married to an American woman and they had a baby girl, but it seemed to me that he treated his books more affectionately than his family. Walter worked by night in a loft, where he cooked his supper, a couple of boiled potatoes. To be able to devote all his time to research, Walter had declined all opportunities to be promoted. Therefore his salary was the lowest in his department, and accordingly he drove a modest Beetle, whereas the trigonometry professor was paid several times more and drove a high-powered Mercedes, which he washed lovingly every Saturday. (Never on Sundays. We were severely scolded, and threatened with being denounced to the authorities, if caught again doing our laundry on the Lord's Day.) At that time, as in the Middle Ages, university teachers were paid according to their number of students. When he got tired of that slave's life, Walter accepted a chair and, to recover the lost time, he quickly became a dean at a different university.

We also were on good terms with Helmut Hönl, my supervisor *pro forma*, to the point that we attended a boring play by his son in law. And occasionally we enjoyed the company of Friedrich Hayek, Siegfried Flügge, and their respective wives. When Marta learned to cook in Corfu, we invited all four for supper. As we should have foreseen, but did not, nothing came from that encounter: Germany is a land of specialists, not amateurs. When the rare generalist dares rear his head, he is called Gottfried Leibniz.

Despite his reactionary ideology, Hayek was charming, affable, and hospitable. He was a bookworm, with an enormous erudition and a sparkling conversationalist about ideas and books. He had retired from Chicago and the LSE, and, according to our common friend Popper, was waiting to be appointed rector of the Freiburg university, one of the oldest in the country. As the man of letters he was, Hayek had not noticed the leftward shift of the student body. How different might the social sciences be if their practitioners had the social sensitivity and political nose of Keynes, Myrdal, Braudel, Bloch and their disciples?

As for the friendly and placid Flügge, he had been a disciple of Max Born and specialized in nuclear physics. Like his teacher, he was knowledgeable in all the branches of physics. He had a hobby explainable in someone who had experienced so many deprivations during the decade following the war: he collected gold coins. Flügge sympathized with my scientific-philosophic project and introduced me to his last doctoral student, Gerhard Vollmer, who a few years later spent a fruitful postdoctoral year with me.

In Freiburg I had no obligations other than doing my thing and attending the occasional talk. In the departmental seminar I gave a talk on "Quantum theory without observers" that was better received by the young than by the old. (The same thing happened some years later in Copenhagen.) Even so, one of professors liked it so much, that he translated it into German. And a group of doctoral students and postdocs asked me to teach a course in the philosophy of physics. Flügge, who was then the departmental chair, appointed me a visiting professor, and I taught the course in Germlish.

At the end of the semester, my students thanked me for having allowed them to ask questions and formulate objections. When I replied that this was only normal, they told me that such a thing was unusual, even in Professor Flügge's courses. Once, when one of his students had asked a question in class, the famous professor looked at him in such a severe way that the infraction was not repeated. The student revolts a few years later swept out those cobwebs.

Marta kept working on her dissertation and meeting periodically with her advisor Bill Lawvere in Zürich. We celebrated every time she proved a new theorem, dining at a fancy restaurant facing the stunning cathedral. She always ordered frog legs, which repelled me, and the waitress felt offended when we ordered water instead of wine. One night we were spotted at that place by Heinrich Pfeiffer, the tireless and charming secretary of the Foundation that had given me the fellowship. He was always ready to help fellows in trouble, which is why I called him "the uncle and aunt of all the Humboldt fellows." Twice a year he gathered us in different university towns, giving us the chance to chat with one another and with the Foundation president, the eminent but unassuming Werner Heisenberg. Germany has never had a better cultural ambassador than Heinrich Pfeiffer.

During that year we hosted my mother, the German-Chilean logician Gerold Stahl, Marta's former classmate Murray Edelman – who took her place in the graduation ceremony at Penn – and the German-Canadian mathematician Jim Lambek, who offered Marta a postdoctoral position.

Jim, whom we had met 2 years earlier at the Jerusalem congress, had been forced to flee from Germany to England at 17. Despite being Jewish, he was regarded as an enemy alien, and was locked in a Canadian prisoners' camp along with some hardened Nazis. While in the camp, he borrowed books from McGill University, where he went to study at the end of the war. He got his PhD in mathematics, and had a brilliant academic career. (To earn a few dollars he wrote for classmates term papers in philosophy, that got better grades than his own.) When Jim visited us in Freiburg, he was taking his first steps in category theory, and had decided to set up a Category Center at McGill, an aim that he fulfilled. Over a couple of decades Montreal was a world center in category theory.

The Two Berlins

In West Berlin we watched the intelligent and funny political cabaret *Die Stachelschweine* (The Porcupines), that made fun of the day's politicians, many of whom had done very well under Nazism. For example, the president of the Federal Republic had made millions building concentration camps.

My day in East Berlin was memorable. In contrast to the western half of the city, the eastern one had not been fully rebuilt. In particular, the holes made by the Soviet artillery had not been filled. As a compensation for austerity, there seemed to be a bookstore at every corner. Fruit was scarce, but the cafés served delicious if fattening *Kuchen*.

When I went to the Academy of Sciences to visit the philosopher Georg Klaus, a *rara avis* among Marxists, I was kept in an office while inquiries were being made. Once in a while a curious head showed up, and I heard that the employees addressed one another ceremoniously as *Herr Kollege* or *Frau Kollegin* rather than *Genosse* (comrade). I also noted that the East Germans had rehabilitated many foreign-sounding words like *Telephon*, which under Nazism had become *Fernsprecher* (far-distance speaker).

After waiting for more than an hour I was told that a car was to drive me to Herr Professor Klaus. This was a mathematics teacher, chess champion, communist militant, and Marxist theorist who wished to update Marxism. In his writings he used and defended mathematical logic and cybernetics. His was a hard fight, because the conservatives were protected by the Soviet ideologues. But Klaus persevered and finally won: he was promoted to the highest academic ranks, and his writings were disseminated. I admired Klaus for his intelligence, culture, courage, and integrity, which had won him 5 years in a Nazi concentration camp. The Spanish Marxist who most resembles him was Manuel Sacristán, who translated my *Scientific Research*.

When I visited him in his house in a pleasant village, Klaus moved about in a wheelchair, spent several hours a day connected to a dialysis machine, and swam by medical order in a swimming pool built for him in his basement. But he had kept his lucidity and sense of humor, so that we had an interesting and lively dialogue. After a couple of hours Frau Klaus came with coffee and *Kuchen* that, of course, he was not allowed to touch. During the long journey back, which bordered the lakes that Einstein and Brecht had loved, the driver complained that the Black Sea brothers preferred selling their grapes to the West rather than to East Germany.

Portugal

The 1965–1966 winter was exceptionally cold, but we sledded in the forest and drove up to sky resorts to watch. At the height of the season we traveled for a respite to warm Madeira. In Lisbon we were kept waiting for a long time by PIDE, the much-feared political police, because we had no visas. After many consultations, our Argentinian passports were stamped.

The once-proud imperial capital, much deteriorated after decades of the colonial wars that Oliveira Salazar' fascist dictatorship had been waging with NATO weapons, seemed to us shabby, old fashioned, and sad. The booksellers looked at us with surprise and suspicion when we asked for works by the great anti-clerical novelist Eça de Queirós, and we got indigestion from a cup of hot chocolate with *churros*. But we were impressed by the views of the estuary of the majestic Tagus, and moved by the *fados* sung by Amália Rodrigues, the *Rainha do Fado*, whom we listened to again in Montreal some years later.

We travelled through Madeira and swam in its angry ocean. We were amazed by the large number of villages sandwiched between tall mountains, which got the sun's rays for only a couple of hours a day. By contrast, the sun is visible all day from the wide and beautiful bay of Funchal, the island's capital. On Year's Eve Funchal was lit up by fireworks. No one seemed to remember the unfortunate soldiers who were fighting the Angolan guerrillas and the Cuban soldiers in the remote African forests.

To Heaven and Back

In the Spring of 1966, Marta and I completed our Freiburg projects: she her PhD thesis, *Categories of Set Valued Functors*, and I my book *Foundations of Physics* (Bunge 1967a, 1969a). As agreed, I mailed my typescript to Clifford Truesdell for his Springer series. Truesdell accepted it right away. Dr. Helmut Mayer-Kaupp, the editorial director of that prestigious academic publisher, invited me to his office in Heidelberg, and we signed the publication contract without further ado. I seized this opportunity to leave with him the bulky typescript of my *Scientific Research*, and proposing that he start with it a new series, to be called *Studies in the Foundations and Philosophy of Science*. This project interested the Springer bosses, and shortly thereafter they invited me to visit them again. This time Mayer-Kaupp invited me to lunch at a fancy restaurant in the company of the business director of the firm. I was in my glory.

My euphoria did not last, however: a few weeks later I got a phone call from Mayer-Kaupp, asking me to see him at his office to discuss an important issue. As soon as I arrived, Mayer-Kaupp informed me in a glacial tone that our contracts were null and void, and that the lead plates for both books had already been melted. He also told me that my stand was was like that of an abandoned dog. A few days later Clifford Truesdell, who edited a book series and two important journals for Springer, wrote me that he had resigned his positions. What had caused these unexpected changes? The arrogance and pedantry of Springer's new technical consultant, a former student of the logician Hans Hermes. That individual had stated that I did not understand the concept of a relation, just because he had learned the textbook definition of a binary relation as a set of ordered pairs, whereas in my opinion this defines only the graph, extension, or field of validity of a relation, while ignoring its intension, sense, or content. Thus, to my critic "heavier than" and "more expensive than" are the same relation $\geq$, whereas to me they only have the same logical form. My critic, a formalist (or structuralist) – as befits a logical imperialist – overlooked senses or contents, the way the structuralist school of Patrick Suppes, John Sneed, Wolfgang Stegmüller, and their Spanish-speaking followers had done.

In the factual sciences there is no form without content. This is why the same mathematical constructs, such as the logarithmic and trigonometric functions, appear again and again in different branches of science, although with different senses. This is why the structuralist (also misnamed semantic) school is just an academic industry that has contributed nothing but confusion to the foundations and philosophy of science, in particular to the debates over quantum mechanics. They deserve the cynical definition *La filosofía è quello con lo quale, e senza lo quale, il mondo resta tale e quale*. ("Philosophy is that which, and without which, the world stays just the same.")

Ultimately my critic lost the battle, not because I was given the opportunity to debate with him, but because of an unrelated event. I had organized the 1966 colloquium of the Académie Internationale de Philosophie de Sciences, to which I belonged, and had proposed that it should deal with the foundations of physics, and be held at the house that the university kept in Oberwolfach, in the heart of the Black Forest. I obtained the university's permission and, together with the Academy's secretary, the Flemish priest Stanislas Dockx, we visited the place, alerted the local inn-keepers, and booked the bus that would take us to the source of the Danube at the end of the meeting.

Nearly all the people whom I approached accepted our invitation, and a few others sneaked in. The list of invited speakers included Peter G. Bergmann, Joseph M. Jauch, Henry Margenau, Patrick Suppes, and above all John Archibald Wheeler, famous for his spectacular diagrams and his philosophical excentricities. Other notables, like Paulette Février, Günther Ludwig, Friedrich Hund, Henryk Mehlberg, André Mercier, and Erhard Scheibe, confined themselves to making brief comments. My communication, "Ghostfree quanta", was commented on favorably by various colleagues.

Back in Heaven

To my surprise, Springer's Mayer-Kaupp attended all the lectures, and met privately with a number of participants. He watched me attentatively when I conversed with colleagues. A few days later, out of the blue, he phoned me, asking me to see him at his Heidelberg office. When I arrived he told me that, having noted the respect that the Colloquium participants showed me, and the familiarity shown by some of them, he had resolved to backtrack: our contracts recovered their validity, both books went back to the printing shop, and I was free to commission the following volumes of the new series I had proposed. Thus, my good luck had returned.

But I still had no job for the coming year, so I asked for a renewal of my Humboldt fellowship, and got it. Actually I did had an offer: the one Yale had made me earlier. But neither Marta nor I were tempted to reside in a country whose government kept burning Vietnamese civilians with napalm. Besides, Marta had just accepted a postdoctoral fellowship at McGill, so I had the duty to follow her. But how to get a job at the same university? I had no contacts with any McGill philosophers, or so I believed.

Looking for a Job

Marta and I completed our respective projects at the same time, and decided to reward ourselves spending a few days in what Goethe had called "the land where lemons bloom." We packed our Beetle and drove to Florence, to visit Michele Ranchetti and his family, whom we had befriended in Corfu a couple of years earlier. Michele and Hazel Ranchetti and their three young children resided in a large, old and dilapidated house on the slope of a small valley from which one could see the house where Galileo – the earliest martyr of philosophical realism – had been forced to spend his last years.

Michele earned his living teaching Church History at the university, but Hazel confided in me that she still did not know whether he believed in God. What Michele really liked was to write poems, make small delicate drawings, read Freud and Wittgenstein – still novelties in Italy – and entertain friends. Their house was in the middle of an orchard cultivated by a sharecropper, who was also in charge of a cow and a pig. No timetables were kept at the Ranchettis's, except for those of the schools. Michele's income was never enough, for he could not afford not to buy the latest book that caught his attention, nor could he deny hospitality to anyone. Like most of his colleagues, Michele supplemented his university salary doing small jobs for several publishers. One evening, when Marta and I were staying with him, there came the editorial director of La Nuova Italia Editrice, an old and prestigious academic publisher. In the course of the conversation he said that his latest publishing adventure was with Raymond Klibansky, a Renaissance scholar who chaired the Department of Philosophy of McGill University, the same where Marta was expecting to go next fall. I asked him for Klibansky's coordinates, and wrote him asking whether he might have a job for me.

Canada Calls

Klibansky's answer came immediately and it astonished us: he had known about me since my journal *Minerva*, and he had even kept a letter of mine duly crossed through by the war censors. He arranged to have me invited to visit McGill, the oldest and most prestigious Canadian university, where Ernest Rutherford – the father of experimental nuclear physics; Hans Selye – the discoverer of stress; and Wilder Penfield and Donald Hebb – pioneers of cognitive neuroscience – had worked. Besides, McGill's nuclear physicist Robert Bell had just been named a fellow of the Royal Society of London.

When I arrived in Montreal I was taken to meet Dr. Rocke Robertson, the principal, and two deans, all of whom seemed eager to hire me. I was also invited to deliver two lectures, and to dine at the best restaurant in the company of Charles Taylor, at the time a rising philosophical star and a socialist activist. Jim and Hannah Lambek drove me around in her red Volvo to see the sights, and I walked around the town, which did not impress me.

After a couple of days I was offered the chairmanship of the philosophy department. I declined this offer, stating that I wished to devote myself to research, but added that I would gladly come as an ordinary professor. My counter-proposal was accepted, and I was assured that they would expedite the lengthy immigration process. I was happy at the prospect of joining a research university as well as the cradle of hockey, the national sport. Besides, McGill was a private university, hence without political strings, and it wanted both Marta and me. Still, I had doubts: Montreal seemed ugly and cold by comparison with Freiburg, I hardly understood the local *patois*, and the recent terrorist attacks – in particular the bomb that exploded in a mailbox – frightened me. On top of all this, a Swiss psychologist was trying to get a chair for me in his country, which Marta and I loved.

To clear my doubts I went to Boston to ask for Quine's opinion, and to London to ask for the Poppers' advice. Quine thought I should go to McGill, since I would be able to work on my projects. Along the way we chatted about philosophers with a program, like himself and Carnap, and those who, like his colleague Hilary Putnam, jump from branch to branch, without mastering any subject in detail and depth. I also met with Robert S. Cohen who with Marx Wartofsky was the tireless organizer of the Boston Colloquium for History and Philosophy of Science and who held a joint appointment in both the Philosophy and Physics departments. In Boston I attended the funeral service for the physicist and philosopher Philip Frank, whose *Foundations of Physics* I had adopted for my seminars for physics students in Buenos Aires and La Plata despite my disagreements with it.

As soon as I arrived in London I went to visit the Poppers. Karl did not show any interest in my problem, for it did not regard him. (He never once helped me get a job or a publisher.) By contrast, Hennie asked me the key question: What do you expect from Montreal? I answered that I expected the best from both America and Europe, to which she replied: What if you find the worst from both? This worried me, but the die had been cast: Canada called us, and it was not at war. However, Marta and I needed some relaxation, and so we drove to Spain for a short holiday.

First Visit to Spain

Before crossing the Pyrenees we bought copies of *Le Monde*, *Le Nouvel Observateur*, and other liberal publications, that we assumed were banned in fascist Spain, and hid them under the Beetle's carpet. We were amazed to see them on sale in Spanish news stands, presumably for the hordes of French tourists who, involuntarily, were doing something much more subversive: driving their own small but sturdy Citröen 2CVs for their long paid-holidays.

In Barcelona we visited our old friend Alejandro Argullós, who showed us the new plant of *Editorial Ariel*, and introduced us to Manuel Sacristán, the unemployed Marxist philosopher, with whom we sympathized instantly. Manolo, whose hair was cropped, turned out to be a friendly and open-minded fellow, who accepted to translate my *Scientific Research*, a task that he performed to perfection. I offered him help to obtain a Humboldt fellowship, so he could to do some research and breathe freely for a couple of years, but he firmly declined my offer, saying: "My place is here. I won't move until we defeat fascism." Neither of us could foresee that fascism would fall 10 years later without a fight. I lost touch with Manolo, except for a brief encounter at a philosophy congress in Morelia. From Barcelona we drove southward looking for a chalet on the sea. We found it only in Altea, an ancient Greek colony on the White Coast. It was an isolated place, where at night we heard the howling of foxes. The beach, of gray pebbles, was deserted and inhospitable, but the Med was as lovely as ever. By night we tuned on the Albanian radio station, which played the best classical music. We experienced only one scare: when visited by a couple of Civil Guards, curious about a couple who stayed away from the tourist masses. We learned much later that our homeland had just suffered a military coup, and that the police had occupied the Buenos Aires Faculty of Science and badly beaten up its professors.

Piaget and Kuhn in Geneva

Back in Freiburg I got an invitation from Jean Piaget, with whom I had interacted at several academic meetings. He invited me to attend a conference on causality. One of the participants was Thomas Kuhn, then basking in the success of his *Structure of Scientific Revolutions*. His presentation disappointed me for exhibiting his poor knowledge of the history of science. For example, I had to remind him about Johannes Philoponos's theory of movement. According to Philoponos, an arrow keeps moving as long as it retains some of the *impetus* that the archer had imparted to it. The rule of this theory from the sixth to the sixteenth centuries shows both the power of causal thinking and the artificiality of science – surely a subject of interest to a colloquium on causality. Kuhn's presentation impressed no one at the meeting, and it confirmed my impression that his history of science was second-hand, his philosophy confused and backward, and his sociology of science non-existent. Fortunately, there were also a few good presentations and, of course, Piaget's interesting observations throughout.

When moving away from familiar places and bonds, one is let free to imagine new projects. This time, in the long but comfortable journey from Freiburg to Genève via Basel, I conceived the subject of my next long-term project: the semantics and ontology of science. That is, I would attempt to exactify and systematize the concepts of reference and sense, which I had used intuitively, and dig up the tacit metaphysical hypotheses of scientific research, such as "The universe precedes the knowing subject." In fact, this project kept me busy over the next two decades. There is nothing like toiling on a vast, intriguing, and incomplete project to live a happy and long life.

Chapter 8
Canada and Scientific Realism

Montreal

We flew from Frankfurt to Montreal on November 27th, 1966. Our arrival coincided with the first snowfall of the season. With our Delaware savings we rented, furnished, and equipped an apartment on the 20th floor of a newly but shoddily built tower near McGill University. From it we could see the mighty Saint Lawrence river and beyond, down to Vermont's Blue Mountains.

My university office was located in the newest building of the university, named after Stephen Leacock, a professor of economics far better known for his funny novels, which he produced at the rate of one per year, than for his economic opinions, now as forgotten as will be those which are now winning Nobel prizes. My university building was modern and elegant, but built with such cheap materials that I overheard the whispers of my neighbors, in particular those of Alistair McKinnon, our Kierkegaard expert. Although he did not know Danish, Alistair employed several computer technicians to build a concordance of Kierkegaard, the precursor of existentialism, that long lamentation of human misery. There I installed my German typewriter, which was promptly stolen. A month later arrived most of the books that I had left in Buenos Aires, and gradually came many new ones, that I could finally afford to buy. I was also reunited with my faithful filing cabinet, which had followed me to Delaware, and where I filed some of the letters that took 20 % of my time. My mail in four languages took so much time that once I decided to write under the pseudonym Raimo Begun, but my publishers objected.

Marta and I wished to integrate in Canada, but this was not easy to achieve in a province bitterly divided into francophones and anglophones, which led us to initially associate with fellow allophones. We interacted mainly with Stelios, a Greek mathematician, and his American wife Joan, a couple of young Romanian mathematicians, Norbert and Dana, and another of Argentines, Bernard and Diana. We also visited with Jim and Hannah Lambek, as well as with the Franco-German Klibansky and his entourage, that included the linguist Michel Paradis and the

M. Bunge, *Between Two Worlds*, Springer Biographies,
DOI 10.1007/978-3-319-29251-9_8

princess Nina Troubetzkoy – the widow of the famous linguist's brother – the most interesting and helpful member of Klibansky' circle. All of these persons regarded themselves as citizens of the world and therefore above the hateful and wasteful Franco/Anglo split. That's where Marta and I belonged at that time: right on the fracture, and therefore doubly strangers, and looked upon with suspicion by the more fanatical of both linguistic bands. However, we slowly integrated into the Québec society, all the more easily because we were poliglots – unlike the vast majority of our colleagues.

Raymond Klibansky

Raymond Klibansky, my senior colleague and the one who had done the most to get me hired by McGill, was an internationally known expert in Renaissance philosophy and science, and founder of the local chapter of the Canadian Society for the History and Philosophy of Science. He had been born in Paris into a family of German Jews, but when the First World War broke out, he and his family were deported to Germany, because his parents were German nationals. It did not help him being named Raymond in honor of the Republic's president, nicknamed "Poincaré la guerre." The child Raymond was enrolled in a famous progressive school, where he had classmates who, would later distinguish themselves, like the Nobel laureate Hans Bethe. When the Nazis seized power, Klibansky, who was then an assistant at the famous Heidelberg University, helped a number of colleagues emigrate, the way Hannah Arendt did but Isaiah Berlin did not. He himself narrowly escaped at the last minute, first to France and then to Oxford, where he got a minor job. When the war started he and several of his fellow philosophers were recruited by the British Secret Service, where he attained the rank of colonel. Several other British philosophers were assigned similar tasks, but Gilbert Ryle was the only one who kept a military demeanor all his life.

No one knows for certain what those many dons accomplished, but Freddy Ayer confided to me that his single contribution had been to catch a spy, a simple sailor stationed in Gibraltar. When Italian fascism fell, Klibansky was sent to Italy, where he got hold of Mussolini's memoirs. He spent the rest of his long life in Montreal, where he taught at McGill and at the Université de Montreal.

Klibansky kept fit, had Old World manners, dressed elegantly, favored blue bow ties that enhanced the blueness of his shiny eyes, and cultivated a long mane that earned him the nickname "The mad Russian" among the students. Although he enjoyed feminine company, Klibansky dodged successfully the siege of at least one of his female graduate students, who saw him as a good catch: bachelor, good company, and wealthy. But as an octogenarian, he surprised us all by marrying his old friend Ethel Groffier, a refined, friendly and quiet Belgian legal scholar and translator who saw to his health and kept house for him till the end. Raymond gave us another surprise when he sprang on us his Italian daughter, a secret souvenir of his days as a British agent. She had inherited his manly profile, hookish nose included.

Klibansky did not publish any original papers or books, but he superbly edited a number of collective works, and had a good nose for identifying and helping productive scholars and promising students. In his large office full of valuable books he offered tea to his visitors from many countries, and showed them his numerous trophies, prizes that kept coming long after his main work had been done. Klibansky spent every summer at his old Oxford college, looking for rare books and manuscripts. He was also very active at the exclusive Institut International de Philosophie, where he soon got me invited, and at whose annual *entretiens* I met many interesting philosophers. The Institut sponsored Klibansky's multi-volume work *La philosophie contemporaine*, which included several papers of mine and of a number of scholars I had recommended.

Klibansky trusted my judgment, and I consulted him on a number of obscure passages in ancient and medieval texts. He tried unsuccessfully to involve me in departmental politics, and introduced me to a number of interesting characters, such as Freddy Ayer, Max Black, Guido Calogero, Leszek Kolakowski, and Chaim Perelman. One night, just disembarked from England, Klibansky phoned me announcing a sensational fact that demanded a radical revision of the foundations of physics: his Oxford friend Alan Montefiore had just informed him about the spoon bender Uri Geller, the Israeli stage magician who would soon be debunked on TV by James Randi. But on his own turf Klibansky was skeptical and more outspoken than most of his colleagues. For instance, when the fourth centenary of Giordano Bruno's murder came in 2000, he was the only Montrealer to commemorate it, inviting a number of people to listen to his scholarly talk on the martyr, and raise our glasses to his memory. Klibansky was then 95, and he lived to be a few weeks short of a century.

Walter Hitschfeld, Scientist and Administrator

I often had lunch at the Faculty Club or at one of the ethnic restaurants nearby. At the Club I met some like-minded colleagues, such as the meteorologist Walter Hitschfeld. While Dean of Graduate Studies, Walter supported my research and sided with me against Robert Vogel, the Dean of Arts – a funny light-weight historian – in our effort to appoint a serious historian of science, such as my candidate Stephen Brush, who had started as a physicist. Walter also helped finance some of the meetings I organized, such as the one where the Society for Exact Philosophy was launched in 1971, and the large meeting on the mind-body problem, which the psychologist Dalbir Bindra and I organized in 1979. That same year he presided over the dinner that my colleague William Shea organized to celebrate my 60th birthday. When it became known that Walter's pancreatic cancer would allow him only a few more months, the university organized a goodbye party for him. Walter went around the large hall cracking jokes while many of us could hardly keep back our tears.

Storrs McCall: Scholar and Farmer

Another convivial luncheon companion was Storrs McCall, a gentleman who, in addition to teaching philosophy, owned and managed a cattle farm in the Laurentian hills. He had studied at Oxford, was very learned in arcane nonstandard logics, and genuinely interested from afar in physics. Storrs' contribution during the last half century was his "branching model of the universe," where every point in an empty and undefined Cartesian space was assigned a fork representing events with complementary probabilities. He also thought that time flows like a river, and when asked what was the speed of time, he answered unfazed: "one second per second."

When the separatists came to power in the province of Quebec, the most fanatic among them introduced mean laws restricting the rights of the anglophone minority, such as banning commercial signs in English, and confining the attendance of anglophone schools to the children of Anglo-Canadians. Storrs, who had not bothered to learn French, was incensed, and organized AQ (Alliance Quebec), a resistance group. I accepted Storrs' invitation to join the governing board of AQ, for I thought that it was criminal to break up a nation that worked quite well just to satisfy the desire for revenge of the francophone majority, whom the Anglos had exploited and despised in the past. We met several times and talked a lot but accomplished nothing: none of us had any political skills or connections, much less the will to devote ourselves fulltime to politics. My only contribution was to debate a few separatists on local TV and radio stations.

In the meantime the separatist government grew weaker because it accomplished nothing constructive for its constituency, and eventually it was replaced by the Liberals. I had joined the Liberal party when it was in the opposition, but left it when it shifted from what Canadians call "progressive Conservatism" to the conservative policies typical of the so-called neoliberals. Before joining McGill, Storrs had taught in Kampala, Uganda, until its ferocious dictator Idi Amin – accused of anthropophagy – expelled all the foreigners. Anne, Storrs' charming wife, had painted in Africa with lively colors; when back in Canada she only used whites.

William Shea: Catholic Historian of Science

A closer colleague and good friend was William R. Shea. Bill had studied chemistry, theology, and philosophy, specializing in Galileo and Descartes. He tried to reconcile his strong Catholic faith with his admiration for Galileo and Descartes. Unsurprisingly, his 2006 book on Galileo with Mariano Artigas, a member of Opus Dei, fails to tackle the incompatibilities between nature and the supernatural, law and miracle, and between research and faith. Bill also failed to realize that, as I argued in the Vatican colloquium on cosmology in 1991, Galileo's realist philosophy was closer to Aquinas's than to Bellarmino's conventionalism, which the Church had adopted to avoid taking sides in the geocentric/heliocentric debate.

I claimed that the infamous Galileo trial would not have happened if the Church had remained faithful to Thomas. The audience, which consisted almost exclusively of priests and novices, warmly applauded my words. In his big book on Descartes, Bill managed to suppress the materialist side of the great "masked philosopher" and painted him as a good Catholic – although he first fled to Calvinist Netherlands and later to Lutheran Sweden for fear of the Inquisition.

Three Psychologists: Hebb, Milner, and Bindra

The colleagues who taught me the most were the experimental psychologists Donald Hebb, Dalbir Bindra, and Peter Milner. All three worked on important problems, held the psychoneural identity hypothesis, and designed ingenious experiments. Hebb taught me through his books and papers, starting with his groundbreaking *The Organization of Behavior* (Hebb 1949). He was the first to advance a specific hypothesis about the neural mechanism of learning, which inspired an avalanche of experiments, starting with his own on sensory deprivation – an early confutation of behaviorism. While I directed the 'Foundations and Philosophy of Science and Technology' series for Pergamon Press, I persuaded Henry A. Buchtel (1982) to collect Hebb's papers scattered among more or less obscure journals.

Dalbir Bindra, aka D.B., born in the Punjab, had studied at Harvard and continued Hebb's work at McGill, as did his wife Jane Steward at Concordia University, who specialized in the physiology of motivation. I often consulted D.B., he included me in his faculty course on pseudoscience, and in 1979 we organized together a very well attended symposium on the mind-body problem, where Hebb acted as the commentator at large. The book that resulted from this meeting (Bindra 1980) contains two pieces of mine: the Introduction and the chapter "The psychoneural identity hypothesis." D.B. died unexpectedly, possibly from a massive stroke. At his memorial service I read a poem by Omar Khayyám, D. B.'s favorite poet.

Last, but not least, I greatly benefited from Peter Milner's patient criticisms while writing my *Philosophy of Psychology* (Bunge 1987a). Like D. B., Peter was friendly and cooperative as well as smart. He and James Olds, a postdoc of Hebb's, had become instantly famous in 1954 when discovering the reward (or pleasure) center in the rat's brain – a serendipitous finding that should have given pause to the psychoanalysts. Peter was far less famous than Brenda, his first wife, but he was also friendlier, at least to me.

Other Valuable Colleagues

Another of my most original and helpful colleagues was Michel Paradis, who devoted his life to studying bilingualism after getting his PhD on philosophy under Klibansky. Michel started by visiting Arab patients interned in Parisian hospitals, and ended up by studying the same subject elsewhere. This problem intrigued him for being at the intersection of linguistics, psychology and neurology. A bilingual subject who becomes aphasic as a result of a brain lesion may recover one or two of his languages. Which one if any will return first, depends on both the lesion site and the manner of learning – e.g., whether at school or at work – which corroborates at once the old hypothesis of the localization of mental faculties and the conjecture of the social embeddedness of speech.

Michael Brecher, a political scientist specialized in armed conflicts, has paid close attention to numbers and sought to find laws of international conflicts. Although a Zionist, he is also a scientist, so he admits many of the errors and crimes of the Israeli governments from their beginning in 1948. Einstein too was a Zionist, but he warned his friend Chaim Weizmann to go to Palestine as cooperant, not as conqueror. Peter Hoffmann, our most distinguished historian, is the world's expert in the anti-Nazi underground movement during Hitler's rule. His book on the Count Claus von Stauffenberg, the crippled and clumsy officer who headed the failed plot against Hitler in 1944, reads as a combination of Greek tragedy with thriller, but it took years of archival search in several countries.

Michael Mackey, a physicist who has worked mainly on mathematical biology and medicine, is a pioneer in the application of chaos theory (nonlinear dynamics) to biological facts such as arrhythmia and the production of red blood cells. When a Chinese student came to do his graduate work with me, I recommended his bright wife, a biologist, to study with Michael. She did very well, whereas her husband, whom I had suggested to look at chaos theory through a philosophical lens, failed to have any original thoughts. He had been taught to admit, not to question. His gullibility was such that he believed the ancient Chinese theory of the five elements – the fifth being wood.

Another McGill notable was Bruce Trigger, the first Canadian anthropologist and archaeologist to study Canadian natives instead of going to other continents. Bruce was a gentle man with whom I shared much philosophy, in particular scientism, realism, and materialism (Trigger 2003; Bunge 2013b). But, of course, Bruce did not ignore the importance of ideas, particularly when combined with power, as in ancient Egypt. That had been Bruce's first subject. When he returned to Egypt after several decades, he found that the *fellahin* interrupted their agricultural tasks at 6:00 p.m. to watch the *Dallas* TV series.

Claudio Cuello and the Status of Pharmacology

Two decades after Marta and I joined McGill, came Claudio Cuello, a pharmacologist who had studied with luminaries like Leslie Iversen, and had collaborated with César Milstein, a compatriot who earned a Nobel prize after having been rejected by his fatherland. Claudio is bright, witty, hardworking, and amiable. He is also remarkably adaptable. In Buenos Aires he had attended the military lyceum, studied medicine, and joined the Argentine base in the Antarctic. There he made his scientific debut by studying the seal's hypophysis (pituitary gland) – which earned him the nickname "Mr Seal." After trying unsuccessfully to continue his scientific work in Buenos Aires, he went to Oxford, where he became an ardent anglophile and got a doctorate in science as well as a British passport. After a few years, he got tired of depending on "soft" funds, and came to McGill to chair the Pharmacology Department, which he upgraded without making enemies. In fact, he could claim to be 'The Man Without Enemies.' I don't envy him.

With Claudio we have been arguing over three decades about the status of pharmacology. Whereas he claims that it is a basic science, I argue that it is an applied one, that is, a bridge between basic science – the search for truth –and technology – the design of artifacts, from rockets, drugs and therapies to legal codes and formal organizations. But Claudio, a laboratory man, is impatient about definitions, so our controversy is unlikely to be productive.

Bernardo Dubrovsky and the Two Psychiatries

For several decades our closest Montreal friends have been the fellow *porteños* Bernardo and Diana Dubrovsky, he a neurochemist and psychiatrist, and she a mathematics teacher. Bernardo had studied medicine at the Buenos Aires University during one of its rare good periods. At McGill he made the mistake of joining the Psychiatry Department, which was then dominated by psychoanalysts, and is now under the influence of "transcultural psychiatry," whose members neither care for the brain nor see patients.

Bernardo taught a few crash courses far from mainstream psychiatry, and has kept the only one-man lab on campus. We have had many interesting discussions over four decades, particularly on the division between scientific and literary psychiatry, and on the relevance of "mapping the mind on the brain," or localizing the mental functions. Many brain scientists and psychiatrists have rejected this work just because the phrenologists' localizations had been speculative. But nowadays such localizations are performed on the strength of electrophysiological experiments like those that brought Wilder Penfield and his Montreal Neurological Institute international reknown, and the brain imaging made possible by the MRI machines.

There are still neuroscientists, like my friends Rodolfo Llinás and Vernon Mountcastle, who disparage this kind of work. Bernardo shares their skepticism, whereas I am enthusiastic about it for two reasons. One is that the first thing we must find out about any fact is where and when it happens. The other reason is that localization refutes both psychoneural dualism and neural holism, the idea that the brain or even the whole body, acts as a unit.

It has been known for more than a century that the main organ of vision is in the back of the brain, that language is produced mainly by the left cerebral hemisphere, that anxiety and fear are functions of the cerebral amygdala, that disgust occurs in the insula, and so on. What is true is that none of these organs acts alone: every one of them interacts with some others, just as, although we walk with our legs and not our arms, legs would not move without the brain, the heart, etc. In short, there is localization without autonomy: the brain is a system of systems, not the collection of independent modules imagined by some philosophers and pop evolutionary psychologists (Bunge 2010a). Although I have discussed the above problems and their cognates with Dalbir Bindra, Rodolfo Llinás, Peter Milner, Ernst Pöppel, and other professionals, Bernard Dubrovsky is the one who has paid more attention to my questions and conjectures about the mind-body problem.

Less Interesting Colleagues

Because of my scientific background, I have felt very distant from my younger colleagues, nearly all of them specialized in one or two philosophers, and none of whom ever taught me anything. For instance Harry Bracken, a Berkeley expert, atheist and crusader for some good causes, did not even attempt to refute Berkeley's absurd if clever ideas. His protégé John Trentman was an expert on St Francis Xavier, more of a martyr than a philosopher. For both of them philosophy was a profession rather than a calling. But both were good persons and neither of them obstructed my work.

By contrast, David Norton, Harry's other protégé and fellow American, was enthusiastic over David Hume, but did not realize that, while his religious skepticism was progressive, his phenomenalism and corresponding rejection of Newtonianism were reactionary. Worst yet, David held that the philosophy of science should be replaced by the history of the discipline, and during his chairmanship he advertised a position for this unborn entity, for which no candidates showed up.

The fourth American, Jim McGilvray, peddled eggs from his farm, and eventually became the world expert on Chomsky's philosophy – whose existence is yet to be established. Jim joined Marguerite Deslaurier in preventing some students from doing their graduate work with me, as well as in forcing me to retire me as soon as I turned 65 – until a new dean reinstated me 5 years later.

Stephen Menn, the fifth American, who had been a precocious mathematician, specializes in arcane subjects, such as the Augustine-Descartes connection. He feels so superior that, to the amusement of the secretaries, he refused to return his colleagues' greetings. Another hostile colleague was Marguerite Deslauriers, who, after failing to become an Aristotle expert, found her gold mine: feminine philosophy, a favorite with girls without any serious intellectual interests other than their fear of science – that well-known tool of male domination.

Another stone on my road was George Di Giovanni, who had written about Hegel's dialectics in terms even more obscure than his hero's and led his only graduate student to write his dissertation on the religion of Kant, who in his first *Critique* had stated that "God is a mere idea." George denied me merit increases throughout his chairmanship. I suspected that his motive for doing so was that I violated his rule

"Publish *and* perish." To test this hypothesis, I once refrained from submitting my annual report on publications, although I usually published more than all my colleagues put together. That was the year that George gave me a merit increase. He was eventually disciplined for having upgraded the academic standing of a female student in exchange for sex.

Finally, there was Philip Buckley, who had studied Husserl in Louvain and written a book on his antiscientific philosophy, phenomenology. He introduced existentialism to McGill, forcing the philosophy students to read *Being and Time*, which Heidegger had written at full speed to qualify for the Freiburg chair he coveted. I used to denounce this book as one of the worst academic swindles of all times: its original sentences, such as "The essence of truth is freedom," are nonsensical, while the meaningful ones are either trivial or false. Unsurprisingly, Phil reciprocated during the 10 years that he occupied the chair to the applause of all my colleagues.

Last, but not least, there was Charles Taylor, aka Chuck, the most famous of Canadian philosophers, who had helped expedite my immigration process. For a while Chuck was a member of the Political Science department, but data collecting was not his thing. Since he was more of a political essayist than a political scientist, he transferred to our department. Chuck was a member of Dilthey's *Verstehen* ("interpretation") school and, like the structuralist Claude Lévi-Strauss, he believed that "social facts are languages or like languages." He was a popular teacher and a public intellectual, who tried to calm the Quebec political waters. But he lacks a coherent philosophy of his own, for he has tried to reconcile Socialism with Catholicism, and Hegel with Wittgenstein. The only constant along his intellectual development has been his antiscientism. But he is an honest person and has embraced some good political causes.

In short, none of my younger colleagues taught me anything. Fortunately, there were plenty of serious colleagues in other departments.

Integration

Although anglo-Canadians are notoriously reticent and frosty, initially we had cordial relations with all our colleagues. They gave us a chance to prove that we had been good acquisitions even though we came from an obscure place and would never learn to behave like self-effacing beavers.

No sooner did we settle in our Danish-style apartment, than the University appointed me as its representative in the assembly of Quebec university professors charged with designing the syllabi of the new *collèges*, or preparatory schools bridging the high school to university gap. I may have been chosen by default, just because I was one of the very few multilingual McGillians.

It was rumored that the real goal of those new schools was to provide jobs for the thousands of priests and nuns who were leaving the monasteries and convents in droves and in a hurry – a smooth process that took only a couple of years. At the same time, the francophone universities were being secularized. This was only one of the many facets of the radical religious reformation led by Paul-Émile, cardinal Léger, one of the most enlightened prelates in history.

Another leader of that cultural transformation was Jean-Paul Desbiens, an intelligent and amiable Marist priest who had studied philosophy. He had written *Les insolences du frère Untel* (1960), a bestseller that infuriated the Quebec nationalists for stating that the provincial public education was poor and for making fun of *joual.* This garbled speech had evolved when the British invasion provoked the exodus of the francophone intelligentsia. Desbiens, who knew that the anglo-Canadians did not face these problems, took me aside and advised me to try and persuade the McGill administration to set up their own *collège*. He even suggested where to place it: at the high school across from the university.

I persuaded the assembly to adopt a philosophy curriculum for the anglophone colleges different from that for the francophone ones, for fear that the latter would embrace either Thomism or the current French irrationalism. My plan for the 2 years included informal logic, analytical philosophy – which should be taught in elementary school, for it only seeks to speak "felicitously," as John L. Austin had put it – and two courses in the history of philosophy.

I took this resolution and Desbiens' advice to my Dean, who placed the dossier in the hands of Robert Vogel, the history professor who soon thereafter maneuvered to dismiss Immanuel Wallerstein, the father of the new historiographic approach called "world system." That individual, soon to be appointed Dean of Arts, discarded all my proposals and suggested a collection of light courses inspired by recent American developments, such as "War and love." The administration adopted his suggestions, and let me do my own thing, as we used to say in those times.

Canadian Winter

Marta and I had successfully braved several winters in the USA and in Germany, but were ill-prepared to face the brutal Quebec winter, that lasts half a year with temperatures that sometimes drop as low as 35 °C below zero, freezing the spit before it hits the ground. Winter dominates Quebec life to such an extent that Quebec's unofficial national anthem is Gilles Vigneault's song "Mon pays," that starts "My country is not a country, it is winter."

We had been escaping winter instead of trying to adapt to it by practicing winter sports. It took us a full decade to try cross-country skiing and spend the winter holidays in the countryside. Marta learned skiing after a few lessons, and was soon able to distinguish the various kinds of snow and the corresponding kinds of ski wax, of which there are several dozen. I followed her clumsily, and never learned how to stop, as a consequence of which I broke several ribs when crashing into snow banks.

Marta and I joined the Viking Club, and during the fall helped to clear the skiing tracks. Our most famous fellow member was Jackrabbit Johansen, the Norwegian engineer who introduced cross-country skiing to our region, and kept skiing until his death at the age of 111.

Once, when trying to keep up with our hardy dog Finn and our Norwegian friend Unni Angel, I became so stressed that I got arrhythmia. Since then, every morning I take an infallible pill that comes with the warning that one of its potential side effects is death.

We soon discovered the beauty of the snow-covered trees – as well as the anxieties of electricity cuts, frozen roads, and snowplows that took too long in coming. Another beauty of Quebec snowfalls is that they clear the atmosphere and leave blue skies and an intense luminosity that reactivates optimism – until about 4:00 p.m., when night comes suddenly. The winter sun bothered one English professor to such an extent, that he returned to his fatherland's *grisaille*. A colleague keeps the blinds down and wears sun glasses all year round – a cheap way of putting into practice Husserl's *épojé* "method" of pretending that the real world does not exist – in order to better know it. He had studied under Hilary Putnam, and got annoyed every time he heard the word 'real'. It would be pointless to ask his opinion about the reality of atoms or of placebos.

The Laurentians, only 80 km North of Montreal, are believed to be the oldest hills on Earth. They consist of a long and narrow East-West cordillera of low hills spotted with half a million small lakes and covered with thick forests of birch, fir, maple, and wild-cherry trees, as well as ferns, fungi, and moss. They are home to bears, moose, beavers, chipmunks, squirrels, deer, herons, giant porcupines, white foxes, and white owls, in addition to many migratory birds – and black flies, the nastiest of the regional fauna.

Ten years after arriving in Canada we bought a wooden cottage in the midst of a forest of maples and birches, and near Lake Cook, which we reached on foot through a wood. There we learned to walk in extreme cold weather, shovel snow, chop firewood, and do without well-water, electricity, or heating every time a storm knocked down the electricity poles. And there I learned to type on the cold windowsill of a dormer.

We had a couple of interesting neighbors, both physicians. One of them was Dr Henry Morgentaler, a Nazi camp survivor, the Canadian champion of reproductive rights and a militant secular humanist. He spent time in jail for deliberately defying the abortion law, thus sparkling a popular movement that forced the change of the law, first in the province and later in the whole country. Morgentaler ended up receiving the Order of Canada and lived 90 years.

Across the road lived Jean Lavigueur, a friendly psychiatrist always overworked and worried because of the dearth of effective treatments for mental diseases. Another neighbor was Grotkop, an airline pilot who was angry at everybody, starting with himself, and who "vented his frustrations", as he put it himself, by cutting down trees with his noisy chainsaw. Totó, a Sicilian who smoked an English pipe and had two children who looked as though painted by Botticelli, kept enlarging his house hidden in the woods. Finally there was amiable Mister Smith, who had a glass eyeball in replacement for the one he had lost at war and who greatly admired our daughter for her swimming at age four. His widow confided in me that she had long talks with God, but did not disclose the subjects they discussed.

First Courses and Talks

When I arrived in McGill I offered to teach a course on the foundations of physics. Only two students enrolled in it, so I taught them in my office. One of them, a native of the Prairies, survived and wrote his master's physics thesis under my supervision. He could hardly concentrate at home because his wife, a music student, kept banging on the piano.

Six students enrolled in my course on mathematical sociology. The only available textbook on this subject was James Coleman's, which had just been published. When I asked the chairman of the sociology department to send me some students who might be interested in mathematical sociology, he asked me "What's that?" In future I chose less exotic titles for my courses.

In addition to courses I gave some talks in other departments and at different universities around the country. The talk I gave to the physicists at the Université de Montréal on a realistic interpretation of quantum mechanics went well, whereas the one I gave to the McGill physicists on the same topic, was a flop.

I do not know the reason for my success on the other side of the mountain, but it was clear that my failure on my home turf was due to the objection raised by Tony Whitehead, a loud and aggressive Brit. He complained that I did not include any "operational definitions," such as Dirac's for the eigenvalues of a dynamical variable, namely, as the values that measurements of such variables would deliver. I did not include such "definitions" because my point was precisely that the theory in question is about nature, not about human operations. An additional reason is of course that precision measurements are indirect: they are performed on indicators or markers of the variable of interest, such as the height of a mercury column in the case of temperature.

My reply did not persuade many because physicists were used to operationism, the variety of empiricism that had been articulated by the eminent experimentalist Percy W. Bridgman (1927). They did not ask themselves whether bodies weighed anything before weighing scales had been invented.

I had my second surprise in the geology department. The department's director, a Professor Saul, did not believe in plate tectonics, which had only recently become geological orthodoxy. I said playfully that I hoped that Saul, like Paul earlier on, would soon find his road to Damascus. This did not amuse him, for he was religious. But worse was in store for me.

Foundations Center Aborts

Soon after my arrival in McGill I wrote down and circulated a proposal for a foundations and philosophy of science center. My proposal was considered at the professor's assembly, where I was asked to explain what I meant by 'foundations,' a word that some took to be synonymous with 'introduction,' and others with 'historical roots'.

When I said that foundations included the axiomatization of scientific theories, the great psychologist Donald O. Hebb, a McGill star, stood up and spoke vehemently against axiomatization, holding that it forces science into a straitjacket, but he did not say why.

I replied that, on the contrary, axiomatics helps advance knowledge, in exposing tacit assumptions that may turn out to be false, and in centering attention on the basic concepts and hypotheses. But my reply carried no weight, for my colleagues trusted Hebb, and only some mathematicians knew what foundations research was, but they did not come to my help. Besides, the psychologist Clark Hull had unwittingly discredited axiomatics with his own axiomatic learning theory, which I had criticized in my Austin seminar as simplistic and too close to behavioral data. In addition, as Hebb admitted to me when we became friends, he had always preferred literature to mathematics. Last, but not least, the news of the flop of my talk to the physicists had travelled fast.

These negative reactions to my work reached the university administration when they had to choose between my proposal and that of my colleague Harry Bracken, of setting up a department of Jewish studies. Harry himself did not care for the Jewish tradition: he specialized in skepticism and in Chomsky. But he wanted McGill to pay for having practised until recently the *numerus clausus* restriction. In any case, the choice between the two proposals was easy and fast: it was a matter of choosing between one bound to be popular with the strong Montreal Jewish community, and another that looked arcane or even wrong.

Apparently, Hennie Popper had been right in warning me that in my land of choice I might find the worst, not the best, of the two worlds, the old and the new. But it was too late to go back on our decision to settle in Canada. In addition, in that same year of 1967 two auspicious events happened, as an astrologer would say, that overshadowed my failure.

First Auspicious Event

The first happy event was the simultaneous publication, in the spring of 1967, of my works *Foundations of Physics* and the two-volume treatise *Scientific Research*. The three volumes came with hard covers in orange, and looked beautiful on our blue couch. Presumably, Van Gogh would have approved.

The heart of *Foundations* was the axiomatization of five theories: point-particle and continuum mechanics, classical electrodynamics, Einstein's theory or gravitation, and nonrelativistic quantum mechanics. In all cases the format was this:

1. *Formal assumptions*, such as predicate logic, the calculus, and Euclidean (or some other) geometry.
2. *Protophysical assumptions*, such as mereology (the theories of the physical addition of things, of time as a sequence of events, etc.)

3. *Physical assumptions*, such as classical electrodynamics in the case of special relativity, and the rest of physics in that of general relativity.
4. *List of specific primitives* (undefined physical concepts), such as the electromagnetic potentials and the current densities in the case of classical electrodynamics.
5. *List of definitions*, such as that of total mass in terms of mass density in continuum mechanics, and of field intensities in terms of potentials in the case of electrodynamics.
6. *Mathematical postulates*, such as the Newton-Euler equations in the case of classical mechanics, and the Schrödinger equation in that of quantum mechanics.
7. *Semantic postulates*, such as Born's about the state function-probability connection, and the postulate that the advanced electromagnetic potentials are physically meaningless because they violate the protophysical postulate that the cause precedes its effect.
8. *Sample of typical theorems*, such as conservation laws.

In addition, all the important propositions were followed by comments intended to clarify or justify them, or to suggest their philosophical implications. The text was dense and bristling with mathematical symbols, hence it was accessible only to those familiar with the standard or heuristic presentation.

There was only one scientific novelty in this book: the confutation of the so-called fourth indeterminacy relation, the one involving time and energy. My augment was that in nonrelativistic quantum mechanics time is not an operator but an ordinary variable, and the "relation" (or rather inequality) in question enjoys only Bohr's authority – a clear case of dogmatism. However, much later I showed that relativistic quantum mechanics does contain time-energy "indeterminacy" inequalities that are not found in the literature (Bunge 2003a).

The main novelties of the book were as follows.

1. *Some important theories, in particular quantum mechanics, were presented in a logically ordered fashion*, that is, in the form Primitive concepts–Defined concepts – Postulates – Theorems. Obviously, this is an artificial logical reconstruction, very different from the rather messy way theories are invented and developed.
2. *Every key mathematical construct was assigned a physical meaning*, in particular a reference class, through what I called a *semantic postulate*, such as "The value of the vector E at the point x in spacetime represents the intensity of the electric field at x."
3. *The book distinguished concepts that are often confused*, such as those of reference frame (a physical thing), coordinate system (a mathematical object), and observer (a biosocial entity), as well as those of proposition (a logical object), operator (a mathematical object), and measurement (an empirical item).
4. *It was the first rigorously realist (objectivist) presentation* of theories that, like special relativity and quantum mechanics, are usually presented in terms of observers and their decisions and operations. In other words, it showed physical

theories as representing chunks or traits of reality, not experiences or actions of scientists.

5. *It placed every construct in its context.* For example, it treated "$E = mc^2$" as a theorem in relativistic mechanics, not as a universal law. And it deduced Heisenberg's inequality, "$\Delta p.\ \Delta q \geq h/4\pi$," from principles and definitions, the way De Broglie had done, instead of trying to justify it with thought measurements and experiments, such as Heisenberg's imaginary microscope. Such contextualization suffices for discarding most of the interpretations of those famous formulas as so many arbitrary opinions wedded to subjectivistic philosophies.

In conclusion, by axiomatizing some of the best-known and most debated physical theories, I sought to help understand them better, as well as to order the discussions about their philosophical implications.

Reception of Foundations of Physics

Few physicists took notice of my *Foundations of Physics*, and a number of them expressed their appreciation for it. Among them were my teacher Guido Beck, who liked in particular my rejection of the classical analogies with particles and waves; Clifford Truesdell, the top authority on classical physics and the slayer of the structuralist (or "semantic") sect of Suppes, Sneed, Stegmüller, and Moulines; Stephen G. Brush, the expert in the foundations and history of statistical mechanics; my student Andrés Kálnay, who had formerly embraced the orthodoxy that he had learned in the famous treatise by Cohen-Tannoudji; Héctor Vucetich and his students Gustavo Romero and Santiago Pérez Bergliaffa, who updated parts of my *Foundations*; the Nobel laureate Willis Lamb, of Lamb-shift fame, who sought my collaboration at a time when I was immersed in social science (see Fondo Bunge); and the quantum theorist Jean-Marc Lévy-Leblond, who shared my realist stance and adopted my neutral term *quantons* to denote the sui generis things described by the quantum theory (Lévy-Leblond and Balibar 1984).

Lévy-Leblond arranged for the French translation of the book and its publication (Bunge 1974a) and invited me to lecture in Paris and Nice. The particle physicist Michel Paty invited me a couple of times to give talks in Strasbourg and in Paris. José Leite Lopes, the best-known Brazilian theoretical physicist, attended my talks and made constructive comments.

By contrast, Bernard d'Espagnat (2006), who attended one of those talks, criticized a popular paper of mine without going into technical details. But he did not have the last word. In fact, as I write, a lively controversy is developing among the French preparatory school teachers, as to whether to teach quantum mechanics using the classical wave-particle duality or my viewpoint and the associated neutral word *quanton*.

Only one physicist, Martin Strauss (1969), attacked my book, pointing out a couple of minor errors and deploring that it had been printed on good paper, unavailable in his country, East Germany. I replied (Bunge 1969a) admitting my mistakes and pointing out many of his own. Also the Dutch mathematician Hans Freudenthal (1970) attacked the book without admitting any of its merits. The British physicist David Salt (1971) counterattacked him.

Only three philosophers took notice of the book: Karl Popper, John Earman, and Erhard Scheibe. Popper thought it was a good book except for my statement that "$E = mc^2$" refers only to things endowed with mass. Earman, who had then just graduated in philosophy, condemned the book as a whole – although he had not learned the physics required to understand it. And Scheibe, a Göttingen professor, stated that I had failed in my attempt, but did not say why. A couple of years later, he invited me to talk at his seminar and to attend a party in his home in Heidelberg along with some of his colleagues.

Neither of my critics offered alternative axiomatizations or analyses. Nor did they discuss my thesis that axiomatization facilitates philosophical analysis by supplying context, unveiling hidden assumptions, and flagging the basic ideas. For example, how can one *prove* that the quantum theory does not *prove* that the universe is mental other than by exhibiting its referents and reminding the reader that physics does not account for the mental?

In conclusion, *Foundations* was not exactly a success. Most physicists and philosophers still bought arguments from authority and remained uninterested in rigorous reasoning outside mathematical formalisms. However, my book may have contributed to rendering the expression "foundations of physics" academically respectable, as 3 years after its publication Wolfgang Yourgrau, a physics professor launched a successful journal with that title.

Treatise on Scientific Methodology

The two volumes of my treatise in the methodology of science, *Scientific Research*, appeared the same year of 1967 and was reprinted a couple of times. Two years later it was published in Spanish, very well translated by the philosopher Manuel Sacristán. The Cuban publishing firm reprinted this translation, adding a preface that warned the reader that I had missed the social aspect of philosophy. And two decades later Transaction and the Mexican publisher Siglo XXI published revised versions of the work.

In my opinion, the main novelties that *Scientific Research* brought to the philosophy of science are as follows:

1. *It is the only systematic book on scientific methodology*, for it examines in orderly fashion all the traits of scientific research, from problem statement to evaluation to philosophical presuppositions and implications.
2. *It examines problems rather than authors*, and it resorts to the history of science as only a purveyor of examples and counterexamples, not as an excuse for

doing amateur, superficial, and biased history of science (à la Kuhn, Feyerabend, and Lakatos).

3. *It analyzes and exactifies key components of the scientific research project*, that had been ignored or poorly treated by other authors, such as the concepts of problem, presupposition, philosophical significance, theory (as different from hypothesis), indicator, experiment (as different from observation), and external consistency or compatibility with the bulk of antecedent knowledge.
4. *It exhibits the inadequacies of the most popular visions of science* – empiricism (or inductivism), pragmatism, refutationism, and conventionalism.
5. *It warns from the start against pseudoscience and pseudophilosophy*.
6. *It alerts against some widespread confusions*, such as those of fact with phenomenon (or appearance), generalization with induction (or generalization from a handful of data), and given (datum) with produced (with the help of theories).
7. *It shows the complexity of real-life science*, in contrast to the simplifications of the popular views of it.
8. *It does not fall into popular philosophical pitfalls* such as the belief in the sacredness of data and the omnipotence of computation.
9. *It highlights the heuristic role of certain philosophical theses*, such as realism, materialism, and systemism, in the search for research problems and the evaluation of research projects.
10. *It completes each section with a list of problems* of different degrees of difficulty.

Reception of *Scientific Research*

Joseph Agassi, a student of Popper's who neither flattered nor betrayed him wrote a long and positive review of *Scientific Research* (Agassi 1969). In Göteborg, Håkan Törnebohm told me that the book was being very well received in Scandinavia, and in Fribourg the psychologist Meinrad Perrez informed me that it was also well known in Austria and Switzerland. In 1969, when I met Guido Beck in Darmstadt, he told me that the student activists had initially barred access to his talk, but lifted their ban when told that he had been my *Doktorvater*.

The North American market had already been won by the excellent if antiquated textbooks by Morris Raphael Cohen and Ernest Nagel, as well as by a plethora of more recent works, much lighter, written for students who devoted only 25 h a week to studying outside the classroom – a time lapse that has shrunk to 15 h since the students' attention has been monopolized by Facebook and electronic gizmos of several kinds. On the other hand, from the start my treatise has had a large readership in Latin American except in Argentina, where psychoanalysts and their sympathetic philosophers, particularly the logic teacher Gregorio Klimovsky, attacked my work because it exposes their credo as a pseudoscience.

My claim that agreement with facts was not decisive in adopting a scientific theory, was not well received by the American philosophical community, as I learned at the conference that met in 1969 in the dismal town of Brockport, Long Island, where I interacted with Brand Blanshard, Ernest Nagel, Tom Settle, and the cosmologist Ted Harrison. In my presentation (Bunge 1970c) I said that, before seeking a bunch of relevant data, a theory had to pass a battery of non-empirical tests, such as compatibility with the bulk of extant knowledge, on top of which it had to be enriched with indicators (theory-data bridges), as well as with a hypothetical model of the object of study – except in the case of a narrow theoretical model or *teorita*, as Richard Braithwaite called it.

To illustrate the latter's crucial role I concocted an ad hoc model whose object of study could be either a sphere or a disk. Two very different results were obtained, which matched the experimental data. So, the latter were inconclusive, and if the general theory was worth saving for having a good track record, a third auxiliary hypothesis had to be invented and tried out.

The lesson I drew from this example was that an inadequate auxiliary hypothesis may ruin a good theory, whereas a good model may temporarily save a poor one. At this point the loud guffaw "Bunk!" came from the back of the room. This loud bray came from Sidney Morgenbesser, a protégé of Nagel's. Morgenbesser, whom I called *Morgennschlechter*, had become a full professor at Columbia without publications, but was the author of some celebrated jokes, and had courageously participated in student's marches and sit-ins in protest of the American war in Vietnam. He later privately sought my forgiveness and confessed, without my prompting, that he had been the hostile critic who had rubbished a submission of mine to Harvard University Press. In some places *chutzpah* will take you further than hard work.

Popper's Reaction

Popper's reaction to *Scientific Research* was characteristic. He told me that he did not like it because it was not centered on the problem of induction. He claimed to have solved this problem by reformulating Hume's objection, that "there can be no good reason to believe a scientific theory" – to which he only added that, on the other hand, there may be decisive reasons for rejecting it, which is why, according to him, scientists are eager to have their own findings falsified. A few years earlier, in my review of his main work on methodology (Bunge 1959b), I had held that Popper's refutationism was nothing but a version of confirmationism, since denying *A* amounts to affirming non-*A*. I could also have added that refutationism involves a seriously wrong psychology of science, as only masochists enjoy punishment.

Three years later, at the Boston University colloquium on Popper, which he attended, I criticized what I called his crypto-empiricism, for it centered on isolated empirical data, whereas in real life every hypothesis is evaluated not only in the light of the relevant empirical evidence, but also in that of the bulk of background knowledge (Bunge 1973b). This is how Einstein had reacted to the alleged experimental refutation of his formula about the mass-velocity relation: his theory *had* to be true because it followed from classical electrodynamics.

I started off my talk saying that I offered my reflections as a constructive criticism. Popper stood up and exclaimed vehemently: "There is no such thing as constructive criticism! Critics always aim straight at the jugular." This reply of Popper's confirmed my impression that he was unfamiliar with the way scientific communities work. Indeed, the first step a scientist takes when completing a draft is to ask colleagues to review it with a view to improve it. This is why most research papers end with a sentence of the form "We thank X for helpful comments." Systematic refusal to contribute constructive criticism may lead to ostracism from the scientific community. But Popper stuck to the idea of science in a social vacuum, just as the new crop of sociologists of science focuses on networks and institutions without creative brains and without scientific findings.

Lakatos' Attempted Extortion

I submitted my Boston paper, titled "Testability today," to the *British Journal for the Philosophy of Science*. Imre Lakatos, its brand-new director, wrote back saying that he would gladly publish it provided I quoted him. I refused to fall for this attempted extortion, because I had never learned anything from him. In fact, my *Scientific Research* had been published 11 years before his much-publicized "methodology of scientific research programmes," (Lakatos 1970) which in my view was just as remote from real science as his teacher's methodology. In any event, my rejected paper ended up as a chapter in my *Method, Model, and Matter* (Bunge 1973b).

It gradually became known that extortion was not Lakatos' only means to advance his own career. Popper told me that Lakatos had gained his confidence by telling him that he held a doctorate in mathematics, while actually his degree was in mathematical didactics, and so he had never published any original papers. His lie, combined with shameless flattery, got Lakatos what he wanted: to succeed Popper in his LSE chair. The moment he fulfilled his ambition, Lakatos blocked his erstwhile protector's participation in the LSE philosophy seminar.

In contrast, John Watkins – who, like Joseph Agassi, was a loyal disciple and a decent person – welcomed Popper into his own seminar. By that time Agassi had settled in Hong Kong, where he mentored two excellent students, Ian Jarvie and Tom Settle, both of whom I befriended in Canada. Settle became a Popper expert, whereas Jarvie is a distinguished film sociologist – which is why I quip that he is much of the time in the dark.

Second Auspicious Event

Marta gave birth to our first child on June 1st, 1967. We named him Eric Russell: Eric because it is a bilingual name, as befits the native of a bilingual region; and Russell in honor of the logician – who wrote us a gracious thank-you note when notified. From an early age, Eric combined technology with art, and jointly with his Vietnamese wife Mimi Hoang he runs a prestigious architecture firm in Brooklyn, N.Y.

A few weeks after Eric's birth I attended the 7th Inter American Congress of Philosophy, which met at Laval University, near Quebec City, the provincial capital. There I met again the Quines, and was introduced to the Mexican philosopher Fernando Salmerón, who was to become a colleague and a dear friend of my family. Fernando, trained by the Spanish exile José Gaos, had passed from phenomenology to lingüistic philosophy, and used his political connections to modernize the Mexican university. A few years earlier he had renewed the Veracruz university so thoroughly, that the provincial government took offense to the point that one night he was visited by a police squad, which ordered him and his family to leave Veracruz within a few hours.

Fernando, his wife Licha and their eight children moved to México City, where they soon put down deep roots. He was appointed director of the Instituto de Investigaciones Filosóficas, which not only gathered some Mexicans, but was also a harbor for a number of Latin American scholars who had lost their positions in their own countries, such as the Uruguayan Mario H. Otero and the Argentinians Ernesto Garzón Valdés, Eduardo Rabossi, and Tomás Moro Simpson. The institute also published two journals, *Dianoia* and *Crítica*, and its library received more than 100 foreign journals. I spent there the academic year 1975–1976.

The Laval Congress

My congress paper, "Quanta y filosofía" (Bunge 1967f), held two theses that I had already defended elsewhere. The first was that the Copenhagen interpretation of quantum mechanics is observer-centered, rather than object-centered, hence inconsistent with the realist philosophy inherent in modern science. Moreover, subjective idealism is incompatible with the scientific method, whose heart is the contrasting of our maps with the territories they represent. Yet, that unscientific philosophy has been defended by some eminent scientists, and only a few have dared challenge it. Worse, in recent years some physicists have claimed that there is no need to subject the observer-centered and the multiple-universe extravagances to empirical tests.

Tolerance and even support for unscientific fantasies among physicists is such, that the eminent cosmologists George Ellis and Joe Silk felt compelled to attack them in *Nature*, to "defend the integrity of physics" (Ellis and Silk 2014) Regrettably, both embrace Popper's equation of scientificity with falsifiability, which led him to admit steady-state cosmology and mind-body dualism, even while recognizing that both violate energy conservation – just one more conjecture according to him.

I think that what I call 'external consistency', or compatibility with the bulk – not the totality – of antecedent knowledge (Bunge 1967b), is necessary and sufficient to indict as unscientific string theory, the participatory universe, parallel universes, cosmological creationism, and the "it from bit" fantasy. Astrology, palmistry, acupuncture, and homeopathy were refuted long ago, but this did not make them scientific.

My second thesis in the Laval congress was that a sober formulation of the quantum theory, one confined to its postulates and typical theorems, and interpreting the mathematical formulas in strictly physical terms, prevents philosophical smugglings like that of the observer-dependence of nature. Using this procedure one can show, for instance, that the so-called uncertainty relations have nothing to do with mental states, such as that of uncertainty. In other words, those inequalities are laws of nature, holding in stars as well as in labs, and in the remote past as well as after the emergence of observers.

To Corfu, Liège, and Amsterdam

When returning from that congress we flew to Corfu, to exhibit our new product and let Marta recover by swimming and rowing in the blue Ionian, still teeming with fish, starfish, and sea urchins. Eric was only 1 month old when we arrived – proof of our parental irresponsibility, as Nina Troubetzkoy rightly told us later. From Corfu we flew to Liège, in French Belgium, to attend my first colloquium of the Institut International de Philosophie. There I was introduced to Nicholas Rescher, the total philosopher, as well as to the Romanian logician Grigori Moisil, the Belgian philosopher Chaim Perelman (later Baron Chaim), and Sir Freddy Ayer, one of the last logical positivists. The latter told me that, when the Vienna Circle met in a classroom of the Vienna University, they locked the door to prevent Popper from sneaking in.

My presentation argued that scientific data are not given (*data*) but sought, and not in conceptual darkness but in the light of theories, among them the ones that explain how the instruments involved in the pertinent observations or measurements work. Hence the need to think of scientific *data* (given) as *peta* (asked for). As an example I mentioned the theory of the weighing scale of equal arms, that my friend, the physicist and philosopher André Mercier, volunteered to explain on the blackboard. As a student I had learned the theory of this instrument in Friedrich Kohlrausch's pioneering book (Kohlrausch 1870) on physical measurements – which of course had nothing to do with the phony "measurement theory" that used to be cultivated by some American scholars, such as Pat Suppes and Duncan Luce.

From Liège we went to Amsterdam, to participate in the Third International Congress of Logic, Methodology and Philosophy of Science. As with the first (Stanford) and the second (Jerusalem) congresses, I was the only Latin American present. I met again some notables, like Tarski and Popper, and was introduced to the Italian Evandro Agazzi, who became the main organizer of international colloquia and congresses over several decades. The Poppers visited us in our hotel room to meet Eric, who was only 3 months old. Karl did not know what to do with him, but Hennie held him affectionately in her arms.

I presented my axiomatization of classical point mechanics, which contained more axioms than that of McKinsey, Suppes, and Sugar, if only because it stated explicitly the physical meanings of the key concepts. On the other hand it did not include the action-reaction "principle," for it is a logical consequence of the theorem of conservation of the linear momentum. Joseph Sneed, who was present, did not comment.

Back from Europe

On returning to Montreal, in the fall of 1967, I resumed my course work. This was the first time I could use a text on scientific methodology that I liked – my own. I could also use to more effect the teaching method I had introduced in Buenos Aires: replacing the formal lecture with the Socratic dialogue with students who have read the preassigned chapter of a book. And I replaced the final exam with an oral presentation plus a term paper on subjects chosen by the students themselves, who were allowed to write in either of the two official languages of the country. In addition, students could consult me in my office. In working in this manner, the student finished the course taking home a monograph instead of the memory of a stressful or even humiliating interrogation.

My colleagues disapproved of my method: they thought it too demanding for the teacher and too lenient for the student. I suspect that most of them feared facing a pack of curious and uninhibited adolescents who took delight in confronting someone afraid of having to answer unexpected questions that might fall outside his/her narrow specialty.

It took me a while to think of problems other than those I had tackled in my recently published books. So, during my first few Canadian years I engaged in filling some holes in those buildings. In fact, I wrote a paper on the heuristic virtues and shortcomings of analogy (Bunge 1967d, 1969e), that favorite of postmodernists; a detailed technical study on the so-called arrow of time (Bunge 1968a); and an update of my work on the new electron coordinate and the corresponding constants of motion (Bunge and Kálnay 1969).

About 1969 I started to concentrate on my new long-term project: the semantics and metaphysics of science, on which I taught some seminars. These attracted two mature students who asked me to supervise their graduate work: Charles Castonguay, who had completed a master's in mathematics, and Roger Angel, who was already a professor of logic and philosophy of science at Sir George Williams University, a small local college that sprang from the YMCA and that was later to become one of the ancestors of Concordia University.

I suggested to Castonguay to work on meaning and truth in mathematics, and succeeded in allowing him to defend his dissertation in the mathematics department, since the philosophers would have required him to take a number of history courses that did not interest him. From Angel I demanded that, before philosophizing about relativistic physics, he learn the mathematics and physics required to understand it. He complied so well with this condition, that a few years later he taught mathematics at his university. Both Castonguauy and Angel got their dissertations approved, and I arranged for them to be published by Springer and Pergamon respectively, although I did not concur with their views.

I befriended Angel, who since then has been reading only physics books and papers. By contrast, as soon as Castonguay got his PhD, he attacked me in public and became a fanatic québecois separatist, devoting his energy to trying to prove that the francophone to anglophone population ratio was declining, although statistics proved the exact opposite. However, all that happened years later.

Summer in Cuernavaca

We spent the summer of 1968 in México, the USA, and Austria. With Eric, who was just 1 year old, we resided for a couple of months in a chalet in Cuernavaca while Marta attended a summer course in Vancouver. When I went to México City for my lectures, Il left Eric in charge of the house owner, a kind French lady whom I conquered by speaking in her native language, and drove to México city in a one-eyed Beetle along the poorly lighted old road.

I retain a few vivid images from that stay in Cuernavaca: the black swarm of flies buzzing in the butcher's shop; the public square where Eric and I ate the ice-cream cones that gave us both "Montezuma's revenge" with a high fever; the exasperated barber who attempted to trim Eric's blonde mane; the avocado vendor whom Eric followed because his dark skin reminded him of his Trinidadian nanny; and the secluded garden where I tried to teach Eric both the English and the Spanish names of the enormous moon, or luna, hanging in the early evening, until he cried triumphantly: "Moona!"

At the Universidad Nacional Autónoma de México I gave a few philosophical talks, and a crash but dense course on the foundations of physics. The most active participant in that course, Holger Valqui, had been Kálnay's student in Lima, Perú, and we met again three decades later in the same town.

The highlight of that trip was listening to the talks that the Mexican physicists Luis de la Peña and his young wife Ana María Cetto gave at a well-attended Latin American physics colloquium held nearby. We shared our criticisms of the Copenhagen orthodoxy, but disagreed about what quantum mechanics is about: whereas they held that it concerns ensembles of particles, I argued that the basic theory is about single quantons, since ensembles are the referents of quantum statistical mechanics. A few years later, while in México City, Marta and I befriended Luis and Ana María.

After Marta returned from Vancouver we flew to Albuquerque, New Mexico, to participate in a foundations of physics symposium I had helped organize. It was held at Colgate University, a green oasis in the middle of the desert. There we read the news of the Soviet invasion of Czechoslovakia, further evidence that "real socialism" did not tolerate reform, much less self-determination.

Salzburg

Our next destination was Zell am See, a pretty Alpine lake resort where I parked Marta and Eric while I went to Salzburg, Mozart's birthplace. There I attended a colloquium organized by my friend, the philosopher and theologian Paul Weingartner, who had also invited Bruno de Finetti, I.J. Good, Adolf Grünbaum, Henry Margenau, Karl Popper, Martin Strauss, and a few others.

Strauss's and my presentations were about the relations among physical theories. Strauss's dealt with this matter in the same intuitive way as Nagel, Tisza, Kuhn, and Feyerabend. In my view this approach is shallow and sloppy, if only because it cannot tell the conceptual (in particular logical) relation from the heuristic or historical one.

In my own presentation (Bunge 1970a), I argued that, to correctly compare two theories, one must first axiomatize them, because only then can one concentrate on their basic concepts and assumptions; in particular, whereas in some cases what is a basic idea in one of the theories is a derived one in the other. For example, Maxwell's formulation of his electrodynamics is deducible from its formulation in terms of electromagnetic potentials.

I had received a mimeographed copy of Strauss' paper and had written unflattering remarks on its margins: "hodgepodge," "sloppy," "false," and the like. Unintentionally I had left that text on my chair, and a well-wisher gave it to Strauss, who returned it to me by mail along with a note saying "Here is my hodgepodge."

Vienna Congress

Finally, from Salzburg we went to Vienna, where I was expected to participate in the 14th World Congress of Philosophers (1968). I gave an invited talk on scientific realism, which differs from both naive realism and the Platonic realism of ideas. Its main theses are that the world external to the knowing subject exists independently from the latter, and that it can be known partially and gradually through the scientific method (see Mahner 2001).

Scientific realism differs from the "critical realism" that Roy Bhaskar and his faithful, most of them social scientists, have been cultivating at Oxford and environs. This school does not use any formal tools and it ignores science proper. Because of these limitations, their members have been unable to participate in the most important scientific-philosophical controversies of our time, such as: appraisal of the Copenhagen school, the reality of species debate in biology, the mind-body problem, and the scientific status of standard economic theory. The low power of critical realism is partly due to its ontological neutrality, in particular its indifference to the idealism/materialism disjunction.

The failure of critical realists to participate in those controversies is characteristic of any epistemology divorced from ontology – a divorce consecrated by empiricists but not by scientists. Indeed, one cannot even start studying an object without assuming something about its nature: whether it is real or imaginary, material or ideal, physical or living, individual or social, and so on. The moral is that we should strive for the fusion of realism with idealism, as Plato had done, or with materialism, as I have attempted to accomplish with what I call *hylorealism* (see Bunge 2006a).

Reactions to My Defense of Scientific Realism

My presentation was received with interest. I saw the famous astrophysicist Viktor Hambardzumyan nodding repeatedly in assent. By contrast, Freddy Ayer, the earliest British logical positivist, accused me of falling for naive realism – the view according to which the world is just as it looks. I rejected this accusation but admitted to preferring naive realism to sophisticated irrealism, for the former can be refined, whereas the latter blocks the exploration of reality. For example, Democritus believed that the sun was the size of a coin, but he did not question its autonomous existence, and eventually astronomers discovered the sun's real size. Ayer had adopted the majority view in the Vienna Circle, according to which the question of the reality of the external world is a pseudoproblem best left to metaphysicians, for neither of the theses in question can be either confirmed or falsified.

I hold that, even if the hypothesis of the reality of the external world were untestable, it would not matter, for whoever endeavors to study a concrete thing takes its existence for granted. The experimentalist who finds a new particle and the field worker who unearths a new fossil feel elated at their successes in the exploration of reality. And the theoretical physicists who postulate or deduce a new formula make sure that it is frame-invariant before concluding that it represents something real. (See Bunge 1968a for the invariance-reality relation.)

The Place of Mathematical Objects According to Realism

Mathematics has always been the nemesis of naive realism, empiricism, and vulgar materialism, for the mathematical objects are neither perceptible nor material, and they would not exist without mathematicians. Nominalists hold that they are mere signs, but these are conventional, whereas the mathematical concepts are constrained by laws that can be represented by different symbols. Nor can they be regarded as possible thoughts, for some of them, like the tenth power of 10^{10}, the continuum, and the category of all sets, are unthinkable. I have suggested that mathematical objects are *fictions*, just like gods and cartoon characters – with the difference that they are subject to inflexible laws. Moreover, to grasp them one need not fall into a trance, but, quite the contrary, one must be alert and ready to detect and correct mistakes. This is why my philosophy of mathematics may be called *moderate fictionism* (Bunge 1985a, 1997a).

Moderate fictionism is unrelated to both realism and materialism, since it does not place mathematical objects in the material world, but in the ideal one of conceptual artifacts. So, it embraces Plato's thesis that the abstract objects are immaterial, but rejects his claim that they exist by themselves, in particular without the help of mathematicians. As well, moderate fictionism is not to be confused with the total fictionism of Hans Vaihinger (Vaihinger 1920), who held that all the sciences are mere fictions. Moderate fictionists may accept the Aristotelian concept of *factual* truth as *adaequatio intellectus et rei*, but they will argue that this is very different from the concepts of truth used in mathematics: as theoremhood (or demonstrability) or as satisfaction of an abstract theory in or by a model or example.

In his famous paper addressed to philosophers, Alfred Tarski failed to distinguish factual from formal truth: he held that the former suffices (Tarski 1944). But he gave neither arguments nor examples, yet he fooled a number of philosophers, Popper, among others. If Tarski were right about truth, experimental science would be either omnipotent or unnecessary, but it is neither.

Materialists as such have nothing original to say about mathematical objects because these are immaterial and have no independent existence. They are imagined by humans and would cease to have (ideal) existence if mathematicians became extinct. Parallel pairs are gods/believers and music/musicians.

Library of Exact Philosophy

In the course of the Vienna congress, Dr Heinz Götze, director of the Vienna branch of the Springer publishing firm, asked me to gather a few philosophers to discuss the possibility of setting up a library of authors imbued with the "Vienna Circle spirit," that is, respectful of both logic and science. The meeting place was to be the famous restaurant in the basement of the Vienna City Hall.

I invited Richard Montague, Richard Martin, and a few others. We dined like kings and approved Montague's proposal, of calling the new book series the *Library of Exact Philosophy* (LEP). The LEP published 13 volumes, among them the English version of the important book on epistemology by Moritz Schlick, the founder of the Verein Ernst Mach, later known as the Vienna Circle (1922–1936). The LEP died when I ceased receiving suitable submissions. Rationality, in particular exactness, was all but banned by the generation seduced by Kuhn and Feyerabend or by Althusser and Foucault.

The Two Faces of the 1960s Student Revolt

The student revolts of the 1960s, in particular Paris' May 1968, had been hijacked by illiterate radicals who rejected not only political authoritarianism but also intellectual authority and even learning. A large wall of the Goethe University in Frankfurt showed in big black letters on whitewash the slogan *lernen macht dumm*: "Studying dumbs down." In some places the new barbarians went even further. The electron microscope at the Buenos Aires medical school was defenestrated; and the computer center at Concordia University in Montreal was burnt down.

In the past, from Galileo's time on, students had demanded the modernization of the university, a closer relationship between gown and town, or the cessation of unjust wars. In the 1960s the student leaders wanted noise, not light, and instant gratification rather than hard work. Although they called themselves progressives, actually they were reactionaries.

German Tour

At the beginning of 1969 the Alexander von Humboldt Foundation invited my family and me to tour German universities. We left Montreal under a white blanket, and arrived in Bonn, the capital of the German Federal Republic, surrounded by green prairies and blooming cherry trees. Heinrich Pfeiffer, the foundation's indefatigable secretary, received us affectionately and showed us the plan of our tour: Bonn-Heidelberg-Göttingen-Freiburg-Düsseldorf-Berlin-Hamburg-München.

We started by picking up the Mercedes 230 that I had bought in Montreal for the equivalent of 2 months' salary. I visited some philosophy professors and gave several talks. In Bonn I praised what since mid-eighteenth century has been called *enlightened* philosophy, the one attached to rationality and willing to interact with science. I also praised Heinrich Scholz, who had been successively a professor of theology, philosophy, and logic, had spent part of the Nazi period under house arrest, and had hosted Tarski when he was fleeing from Warsaw to London.

My praise did not go down well. I contrasted Scholz's integrity and search for clarity with Heidegger's political servilism and obscurantism, and called him a *Kulturverbrecher*, that is, a cultural delinquent. This remark too was received icily, although nobody dared say anything. A few years later the Foundation felt obliged to organize a public homage to that charlatan. On that occasion Gadamer and Derrida praised him and had the nerve to deny that Heidegger had been an accomplice of Nazism.

In Göttingen, at Erhard Scheibe's invitation, I spoke about exact philosophy. I said that it is not a doctrine but an approach or method, hence ontologically as well as epistemologically neutral, so that it may be adopted by anyone except irrationalists. Therefore, the basic split of the philosophical community is not that between materialists and idealists, but between the exact and the inexact: there can be fruitful rational debates among the rigorous but not among the confused. Francis Bacon would have concurred, for three centuries earlier he had said that confusion is worse than error.

In München we visited the rich Pinakothek and the Deutsches Museum, so amazing that Eric, not yet 2 years old, refused to leave after looking at its artifacts during 8 h. I also went to visit Werner Heisenberg, whom I had met several years earlier in Bonn and Göttingen. He was reading proofs of his memoirs, *Der Teil und das Ganze* (1969), which I reviewed in *Physics Today* (Bunge 1970b). On that occasion Heisenberg told me that he had always sought to understand, not to either compute or design experiments. A few years later I was invited to write a paper for his latest Festschrift (Bunge 1977d), just as my teacher had contributed to his first. I also had a brief conversation with his disciple and friend Carl Friedrich von Weiszäcker, who was trying to do philosophy and impressed me as arrogant – quite unlike his teacher.

When we had enough of Northern humidity we drove to Corfu via Brindisi. There we stayed at the small chalet of Rápanos Mikhalás, a man who every day went up and down Lakones, a village on a hilltop, in the company of his beautiful donkey, whom he kept well fed and shiny, and never loaded with the tools and things he carried. Rapanos' daughter Eleni, unhappily married to a brute, did the cleaning and cooking for us. (Half a century later, I wrote a short story about them, which was published by the oldest Argentine daily, and reproduced in Bunge 2012c.) From the bottom of a ravine I rescued a whining puppy whom, after bathing and feeding, we named Snoopy II and gave to Eric. The two young animals immediately became inseparable playmates.

Reality Is There, but the Reality Concept Is Still Elusive

I had spent three decades battling subjectivism, but my goal of vindicating philosophical realism was still unfulfilled. Indeed, although I could boast of having slain the old irrealist dragon in its most sophisticated lair, namely, quantum mechanics, I still did not have a satisfactory definition of "real." Of course, there was the old reality criterion of experimental scientists: "something is real if and only if it kicks back." But it does not apply to the past, space-time, or thought. And in any case, criteria cannot double as definitions, which are identities, that is, formulas of the form "$A = B$," not equivalences like "A if and only if B." The solution to this problem came years later and unexpectedly, namely, through reconstructing materialism with some elementary modern formal tools. Hence the next two chapters: exactness without barren formalism and materialism without Hegelian fog (Bunge 1981d).

Chapter 9
Exact Philosophy

From Intuition to Rigor

Philosophies may be exact or rigorous to different degrees, from the ones cast in mathematical terms to those that openly spurn rigor. Among the latter are the doctrines that place feeling or intuition over reason (like Rousseau's, Vico's, and Bergson's), exalt everyday language (like Wittgenstein's), praise "weak thought" (like Gianni Vattimo's), attempt to pass off obscurity as depth (like Hegel's and Husserl's), love the absurd (like Heidegger's and the postmodernists), or tell us solemnly that matter and mind are eternal mysteries (like the later Chomsky's).

Spinoza' *Ethics* is likely to have pioneered exact philosophy. However, he applied the *more geometrico* only to the organization of ideas in Euclid's format: Axioms and definitions – Theorems – Commentaries. Spinoza could not question the inexactness inherent in the ordinary language he used to cast his ideas, because mathematical logic was invented two centuries later, although the basic idea, of replacing intuitive discourse with calculation, had occurred to Leibniz shortly before 1700.

It is generally assumed that Bertrand Russell's "On denoting" (Russell 1905) was the earliest paper in exact philosophy, and Rudolf Carnap's *Logical Construction of the World* (Carnap 1928) the first book, in the same vein. These were indeed the earliest philosophers to make full use of modern logic and naïve set theory to philosophize. Regrettably, both works were flawed. Indeed, Rusell's analysis of expressions like "The time unit" and "The dog next door" involved the existence requirement, which is unnecessary because we often deal with fantasies, such as "The mother of God" (Bunge 1974b). Besides, Russell misread "$\exists xPx$" as "There exist Ps," while actually it is short for "Some individuals possess P", as is obvious from the standard definition of $\exists$ as "not-all-not" (Bunge 2012a). As for Carnap's "logical reconstruction of the world", it remained shackled to Hume's and Mach's phenomenalism, and consequently did nothing to advance scientific knowledge,

M. Bunge, *Between Two Worlds*, Springer Biographies,
DOI 10.1007/978-3-319-29251-9_9

Table 9.1 Examples of exactification

Intuitive idea	Exact idea
Some	$\exists$
All	$\forall$
Property	Predicate
State of a thing	List of properties of a thing
Species	Set of things sharing some essential properties
Dependence	Mathematical function
Size	Volume, area, or length
Physical addition	Mereological sum $\oplus$ of things P
Free will	Ability to overcome external constraints
Liberation	Expanding the subject's state space
Confinement	Restricting the subject's state space
Idea	Concept, proposition, or theory
Many	Not available
Credence	Not available
Subjective utility	Not available
Sacred	Not available
Transubstantiation	Not available
Triune	Not available
Dialectics	Not available

which locates phenomena (appearances) in brains and admits plenty of unobservables.

Much the same holds for Nelson Goodman's brilliant *Structure of Appearance* (Goodman 1951): it is exact but, because of its phenomenalism, it is also shallow. Not even its title is correct, for appearance is structureless: we experience sensory impressions as haphazard sequences of unrelated events. And when claiming that simplicity is the seal of truth, Goodman did not even attempt to analyze the two concepts at stake, so he combined falsity with inexactness (Goodman 1958). I have analyzed and clarified both concepts (e.g., Bunge 1963b).

Exactification

I claim that that any interesting idea, if passably clear, can be exactified – a neologism I had introduced while writing on values. See Table 9.1.

Exact philosophy may be defined as philosophy done with the explicit help of mathematical logic and mathematics: it is disciplined rather than wild, contextual rather than context-free, systematic rather than fragmentary, and capable of being argued about rather than oracular.

From the moment I learned some modern logic and got rid of Hegelianianism, I preferred exact to inexact or intuitive philosophies. This conversion started around

1950, but it was a very slow process because of my mental inertia and because the libraries and bookstores accessible to me in my remote corner of the world offered very few publications of that genre. For example, there were only about a dozen pre-war issues of *Erkenntnis* in the library of the Institute of Philosophy of the Buenos Aires University. In addition, even the *Journal of Philosophy*, perhaps the best at that time, published very little exact philosophy.

In extreme cases I wrote to philosophers asking them for reprints. Bertrand Russell answered my request for his pioneering paper on vagueness by mailing to me in Buenos Aires a copy of the issue of the *Australasian Journal of Psychology and Philosophy* that had published it (Russell 1923). This is how I got to know about the Australian philosophers who dared call themselves 'materialists' rather than 'naturalists' or 'physicalists', and defended psychoneural monism. Eventually I met Jack Smart, Ullian T. Place, and David Armstrong, as well as John Passmore, whom I found as delightful as learned when we first met in Philadelphia in the early 1960s and discussed the possibility of my moving to his country. Later I learned that the position I had applied for was filled by two unproductive but far cheaper scholars. This is an example of the formula "1 + 1 = 0," a cornerstone of what I call 'dean arithmetic'.

I sympathized with the Australian materialism project but thought it had no future because of its inexactness, radical reductionism, and detachment from contemporary science. And I found Smart congenial but Armstrong haughty and aggressive, and Ullian – who gave a lecture at my seminar – more interested in farming than in philosophy. Regrettably, around 1980 the American David Lewis and his extravagant possible-worlds metaphysics seduced overnight the younger Australian philosophers, and consigned the native materialists to oblivion. However, back to exactness.

In my opinion (Bunge 1973e, Foreword) exactness, though highly desirable, is only a means for enhancing clarity and systemicity. Exactness without substance is just intellectual gymnastics, as Einstein might say. For example, Max Black, Héctor-Neri Castañeda and Jaakko Hintikka, the sharpest analysts I have known, produced little worthwhile because they did not tackle any deep philosophical problems. In other words, we should strive for exact *and* deep truths.

These ideas prompted me to convening the first symposium on exact philosophy, which met at McGill University on November 4th and 5th, 1971 with the support of both the university's administration and the Killam Trust, administered by the Canada Council, a government agency. That meeting was also the origin of the Society for Exact Philosophy, which has been holding annual meetings since then. I collected the papers read and discussed at that meeting in *Exact Philosophy* (Bunge 1973e). This volume includes the papers read by Carlos E. Alchourrón, Nuel Belnap, Héctor-Neri Castañeda, Peter Kirschenmann, Hugues Leblanc, Lars Svenonius, Raimo Tuomela, Bas C. Van Fraassen, and their commentators, as well as my own "A program for the semantics of science." The latter summarized the semantic theories I was working on (Bunge 1974c, d).

Misunderstandings About Exact Philosophy

As stated above, exact philosophy uses some of the tools forged by logicians and mathematicians. For example, it uses naïve set theory as expounded in Paul Halmos' beautiful book on the subject (Halmos 1960). Philosophers can also make good use of some abstract algebra, but they have little use for model theory, the branch of mathematics that deals with the models or examples of abstract theories, such as the set of positive integers as a model of semigroup theory (but not of group theory). Nor are Gödel's theorems of much use to exact philosophy, for the same reason that they do not occur in physics, namely, they refer to abstract theories of interest to only a tiny subset of the mathematical community that works in the foundations of mathematics.

Of course, theologians, possible-worlds philosophers, and other fantasists are free to help themselves to logic and mathematics. As a matter of fact, in the 1980s there were attempts to do analytical theology, a cross between Aquinas and Wittgenstein, like those of Elizabeth Anscombe, Peter Geach, and Paul Weingartner – as well as analytical Marxism – like those of Gerald Cohen, Jon Elster, and John Roemer. There was even some talk of analytical existentialism. But none of them solved the mysteries of religion, dialectics, *Dasein*, or *Aufhebung* (sublation or double negation). Correct form cannot make up for wrong or confused content. For example, a purely formal analysis of "Is a good teacher." equates this predicte with the conjunction of "is good" and "is a teacher," which is obviously false. The correct analysis of "good teacher" is the conjunction of "teacher" with "competent" and "dedicated." Even clearer evidence in favor of the thesis that correct conceptual analysis sometimes requires substantive knowledge is the frequent occurrence of oxymorons such as "Christian science."

Analytical theology is not superior to its ordinary-language counterpart, for exact fantasy is still just fantasy. In my *Scientific Research* (Bunge 1967b, p. 477–8) I offered a mathematical theory of ghosts, with ectoplasm and all, just to ridicule the myth that doing science amounts to mathematizing. Likewise, my mathematical theory of secrecy (Bunge 1979g) is a cartoon of mathematical politology.

The least that exact philosophers aim at is analyzing some intuitive but important concepts, such as those of space, time, causation, value, and justice. The more ambitious ones will attempt to construct theories – that is, hypothetic-deductive systems, involving exact concepts. Moreover, contrary to the so-called analytic philosophers who swear by Wittgenstein, I hold that *the best analysis is synthesis*, or the inclusion of the idea to be analyzed into a theory. Giuseppe Peano set a prime example of this strategy when he built his axiomatic theory of the natural numbers, which begins with the humble-looking postulates "Zero is a number" and "The successor of any number is a number."

The same strategy can be used in the factual sciences. For example, the physical law that the electric charge of two or more bodies equals the sum of their individual charges involves the notion of mereological sum or physical addition, as in "The charge of $a \oplus b$ equals the charge of a plus that of b," or "$Q(a \oplus b) = Q(a) + Q(b)$".

The concept ⊕ is ontological because, like the part-whole relation, it underlies all the factual sciences. This case sheds light on the science-philosophy connection: it shows that science presupposes some exact metaphysics (or ontology). In turn, this falsifies the vulgar belief that science is incompatible with metaphysics. However, let us go back to the analysis-synthesis connection.

Philosophical Analysis Unveils Hidden Complexity

We stated above that conceptual synthesis, as in classing and theorizing, provides the best analysis. By contrast, talk of concepts out of context may lead to confusion, as in the cases of 'entropy' and 'information', which do not mean the same in thermodynamics as in information theory. In the former, "entropy" occurs along with the concepts of temperature and heat, whereas in information theory it is associated with the concepts of signal and information channel. Logical analysis helps evaluate opinions and conjectures of all kinds. For example, the famous psychiatrist Sir Michael Rutter wrote once that resilient individuals recover easily from adverse experiences: he did not notice that he was just stating a tautology, that is, a logical truth, not a factual one, for he was just defining the predicate "is resilient."

A more important case of failure of analysis is the individualist (or radical reductionist) thesis that, to account for complexity, we must start from the simple, in this case the isolated individual. This programmatic thesis is false for two reasons. First, individuals are partly shaped by their environment. Second, whereas relations cannot be built out of intrinsic properties, some of these can be analyzed in terms of relations. For example, "x is blind = there is nothing that x can see." Note that we are construing "to see" as a binary relation: there is something that x sees. But the vision psychologists use a quaternary relation: a is seeing b under illumination c during the time period z. They would laugh if told that Gilbert Ryle, a prominent practitioner of soft or Wittgensteinian philosophical analysis, decreed that conceiving of vision as a process is wrong, because "the grammar of the verb *see*" demands that seeing is a point event. The ontology of soft analytical philosophy may be summed up in the formulas: *To be is to be talked about*, and *Knowing is mastering ordinary language*.

Conceptual analysis is more useful in the "soft" sciences than in the "hard" ones, since the latter make intensive use of mathematics, the most exact of all sciences. For example, in his famous study on nationalism, Ernest Gellner defined it as the doctrine that promotes the "congruence" of the political with the national; he also asserted that the nationalisms emerge as the "crystallizations" of new cultures (Gellner 1983). But he did not bother to clarify what he meant by "congruence" or by "crystallization." When I asked him in public how these intuitive ideas would apply to the American Independence movements, Gellner admitted frankly that they don't: he was thinking of the nationalist movements born in the Austro-Hungarian empire, where he had been born.

Annoyed by such narrowness and sloppiness, I defined in my *Social Science Under Debate* (Bunge 1996a) the concepts of country, people, nation, and their kin. In addition, I distinguished dozens of varieties of nationalism, that the experts have conflated, hence have failed to help the people who face the problem in the political arena. For example, all aggressive nationalisms are morally reprehensible, whereas defensive nationalisms are not; and cultural nationalisms are self-destructive, whereas moderate economic nationalisms are constructive in the face of imperialism. I admit that all these evaluations of mine are controversial. But this is precisely my point: that serious conceptual analysis is relevant and controversial, whereas Wittgenstein's aphorisms are at best quotable witticisms, and at worst boring flippancies.

From Parmenides to Hegel

As stated above, exact philosophy is not confined to philosophical analysis, which at best corrects, but it may engage in synthesis, which is constructive, for it involves the creation of new ideas. A clear case of error induced by conceptual poverty is that of the Eleatic philosophers, who denied change just because they lacked the concept of speed. In fact, when thinking of the arrow passing in front of them, they reasoned thus: the arrow *is* and *is not* here at the same time. But since this is a contradiction, the arrow cannot move. Hence motion, and change in general, is illusory.

Parmenides, Zeno, and their likes would not have denied change if they had had the concept of speed, for this allows us to say that the arrow is now moving in front of us at such and such a speed. Rest is thus just the particular case of null speed. By contrast, the Eleatic stand on change is not just contrary to experience but sophistical.

Taken literally, Parmenides' famous thesis, "What is, is, and what is not, is not," is meaningless. A modern logician would say that it is an ill-formed sentence, for "is" is a syncategorematic particle, that is, an expression that lacks meaning in and by itself. We must add some predicate *P*, such as "is poor," and the universal quantifier $\forall$, or the original sentence to make sense. If this is done, we get something like this: "For all *x*: if *x* is a *P*, then it is false that *x* is not a *P*", or "$\forall x(Px \Rightarrow \neg\neg Px)$," which in classical logic amounts to "$\forall x(Px \Rightarrow Px)$," which is a tautology or logical truth, not an ontological thesis.

Maybe Parmenides intended to state the principle of non-contradiction and/or the excluded middle, something that Aristotle did clearly much later. However, we may never find out what Parmenides really had in mind. As Guido Calogero, the great expert in the pre-Socratics, told me when I consulted him about another pre-Socratic fragment: "Do not worry if you fail to understand those thinkers correctly, because they only babbled, so we can only guess what they thought. All we know is that they were the first to pose interesting but disinterested questions: they engaged in purely spiritual quests.

Two millennia later, Hegel and his disciples repeated the Eleatic contention that change is contradictory but, instead of denying it, they praised contradiction (or conflict) as the ultimate source of novelty. One century later, Mao Zedong added that conflicts are not to be resolved but, on the contrary, exacerbated, to avoid stagnation. This idea seems to have resulted in the so-called Cultural Revolution, one of the worst man-made disasters in history.

While the self-styled dialectical philosophers got confused over change, philosophers of other schools and scientists in all fields, from physics to history, have studied it rationally, that is, without giving up the logical principle of non-contradiction.

Clarity Without Triviality

Analytic philosophy comes in two main shapes: soft and hard. The former, and by far the most popular, is the linguistic philosophy launched by the latter Wittgenstein, the author of *Philosophical Investigations* (Wittgenstein 1953) and other equally shallow works by the same author and his closest disciples, Elizabeth Anscombe, John L. Austin, Peter Strawson, and a few others.

These authors did not tackle any serious philosophical problems, whether old or new: they denied their very existence or declared them to be just linguistic mistakes. They did what Bertrand Russell called "philosophy without tears," that is, simplistic or even childish, for it does not question vulgar knowledge and the associated ordinary language, and so it cannot advance knowledge. The highest praise Austin pronounced about a text was not 'true', 'deep,' or even 'thought-provoking,' but *felicitous*: for him, style trumped cognitive value.

Clarity is necessary for any rational discourse, but it is insufficient, as shown by the mounds of clear trivialities and falsities. In academic matters what we say should not only be clear and thus capable of being examined, debated, and corrected if need be; we should also say something worthy of being discussed, and thus either being improved or corrected. Because they focus on words rather than things or ideas, the soft analytical philosophers are bound to remain marginal to the advancement of knowledge, just as the lexicographers and grammarians do not help creative writers. In the best of cases, the soft analytics are irrelevant. In the worst, they actively oppose the advance of knowledge, as when Wittgenstein declared it to be wrong, and even dangerous (sic), to mention the brain when talking of the mind, or when his follower Peter Winch rejected scientific social science.

The soft conceptual analysts are likely to be openly hostile to the hard ones. For example, Eduardo Rabossi, a co-founder of SADAF (Sociedad Agentina de Filosofía Analítica), who got his Wittgenstein and his pipe at Oxford, became very agitated when he heard me talk in México about exact philosophy, which he regarded as "just like a tic." Incidentally, in 1985 the same man invited me to talk at the Faculty of Philosophy of Buenos Aires University. He was so angry at my criticisms of linguistic philosophy for being superficial, that the same Faculty did not invite me again until 2012, when I addressed a crowd of youngsters whose brains had been severely damaged by the compulsory reading by Hegel, Nietzsche, Husserl, and Heidegger.

In 1999, when the Rector's office of the university invited me to occupy the prestigious Bernardo Houssay Chair, that Faculty found no available classroom, so I lectured at the traditional newspaper *La Nación*, whose secretary, Germán Sopeña, ably moderated the discussion. That is how dissidents are dealt with in authoritarian circles: when they cannot be eliminated, they are ignored or rubbished instead of debated with.

Linguistic Philosophy and Phenomenology

I am critical of both Wittgenstein's linguistic philosophy and Husserl's phenomenology. However, I do not put them in the same bag, for whereas linguistic philosophy is alien to science, phenomenology is its "polar opposite" (Husserl 1995, p. 31), even though Husserl called it a *strenge Wissenschaft* (rigorous science). The phenomenologist must also feign to understand sentences such as this: "As primal ego, I constitute my horizon of transcendental others as cosubjects within the transcendental intersubjectivity which constitutes the world" (Husserl 1970, p. 184).

Unsurprisingly, phenomenology has had an impact on the speculative studies of man, though without inspiring any new findings. In particular, Merleau-Ponty's phenomenological psychology has not enriched our knowledge of either perception or intelligence, and Alfred Schütz's phenomenological sociology could not compete with the scientific sociology of Durkheim, Merton, and Coleman, or even Weber's, Parsons', or Sorokin's. Nor has phenomenology inspired any serious work in political science. In particular, it has not unveiled the kinds and sources of power. Such barrenness should not be surprising, for phenomenology encourages groundless speculation and does not demand evidence. On the other hand, Husserl taught Heidegger, one of the most harmful charlatans of his time, and a main source of the French postmodernists, from Sartre to Foucault, Derrida, and Deleuze.

By contrast, analytical philosophy, even in its soft Wittgensteinian version, had the merit of acting as an emetic to detox a number of thinkers, like my friends José Ma. Ferrater Mora, Fernando Salmerón, and Luis Villoro, who in their youth had been dazzled by the verbal tricks of Husserl and Heidegger, just like their teachers José Ortega y Gasset and José Gaos before them. In addition, SADAF kept burning a small flickering flame of reason in dark Argentina between 1972 and 1985. Although science beats commonsense, the latter trumps nonsense.

Exactness Is Not Enough to Philosophize

To work in exact philosophy, one needs a modicum of logic and abstract (nonnumerical) mathematics, in particular, naïve set theory. The logical imperialists, like Carnap, Montague, Putnam, Hintikka, Tuomela, Mosterín, and his teacher Hans Hermes, seem to have believed that those disciplines suffice to philosophize and even to write about science. But none of them has exactified any key philosophical notions, such as those of existence (not to be confused with someness), property, reality, matter, mind, life, society, justice, power, justice, knowledge, reference, sense, representation, or factual truth.

Formal tools suffice to detect formal errors as well as to exactify intuitive but clear ideas. But they are not enough to philosophize, because they provide neither philosophical problems nor the substantive knowledge required to handle them in accordance with current science. For example, although the great Alfred Tarski has been called "The man who defined truth," he did not distinguish formal from factual truth, let alone exact from approximate truth. And Rudolf Carnap, who wrote extensively on both semantics and probability, missed the concept of reference and did not relate probability to either real possibility or randomness, so that his writings are of no use to the scientists intent on finding out what their theories are about or who are not sure how to interpret propositions of the form "the probability of x is y." Aristotle had rightly advised to start every discourse by stating clearly what it would be about. Regrettably not everyone followed his advice. For example, the constructivist Bruno Latour believed what he had read in vulgarization writings, that special relativity dealt with the measurement of distances and times, and concluded that the said theory dealt with everyday affairs. And when the eminent evolutionary biologist Stephen Jay Gould visited McGill University in 1970, he claimed that paleontology was a branch of geology, because most fossils were found by prospectors – which is like saying that biology belongs in microscopy. I wrote him criticizing his statement, and for a while we kept a lively correspondence.

These and similar equivocations about the proper object or referent of scientific theories led me to writing *Semantics*, which a few years later became the first volume of my *Treatise* (Bunge 1974a). Carnap too had written on this subject but, by combining it with modal logic, he produced a confused theory incapable of identifying the referents of even the simplest proposition, such as "Abelard and Héloise are in love," much less those of complex theories. To my knowledge, my theory of reference is the only one equal to the task, in particular capable of *proving*, not just asserting, that the quantum theory is about physical things like electrons or photons, rather than about the experimenter's mind or his lab operations.

In short, the logical imperialists have not produced a useful semantics, in particular one allowing one to distinguish the mathematical formalism of a scientific theory from its factual referent. Nor has any of them participated in any of the philosophical controversies generated by scientific break-throughs, such as the emergence of field theories, atomic and nuclear physics, physical cosmology, evolutionary biology, cognitive neuroscience, mathematical economics, or political science, in particular the analysis of the key concepts of liberty, equality, and justice.

The logicians who have not thought deeply of science may do worse than being irrelevant to it: they may make elementary errors, like confusing quantitation with measurement, and a scientific model (or circumscribed theory) with a model in the model-theoretic sense, that is, an example of an abstract theory – which is what the structuralist school of Suppes, Sneed, Stegmüller, Moulines and others have done (Bunge 1969f, 1978a; Truesdell 1984).

Another conspicuous mistake of structuralism is to overlook the difference between conceptual existence (e.g., that of infinitely many fractions lying between any two numbers) and real existence (e.g., that of molecules in a drop of water). This mistake induces another: that of identifying things with the concepts representing them, as Patrick Suppes did when he defined a material body as a region of Euclidean space, thus forgetting that mathematical constructs are massless and cannot move.

Such errors and confusions are typical of magical thinking, and they discredit the noble endeavor of using mathematics as a tool for constructing and analyzing theories. No wonder that the structuralists have not contributed to resolving, or at least clarifying, any of the great scientific controversies of the day: they have failed the fertility test.

What Exact Philosophy Can Deliver

Exact philosophers may not only contribute to clarity, precision, and consistency: they may also perform semantic and methodological analyses, and they may build exact philosophical theories. Let us peek at these tasks.

Semantic analysis consists in uncovering the meanings, that is, the referents or denotations, as well as the senses or connotations, of key concepts, statements, or theories. For example, when trying to read quantum mechanical formulas in physical terms, one should not remain satisfied with the opinions of their creators, for they are often extravagant and borrowed from philosophers. Instead, one should start by finding out through analysis what the formulas are about. When this is done, one finds that the referents of the key concepts of quantum physics, such as those of Hamiltonian and state function, are physical entities, such as atoms, or physical processes, not laboratory operations, much less an experimenter's ideas. (Recall Chap. 8.)

The *methodological analysis* of a statement seeks to find out whether it is empirically testable and, if so, how. For instance, the stunning conjectures of the pop evolutionary psychologists are hard to test, because fossils do not show the traces of their original owners' behavior. But if it is assumed, as many of those fantasists do, that we ceased to evolve around 50,000 years ago, and are now walking fossils, then those stories can be checked by studying ourselves. For example, the popular ideas that we are born violence-prone, and therefore primitive tribes lack social brakes and live in permanent war, has been tested on contemporary primitives, and found to be false (Fry 2013).

Likewise, Samuel Huntington's popular hypothesis that modern imperialist adventures are "culture clashes" does not fit history. In fact, nearly all military aggressions, at all times, have been motivated by the wish to grab land, oil, or other natural resources, and in recent times also by the wish to keep the arms industry in business. Another fact at odds with Huntington's account is that of the different cultures that coexisted peacefully for centuries under Roman, Arab, and Ottoman rules, as well as the Hindu until the Moghul invasion. But Huntington's simplistic and profitable ideology will not be budged by ugly facts (Bunge 1998a).

I got an early warning about Huntington's political pseudoscience when listening in 1986 to a lecture by Serge Lang, the distinguished Yale mathematician who waged the campaign that blocked Huntington's admission into the US National Academy of Sciences for fabricating data and writing bogus mathematical formulas. So much for methodological analysis: the previous lines suggest that, far from being a neutral philosophical exercise, such analysis may benefit or hurt vested interests.

As for *exact social or philosophical theories*, there are many of them. In particular, there is no dearth of theories using decision theory to account for almost anything, from disease and war, but they do not fit the facts because the processes in question are causal, not random, and because the probabilities and utilities – the key concepts of the theory – are assigned arbitrarily (Bunge 1996a, 2013a). Ideally, a philosophical theory will be not only rigorous but also compatible with the science of the day.

First Encounters with Exact Philosophy

The first exact philosophers I read were the great mathematician George Boole, interesting positivists like Bertrand Russell, Hans Reichenbach, and Nelson Goodman, as well as some Bayesian (or subjectivist) probability theorists, like Harold Jeffreys. I tried to learn from them while jettisoning their empiricist baggage. When I first met Goodman at Penn, in 1960, I told him that I admired his book *The Structure of Appearance* (Goodman 1951), but rejected his phenomenalism or confinement to appearance. He replied that content did not matter: that only formal precision mattered. I do not share his indifference to truth. Like Euler's dog, I bark at strangers, thus tacitly admitting that they exist outside of my mind.

My earliest exactification trial was my definition of philosophy as the discipline whose philosophy is itself philosophical (Bunge 1944a). I wrote something like "$F^2 = F$," and added that philosophizing is like an idempotent operator. However, that was an isolated episode. My first exact pages were the ones about induction (Bunge 1960b) – which I no longer countenance – and about value (Bunge 1962b). Later came my first theory of partial truth (1963b) and my exactifications of the concepts of problem, and of explanatory and predictive power in *Scientific Research*

(1967b). However, my earliest long-haul project in exact philosophy was my work on semantics (Bunge 1974c, d), a preview of which I presented at the founding meeting of the Society for Exact Philosophy (Bunge 1973e).

The only statement of mine that raised objections on that occasion was that truth cannot be elucidated in probabilistic terms because it neither emerges nor submerges by chance. In my view, the attempt to use probability as the panacea for all philosophical ills, the way I.J. Good did, is just a symptom of laziness. The same holds for the alleged universal applicability of Boolean algebra, modal logic, model theory, and Turing machines.

Semantics

In 1969 I taught a course on semantics, the theory of meaning and truth. I had read with profit Russell's stimulating book on the subject (Russell 1940), but had been utterly disappointed by Carnap's and van Fraassen's books. None of the three philosophers, not even Russell, had taught me how to find out the referents of a theory, nor how to search for its primitives (undefined concepts), nor why mathematical truth is a priori. In addition, both Carnap and van Fraassen made much of modal logic, which is never used in science, not even to elucidate any of the various possibility concepts (real, conceptual, practical, legal, etc.).

Since I was improvising, I had some interesting discussions with my students. For example, we discussed the problem of whether not only propositions, but also the concepts that constitute them, may be meaningful, and the problem of whether meaning is context-dependent. Of course, we started by questioning the nominalist principle that there are no ideas, so that we should talk about terms and sentences instead of concepts and propositions. This point shows, incidentally, that semantics is not done in an ontological void.

Gottlob Frege had emphasized that form is insufficient: that meaning too is needed. But I rejected his extravagant assertion that two propositions mean the same if they have the same truth-value. I thought that meaning precedes truth, if only because we must understand a proposition before attempting to finding out whether it is true. For example, "The Earth is flat" is false but meaningful, whereas Heidegger's "The essence of being is itself" cannot even be false because it is meaningless. And yet Carnap's "Testability and meaning," of 1936, is still being cited as authority, although it is only an embellishment of the operationist dogma that meanings are found by performing measurements and experiments rather than conceptual analyses (Carnap 1936a, b). In real science, the stipulation and discovery of meaning are conceptual operations, not empirical ones. For example, in thermodynamics we stipulate that dQ/T is the entropy increase that accompanies the increase of the quantity of heat in a system by dQ at temperature T; and we find that kT is a thermal energy by performing a dimensional analysis of Boltzmann's constant k in relation to any of the formulas where kT occurs.

Analysis of Predicates

We learned in Gottlob Frege's *Sinn und Bedeutung* that predicates, such as "is habitable", are functions. This is certainly correct, but we cannot admit his thesis that the values of such functions are the truth values 1 (true) and 0 (false). Obviously, the value of "is habitable" for "Earth" is the statement "Earth is habitable". Thus, the function H (for "is habitable") maps the domain P, the set of all places, into the codomain S, the set of statements of the form "x is habitable." In short, $H: P \rightarrow S$ (Bunge 1974b, 1975a). Even shorter: predicates are functions that map individuals into propositions – not truth values.

Thus, reference or denotation has nothing to do with truth. To get to truth we need a second function V: one from propositions to truth-values. Thus the composition of H with V yields Frege's map F from individuals to truth-values. In short, $F = V \circ H$.

Presumably Aristotle, who had warned that every discourse ought to start by saying what will be talked about, would have approved. Of course, we cannot know what Frege would have thought about my emendation. But we can be sure that he would have objected to Anscombe's translation of *Bedeutung* (meaning) as *reference*. As for the *sense* or *intension* (with *s*, not *t*) of either a predicate or a proposition, I equate it with the union or local sum of both its logical consequences – what I call its *import* – and of the constructs that precede it logically – its *purport*. Since primitives precede definienda, and axioms precede theorems, we may stipulate that the *potency* of a construct equals its import.

Finally, I define the *meaning* of a predicate or a proposition as the ordered pair <*reference*, *sense*>. In particular, the meaning of a primitive concept or a postulate p is <reference of p, import of p>. Again, truth does not occur here: we must grasp the meaning of a proposition before trying to find its truth-value.

My approach to semantics, by contrast to its alternatives, makes intensive use of the mathematical concept of a function, in particular that of set-valued function. This clarifies and simplifies a lot. For example, the definite description "Clare's mother" may be analized as the first part of the function "mother of," which refers to mammals, and whose values are propositions of the form "The mother of x is y." Calling M this function, and abbreviating Clare as c, we may condense "Clare's mother" as $M(c)$, or just Mc, which in turn is the first part of propositions like "Clare's mother is dark."

I submit that my analysis is far simpler and clearer than Russell's in his famous 1905 paper "On denoting," which passes for being the earliest specimen of exact philosophy (Russell 1905). In particular, I do not presuppose the existence of the description's subject, the way Russell did. My exactification allows us to keep using descriptions such as "The Holy Ghost", "God's mother," and "The greatest number," without assuming that their referents exist outside certain discourses.

Existences and Someness, Factual and Formal Truths

Nor do I accept the theses that there is a single kind of being or existence, and that the particularizer ∃ is to be interpreted as existence. For example, numbers are in number theory but not in nature, whereas fish are in nature but not in mathematics – whence the very project of "naturalizing mathematics" is wrong-headed. Only objective idealists, such as Plato and Hegel, can afford to ignore the polisemy of the word 'exists'. And only naive (or crass) materialists can deny that all the sciences, even the natural ones, are unnatural.

In my view Alexius von Meinong was right when, while dealing with logic, to use a single word, namely *Gegenstand* (object), to denote both existents and non-existents, since from a logical point of view they are the same. Meinong was wrong only in supposing that there could be a single theory about both. It follows that the Meinong-Russell controversy over denotation, which occupied both philosophers at the beginning of the twentieth century, was a storm in an empty teacup.

Logicians have nothing to say about the present King of France because logic concerns the validity of arguments, not the truth or falsity of the propositions referring to that imaginary object. Obviously, the logical imperialists could not be pleased by my opinion, that they have no credentials to pontificate about either existence (whether real or imaginary) or truth (whether factual or formal).

As for the operator ∃, it is defined as *not-all-not*, that is as *some*. Hence using ∃ does not involve any "ontological commitment" (Quine's phrase). Indeed, the definition in question is this: For any predicate *P*, $\exists xPx = \neg\forall x\neg Px$. For example, "Some people are literate = Not everyone is illiterate." Incidentally, Nicholas Rescher informed me that the great Charles Peirce is to be blamed for the interpretation of ∃ as existence. Hence, if we wish to exactify the intuitive notions of real (or material) and conceptual (or ideal) existence, we must define an existence predicate independent of ∃. This I did four decades ago (Bunge 1976d, 2012a). For example, one may say that some gods are evil without admitting that there are gods. As is well known, mathematical existence is either postulated or proved, whereas in physics and other factual science existence may be conjectured until some experiment confirms or refutes the hypothesis.

Bloopers in the Name of Logic

Mathematical logic is a powerful exactification tool, but it is not the master key its enthusiasts claim to be, and its misunderstanding has generated some wrong philosophical doctrines. I claim credit for having detected and exposed four of them: those of the ontological commitment of the wrongly called existential quantifier; the view that modal logic is the master key for opening all philosophic al doors; the single truth doctrine; and the so-called structuralist view of scientific theories.

The first doctrine, that "$\exists xPx$" must be read as "There are *P*'s", was analyzed, and found wrong, in the previous paragraph, where we showed that its correct interpretation is "Some individuals are *P's*", because this is what the implicit definition of $\exists$ says.

In volume 3 of my *Treatise* (Bunge 1977c) I make my case against the claim that modal logic is an all-purpose theory: it is not, because $\diamond$ is both imprecise and polysemic. Indeed, "$\diamond p$", or "*p* is possible", makes clear sense only if '*p*' stands for a fact, which is not the case of modal logic, where it is assumed to stand for an arbitrary proposition – so much so, that it is universally admitted that every system of modal logic is generated by enriching predictate logic, or even propositional logic, with some axioms or definitions concerning $\diamond$, such as "*p* is necessary = $\neg\diamond\neg p$." That $\diamond$ has multiple interpretations (conceptual, physical, technical, moral, legal etc.) is obvious, and it has been underlined by Reichenbach (1959) and a few others. In particular, the definition of "physical possibility" as "compatible with the physical laws" has nothing to do with "conceptually possible" as identical with "noncontradictory". Therefore the project of building metaphysics of possibility with the sole help of modal logic is simply foolish. Diamonds may be forever, but they are not for every occasion.

As for the vulgar idea that there is a single kind of truth, in particular that of correspondence or adequacy of idea to fact, it is wrong even though it was adopted by the great Tarski. Indeed, it leaves out, among others, the mathematical identification of "*p* is true in theory *T*" as "*p* is provable in *T*," which is common in mathematics even after Gödel's results, which have excited philosophers but has hardly affected mainstream mathematics.

Finally, the "structuralist" idea that "a theory is the totality of its models," championed by Suppes and his disciples, involves the elementary confusions of 'theory' with 'abstract theory', and of 'model' as an example of such with 'model' as a '*teorita*' (Braithwaite 1953), or theory about a narrow scope or reference (Bunge 1974a). In short, mathematical logic is a potent lamp, but when it shines alone it may dazzle to the point of darkening the field of knowledge instead of illuminating it. So much so, that logicism has given birth to a number of serious mistakes. One of them, the "one truth concept fits all," thesis, has blocked research into the investigation of approximate factual truth, of which more anon.

Partial Truth

I hold that the extralogical propositions are not intrinsically true or false that their truth-values are assigned in the light of conceptual or empirical operations. Before these are carried out, the propositions in question are neither true nor false. In biological terms, factual truth values are acquired properties, not inborn ones. And untested propositions are like human embryos: they are human but not persons. This explains why factual truth-values may change in the course of time.

But if truth-values are time-dependent, then some new truths may retain something of their predecessors. In turn, this suggests that there are partial truths, or more or less approximate ones, such as "$\pi = 3$", "$\pi = 3.1$", "$\pi = 3.14$," and so on. All the sciences and technologies use an intuitive concept of approximate truth – to the annoyance of logicians. I have tried several times to exactfy this slippery concept (e.g., Bunge 1963b, 1974b, 1981b, 2012a). My doctoral student Jean-Pierre Marquis (1991, 1992) worked on the same problem with greater mathematical sophistication, and invented what he called a *Bunge algebra.*

I am still dissatisfied, and I do not even know when a theory of truth, or of any other philosophical category, is true. Of course, one may say that a given theory of this kind is true if "it adequately captures the underlying intuitive ideas." But how might one formalize this intuitive notion? So far as I know, both problems are still open.

Books

My semantics course got gradually transformed into a book that I declared completed in the summer of 1972. In the fall of the same year I travelled from Aarhus – where I was spending a sabatical semester – to Cambridge, England, to attend an Institut colloquium. There Max Black asked me about my semantics over tea. As expected, he posed some penetrating questions.

On my way back to Aarhus I stopped by the Routledge publishing firm and foolishly bragged that my semantics book was original, and in particular owed nothing to Wittgenstein, Carnap, Tarski, or Popper. The editorial directors asked me "But then, who is going to read it?" They were right, of course: originality is not a market value, and is not even appreciated in the philosophical community, unless combined with outrageous error. Some time later, the prestigious Dutch publisher Anton Reidel – perhaps encouraged by the success of my *Philosophy of Physics* (1973a) that he had published recently – dared publish that work as the first installment of my *Treatise on Basic Philosophy*. His successors, Kluwer and Springer, kept old Reidel's trust in my judgement.

The treatise was originally planned to occupy seven volumes, and to be completed in the same number or years. The actual result was nine volumes published in the course of 13 years (1974–1987). My *Political Philosophy* (2009) should be regarded as the tenth volume of this work. So, altogether it took me four decades (1969–2009) to complete that project in addition to several others.

Foundations and Philosophy of Science Unit

As I was teaching my first semantics course at McGill, I proceeded to set up the Foundations and Philosophy of Science Unit despite the unwillingness of the university administration. I wished to invite a couple of postdocs a year, as well as visitors for shorter periods and the occasional lecturer, to keep a lively seminar going. To carry out this plan I needed funds and a locale. I got both during the 1969–1970 academic year. I started by applying for the prestigious and demanding Killam fellowship, administered by the Canada Council, a government funding agency in Ottawa, the federal capital. One winter day of 1970 I got an invitation to visit the Canada Council, to describe my project to a panel of scholars. Several decades later, John Polanyi, a Nobel laureate in chemistry and the son of the famous polymath Michael Polanyi, told me that he had been a member of that panel, and had strongly supported my application. My university diligently administered the funds without the discount usual in American universities.

Once I secured the funds I went to see the university official in charge of buildings, asking him for a set of offices. The man, an expeditious war veteran, gave me all I needed without further ado and behind the backs of my chairman and my dean, perhaps because I addressed him in French. During three decades I occupied half of a furnished building constructed a century earlier, where my visitors worked, and where I taught my courses and held my seminar. Finally I went to the printing shop and ordered stationary with the name and address of my Unit, which had no official standing.

Though ghostly, the Unit was even listed in the entrance nameplate. All of this was achieved without any bureaucratic hassle. But our building was in such a bad state of repair that once the roof over my office caved in under a heavy snowfall and I lost countless documents and books.

When my visitors started coming from a dozen different countries, I got them an office at the Unit, a library card, and access to the mail system. The first postdoc was David Salt, a physicist with a Cambridge doctorate. Later came Gerhard Vollmer, whom I had met in Freiburg through Siegfried Flügge, and the Japanese Hiroshi Kurosaki, who came highly recommended by a wellknown Japanese philosopher. Both men had studied physics and philosophy. We met at least twice a week: at the seminar and in my office, where we discussed the problems they encountered in their research projects. Both withstood stoically the first draft of my 'scientific metaphysics' – an expression coined by the incomparable Charles S. Peirce.

In the following years I received the Finn Raimo Tuomela, with a brand new Stanford doctorate and recommended by Jaako Hintikka. He was an empiricist on arrival, but a realist when he left. While at the Unit, Tuomela wrote *Theoretical Concepts*, which I published in the Library of Exact Philosophy. However, he covered his tracks: nowhere in his writings is there any mention of his three visits to the Unit. Other guests of the Unit were the German Peter Kirscheman, an expert in Soviet philosophy recommended by Józef Bocheński, OP; the Salamanca professor Miguel Angel Quintanilla, who during his tenure at the Unit decided to specialize in the philosophy of technology; the Venezuelan quantum chemist Máximo García-Sucre, introduced to me by my former physics student Andrés Kálnay, and with whom I wrote several quantum physics papers; the American mathematician William Hartnett, whom I "discovered" by reading one of his crystal-clear books; the Austrian philosopher, logician and theologian Paul Weingartner, whom I had befriended in Salzburg; the Mexican physicist Guillermo Covarrubias, whom I had met in México, and who updated my axiomatization of general relativity; the Argentine-Canadian mathematician Arturo Sangalli, who helped me craft my theory of properties and kinds (Bunge and Sangalli 1977); the Argentine-Venezuelan physicist María-Esther Burgos, who persuaded me that the "collapse" (or decoherence) of a state function was a natural event; and the Argentine-Cuban biochemist Ernesto Mario Bravo, who had been tortured by Perón's political police, and who taught me some biochemistry.

Six of the 20 something postdocs who passed through the Unit were unproductive: the Austrian recommended by Popper, and who every week asked me for a new project; the brewery-heir David Probst, recommended by Prigogine; the Argentine logician recommended by a former teaching assistant, who kept typing furiously but never showed me any of his productions and never participated in our seminar; the Swede who started out thinking about quantum mechanics and ended up as a neoconservative ideologue; the Brazilian who had grown up in Goa and worked in Uganda until being expelled by Idi Amin, and spent all his philosophical energy ably criticizing Feyerabend's nonsense; and the Serbian physicist who, instead of working on the philosophy of physics as agreed, spent all his time reading Marxist books that he borrowed from our library and took back home. What a weird profession ours is!

Society for Exact Philosophy

The inaugural meeting of the Society for Exact Philosophy (SEP) coincided with the first big snowstorm of the 1971–1972 winter. This association, constituted by American and Canadian philosophers, is holding its 42nd annual meeting as I write. The first meeting was attended by the Argentine Carlos Eduardo Alchourrón, the Guatemalan-American Héctor-Neri Castañeda, the German Peter Kirscheman, the Quebecois-American Hugues Leblanc, the Dutch-Canadian Bas van Fraassen, and the Finn Raimo Tuomela. I had also invited Quine, who excused himself, and Richard Montague, who promised to attend but was murdered by one of his occasional partners a few days before the meeting started.

We founded the SEP during the dinner that McGill offered in the restaurant Le Caveau, decorated with curtains with prints of the beautiful Bayeux tapestries. When it came to electing authorities, somebody proposed me for the presidency, but I declined and proposed Castañeda instead. He accepted but accomplished nothing. By contrast, Bas van Fraassen, voted secretary general, organized the second meeting, held successfully at the University of Toronto. Bas was a good organizer who in his youth had volunteered for the Dutch Salvation Army, where he saw corruption. Incidentally he, the Chilean Roberto Torretti, and the Argentinian Riseri Frondizi called themselves existentialists, but I never detected any nonsense in their writings, on top of which they are not about metaphysics. We decided to charge a membership fee of $2, but Jim Lambek paid only $1, a prank that gave our treasurer a headache. I collected the main presentations in the volume *Exact Philosophy*, published by Reidel in 1973 (Bunge 1973e). My own presentation, titled "A program for the semantics of science," was also published by the *Journal of Philosophical Logic*, which van Fraassen had just launched (Bunge 1972).

Scientific Metaphysics

The idea of scientific metaphysics seems to have been conceived by the amazing American polymath Charles Sanders Peirce, who started to write a whole book on the subject (Peirce 1935). Regrettably this book is rather chaotic, and Peirce sought counsel more in Suárez's late scholasticism than in science, although he knew a lot of it. I started from science rather than from dusty books. In particular, I started with what I regarded as the tacit metaphysical assumptions of all scientific research projects: Heraclitus' *panta rhei* and Lucretius' *ex nihilo nihil*. I also assumed that, although some metaphysical principles or postulates may not be checked with empirical data, they should be both clear and consistent with the best available scientific knowledge. Consequently, a theory that holds that "time flows," instead of "things flow," is unacceptable because it invites the absurd question: "At what speed does time flow?" Likewise any philosophy of mind that detaches mental processes from brains is to be rejected because it is a variance with cognitive neuroscience.

Corsica

We spent a long 1971 summer in Corsica, where we arrived just a few weeks after my tennis elbows were operated on. We drove all around that very beautiful island until finding a suitable house for rent near Porto Vecchio, the seat of the redoubtable Foreign Legion. To reach the best beach, we drove across the island in half an hour. Although in France beaches are public property, the best portions of Palombaggia had been privatized.

The house we rented was in the middle of a large villa owned by a Madame Cortial, an avaritious and domineering woman who shouted at me for having allowed a family of spiders to cover the huge fireplace with cobwebs. I could never bring myself to destroying a spiderweb, least of all its creator. Nor could I stop the resident she-goat from jumping onto the mantelpiece of a large fireplace and taking possession of it by peeing vigorously from it. Who was I to challenge such a persuasive notarial procedure? After all, "naturalized legal work" does not sound more absurd than "naturalized mathematics."

Madame Cortial had two sons, one of whom a Trotskyist fanatic who reminisced with relish about his vandalisms in the May 1968 student rebellion, which had demanded *Tout, tout de suite*, but only achieved President De Gaulle's resignation and the creation of the Vincennes university campus, where students were allowed to insult their professors.

The villa's caretakers were a timid and friendly Frenchman and his formidable Polish wife, who bred rabbits whom she called *mes bébés* but sent the oven without compunction. One day Eric, then 4 years old, and his gang "liberated" the rabbits, to their owner's indignation. On another occasion I announced to the caretakers, both devout Catholics, that a miracle had just occurred: the Lord had flooded the street with wine. They looked with disbelief me, until they found that the wine had been spilled by a tank truck.

Eric was admitted in the local *école maternelle*, and treated with loving care by his teacher. He soon mastered the language he needed to recruit a gang that kept inventing new games, while Marta and I either studied or read some great French novels, like Marguerite Yourcenar' masterpiece *Mémoires d'Hadrien*.

That summer I concentrated on learning some molecular biology. Back in Montreal I wrote to Jacques Monod criticizing his teleonomy, which I regarded as a disguise of the old teleology ridiculed by François Rabelais half a millennium earlier. Monod answered politely but declined to engage in a discussion due to work pressure. In fact he was dying from cancer. It seems as though the French climate were just as inhospitable to Darwinians, like Monod, as to exact philosophers. Near the end of our long Corsican holiday we flew to Bucharest to attend a congress.

Bucharest Congress

I had discussed my preliminary ideas on exact metaphysics at the philosophy seminar in Göttingen, in 1969, and subsequently in a paper published by the *Journal of Philosophy* (Bunge 1971). I also presented them at the 3rd International Congress of Logic, Methodology and Philosophy of Science held in Bucharest in 1971. The only objection I heard was that I used the word 'metaphysics'. I answered that the old whore cannot be ignored, and might yet be rehabilitated.

In Bucharest I met again with Tadeusz Kotarbinsky, Alfred Tarski, Bonifati Kedrov, Azarya Polikarov, Nicholas Rescher, Hiroshi Kurosaki, and Tom Settle – a former Methodist missionary whom Agassi had converted to philosophy in Hong Kong. I also chatted with the Romanians Athanase Joja–a philosopher whom I had met a decade earlier in Buenos Aires – and the logician Griori Moisil, whom I had met in Liège.

I was also introduced to Carl Hempel aka Peter, the Russians mathematician Andrei Markov Jr, and the charming and open-minded Marco Markovic – son in law of the Serbian dictator. Hilary Putnam astonished the audience with his vehement denunciation of the Ceaucescu regime from a Trotskyist viewpoint – which he soon jettisoned. And Nicholas Rescher surprised some of us for his legal acumen.

Joja and Moisil were survivors of the old regime. Joja had lasted due to his mimetic skills, and Moisil by taking refuge in the history of logic. When I invited

Joja to address my philosophy students during his State visit to Buenos Aires in 1958, he spoke about his studies in pre-war Italy and, to my disappointment, he referred to phenomenology as a respectable doctrine. Moisil had spoke in Liège about the need to adapt logic to electronic computing, for instance adding the symbol ':=', to distinguish value assignment (as in "let x equal 3", or $x := 3$) from identity as in "$x = x$". Presumably, the ruling party was not interested in such subtleties. And during the congress it decided that having two sets of courses in the history of philosophy, one taught at the party school and another at the university, was an unnecessary duplication. The reader will guess which one was eliminated.

I also got to know several young Romanian philosophers interested in updating Marxism, though not to the point of trying to do something about it. One of these young scholars translated my long paper on social structure, which first appeared in Romanian. Another, Radu Bogdan, soon migrated to the USA, and invited some North American philosophers to write their autobiographies, which he got published. I declined his invitation, whereas Pat Suppes accepted, and had his autobiography published at the tender age of 57 years. There we learn that he got the National Medal of Science for his many publications, particularly the ones on "measurement" theory, a specimen of pseudoscience since it claims to account for measurement, an empirical procedure, in purely mathematical terms (Bunge 1973d). The origin of this confusion was the mistranslation of the German *Maß* (measure) by Ernest Nagel, Suppes's teacher.

The most picturesque character I met in Bucharest was a friendly chainsmoking analytic theologian who claimed to have proved the existence of God using only set theory – a case of pseudoscientism. He worked as a janitor at the local university, but got all his theological fantasies published in a scholarly journal because they bristled with mathematical symbols that no one bothered to check.

The applied mathematician Mircea Malitza, who had published a paper in mathematical history, invited us to a party in his flat along with Tarski, Kedrov, and others. When I told Kedrov that one of my discrepancies with Marxism was that I rejected dialectics, he replied: "Don't worry, *tovarich* Bunge, for Marx mentions dialectics only six times in his *Kapital*."

Another memorable encounter was the one I had with the dictator Nicola Ceaucescu – then at the peak of his career but with only 17 years left to live – at the official party for the congress participants. He seemed annoyed when I told him that I was glad that a congress in my field was being held at a socialist country – but of course this may have been the interpreter's fault. Ceaucescu may be remembered as the second communist leader, after Marshall Tito, to defend his country's interest against Soviet imperialism. It must also be admitted that, right up to the said congress, there had been considerable cultural freedom in Romania. The Wild-West style murder of the Ceaucescu couple, without a proper trial, brought neither honor to his executioners nor stability to his country.

At that party Marta and I first met the charming Salamanca couple Miguel-Angel and Ana Quintanilla, with whom we have been close friends ever since. Shortly thereafter Miguel-Angel spent some time at McGill as my research associate, and a few decades later he succeeded in setting up at his university a unit for Hispano-

American students in the sciences of science and technology, which José Luis Pardos and I had planned but never got beyond the initial ceremony.

Essentials of My Scientific Metaphysics

The essentials of my scientific metaphysics are found in volumes 3 (1977) and 4 (1979) of my *Treatise*, on which I had been working in Montreal, Aarhus, Zürich, and México City. Volume 3 deals with concrete things and their properties and states, as well as their changes, space, and time.

One of the novelties of scientific metaphysics is that, because it makes use of some elementary abstract mathematics, such as semigroup theory, my mereology, or theory of physical addition and the part-whole relation, is far simpler and briefer than its Polish precursors – the work of Lesnievsky, Sobocinski, and Tarski – which accomplishes very little in an extraordinarily complicated and therefore counter-intuitive formal apparatus.

I postulated that the set S of concrete things, together with the binary operation $\oplus$ in S, constitutes a semigroup. I interpreted $\oplus$ as the physical or mereologiacal sum, and stipulated the following definition of the part-whole relation: $a \leq b = (a \oplus b = a)$. The ontological concepts of physical addition and part-whole relation are presupposed by all the factual sciences.

Another novelty of my ontology is that it includes a theory of properties as distinct from (but representable by) predicates such as "is new" and "is newer than." It is usually taken for granted that properties are the same as predicates. My rationale for the property/predicate distinction is that, whereas the logical negation, disjunction, and conjunction of predicates are proper predicates, this does not hold for properties. For instance, commonsense resists reifying the predicates "is prime or vertebrate," and the quantum theory has no use for "the electron passed here at such and such a speed." My paper was rejected by several leading philosophy journals.

The concept of a property of a thing allows one to define that of state, which in turn leads to that of state space, or totality of states a thing may be in. (David Armstrong's proposal, that the world is made up of states in themselves, would be laughed at by any science student.) But only some of all the conceptually possible states are possible: those that fit the relevant laws. In other words, a realistic ontology, unlike that of possible worlds, will focus on the lawful state spaces.

An *event* is representable by a pair of points in a lawful state space, and a *process*, or sequence of states, by a trajectory in a lawful state space. These basic ontological ideas are useful to exactify and organize the material, often vague and chaotic, of an emergent discipline in the social and biosocial sciences. For example, they have helped me understand cognition, disease, social development, and liberty.

The same volume 3 of my *Treatise* also contains a mathematical theory of spacetime as the basic structure of the set of changing entities. I adopted the triadic relation of interposition (as in "the torso lies between the head and the limbs") as primitive or defining. And, of course, I had no use for the relational (or adjectival) theories of space starting from perception.

This third volume (Bunge 1977c) took me far more time than expected, as I started it in Montreal in 1969 and completed it in México City in 1976 with the help of my research associates, the mathematicians Adalberto García Máynez and Arturo Sangalli.

I went to both Aarhus (1972) and Zürich (1973) on a Guggenheim fellowship, which I had won on the surprising recommendation of Richard Rorty, whom I met only three decades later at a congress in Moscow. I chose those towns because Marta was then working with mathematicians based at both cities. But before going there we went to southern Spain for the summer.

The *Treatise* Plan Is Born in Spain

We flew from Montreal to Madrid and went straight to the central railway station, with the naive idea of catching the next train to the Costa del Sol. (Internet was not yet born, and information about Spanish tourism was hard to get in Canada before our friend Jo Lu Pardos went there as Spain's ambassador.) Since train departures were infrequent, we went by taxi to Málaga, where we were visited by an economics professor who assured us that Spain's economic recovery was bogus. I still had to learn that economics is not yet a science.

The next couple of days we explored Marbella and its environs until we found a suitable house for rent. On the nearby beach we made friends with a limping French giant and his family. One night they drove us to a disco. The so-called music was playing so loud, that it made me sick, and I had to be driven back home. In the backyard there was a green lawn without trees and surrounded by a high wall, so there was nothing to do but think. There I had the idea of expanding my work to encompass all the main branches of philosophy.

Since Morocco was just across the Gibraltar strait, one day we drove our small Seat to Algeciras, where we boarded the ferry to Tangiers. We disliked the rude immigration and customs officers, but were seduced by the views, the beach, the common people, and the *zoco* or market, where we got delicious dates. We were also treated to mint tea, without charge, by an old Jewish shopkeeper who remembered his Spanish from the days of Spanish rule.

On our way back we were dazzled by the blue majolica in mosques and palaces, as well by the green valleys in the Riff, where the legendary Abd el Krim had defeated many a Spanish regiment in the 1920s. But that had also been the place where General Franco had recruited his ferocious Moorish troops, which helped him terrorize Spain. Remember the stirring song that starts: *Los moros que trajo Franco/ En Madrid quieren entrar./ Mientras queden milicianos los moros no pasarán. /¡No pasarán!* The monument that records the gratitude of the Crusader of Christendom to those worshippers of Muhammad is placed in the middle of a prairie in distant Asturias, and it is distinguishable only by a bronze crescent.

From Spain we flew to Paris, where we watched a Charlie Chaplin retrospective that captivated Eric and, of course, we admired once again Notre Dame and the Louvre.

Denmark

Finally we flew to Aarhus via Copenhagen, and settled in a cottage near the beach, where we stayed till the first cold days. Then we moved to a small apartment near the harbor, from which we heard the foghorn. Our lodging was situated above a bakery, whose delicious aromas woke us up every morning with gluttonous desires that we did not always suppress. Our nearest neighbors were the Swiss mathematician Michel Thiébaud and his wife, for whom I carried upstairs the mattress where a few days later she had her first child. (Michel, working hard in his office, could not be bothered.) Fifteen years later we met them again in Geneva.

In Aarhus Eric was admitted to a kindergarten that after a couple of months transformed him into a Danish boy. He learned to speak fluently the local dialect and built large ships with Lego. I studied several articles and books, among them David Marr's intriguing mathematical theory of vision (Marr 1969/1982), as well as Evelyn Piélou's beautiful *Introduction to Mathematical Ecology* (Piélou 1969). I also wrote my first long paper in mathematical sociology, on the concept of social structure. This subject, that ought to lie at the very heart of sociology, had been studied in detail only by the eminent Peter Blau. But, although Blau loved theories more than data, he had only an intuitive idea of social structure.

By contrast with the vast majority of papers in that field, which start off with statistics or equations, mine moved from quality to quantity: I began with equivalence relations like the one occurring in the statement form "*a* is equivalent to *b* with respect to *r*, or $a \sim_r b$." Using equivalence relations such as "same occupation," I built social equivalence classes, such as cohorts and income brackets, which I gathered in matrices. Finally I defined the structure of a society as the family of the given equivalence classes – like a heap of pizzas.

Measurement is thus reduced to its simplest form, namely counting heads in every equivalence class. In this approach, numbers come last but they do come, by contrast to the purely qualitative approach, which is useless to statesmen and government officials, who need to know how many need what and how much. My paper was rejected by various social-science journals. One of the referees confessed that once he had seen a Boolean theory (sic) that he had not understood. It was finally included in an edited volume (Bunge 1974b).

Aarhus is a delightful old town on the Baltic, where all summer we read in the garden till midnight in the sunlight, and swam in tepid if polluted waters. One afternoon I bumped on the beach into my old friend Andrés Raggio, the Argentine philosopher, who also visited me in Zürich a few months later. He introduced me to the delicious müsli, and I introduced to him to Paul Bernays, whose seminar we attended. Regrettably, we did not understand that day's speaker, a young man who spoke the Swiss dialect of German.

One early evening we dined at one of the beach mansions, as guests of a remarkable Danish diplomat who smoke cigars and served us raw eggs that I did not touch. But she entertained us with her experiences in West Africa, where Scandinavians are welcome because they had never raided it. We spent another interesting evening at a colleague's, whose wife wore boots in summer and treated him as an obnoxious child although he was very well behaved. We discussed his project on the so-called relevant logic of Anderson and Belnap, and agreed that it fails because it admits the addition postulate "If *A*, then *A* or *B*," a breach through which any irrelevance can sneak. The man committed suicide shortly thereafter. I never knew whether his wife or his philosophy had driven him to despair.

We had rented the cottage from an impecunious medical student, who had gone to Sweden for the summer to earn a few extra kronen. One late afternoon we were visited by an employee of the electricity company to check the meter – or so I understood. Only at midnight, when it was really dark, and the birds had finally stopped chirping, did we realize that the man had cut the electric power – for lack of bill payment, as we learned the next day. I had suspected nothing because I had understood nothing of what the worker had told me in a very polite way, though of course in a language that I did not understand.

Arguing with Danish Academics

I gave talks at all the Danish universities, and was always invited to long lunches accompanied by solemn short speeches, as well as by alcoholic beverages of seven different kinds, each in a different crystal glass. The most successful of my talks was the one I gave at the Niels Bohr Institute, which my teacher had visited every summer during the halcyon days of the quantum theory, when the whole of physics seemed to be undergoing a radical reconstruction by about 30 enthusiasts, only one of whom, grandpa Niels, dared ask skeptical or at least cautious questions. My criticisms of the standard or Copenhagen doctrine did not please the professors, but was enthusiastically applauded by their students.

At Aarhus I was visited by my former student Andrés Kálnay, with whom we wrote a paper on the peculiarities of quantum physics (Bunge and Kálnay 1975). I also had interesting conversations with a Danish physicist interested in foundations, as well as with the historian of science Olaf Pedersen.

Marta and I made friends with Uffe (pron. Oove) Juul Jensen, the most expansive and helpful of my philosophical colleagues. Uffe called himself a Marxist, and had authored a sophistical philosophy of mind without brains while campaigning against the European Union – which I regarded as a valuable wedge between the two main international blocs. He visited us often, always enthusiastic about something, as well as sweaty and smelly from running. (Less than half of the Aarhus homes had bathing facilities, so most people went to public baths only when absolutely necessary.)

The other local philosophers, though mostly friendly and helpful, were a lost cause: they had no definite research projects, and at that time were fascinated by the Wittgensteinian Peter Winch and Jonathan Rée, the British promoter of "radical philosophy." The self-styled radical philosophers were a motley group of backward-looking individuals: existentialists, critical theorists, unreconstructed Marxists, and academic feminists. I found it impossible to engage Rée in a rational debate, for all he was interested in "alternative thought," the scholarly counterpart of alternative medicine. And Winch, a Wittgensteinian, was admired in Aarhus and elsewhere because he had attacked the very idea of a social science. I learned nothing from my amiable philosophy colleagues, and I only talked to them, over delicious pastries and coffee, about matters that did not interest them. Clearly, I had arrived at my antipodes. Eric and Marta, by contrast, belonged there: building blocks, whether Lego's or conceptual, where their thing.

The Philosophy Department assigned me a comfortable office with a single defect: it was next door to that of a colleague who, as soon as he arrived, fell asleep and snored very loudly. I bought a carpet and hung it over the door that joined our offices, but to no avail. The snorer had studied theology and philosophy and, since he could not produce anything, was waiting to succeed his father-in-law as the pastor of a local church. But when the old man retired, the Bishop appointed his own protégé, apparently without consulting his ultimate Superior. I never heard again about him, and saw only one of his colleagues when I returned to Aarhus a quarter of century later for a meeting of the Académie Internationale de Philosophie des Sciences.

The university libraries kept buying books and were up to date with the main scientific journals. The borrowing procedure was simple: one took out the desired volume and left a note with one's name and office number, and left the door open. Denmark was one of the very few countries where such trust was justified and worked. Behave honorably and you may be treated honorably.

The mathematicians's cafeteria – where I met with Marta and some of her colleagues for lunch– offered a delicious smorgasbord, in fact the only edible and healthy meal of the day. The strongest evening dish at the local restaurants was a chunk of meat drowned in a suspicious brown sauce accompanied by baked, fried, and boiled potatoes – with neither greens nor salads. (How is it possible for Danes to make some of the most beautiful children in the world on such a poor diet?) Towards the end of our stay I discovered a greengrocer who wore a philosopher's beard, spoke flawless English, and sold exotic stuff such as tomatoes and eggplants.

While at Aarhus I applied decision theory – which I had studied in Corfu from a book by Anatol Rapoport – to the American war in Vietnam (Bunge 1973c). I concluded that the American leaders were not acting rationally, as became evident 2 years later when they and their South Vietnamese accomplices fled the country in a hurry. However, I failed to point out that my own calculations, just like those of the US War Secretary Robert McNamara, involved invented probabilities and utilities, so that both were pseudoscientific fantasies. I hope to have redeemed myself in my later work (Bunge 1989f, 1996a, 1998a).

Zürich

When Aarhus had nothing further to offer, we decided to spend the balance of my sabbatical year in Zürich, where we had some interesting contacts, such as Paul

Bernays. But first we engaged Dawn, a drifter who had left her native South Africa disgusted by the Apartheid regime, as an au pair to look after Eric.

We drove through Germany and arrived in Zürich on the night of the same day we left. Over the next semester we stayed at the apartment assigned us by the famous ETH, or Polytechnic, that had invited me as a visiting professor for the first half of 1973.

The office I was assigned was on the second floor of the building occupied by the team that was studying the work of Etienne-Louis Boullée, an original French architect of the Enlightenment whose grandiose designs were never built, yet had greatly influenced the modernist ("rationalist") architects. Are there any twentieth century philosophers likely to help the twenty second century ones?

One quiet Saturday afternoon, when I tried to leave my office, I found the building door locked. Since no one answered my phone calls to the buildings and grounds office, I jumped to the garden, and as a consequence injured my hip. This injury kept me limping for several years. After this incident I was moved to a safer office in the Clausiusstrasse. There I kept filling pages with my exact philosophy, but threw most of them into the wastepaper basket when I realized that they were wrong. How right had been John Maynard Keynes when he advised a friend never to work in solitude, without receiving feedback! Thinking is private, but quality control is public.

During my ETH tenure I gave a talk on the semantic concept of intension or sense, taught a course based on my *Scientific Research*, and filled in a couple of times for a sociology professor at the nearby Zürich University. My formulas "$I(a \,\&\, b) = I(a) \cup I(b)$" and "$I(a \vee b) = I(a) \cap I(b)$" for the intension of the conjunction and the disjunction of two constructs a and b provoked questions from Paul Bernays, that were interpreted as refuting me. Actually he may have been just intrigued, for intension or sense flows against the truth current: the more one says, the more likely is it to be false. Bernays and his sister invited Marta and me to tea, and we reminisced about the time, a decade earlier, when we had interacted at the universities of Pennsylvania and Delaware. Back then I had shocked him by asserting that Freud, who had married his close relative Martha Bernays, had only been a successful charlatan.

In Zürich I befriended the historians Jean-François Bergier and Carlo Cipolla. Bergier regarded himself as a disciple of the great Fernand Braudel, famous for his work on the Mediterranean at the time of Philip II – a gem of the systemic approach. He had just married a nice, plump and pink librarian, who called him *Du*! ("Hey, you!") from the bottom of the high hand-ladder in the huge library where he was working. Despite his timidity, Bergier presided over the commission of experts that denounced the complicity of the Swiss power elite with Nazism. As for Cipolla, an economist by trade, he had written with erudition and elegance about the clock and other artifacts, as well as about "The basic laws of human stupidity."

Towards the end of the semester I flew to Washington D.C., to participate in the crash course for high-school physics teachers sponsored by the National Science Foundation. At the Catholic University of America, where the course was being taught, I renewed my cordial relation with the historian of causality William Wallace, OP. There I also first met Augusto Durelli, a world expert in photoelasticity, and one of the handful of Catholic antifascists in Argentina. He emigrated after having been jailed for his opposition to emergent Peronism.

Two Catastrophes

Before returning to Montreal we spent a month in Corfu, where we would not return for more than a decade. We stayed at a sun-drenched house owned by our old friend Spiros Bourbos The only good thing about that dwelling was that it overlooked a small cove with waters so clear, that from the top we could see the fish, shells, and pebbles on its bottom.

I had brought only one book to study: René Thom's highly anticipated book on what he called a general theory of models, which in the grapevine was referred to as mathematical biology (Thom 1972). How disappointing! The famous inventor of "catastrophe" (singularity) theory had promised to mathematize biology but did nothing of the sort. He was only interested in classical morphology, and disbelieved all of modern biology, from evolution to genes. Moreover, I heard him say in public that his a priori speculations made it possible to close down all biology laboratories.

In 1991 I had an exchange with Thom in the journal *New Ideas in Psychology*. I made fun of his complaint that, when discussing billiards balls, the physicists adopted the viewpoint of the ball set in motion by the billiards player, instead of the target's, which he regarded as unfair. While on a sabbatical in Genoa I wrote a paper for his Festschrift, on the chasm between mathematics and reality (Bunge 1994). My thesis is summarized in Charlie Brown's little sister's complaint: "Only in math problems can you buy 60 cantaloupes, and no one asks what the hell is wrong with you."

Return to Montreal

One day, while in Zürich, I escaped to Montreal from Zürich, and bought our first Canadian home. It was situated on the slope of a hill, and had a garden that our children would greatly enjoy – but it also was on a very busy street, which gave me nightmares. The kitchen had to be renovated, a job we gave to a team of handy-men who came at sundown after having worked at another site and having quenched their thirst with several beers. One evening they were so tipsy, that they got the kitchen floor wrong, so it had to be redone. Another evening they held a party together with some friends, and raided our liquor cabinet. While the bookshelves were being built, our living-room was empty except for a large Afghan rug, which made some visitors wonder whether it was our meditation room.

Shortly after making the house habitable I flew to Varna, the Bulgarian harbour on the Black Sea, to participate in the 15th International Congress of Philosophy, as well as in the annual colloquium of the Institut International de Philosophie. In the former I read a paper on my analysis of meaning, and in the latter, devoted to dialectics, I first acted as interpreter of the Mexican Eli de Gortari, who had written a whole book on "dialectical logic." Then I presented a detailed and harsh criticism of the dialectical ontology of Hegel, Engels, and their Soviet followers (Bunge 1975c). My main objections to this doctrine are that some of its theses – such as that everything is a unity of opposites – are false, while the rest is so muddled, that no one had managed to exactify it. For example, why repeat the Marxist formula "The form of production contradicts the relations of production"? if what is really meant is that "Whereas the production of wealth is social, its appropriation is individual"? (Incidentally, why reduce all social relations to relations of production?)

My paper provoked nearly 20 replies of several kinds. A Dutch phenomenologist claimed that exactness is not a virtue but a positivist prejudice, and the orthodox Marxists, like I.S. Narsky, repeated their catechism. However, there was a surprise: about 15 philosophers from the periphery of the so-called socialist camp admitted that the dialectical texts are obscure, and a few of them, in particular the Romanian Pavel Apostol, promised to mathematize dialectics within a year. Apostol's reply (1985) was published a decade later, but did not meet my challenge: he had failed to exactify dialectics (Bunge 1985c). Worse, he died that year.

Our daughter Silvia Alice was born a few months later, in the midst of a snowstorm so severe, that it paralyzed public transportation, which in turn reduced the numbers of hospital personnel. I had to walk to the hospital, sinking in the snow, but in the end everything turned out all right. A few days later we took Silvia home, though Marta slipped on the ice on our sidewalk.

Whereas Eric had cost us $1,000, Silvia came without charge, for in between the whole of Canada had adopted the Medicare health plan pioneered by the great socialist statesman Tommy Douglas, voted by a Canadian Broadcasting Company poll as "the greatest Canadian."

A few months later we got our Canadian citizenship without having to relinquish the Argentine one. It came just in time, because the military dictatorship had deprived us of our Argentine citizenship.

Another Congress in Israel

At the end of 1974 we flew to Israel, where I participated in a large international meeting about "ethics in a society dominated by technology." All four of us stayed at our own cost at a fancy hotel on the famous Lake Tiberias. Our children were much admired for their good behavior. I, by contrast, was rudely scolded by a rabbi who caught me warming up Silvia's milk in the oven reserved for meat. Eric, just 8 years old, wrote in pencil a journal that he sold for a few shekels. His paper folded when he wrote that the Palestinians too were entitled to land and water.

The congress was held at the prestigious Technion, the Israeli MIT, in the coastal town of Haifa. The very first day, an engineering professor warned us not to believe we were in Asia: Israel was a European enclave. The congress was full of celebrities, like Daniel Bell, Sir Isaiah Berlin, Max Black, Melvin Kranzberg, Hans Jonas, Robin Fox, and Lionel Tiger. I was asked to summarize the presentations. When Berlin's turn came, I excused myself saying that I had not understood his pronunciation, which imitated that of the English upper classes.

México Calls

Something made us accept Fernando Salmerón's invitation to move to UNAM (Universidad Nacional Autónoma de México) in mid-1975. We got the requisite documents, sold our house, asked for leaves of absence without pay, and flew to México City, where we were received with open arms. Marta worked at the Institute of Mathematics and Systems, and I at Salmerón's Instituto de Investigaciones Filosóficas, and soon also at UNAM's Faculty of Philosophy, as well as at the Metropolitan University. Inadvertently, I had acquired the Latin American taste for multiple *chambas* (jobs).

We quickly made several new friends, with whom we visited frequently. These gatherings, held at various homes, joined colleagues from several fields, as well as a few people with direct phone lines with the Republic's president. We met, among others, the Uruguayan philosopher Mario H. Otero and his wife the biologist Lina Bettucci, the seismic engineer Emilio Rosenblueth, the seismologist Cinna Lomnitz, the physicists Tomás Brody, Rafael Pérez Pascual, Luis de la Peña and his wife Ana María Cetto, the mathematicians Adalberto García Máynez and Tomás Garza, the social anthropologist Larissa Adler, and later on my son the physicist Carlos and his wife the chemist Annik Vivier, both of whom moved to Mexico from Brazil for good along with their four children.

The variety of interests of our Mexican friends made our parties interesting and fun. We were also frequent guests of Fernando and Licha Salmerón, the most generous and affectionate of hosts. We played volleyball with Fernando and his eight well-educated children. Marta taught one of them, Leonardo, who became a productive mathematician, and I learned about Mexican history and politics from the eldest, Fernando Jr. We also spent several weekends in warmer Cuernavaca, at old ranches transformed into hotels with old gardens, large swimming pools, and delicious food.

In my office at the Instituto de Investigaciones Filosóficas I received researchers in various fields: mathematics, physics, biology, psychology, engineering, and literary criticism. At that time all the local psychologists were behaviorists. When I talked to them, they were so stunned by my statement that brainless psychology was unable to explain anything, that a strong delegation of them came to my office demanding an explanation.

None of the self-styled social scientists sought my company: except for Larissa Adler Lomnitz, they had read only Marx and Althusser, and none of them had done fieldwork. Larissa's excellent book on how marginal people survived, based on her visits to shanty towns, was summarily rejected as being "functionalist-structuralist," and her critics went on ignoring the society in which they were immersed. They did not know that Marx and Engels, their top heroes, had studied the social before writing about it. And they had no idea of scientific research: they only commented on other people's works – in Spanish translation, of course.

One of my visitors was Joaquín Sánchez McGregor, a specialist in the philosophy of literature, who tried to interest me in his field. He only succeeded in transmitting to me his contempt for the romance novels that Corín Tellado was publishing in industrial quantities. I even wrote about "Corín Tellado's workshop" in the main Spanish daily. She wrote me a respectful letter that put me to shame: she admitted to cultivating what she called the "sentimental-commercial" genre, but informed me that she had typed every one of her weekly novels without help on her old Underwood. I wrote her back apologizing for my irresponsible accusation.

Besides teaching regular courses in México City, I gave talks at various provincial universities, and participated in the philosophy congress held in Morelia. The star at that meeting was Carol Gould, an American Marxist who talked about Marx's *Grundrisse*, a word she mispronounced. That was a manuscript of Marx, a precursor of *Capital*, that had come to light only in 1939. A case of necrophilia? My own presentation (Bunge 1976a) dealt with the advantages of mathematical over verbal social science: a greater clarity and therefore avoidance of interminable debates, and a more precise empirical testability. I wrote a few elementary examples on the blackboard, but made no converts: Marxist schoolmen do not use blackboards.

I spent the long bus drive back to México City conversing with José-María Ferrater-Mora, with whom we made friends for life. He had taken up arms for the Republic, exiled himself to Chile, and was now teaching at Bryn Mawr, the Quaker college near Philadelphia. His philosophical trajectory had been just as complicated – from existentialist darkness to mathematical logic – and he was now a philosophical materialist very near my own philosophy.

Ferrater was a walking philosophical library, as shown by the four thick volumes of his *Diccionario de filosofía*, the 6th edition of which appeared in 1979. His sense of humor was just as amazing as his prodigious memory. He said that the Mexican bureaucracy was so Kafkian, that to initiate any procedure one had to exhibit not only one's birth certificate but also one's death certificate. (In fact, there are professionals, called *coyotes*, who live off conducting transactions in state offices.) Ferrater used to send me hermetic texts by White House clowns such as General Haig, which competed with the latest nonsense imported from Paris's Latin Quarter.

I also took part in several colloquia. One of them was convened by the Mexican Academy of Medicine to discuss problems in what I called *iatrophilosophy*, or the philosophy of medicine. In my presentation, later included in my *Epistemología* (Bunge 1980e), I introduced the health-state space spanned by such biomedical markers as blood pressure and metabolic rate. Three decades later the Swiss biomedical researcher Gerald Thurler and I applied this approach to a particular case (Bunge 2003e).

Another lively colloquium drew physicists and engineers interested in applied probability. As usual, the speakers favored either the frequency or the subjectivist interpretation. I was the only defender of the realist view that, when applied, probability measures real possibility or randomness. I was appalled to learn that some seismic engineers, when specifying rules for earthquake-safe buildings, interpreted probabilities as belief intensities.

Seminar and Association

One day in 1975 I announced to my Mexican colleagues the inauguration of the weekly philosophy seminar. I ruled that there would be short talks, about 15 min long, and long ones, lasting about 45 min. I invited Hugo Margáin, Enrique Villanueva and their friends, just back from Oxford – where they had spent several years at the Mexican taxpayers' expense but without even getting B.Phils – to give short talks. They felt offended, even though they could only comment on such marginal stuff as private language. They were only interested in sinecures and in the power to censor the submissions to the two journals published by my institute. They soon fulfilled their political ambitions. In particular, Margáin was appointed director of his institute, only to be assassinated a few months later under mysterious circumstances.

The seminar ran smoothly without them. The physicist Rafael Pérez Pascual gave the first talk, where he described the sensational experiment by Hubel and Wiesel showing that the visual cortex of kittens fails to develop depth perception unless subjected to visual stimulation. The reaction of the Oxtecs, as I called the local Wittgensteinians, was a mixture of contempt and indignation: they could not understand that the age-old nature vs. nurture dilemma was finally amenable to lab experiment. Six years later the Nobel committee awarded Hubel and Wiesel a Nobel Prize for proving that biological development is not predetermined by the genome, but results from the intertwining of nature with nurture. Obviously, neither Richard Dawkins nor Noam Chomsky, nor the self-styled evolutionary psychologists had ever heard of that breakthrough.

At the beginning of 1976 I invited a number of researchers, in particular my friends the philosopher Mario H. Otero, the urban anthropologist Larissa Adler Lomnitz, the physicists Luis and Ana María de la Peña, Tomás Brody and Rafael Pérez Pascual, the economist Enrique Leff, the neurophysiologist Augusto Fernández-Guardiola, and the engineer Emilio Rosenblueth, to meet in order to set up the Asociación Mexicana de Epistemología (AME). Shortly thereafter we celebrated this event with a party at the residence of Manuel Sandoval-Vallarta, a pioneer of cosmic-ray physics and former teacher at MIT, who two decades earlier had visited *El Ombú* along with Marcos Moshinsky.

The AME remained active for some years. In particular, it held a colloquium on environmental protection, organized by Leff, and another on consciousness, convened by Fernández-Guardiola, at which I read a paper. But it did not topple the wall between the sciences and the humanities that had been erected around 1800 by Kant's followers.

From México City to Boston

In the spring of 1976 I traveled twice from México City to Boston, the first time to attend an AAAS conference on anthropological theory, and the second to participate in a meeting convened by the Nobel laureate Eugene Wigner to discuss researchers' motivations.

At the first meeting I was included in a panel along with Margaret Mead – famous for having described the sexual freedom of the Polynesian adolescents–, Marvin Harris (the father of cultural materialism), and Burrhus Skinner, the most radical of behaviorists. When my turn came, I stated that anthropologists had plenty of loose hypotheses but not a single theory, or a hypothetic-deductive system binding some hypotheses into a well organized whole. Harris did not complain, although he had written his monumental and unsurpassed *Rise of Anthropological Theory* (Harris 1968). By contrast, Margaret Mead got so furious that she made threatening gestures at me with her Polynesian ceremonial mace.

A few years earlier, Harris had written to me from Columbia University praising my materialism without dialectics, but now was offended by the satirical doggerel on the emic/etic distinction, that I improvised for the occasion. He mellowed at lunch after a few beers, but we never communicated with me again. As for Skinner, he was very pleased to hear that he was the idol of Mexican psychology students (smile), but did not like hearing that, because of that, they were allergic to psychobiology (grimace of distaste).

Wigner liked when I started by recalling Aristotle's view, that what motivates the serious student of anything is curiosity. (Robert Merton rightly added peer recognition.) This was the third time that Wigner had invited me. I had declined his two previous invitations because one of them involved the infamous Reverend Moon, while the other was funded by NATO. Unlike Wigner, I believed that *pecunia olet.*

In sum, our long Mexican year was one of the most productive, happy, and memorable of our lives. But our residence permits were precarious, and every time we wished to travel abroad we had to get a permit from the Ministry of the Interior. We felt like prisoners, not visitors, although at the University we were treated like old friends.

Back in Canada

We were not allowed to board the plane back to Montreal, because we had forgotten to obtain the exit permits. I phoned our close friend the under secretary of education but, because it was a Saturday, his minister was out of reach, so we had to postpone our trip back, as well as the signing of the title deed for the house and the cottage that I had bought during a brief escapade in Montreal. At the same, time I wired all our funds in Mexican pesos to our Canadian bank. Three days later our family, along with our competent housekeeper Lupe Luna and her delightful son Gabriel, were safely installed in Montreal. Two weeks later the Mexican peso was drastically devalued, so I was not ruined once again. But our good memories of México were not devalued, and once in a while Marta and I wonder whether we did well in returning to our safe but cold and lonesome harbor. Oh Canada!

Chapter 10
Systemic Materialism

The previous chapter was a paean to exactness. But it was also a reminder that form without content is philosophically and practically insignificant, which explains the barrenness of logical imperialism. To make a difference, philosophers must say something original, reasonable, and inspiring about matter, change, mind, value, knowledge, action, justice, or their cognates.

In Chap. 8 I said something about epistemology, in particular scientific realism; in the present I will tackle ontology or metaphysics, and the subsequent chapters will touch on the philosophies of life, mind, and society, as well as on practical philosophy and aesthetics.

Spontaneous Materialism

Children are spontaneous materialists: they believe in ghosts, devils or gods only if they have heard adults talking about them, or when readings stories, like the Grimm brother's tales, that seem designed to tame them. I grew up without any such beliefs, and so was able to walk in the dark without fear, except when I heard the resident owl's hooting – but only because I had heard from superstitious neighbors that it is an omen of evil. I used to wait expectantly for nightfall, because it was the time when my father might come back from work. In Pavlovian terms: from an early age I was conditioned to love the dark. And of course my children loved nocturnal games involving the search for things or playmates with a flashlight, which they had learned at the YMCA camps. In short, I was predisposed to a materialist worldview long before I read about it. And, because I had not received any religious instruction, when I started reading some materialist philosophers they did not cause me any inner conflicts: I was spared the anxieties of ideological conversion. But, by the same token, I had no solid reasons for my spontaneous materialism.

M. Bunge, *Between Two Worlds*, Springer Biographies,
DOI 10.1007/978-3-319-29251-9_10

The idea of the unity of the world came to me much later, when I first encountered Newton's law of gravitation. In fact, the idea that all things were held together by gravity was the first cosmology inspired by science. Far from being ancient and obvious, this idea first came to light in 1687, and it was ignored by most of Newton's contemporaries, even Locke, and explicitly rejected by great thinkers of the next generation, like Leibniz, Berkeley, and Hume. Half a century later, when Voltaire learned of Newtonianism, he became so enthusiastic, that he wrote some of his most inspired pages, the *Lettres philosophiques* of 1734, which introduced his compatriots to Newton. Voltaire's praise for the new worldview was triply heretical: because the French had been Anglophobes since the time of Joan of Arc; because Voltaire was only a man of letters; and because Newton had countenanced action at a distance, anathema to the Cartesians as well as to the Aristotelians.

I was similarly moved when I learned that the inverse-square force was the cement of the universe. Hume and the other phenomenalists, for whom the world was a patchwork of disjointed appearances, and who were not interested in worldviews anyway, never experienced that emotion. And so, I faced my physics studies with a huge and luscious apple in hand. But it took me some years to discover the worm in it, namely dialectics – as I said in the previous chapter. Let us now move from spontaneous to crafted materialism.

Philosophical Materialism

Philosophical materialism, born simultaneously in ancient Greece and India, holds that all existents are material. Contrary to what its detractors have been claiming, Western philosophical materialism is unrelated to hedonism or the search for pleasure, which is a moral stance, not an ontology. (Only the pop philosophers Ayn Rand, of rational egoism fame, and the Lacanian Marxist Slavoj Žižek support both hedonism and crass materialism.) Moreover, far from being indifferent to the search for the good and the right, materialism facilitates it because it places them within our reach rather than above the material world.

Actually, philosophical materialism is a whole family of doctrines, from physicalism to organicism, from vulgar (or crass) materialism to scientific materialism, and from dialectical to systemic materialism. And all of these varieties of materialism are akin to naturalism, the worldview according to which everything real is natural. Materialism and naturalism are often identified, but the former is wider than the latter, as it admits the artificial or made in addition to the natural or given. For example, sociobiology and its offspring, evolutionary psychology, are naturalistic, whereas most anthropologists explain the social as a result of two different but strongly intertwined evolutions: the biological and the social. For instance, human communication, in particular speech, could not have emerged without the emergence of certain anatomical features of the speech organs, from the larynx to the cerebral cortex and, once present, it contributed powerfully to all social interactions. The social has emerged from the natural but is not natural itself, because the social norms are artificial, and not all of them favor adaptation. For example, sex is natural, but sexism and the fight against it are social; again, eating is natural, but the dietary permissions and prohibitions are social –and some of them are harmful.

Materialists have been accused since antiquity of denying the spiritual. This charge certainly applies to the nominalists and the so-called eliminative materialists like Paul and Patricia Churchland, but it is false with regard to all other materialists, in particular Epicurus and his followers. What is true is that materialists disbelieve the fantasies about spirits, ghosts, souls, and the other immaterial entities inherent in religions, esoteric cults, and alternative medicines.

Scientific materialism brings the spiritual down from the clouds and inserts it into material things: brains and artifacts "embodying" ideas, such as texts and diagrams. Moreover, materialism recasts the spiritual: it considers that poetry, music, mathematics, the basic sciences, mathematics, and the humanities, are authentically spiritual because they are disinterested, unlike religion, which seeks either personal salvation or social control, technology, which seeks profit or power, and business and politics, which aim at power.

The fact that Michel Foucault placed knowledge and power in the same bag, that Bruno Latour stated that "science is politics by other means," and that some Marxists are still debating whether science belongs in the infrastructure or in the superstructure only show their ignorance of both science and praxis.

Materialists oppose spiritualism, whether religious, philosophical, or New Age, as lacking in empirical roots, and sometimes even in sense. A Zen *koan*, a typically existentialist or Lacanian sentence, and many a religious dogma make sense only to the initiated: they do no resist conceptual analysis or confrontation with reality, and are hardly amenable to rational debate. For example, how should Hegel's dictum "The absolute is the spirit" be understood and justified? And who but a shaman or a priest can treat mental diseases if "they are only in the mind"?

Systemic Materialism

I have proposed a particular kind of materialism, which I call *systemic materialism*, or *systematerialism*. This is the fusion of materialism and sytemism. The former concerns the "stuff" reality is made of, namely concrete entities like photons and families, whereas systemism regards the way they are organized. Contrary to both individualism and holism, systemism holds that the universe is the system of all systems. To put it in negative terms, according to systemic materialism, immaterial objects, such as numbers, are imaginary, and the same holds for isolates with the sole exception of the universe. Thus, ancient atomism was half-true: the ultimate constituents of the universe are indeed material rather than spiritual, but they are actual or else potential constituents of structured wholes or systems rather than isolates.

How does my systemic materialism differ from the one proposed by Thiry d' Holbach in his original and influential works *Système de la nature* (1770) and

Système social (1773)? The only differences are that I am less wordy and make more explicit use of science and logic, and this simply because two centuries of scientific research have provided much additional evidence for both the materiality and systemicity of reality. This is why 'scientific materialism' (Bunge 1981d) is a synonym of 'systemic materialism'. Let us take a quick look at evidence for this.

There is scientific evidence for systemic materialism in (a) scientific projects, (b) the scientific method, and (c) scientific findings. Indeed, a scientific research project is a plan to study some material entities rather than freely floating spirits; every scientific activity starts off by either isolating the object of study from its environment or, on the contrary, by looking for previously overlooked links between it and other things; and every scientific finding describes some features of concrete things embedded in some system or other – with the exception of the parts of physical cosmology that describe global features of the universe, such as its expansion and corresponding thinning.

In short, to paraphrase Hegel's confused *Das wahre is das Ganze* ("The truth is the whole"), all existents are material systems. Everything else – in particular the world spirit, the greatest number, and the set of all-powerful Great Men – is imaginary. This is why scientific research is materialist and systemic all the way.

From Billiard Balls to Fields and Quantons

The medieval nominalists, who were naïve materialists, were also individualists. Likewise, many great modern scientists, from Mach and Ostwald on, have ignored systems altogether, and have claimed that science has overcome materialism. Let us take a closer look at this claim.

It has been claimed that, by conceiving of particles as wave packets, quantum physics has effected the vanishing of reality, since the waves in question would be nothing but symbols representing our state of knowledge. In other words, since quanta are irreducibly probabilistic, and since probabilities are identical with belief intensities, quantum physics would be about the mental, not about independently existing material things.

Yet, experimental physicists measure physical traits of concrete things, such as photons and molecules, not mental processes. Psychologists, not physicists, are expected to study the mental. And when someone suffers a mental disease, one resorts to a clinical psychologist or a psychiatrist, not to a physicist or an electrician.

In addition, the subjectivist view of probability as credence is not used in science, where a probability is regarded as one more real trait of a concrete thing (see Fréchet 1955; Bunge 2008a). For example, in the realistic interpretation of quantum mechanics, $|\Psi|^2 \Delta V$ is read the way Louis de Broglie proposed, namely as the probability of the presence of the quanton in the state Ψ and in the volume element ΔV. There is no reference to mental states or even measurements in the preceding (see Bunge 1967a). We learn this by inspecting the axioms of the theory instead of reading the opinions of famous quantum physicists (Bunge 1967a, 2010a).

The mechanistic identification of materiality with hardness and impenetrability became obsolete the moment, in the early 1830s, Michael Faraday discovered the electric and magnetic fields, which lack the properties characteristic of concepts, propositions, and theories, such as meaning, consistency, and truth. Yet most philosophers are still stuck in a pre-Faraday conception of matter. Worse, few recent dictionaries of philosophy contain definitions of 'matter' (or 'material') and 'system' (or 'systemic'). It would seem that philosophers, unlike lowly workmen and traders, deal only with "lofty matters."

What do scientists have to say about matter? Strangely enough, although they rarely use this word, all scientific research projects deal with concrete or material objects, from fields to cells to brains to social organizations. And, though scientists do not define 'material', they tacitly use Plato's equations: "material = mutable," and "ideal = immutable." This is why the central hypotheses of scientific theories are either equations of motion of bodies or fields, or reaction equations. No scientist would propose to study how numbers or triangles change over time, or to suggest that molecules have always been the same – although Noam Chomsky ridiculed the very ideas of molecular evolution and the evolution of languages, which happen to be "hot" topics of contemporary scientific research. When will we learn that political stance is no indicator of scientific status?

Building Blocks

The history of philosophy is a cemetery of doctrines that perished from scientific malnutrition. Let us quickly remember those of Wittgenstein, Popper, Kripke, and Armstrong. Wittgenstein started his famous *Tractatus* of 1921 asserting that "the world is the totality of facts, not of things" (Wittgenstein 1921). Now, in the factual sciences 'fact' denotes either a state of a thing or a sequence of states of a thing: there are no facts in themselves, without material things. For example, there would be no car collisions without cars, or changes of government without rulers and their subjects. Remember Aristotle's criticism of Plato's idea of movement in itself rather than moving things.

Popper was not satisfied with one *uni*verse. Like Hegel and Lenin before him, he wanted three "worlds": those of material things, mental processes, and ideas in themselves along with their materializations – books, records, paintings, and the other "inmates of world 3." After a nuclear juggernaut, only world 3, or at least his own books, would remain. But how would a printed book differ from a blank book if no one capable of reading it were around? And how does such trialism help advance knowledge?

Saul Kripke, David Lewis, my colleague Storrs McCall and the rest of the possible-worlds faction place imaginary "worlds" on the same footing as ours, as if there could be imaginary things subject to laws inherent in them. People endowed with a minimum of common sense, and without the urge to defile blank sheets of paper, know that real entities are very different from imaginary ones, and that real possibility is not arbitrary but subject to laws or norms. For example, it is possible for an ice cube to melt, but impossible for it to transmute into a bird.

Then there is David Armstrong's proposal of replacing things and facts with states – as if there could be states in themselves rather than states of things (Armstrong 1997). An early attempt of this kind was the thermostatics of Gottfried Falk and Herbert Jung contained in a volume of the *Handbuch der Physik* published in 1959 (Falk and Jung 1959). This attempt was wrong-headed because it is well known that thermostatics, a primitive form of thermodynamics, is about macrophysical systems, and that these, like all real things, are in definite states and change states over time.

In the simplest case, that of a gas in a container, its state at any time is the list of values of its volume, internal pressure, and temperature. If the container is evacuated, the gas pressure and temperature vanish. However, the fantasy of states without matter was used by both the "structuralist" Carlos Ulises Moulines (1977b) and the erstwhile Australian materialist David Armstrong (1997) to "refute" materialism. Moulines also stated that no one, not even physicists, know what matter is. But what do physicists (and chemists and biologists and social scientists) study other than material things?

Much the same holds for *energetism*, popular *circa* 1900 but the now forgotten. This is the doctrine according to which the stuff of the world is energy, not matter. The energetists, such as the eminent chemist Wilhelm Ostwald, should have known better, for they measured or calculated only the energies *of* some thing or other, never energies in themselves. What happened is that they overlooked the object variable in their formulas just because it was taken for granted. For example, if one writes "E = 100 cal," one assumes that it refers to some thing b such as a chocolate bar, so that a more correct formula is "$E(b, cal) = 100$."

Finally, informationism, which came into vogue a few decades ago, is similar: although it claims that the basic blocs of the universe are bits, not "its," they are attributed to things. When I asked John Archibald Wheeler after a talk of his how many bits a hamburger contained, he made a quick mental calculation and replied unfazed: "A typical hamburger has so many bits." Nutritionists might not appreciate this reply. And information engineers might remind us that information circulates only around information systems, which are macrophysical things like TV circuits. No bits without matter in bulk (More in Bunge 2010a).

In short, none of those six attempts to free ontology from the concept of matter have succeeded: all four have turned around ill-defined notions, have been alien to science, and have been so sketchy, that they have not helped clarify the key concepts of entity, property, state, process, space, time, or law. And so, our initial assumption holds: the world is the material thing composed of all the material existents. Notice that this is not the same as "The world is the *set* of all material things," for sets are conceptual, and the composition of a thing is only one of its basic properties.

Systemism

Materialism answers the question about the constituents of the universe, but gives no hint about its structure or organization. Systemism answers this question as follows: All material things are systems, or actual or potential components thereof; nothing, save the universe, is an isolate. Thus, an electron or some other elementary particle moving through empty space towards our planet will eventually be captured by a terrestrial atom. The universe as a whole is the only lasting isolate.

I adopted systemism explicitly when I read Ludwig von Bertalanffy's foundational paper on "general systems theory" (von Bertalanffy 1950). I prefer to call it *systemics* because the polymath Anatol Rapoport cautioned that systems theory is not a theory proper, but something even more important, namely a fruitful *approach* or viewpoint: the one that enjoins us to think out of the box, and to remember that every item is either a system or a part of one.

The systemic hypothesis may be reworded thus: *All the properties of a thing come in bundles or systems rather than one by one* – as shown by the very attempt to find laws, which are invariant relations among essential properties. The epistemological partner of this ontological thesis is that the important problems of knowledge too come in bundles (Bunge 2014e). Hence the superiority of theories over isolated empirical data.

In the early 1970s I joined the Society for General Systems Research (SGSR), that had been founded in 1955, and attended several of its meetings. There I met such interesting people like the economists Kenneth Boulding and Amitai Etzioni, the politologist Karl Deutsch, the biologists Ralph Gerard and Robert Rosen, the engineer Mihahjlo Mesarovic, the mathematician and social scientist Anatol Rapoport, and the systems engineer George Klir, who was my main contact with the movement over many years.

There was no shortage of excentrics in that crowd. Forced by his mother, Norbert Wiener had married a Nazi girl who kept two copies of Hitler's *Mein Kampf* at home; Robert Rosen used to play with his twenty-in. penknife, for it was "a conversation piece"; Kenneth Boulding could not control his stuttering; Gordon Pask went around dressed in tails, and Humberto Maturana, who had started as a physiologist, ended up mounting psychodramas (circus-like dialogues) with his wife – a show Marta and I watched in Quito.

I contributed several papers to the society's journal, such as "General systems and holism" (Bunge 1977e), where I warned against confusing systemics with holism. The latter is the doctrine of Aristotle, Hegel, and Parsons, according to whom the whole precedes its parts and is to be understood in and by itself rather than by analysis. Holism is a part of New Age, and it has been popularized by conspicuous members of the SGSR, particularly Ervin Laszlo. I regard these people as constituting the soft or amateurish wing of the systemics movement. I repeated that warning against pop systemism in the Introduction to my book on systems (Bunge 1979c), and proceeded to expound my own version of systemics, based on contemporary science and using some formal tools such as the state space of a system.

I also served several years as the philosophical editor of the society's journal, but after some years I felt obliged to reject the majority of submissions for being just rehashes of old holistic slogans – a thankless task. Since I dislike the job of executioner, I eventually resigned that position.

My latest participation in the general systems movement was as the keynote speaker at the Vienna congress in the spring of 2014. My talk was titled "Big problems call for a systemic approach" (Bunge 2014e). On that occasion I was given the Bertalanffy Award on Complexity Thinking, and had the pleasure of meeting in person the German sociologist Heinz Droste and the Taiwanese sociologist and philosopher Poe Wang, both of whom share my main ideas.

In Vienna I also gave a talk on the so-called Aharonov-Bohm effect at the Institute of Quantum Optics directed by Anton Zeilinger, famous for his experiments on long-distance correlations among quantons. The main point of my talk was that the said effect is no such thing: that electromagnetic potentials are just as meaningful as field intensities since both refer to the same physical entity, namely the electromagnetic field – but they are also conceptually richer (Bunge 2014c). Aharonov and Bohm might not have made that mistake if they had performed a semantic analysis of the formulas in question. Incidentally, these ideas came to me while writing a draft of the present memoirs on the Greek island of Euboea during the summer of 2013. Presumably, that will be my last contribution to physics, 70 years after my first.

The Individualism-Holism-Systemism Trilemma

Most philosophers have been holists, like Aristotle and Hegel, or individualists like the Greek atomists and Thomas Hobbes. Individualism has ruled in the social sciences, from Adam Smith and Alexis de Tocqueville to the rational-choice theorists. The great exceptions have been Ibn Khaldûn, François Quesnay, Karl Marx, John Maynard Keynes, and the *Annales* historiographic school, all of whom may be regarded as systemists. (Let us pretend that Marx did not take literally his own statement that Britain was the tool History used to modernize India).

I have argued (e.g., Bunge 1979a, 1996c, 2000a, b) that both holism and individualism are untenable: holism because, in rejecting analysis, it is obscurantist; and individualism because isolates are fictitious. Besides, both extremes are politically undesirable: holism because it leads to statism, and individualism because it leads to egoism. But my strongest argument against both of them comes from science, which goes now from wholes to their parts (top-down analysis), now from individuals to wholes (bottom-up synthesis).

So, what if any is the correct alternative to both individualism and holism? I have argued in different ways and at several places (e.g., Bunge 1979a) that it is systemism, for it places all items in their contexts and, if they are complex, it encourages us to analyze them. That systemism is the correct alternative, can be seen in both my proposed analysis of systems and in its axiological and political implications. Let us start with the former. I have proposed (e.g., Bunge 2003d) that any system σ may be schematically modeled by the ordered quadruple composition-environment-structure mechanism, or

$$\mu(\sigma)=\langle C(\sigma),E(\sigma),S(\sigma),M(\sigma)\rangle,$$

where

$C(\sigma)$ = Set of constituents of σ at the given level of analysis;
$E(\sigma)$ = Immediate environment of σ;
$S(\sigma)$ = Set of bonding and non-bonding relations among the system's constituents;
$M(\sigma)$ = Set of processes that keep σ going.

For example, the composition of an organism at the cellular level is the set of its cells, whereas at the organ level it is the set of its organs; the structure of an organism is the totality of its bonding relations, such as connecting tissues and hormonal fluxes, and nonbonding relations, such as those of position and succession; and the mechanism of a system is composed of the processes that keep it alive, in the first place metabolism and interaction with its environment.

The distinction between a system σ and any of its models μ(σ) escaped Ross Ashby in his 1956 classic. Indeed, one and the same concrete system may be modelled or conceptualized in different inequivalent ways.

Projections of the Quadruple

The above quadruple not only helps in reminding us that wholes are analyzable: it also shows that projecting the quadruple onto any of its components involves an unacceptable oversimplification. Let us take a quick look at any of the four possible projections.

1. *Leave only the composition*, that is, deny the reality of wholes and accept Margaret Thatcher's famous dictum "There is no such thing as society: there are only individuals." Radical individualism ignores not only the difference between living things and their molecular components, but also allows right-wing statesmen to relinquish their duty to efficiently manage all the components of the commonwealth, from defense and public health to education and science.
2. *Single out the environment* at the expense of the other components of the quadruple: adopt the stance of the radical behaviorists, and overlook the components's activities, in particular the spontaneous ones. Structuralism is also obvious in the radical fringe of the environmentalist movement, who accuse of "speciesism" those who care for the welfare of humankind.
3. *Replace the whole by its structure*, the way radical structuralists do. This move is logically impossible, because there are no nets or graphs without nodes. In other words, every structure is the structure *of* some thing. Yet structuralism is being preached by the psychologists and philosophers of mind who claim that the mental consists of information flows, as well as by the social philosophers,

like Niklas Luhmann, who claim that individuals do not matter: that only communications matter. In his technologically path-breaking *Cybernetics*, Norbert Wiener had claimed that cybernetics had solved the materialism/dualism dilemma, by showing that information is neither material nor spiritual (Wiener 1948). He overlooked the fact that the information source, channel, and receiver of any information system are parts of material systems such as TV networks. So much so, that it takes only a hammer to smash an information system.

4. *Eliminate mechanisms*: adopt the Eleatic thesis that nothing ever changes, or that there are changes without any underlying mechanisms. This is the view of the scientists who remain satisfied with rate equations, without inquiring into the change sources. That is also the view of idealist process-philosophers, such as Alfred North Whitehead (1929), for whom a thing was a bundle of processes.

Values Too Come in Bundles

When applied to value theory and ethics, our model suggests that values, too, come in clusters. For instance, whenever we make a choice, and do not act on impulse, we weigh several traits of the object of choice and our means to attain our goal. Likewise, when confronted with moral dilemmas and wish to do the right thing, we take into account duties along with rights – unlike the bully portrayed by standard economic theory, who only cares about his own interests.

Much the same applies to politics. For example, the oldest and most famous of all political slogans is that of the French revolutionaries of the 1790s: *Liberté, égalité, fraternité*. This was a wise slogan, because emphasis on any one member of this triplet to the detriment of the other two results in unfairness. For example, if freedom is prioritized, we get libertarianism, which consecrates selfishness and the rule of the few who can afford to make free choices. Likewise, overemphasizing equality, as the utopian socialists did, leads to deriding excellence, suffocating initiative, and relinquishing individual responsibility. And prioritizing solidarity, the way communitarians do, favors the group of people who are willing to be solidary with those who share their values – family, party, church, masonic lodge, or mafia – at the expense of all other social systems.

Single-issue political parties, such as libertarianism, communism, feminism, and environmentalism, adopt non-systemic strategies. This is why they seldom mobilize majorities: because individuals and organizations are polygonal, and a polygon collapses when any of its sides are removed. All the environmental and social issues, such as those of energy scarcity, environmental degradation, housing, income inequality, public health, and public education, and now also anthropogenic climate changes, are systemic, hence they cannot be solved by "technological fixes." Yet, they are seldom studied in this way, and they are often put in the hands of economists – the overwhelming majority of whom are one-sided and biased in favor of the powerful – or of engineers, who are seldom aware of the social and moral dimensions of macroprojects such as dams and large housing projects.

Emergence and Levels

Whenever two or more things bind together, they constitute a system with new or *emergent* properties and, in the process, lose their individuality. Obvious examples: the synthesis of a molecule out of its precursors, of a zygote from an ovum and a spermatozoid, and of a battalion of soldiers out of a bunch of recruits.

The precursors of an emergent system may or may not be its components. They are in the cases of a large magnet, a polymer such as a plastic toy, and a formal organization; but they are not in the case of chemical compounds such as a water molecule, a fertilized egg, or a business firm resulting from the merger of two companies.

In short, systematization involves emergence. This is why systemic materialism is emergentist, so that it may also be called *emergentist materialism*. Yet most philosophers dislike the very idea of emergence, and few philosophical dictionaries include it. In mine (Bunge 2003c), by contrast, emergence has pride of place. On the other hand, the fuzzy notion of "supervenience," used by some philosophers but not by scientists, plays no role in my philosophy. The word 'emergence' has been appearing with increasing frequency in the scientific literature, and the papers on emergence and levels by the Nobel laureate Philip Anderson (2011) are often cited by fellow scientists.

I have proposed two mutually complementary definitions of 'emergence', which I call 'static' and 'dynamic' (see, e.g., Bunge 2003c). The static concept is this: A property of a system is emergent if is not possessed by any component of the system. The dynamic concept: A property of a system is emergent if it is acquired by the system at some time during the process of its synthesis or evolution. Some cases, such as those of the origins of life and the state, call for both the static and the dynamic concepts of evolution. Just think of speciation and species extinction, or of the novelties that emerge or submerge in the course of a person's development or of human history.

Integrative Levels

The concept of emergence is closely related to that of integrative level, or level of organization, which intrigued me from the moment I first met it in Nicolai Hartmann's 1948 paper on categorial analysis, which I read in Spanish translation in *Notas y estudios de filosofía*. Only much later was I able to obtain a couple of books by that interesting apostate of phenomenology, who survived Nazism by keeping silent and avoiding to be seen, even abroad, conversing with a non-Nazi. (Georges Gurvitch spoke of his secret encounter with Hartmann in the Bois de Boulogne during the 1937 Descartes congress in Paris: they went separately and each of them spoke from behind a tree.)

However, Hartmann's concept of a level – which he called *Schichte*, or layer – was unrelated to that of emergence, which he called *categorical novum*: it was but an elaboration of the old notion of "layer of being" occurring in the hierarchical worldview of Dante's favorite philosopher, Pseudo-Dionysius the Areopagite, who had placed God at the top of the ladder. In the mid-1950s, while researching my book on causality, I came across Arthur Lovejoy's learned book *The Great Chain of Being*: *A Study of the History of an Idea* (Lovejoy 1936).

But only reading the 1945 pioneering paper by the biologist Alex B. Novikoff, who had stressed the non-physical peculiarities of organisms, did I realize the centrality of the level concept to a scientific philosophy. I wrote a lot on the twin concepts of emergence and level, and gave papers on them at several congresses, adding some refinement every time (e.g., Bunge 1960a, f, 1980a, f, 1989g). My most frequently cited paper on this subject was titled "The metaphysics, epistemology and methodology of levels" (Bunge 1969b) which I read at the conference on hierarchical levels held in 1968 at an unlikely place: the Boeing factory. My latest book on this subject was *Emergence and Convergence* (Bunge 2003d).

Reduction and Reductionism

Physicalism is radically reductionist or leveling, for its pivotal principle is that every existent is physical. Systematerialism is not, for it distinguishes various kinds of matter: physical, chemical, living, thinking, social, technical, and semiotic (e.g., Bunge 1969b, 2004a). In other words, systematerialism distinguishes several levels of reality, every one of which is assumed to have emerged from previous levels in the course of natural or social evolution. One decade earlier (Bunge 1960a, f) I had proposed this double ladder (Fig. 10.1).

A physicalist might concur with this double layering, yet might argue that every level is "ultimately" physical or reducible to the physical one. Before addressing this problem, let us review the meaning of 'reduction'. The extant definition, in terms of definability and deducibility, applies to levels of analysis, not to levels of being. I have proposed the following definition, which is expected to cover both the

Mind	→	*Culture*
↑↓		↑↓
Life	→	*Economic & social life*
↑		↓
Inanimate matter	→	*Artiphysis*

Fig. 10.1 The level structure of reality

ontological and the logical aspects (Bunge 1991e). Let *A* and *B* designate two objects. Then, to reduce *A* to *B* is either to identify *A* with *B*, or to include *A* in *B*, or to assert that every *A* is either an aggregate, a combination, or an average of *B*s, or else an effect, manifestation, or image of *B*.

My earliest paper on reduction, written in the late 1940s, dealt with the alleged reduction of thermodynamics to mechanics. I claimed that there is no such reduction because statistical mechanics, the reducing theory, contains the concept of randomness, which is alien to classical mechanics. I submitted my paper to *Philosophy of Science*, whose editor rejected it because it was poorly written, but encouraged me to have it copy-edited by some colleague.

Over the years I have examined in detail several cases of real or bogus reduction, particularly in physics (Bunge 1973a) and psychology (Bunge 1990b). The most detailed of such analyses concerned the alleged reduction of chemistry to physics, which physicists take for granted. I analyzed quantum chemistry, which treats chemical reactions as collisions among reagents, and looked in particular at some papers by the Nobel laureate John Polanyi (Bunge 1982a). I concluded that quantum chemistry does not follow from quantum mechanics alone because it borrows a key equation from chemical kinetics, which belongs in pre-quantum chemistry, as it refers to matter in bulk.

The two main journals of philosophy of science rejected my paper, presumably because they had never come across quantum chemistry. One of the referees objected to my discussing chemical kinetics because he/she mistook it for the kinetic theory of gases. This is the flaw of the peer-review mechanism: that it protects ignorance, conservatism, and malice.

Finally, note the difference between reduction, a conceptual procedure, and *reductionism*, a research strategy based on an ontological conjecture. A reduction project may or may not succeed. The idea that ultimately it must succeed is a dogma based on physicalism. Physicalist philosophers will applaud when a reduction project succeeds, but immaterialists will be glad when it fails. By contrast, systematerialists will view reduction failures as confirmations of their worldview, according to which reality is many-layered like an onion, and that, just as with onion peeling, the uncovering of integrative levels may exact some tears.

A clear example of the effects caused by such unveiling is the commotion caused by the discovery that the Brownian motion of pollen particles was the visible effect of invisible molecular collisions. This discovery (1905–1908) was at once a victory for both atomism and emergentism, a defeat of the anti-atomism held up until then by most physicists and preached by the positivists, and the occasion for Einstein's promotion from clerk of a patent office to the scientific genius of the decade.

When attempting to reduce a given level to a lower one, one may fail for missing the intermediate level. The attempt to reduce the mechanics of extended bodies to particle mechanics is a case in point. In fact, a macrophysical body is not just a clump of particles: there is also the mesophysical level to be reckoned with. For example, a drop of water is certainly composed of water molecules, but these are three-component systems that, far from clumping into a structureless aggregate, assembly into crystal-like systems held together by hydrogen bonds, which are alien to mechanics and were discovered by chemists.

In short, water is not identical with H_2O: this formula exhibits the composition of water, not its structure, let alone its mechanisms, such as its rapidly changing bonds. And, while X-ray crystallography has uncovered the molecular structure of the water molecule while in the solid state, it has not done the same for the intermolecular forces. In short, the analysis of water has exhibited systems within systems within systems – superficially looking like a three-layer Russian doll.

Space and Time in Materialism

The immateriality of space and time has been used as an argument against materialism. Since neither space nor time are tangible things, yet are in the world, Newton thought that they constituted the *sensorium* or sensory apparatus of the deity. By contrast, Leibniz, his contemporary and formidable critic, revived the materialist idea of Aristotle and Epicurus that, far from existing by themselves (the absolutist view), space and time are relational: space is the order of coexistents, and time that of successives. Hence, there would be no space without things, and no time without change: the hollow and unchanging universes are imaginary.

In other words, spacetime would be the basic structure of the totality of changing things. Therefore, it is an objective trait of matter, or the set of all mutable entities. This view is countenanced by Einstein's theory of gravitation, according to which the geometry G of spacetime is determined by the distribution T of matter ($G = \kappa T$). Volume 3 of my treatise (Bunge 1977c) includes a formalization of this relational and therefore materialist conception of spacetime.

Ideal Objects in the Light of Systematerialism

Another classical objection against materialism is that it fails to account for ideal objects, in particular the mathematical ones. Granted, the latter have no room in a materialist ontology, if only because, as Plato noted, they are immutable or, if one prefers, timeless. In an early paper (Bunge 1960g) I examined the question of timeless things. However, this argument affects only nominalism, the "vulgar" materialism professed in recent times by Tarski and Quine (see Bunge 1975b). An emergentist materialist has the choice of *fictionism* for ideal objects: he will maintain that they are fictions, and will add that there would be no fictions without brains capable of inventing them, that is, mathematicians, theologians, rhapsodes, cartoonists, and the like (Bunge 1985a, 1994, 1997a). Without them we would have neither numbers, nor gods, nor talking ducks.

Consequently materialists, as such, have nothing to say about mathematical objects: only mathematicians are equipped to talk about them – to say, for instance, that 7 is a prime number. Only psychologists can investigate in vivo the processes of mathematical invention and the learning of mathematics; and only social scientists are equipped to investigate mathematical communities and their relations with society at large. A materialist theory of mathematical objects would be laughed out of court at as a case of hysterical materialism. On the other hand, materialists are equipped to discuss the idealist theses that ideal objects exist by themselves, and that there is no difference between ideal and real existence: that both types of existence are formalized by the wrongly called "existential" quantifier. Likewise, systemists are expected to emphasize that, far from being isolates, mathematical objects belong in systems, such as number systems, groups, manifolds, and spaces.

Modernity of Systemism

The idea of a conceptual system first occurs in Euclid's *Elements of Geometry* (ca. 300 B.C.). By contrast, that of a material system is modern: William Harvey introduced it in his revolutionary *De motu cordis* (1628), where the heart occurs as a component of the cardiovascular system. True, one century earlier Copernicus had written about the solar system, but only Newton, two centuries later, justified this idea by showing that the said system is held together by gravity. And two centuries after Newton some of the global properties of the solar system, such as its stability and its rotation around the center of the Milky Way, started to be investigated. These questions could not even be posed before Copernicus, when the planets were being studied one by one because they were thought to constitute a collection of isolates. It took one more century for philosophers to take note of the centrality of the system concept. Indeed, the earliest systemic philosopher was Thiry d'Holbach, the author of two very influential books: *Le système de la nature* (1770) and *Le système social* (1773). But of course the philosophical establishment, dominated until midtwentieth century by Kant, Hume, Hegel, and their successors, ignored Holbach and his friends of the radical wing of the French Enlightenment. Hume and Kant were individualists, in so far as they held that the world is the totality of appearances. Hegel was a notorious holist. So was Marx, his admirer, when he prophesied that socialism would arise spontaneously from economic changes, and when he claimed that social classes, not individuals, generate ideas. But Marx was a systemist when he held that individuals make society, which in turn shapes individuals, and when he participated in the formation of trade unions and political parties.

While scientists were busy studying concrete systems, from atoms and crystals to galaxies to nervous systems to immune systems, to business concerns to whole societies, most philosophers aligned themselves with either individualism or holism. Philosophical systemism, which had made a short appearance around 1770, reappeared independently only two centuries later.

The first contemporary to reinvent systemic philosophy may have been the theoretical biologist Ludwig von Bertalanffy (1950). In 1954 he, along with the economist Kenneth Boulding, the biologist Ralph W. Gerard, and the mathematician and social scientist Anatol Rapoport, founded the Society for General Systems Research (SGSR). I was struck by Bertalanffy's initial paper, but became disappointed when he started to mix systemism with holism and even with the hierarchical worldview pushed by some Christian thinkers.

I found Anatol Rapoport and Russell Ackoff particularly congenial because of their interest in real-life problems and their use of mathematics, as well as by their avoidance of holistic blabber. I also benefited from the writings of the mathematical biologist Robert Rosen and the systems engineer George Klir, who was my main contact with the systems movement. In the 1970s and 1980s I participated in several meetings of the SGSR, published some papers in its journal, and served as its philosophical editor for some years. I resigned this position when the serious submissions began to be outnumbered by the light essays of amateurs like the concert pianist Erwin Laszlo.

Something similar happened to me with *Mathematical Reviews* as the editor of its foundations and philosophy of mathematics section. Most of the submissions I had to evaluate were ridiculous, like a paper on semiotics by the great novelist Umberto Eco, or the formally correct products of the Suppes-Sneed-Stegmüller-Balzer-Moulines structuralist school, which did not contribute anything useful to either science or philosophy. After a quarter of a century I got tired of manning the guillotine, published a harsh criticism of a book by Stegmüller (Bunge 1978a), and resigned my position as a gate-keeper.

My Main Systematerialist Work

I expounded my systematerialist ontology in volumes 3 and 4 of my *Treatise*, published in 1977 and 1979 respectively. In this work I elucidate the key concepts of substance, property, state, process, thing, system, emergence, submergence, space, time, causality, randomness, space, time, chemism, life, evolution, mind, society, social structure, participation, marginality, social cohesion, and history among others.

To carry out this task I make use of such elementary formal tools as set and function. In addition, the exposition follows the axiomatic format: primitive (undefined) concepts and defined ones, axioms (or postulates), theorems, and comments. However, the motivation and justification of my principles (axioms) originated in the sciences.

My ontology was ignored by most philosophers, in some cases because of allergy to metaphysics, in others because of mistrust of science, and in still others because of neophobia. There is also the preference for short formulas, such as Kant's "space is an a priori form of sensibility", and "the world is the totality of phenomena."

My ontology was well received by the polymath Anatol Rapoport, and the theoretical physicist Héctor Vucetich and his student Gustavo Romero. My materialist theory of the mind attracted neuroscientists like Vernon Mouncastle, and psychologists such as Dalbir Bindra, Peter Milner, Ernst Pöppel, and Viktor Sarris. And several journals, like *Theory and Decision*, *Philosophy of the Social Sciences*, and *Journal of Socio-Economics*, have published articles of mine about systemism (e.g., Bunge 1979a, 1981c, d, 1989a, b, 1993b, 2000a, b).

Systemic vs. Dialectical Materalism

Dialectical materialism, which had seduced me as an adolescent, has seemed to me from about 1950 a coarse work of amateurs expounded and defended in a dogmatic fashion, from which a single nugget remains, namely the thesis that the world is material and changeable. The rest is either unintelligible, too sketchy to be useful, or just plain false. For one thing, dialectical materialism does not contain a clear idea of materiality. For example, Lenin (1972) stated that material is whatever exists outside consciousness. Obviously, he defined *real*, not *material*. The conflation of the two concepts is anthropocentric, hence of no use to those who study things that existed before the emergence of animals capable of being in conscious states. Ontology is about things in themselves, not things for us. The latter are studied by epistemology, and the thesis that the external world is real is the kernel of realism, not of materialism. Regrettably, dialectical materialists confuse both doctrines.

Second, historical materialism, the part of dialectical materialism that deals with human society, is not monistic, as materialism is supposed to be, but dualistic. Indeed historical materialists split society into two parts: the "material infrastrucure," or economy, and the "spiritual superstructure," or culture. Such a divisi6n is not only inconsistent with materialist monism: it is also incorrect, for deliberate human action is guided by conceptual considerations, which is why business concerns employ managers, engineers, and consultants of various kinds. Consistent materialists will hold that everything real is material. But, of course, they will distinguish several subsystems of any society – the economic, political, and cultural: they will be systematerialists.

My third objection is that dialectics is contaminated by key opaque terms, such as "opposite", "dialectical negation," and "contradiction," as well as by false and even socially dissolving principles, such as "Every change is generated by some conflict," which damns cooperation and peace. As I said repeatedly in my Beijing lectures in 2011, all leaders know that one of their main tasks is to avoid or resolve conflicts instead of exacerbating them.

Fourth, dialectical materialism is seriously deficient in epistemology, methodology, axiology, action theory, ethics, and even political philosophy. Indeed, most of Engels's occasional remarks on knowledge evoke empiricism, and his statement that praxis puts theories to the test is pragmatist, as is Lenin's that "good is what favors the communist party". On the other hand, Marx's statements about the unfairness inherent in capitalism stand. Suffice it to recall that over the past half century industrial productivity has doubled while wages have remained stagnant.

Regrettably, this moral advantage of Marxism has been cancelled by the immorality of the dictatorships of the "proletariat," which actually was that of the *nomenklatura*.

Fifth, dialectical materialists have been busier criticizing and attacking than in learning and building. For example, the Soviet even philosophers initially opposed all the scientific innovations of their time, in particular mathematical logic, the two relativities, quantum mechanics, quantum chemistry, genetics, the synthesis of Darwinism with genetics, and even biological psychology. Nor is there a Marxist sociology or an alternative to standard mathematical economics.

Sixth, Marxists have not participated in the battle against the pseudosciences and alternative medicines. Worse, Marx was the first social constructivist (see Bunge 2000c), and it is well known that not only Stalin, but also the eminent British scientists J.B.S. Haldane and John D. Bernal defended the charlatan Trofim Lysenko in the name of dialectical materialism.

Lastly, dialectical materialism is not a research project but a dying creed. Its latest prophet, Slavoj Žižek (2014), keeps reheating the same stew of Hegelianism with postmodernism dressed in radical disguise. He rightly notes the reluctance of Marxists to even admitting scientific novelties – among which he includes the psychoanalytic fantasies, in particular Jacques Lacan's – but he himself offers none, and even praises endless repetition.

In short, in the twentieth century dialectical materialists have hindered all the sciences except history. Indeed, there have been some important Marxist historians, such as Eric Hobsbawm and Edward Thompson, as well as para-Marxists ones such as the original members of the *Annales* school (Fernand Braudel, Marc Bloch, and Lucien Febvre), and more recently Pierre Vilar, Robert W. Fogel, Eugene Genovese, and Josep Fontana. But there have been no original Marxists in the other social studies.

My Philosophy in the Cradle of the Enlightenment

After World War II German philosophy buried the remains of French classical rationalism. Indeed, the French postmoderns, particularly Louis Althusser, Michel Foucault, Jacques Derrida, and Bruno Latour, demonized the French Enlightenment invoking Hegel, Nietzsche, Husserl, Heidegger, and the Frankfurt school.

The only visible reaction against that irrationalist movement is the group around the journal *Matière première* and the "Mátériologies" book series, led by the publisher and amateur philosopher Marc Silberstein. My book *Le matérialisme scientifique* (Bunge 2008a) appeared in that series. In 2004 Silberstein organized a 4-h lecture of mine. The event was held at the Ligue pour les Droits de l'Homme, and was attended by the biologist Pierre Deleporte, the mathematician Patrick Teller, and others. However, the French materialists are few, and the systemists are hardly visible in the midst of the holistic cloud, and shortly after the journal *Revue Internationale de Systémique* folded.

Metaphysics of Quantum Mechanics

The only philosophers who took notice of quanta in the first half of the twentieth century were the Marxists – who rejected it – the neo-Kantian Ernst Cassirer, and the logical positivists. The latter, in particular Philipp Frank, adopted and enlarged the so-called Copenhagen interpretation that had been sketched by Niels Bohr. The kernel of this interpretation is that the microphysical entities acquire all their properties only if and while observed. At one point, Bohr and Heisenberg claimed that this holds also for macrophysical things like streetcars: they would acquire their properties through repeated observations. These great scientists did not think of their own innards, the interior of the stars, or the past.

Einstein, Planck, and a few others, among them my professor Teófilo Isnardi (1927) criticized the standard interpretation as a piece of idealistic philosophy, but were not listened to. Besides, Einstein confused realism with what I have called *classicism*, that is, the view that micro-things have all the properties assigned them by classical physics – a view that had no empirical support (Bunge 1979f). David Bohm (1952) and others invented alternative theories, involving "hidden variables," that is, dynamical variables that had precise values all the time. But these theories made no new predictions, so they were regarded as having failed (Recall Chap. 8).

The view of Bohr and his followers was that the quantum theory had been successful where classical physics had failed – for instance, in accounting for the very existence of atoms, which, according to classical electrodynamics, would collapse immediately after being formed. The Marxist criticism was even less effective, for it confined itself to pointing out that the philosophical pronouncements of the members of the Copenhagen party contradicted the teachings of Engels and Lenin. None of them proposed a realistic alternative that kept the usual mathematical formalism of quantum mechanics, and only changed its physical interpretation, as I did (Bunge 1967a, Bunge 1967g, 1973a, 2010a).

The quantum theory should have challenged the metaphysical community to update their antiquated ideas about material things as hard and passive entities. But metaphysics was still lying under the rubble left by the Lisbon tsunami, that swept away the last vestiges of Leibniz's metaphysics that Voltaire ridiculed in his *Candide*.

Between 1925 and 1935, while the basics of quantum theory were being built, there were no metaphysicians competent to understand that it involved radically new ideas worth exploring. Among these new ideas were the following: the building blocks, or fundamental particles, are endowed with counterintuitive properties such as spin and parity, and some of them are very short-lived; microphysical things are most of the time in states that are superpositions (linear combinations) of elementary states or eigenfunctions; microphysical systems may persist even after breaking up into parts that fly apart at great distances; the elementary particles are shapeless, and geometric shape emerges gradually as a system grows; when two or more atoms combine into a molecule, they lose their identity, and their elementary constituents are redistributed in nonclassical ways; when fundamental particles collide at high energies, new particles emerge, such as mesons and neutrinos, that are not present in the incident beam; atoms are mostly hollow; Bohr's neat elliptical orbits turned out to be imaginary; classical chunks emerge gradually, as typically quantum bits of matter go into states characterized by large quantum numbers and interact with macrophysical things; many changes occur spontaneously rather than as results of external stimuli; quantum randomness is irreducible and very different from the disorder of a bunch of dice in a dice cup; causality is alive and well, but it often combines with randomness; a total vacuum is impossible: even in empty containers, the quantum vacuum remains, that will exert a force on any electron inserted in it; and *quantons*, as I called the referents of the quantum theory, are neither particles nor waves, but under special environmental conditions either the corpuscular or the field aspect is bound to dominate. In short, material things are not the hard and inert marbles imagined before the rise of modern physics.

From the start, quantum mechanics was interpreted as involving the downfall of causality – to the delight of the positivists, who had kept Hume's idea that the causal connection cannot exist outside the mind because it is imperceptible. In my *Causality* (Bunge 1959a) I sought a partial rescue of causality. I showed that it occurs in quantum physics when it accounts for the scattering of a particle beam by a field force: the latter bends the beam constituents, and its components are distributed at random, so that, by contrast to the games of chance, not even an omniscient being could predict where each of them would go In short, the quantum theory combines causality with chance. In fact, one often calculates the probability that a given cause *C* will have the effect *E*.

In addition, quantum mechanics introduces a new type of randomness. In classical physics, randomness is a collective property, for it amounts to the disorder of a collection of mutually independent events of the same type, such as births and car accidents. Quantum randomness, by contrast, is inherent in individual entities, such as radioactive atoms. Furthermore, the degree of randomness, that is, probability, can be regulated. For example, the random spin directions of an iron bar, initially random, will tend to align themselves in a given direction on introducing the bar into a magnetic field. This kind of randomness is ontological, not epistemological, and therefore Bayesian (or subjectivist) probability theorists have nothing to say about it. Not even God could foresee exactly where each particle in an incident atomic beam will end up.

After the 4th Volume

After finishing volume 4 of my *Treatise* I worked on volumes 5 and 6, devoted to epistemology. But at the same time I also had several other experiences. One of them was my presentation of a materialist conception of culture to the McGill anthropologists. I said that the culture of a society is a subsystem of it composed of the persons who produce or consume such cultural products as poems and cooking recipes, theorems, and theological fantasies. This view was not new: it is the sociological conception of a culture. The novelty resided in my contention that such a system is just as material as the economic one, for being composed of flesh and blood animals, not of pure ideas. This conception of culture matches my systematerialism, and disagrees with idealist monism, Marxist dualism, and Popper's trialism. I made the mistake of not clarifying the point that my conception of culture agreed with the sociological one and disagreed with the anthropological one, which identifies "culture" with "society." I should also have remarked that this confusion had originated in German idealism, and that it leads to talk of "the culture of a culture," by analogy with "the economy of a culture."

By omitting these clarifications I incurred the fury of Professor Jérôme Rousseau, who had specialized in certain tribes in Central Borneo, who shouted: "Shut up! Digging up a latrine is just as cultural an activity as proving a theorem." This interjection, as unusual in a university as in a primitive tribe, left us all speechless. I resumed my presentation and assumed that the incident was over, but every time I crossed paths with Rousseau he stared at me with hatred. Maybe he shared his Borneo tribesmen's belief in black magic.

By contrast, Bruce Trigger, the only well-known member of the anthropology department, held my work in high esteem, and shared my systematerialism. I persuaded him to collect his theoretical papers, and urged Transaction Publishers to publish the volume (Trigger 2003). After his death from cancer, I wrote a paper on his philosophy (Bunge 2013b).

Exchanges with Spanish Philosophers

In 1981 Manuel Garrido, the founder of the journal *Teorema*, organized a colloquium on my work, that was held in Peñíscola, a seaside town near Valencia. Among the participants were the philosophers Jesús Mosterín and Miguel Angel Quintanilla, the logician Mara Manzano, and the physicist Manuel Sánchez Ron. I was not particularly lucid because my cardiologist had forbidden me to drink coffee. While climbing a hill, Mosterín, an enthusiastic naturalist, scolded me for mistaking swifts for swallows – a taxonomic mistake he never forgave me. I retaliated by scolding him for confusing two very different concepts of model: a limited range theory, and an example of an abstract theory.

The following year I attended the first Spanish Congress of Theory and Methodology of the Sciences, held in Oviedo, the capital of the Principality of Asturias and cradle of the revolt against Arab rule. The efficient organizer, Alberto Hidalgo, met my flight in Madrid and put me up in the magnificent Reconquista hotel, where the congress was held. There I met a number of friendly people, among them the physicist Antonio Fernández Rañada and the philosopher Gustavo Bueno, the local philosophy boss. Bueno, who believed to be the founder of philosophical materialism, explained to me his "categorical closure" theory, which had made him famous in Spain. I did not understand it, and every one of the dozen philosophers whom I consulted understood it in their own way. That is the great virtue of confusion in the Republic of Letters: it generates endless commentary.

I listened to a number of good presentations, several of them by scientists. I took part in the discussion of many of them and read two of my own: one on quantum mechanics and reality, and another one on the standard economic theory. I accused the latter of assuming free competition, while actually the most important industrial sectors are owned by half a dozen oligopolies. I also noted the falsity of the general equilibrium postulate.

My presentation angered two local economics professors. One of them became so agitated, that he unwittingly knocked down the standing blackboard. This made me exclaim: "The downfall of standard economic theory!" His exasperated colleague asked the public: "What are we to teach then?" He did have a point: the ruling theory was wrong, but no one had proposed an alternative microeconomic theory. The economics students at Cambridge University applauded the correct and witty criticisms of Keynes's disciples, in particular Dame Joan Robinson and Luigi Passinetti, but the master had not crafted an alternative microeconomics. And the Marxist economists, like Michal Kalecki, only studied global economies: they were not interested in real people.

Stopover on the West Coast

As soon as I got back from Spain I flew roughly the same distance to beautiful Vancouver, the seat of Simon Fraser University. Donning a blue toga and a wide-brimmed hat with a Renaissance air, I gave the Commencement speech and got the first of my 20 honorary doctorates while a kilt-wearing man blew his bagpipe. I also gave talks at several departments and interacted with the physicist Leslie Ballentine, the linguist James Foley, and the Argentine professors Martha Santi and her husband Ricardo Foschi. Martha had been my T.A. when I taught philosophy in Buenos Aires, and while a doctoral student at Stanford she had emigrated from philosophy to social psychology.

From Vancouver I flew to La Paz, in the Mexican province of Baja California Sur, as a visiting professor at its Center of Marine Biology. This was a multidisciplinary institute headed by the biochemist and immunologist Félix Córdoba Alba. My family and I stayed at a hotel on the beach on the Gulf of California. This is said to be the home of about 850 marine species, including whales and sea lions, which makes it the richest ecosystem on Earth. The children spent the day at the swimming pool. Silvia, who at 4 was an accomplished swimmer, complained every time someone used a flyswat, saying "They are my friends!"

My task at the center was to teach a course on systems and talk to young researchers. Some of them where trying to measure and explain the migration of sea lions, but they were on their own: they had not been taught theoretical ecology, and no one had told them what to measure to explain the depletion of the sea-lion population.

Deficiencies like this are endemic in Latin America, which abounds in research centers without enough experienced researchers capable of guiding doctoral students.

Epistemology

Back in Montreal I read the page proofs of the 5th and 6th volumes of my treatise, which concerned the exploration and understanding of reality respectively. Contrary to both the subject-centered and the subject-free theories of knowledge, I started with the brain and what happens to it as it learns – a hot topic in cognitive neuroscience at that time, and even more so nowadays, as the brain-imaging techniques allow us to discover some of the alterations in the brain as it learns or forgets.

I did not forget the social environment, but I did not shift learning from brain to society, as Marx and Vygotsky had done. My thesis was simply that learning is a specific (peculiar) function of the socially embedded brain. Hence epistemology must be developed in close contact with the psychology and sociology of knowledge.

In the 6th volume I also dealt with the various kinds of knowledge, as well as with the demarcation between them and the pseudoknowledge exemplified by theology and pseudoscience. However, far from looking for the borders between philosophy and science, I argued that they have a non-empty intersection, since the exploration of a bit of the world presupposes both its existence and the possibility of knowing it. Decades later I returned to epistemology, first in *Chasing Reality* (Bunge 2006a), and more recently in *Evaluating Philosophies* (Bunge 2012a). In the former work I claim that there is no sound epistemology without ontology, because whoever endeavors to explore a piece of reality starts by making some assumptions about its nature, such as that it is perceptible or imperceptible, alive or not, and so on. I called *hylorealism* the fusion of materialism with realism.

Reality Equals Materiality

The most important postulate of hylorealism is that everything real is material and conversely. In standard symbols, $\forall x(Rx \Leftrightarrow Mx)$. In other words, the concepts of reality and materiality are coextensive. An even stronger postulate is that the concepts in question are also cointensive. That is, $R = M$. For example, ideation is real, hence also material, since it is a brain process. But of course ideas in themselves are not material since, by definition, "in itself" denotes a fiction obtained by stripping the object in question from all the properties associated with change. Admittedly, the claim that reality is the same as materiality, which in turn equals mutability, is counterintuitive. But so is all science. Ordinary knowledge and its companion, ordinary language, are only skin-deep and can be misleading.

In the second of the above-mentioned books I state my preference for the philosophies that help advance knowledge over those that are only games. There I also ask why induction cannot supply deep theories such as those of nuclear physics, cognitive neuroscience, and economic sociology. My answer is that inductive generalizations involve only concepts that occur in the relevant empirical data. And these data, in particular instrument readings, are mostly indicators of things or facts inaccessible to direct observation. This radical limitation of induction implies that the inductive logics of Carnap, von Wright and others are useless for advancing knowledge.

So, Popper was right in opposing inductivism, but his reasons for doing so were not, for he shared the assumption of the inductive logicians that propositions can be assigned probabilities. This groundless assumption had led Popper to asserting that the probability of any universal statement is null. Nor was Popper right in adopting deductivism as the right alternative to inductivism. The most interesting reasonings in scientific research are neither inductions nor deductions: they are what Peirce called *abductions*. And these are not inferences proper but inventions: they are the conjectures we imagine when trying to solve inverse problems, such as explaining observed behavior in terms of intentions or the corresponding processes in the prefrontal cortex (behind our eyes).

Back to Spain: Environment and Carnivory

Tito Drago is an outstanding journalist and social activist who had to flee for his life first from the Argentine military dictatorship and later from the Chilean one. He has been living in Madrid for the last three decades, working at a cooperative news agency, and organizing meetings about social issues. In the 1980s he invited me to attend a meeting on environmental issues to be held in hot and humid Seville.

In my presentation I stated that cattle raising greatly contributes to environmental degradation, since every kilogram of beef requires about 20 kg of fodder, the cultivation of which in turn degrades the soil and accelerates desertification. I further suggested that the sustainability of human life calls for a return to the vegetarian diet of our remote ancestors, which still prevails in most regions of the world. I added that we should start by banning bullfighting, not only because it is cruel and brutalizes its fans, but also because the raising of fighting bulls is even more wasteful than that of ordinary livestock. The vast majority of my listeners, all of them supposedly cultured people, became so angry that rational debate became impossible. But only three decades later bullfighting was banned in Catalonia, partly thanks to the campaign led by the philosopher Jesús Mosterín.

When returning to Montreal via New York, the rude American customs official, asked me the purpose of my trip to Spain. When I told him that I had gone to discuss environmental issues, he asked me what I did for a living. My answer, that I taught philosophy, reddened even more the crest of the fighting cock. So, I added that nothing human is foreign to the philosopher – which only rendered the patriotic civil servant even more suspicious. Presumably, he shared the view of my colleagues, that a philosophy is the better, the more useless. In my opinion, the least a philosopher can do, as Socrates said, is what the horsefly does to the horse: sting it to keep it awake.

Festschrift

In 1982 I was honoured with my first *Festschrift*: a collection of 26 philosophical essays, most of them about my philosophy, edited by Joseph Agassi and Robert Cohen (1982). The editors complained that I rarely answered to my critics, which is true and I explained thus. When one has a long-term project, one fears not having enough time to complete it. In 1982 I still had half of my treatise to write. Nearly a decade later, when my second *Festschrift* appeared (Weingartner and Dorn 1990), I had just completed that project, so I had time and patience to comment in detail on the 31 contribution to that volume. Yet, some of my replies were admittedly too curt. For example, to my former postdoc Hiroshi Kurosaki's assertion that he "wanted to know more" about the way the brain ideates, I replied that all he had to do was to read some of the relevant scientific literature. Apparently my critic expected personal tutoring.

A decade later came the third and fourth *Festschrifte* edited by Guillermo Denegri and Gladys Martínez (2000), who were based in the Mar del Plata seaside, and a fifth by the Grupo Aletheia of Vigo, headed by Avelino Muleiro. I was not asked to comment on them. Nor was it appropriate to comment on the interviews by Laurent-Michel Vacher (1993) or by Raúl Serroni-Copello (2011). Each of them took 2 weeks of fluid daily dialogues in my university office. As I write, Denegri's compilation of 18 essays on my philosophy by 26 scholars is being released under the title *Elogio de la sabiduría* (Denegri 2014).

I got to know quite well four of the above-mentioned editors. Denegri, now a professor of biology and philosophy, had graduated in parasitology, and in the mid-1980s had applied to McGill to do graduate studies in philosophy under my guidance. My McGill colleagues rejected him, along with a Colombian engineer and two Canadians, one an economist and the other a graduate in agriculture, for not having a philosophical background. Motivation and knowledge of a science or a technology were deemed to be irrelevant to philosophical studies.

Muleiro is an enthusiastic philosophy teacher in Vigo, an important fishing port, and presides over the Aletheia group, an association of amateur philosophers who invited me to talk to them, as well as to attend a meeting discussing my philosophy. Vacher had studied philosophy in his native France, and had come to Canada on a Franco-Canadian Exchange program, had liked the country, and stayed here. He was one of a handful of French philosophers to fall in love with science. Serroni-Copello too was a *rara avis*: one of the very few Argentines who had not fallen for psychoanalysis. He directed the only psychology laboratory in the country, established in 1901, when "positivism" (scientism) was at its height in Argentina.

Since a decade ago, after nearly a century of uncontested psychoanalytic supremacy, Facundo Manes and Mariano Sigman imported cognitive neuroscience into the country. Sigman heads the Integrative Neuroscience Laboratory at the Buenos Aires University, and Manes the INECO research and treatment center, as well as the INECO Foundation, which gave me its 2011 prize. While Manes, Sigman and their students carry out international-level research in cognitive neuroscience, the so-called faculties of psychology keep teaching psychoanalytic fables, particularly Lacan's.

Toledo/Madrid: Meeting in the Young Spanish Democracy

In the spring of 1983 I participated in the Meeting for Democracy, a gathering of more than 100 Spanish and Hispanic-American politicians, intellectuals, and writers. We met first in Toledo – where Moors, Christians, and Jews had lived in peace before the Spanish reconquest – and then in Madrid. In Toledo we stayed at the national hotel, that commands a splendid view of the rapid if polluted Tagus river, that flows down to Lisbon. There I had the great pleasure of meeting Raúl Prebisch, my secular godfather, whom I had not seen since I visited him three decades earlier in Santiago de Chile. Raúl, a distinguished economist and senior UN official, was accompanied by his second wife Eliana Díaz, a charming Chilean lawyer.

In Toledo I also met my old friend Jorge Sabato, Ernesto's nephew and a nuclear engineer, who had often visited me at *El Ombú*. There was also the historian Nicolás Sánchez-Albornoz, who had escaped from a Franquist labor camp, and with whom Marta and I had often gone horseback riding in the Palermo gardens. I also chatted with the great Paraguayan novelist Augusto Roa Bastos and his compatriot Domingo Rivarola, the editor of the best Spanish language journal of sociology, whom I met again three decades later in Asunción. The famous Gabriel García Márquez was pleased to learn that my former Mexican student Rafael Vidal had dedicated his master's thesis to the people of Macondo.

Another agreeable surprise was meeting Raúl Alfonsín a few months before he was sworn in as the first constitutional president of Argentina after 17 years of military rule. We met again at the presidential residence and at the Club del Progreso, the oldest Argentine club, of which both of us are honorary members. Alfonsín was a warm, honest and courageous person, in charge of a mediocre team and beleaguered by fanatic and violent groups of many colors. He appointed Manuel Sadosky, an enthusiastic mathematics teacher and my early mentor, as the first minister of science and technology. Regrettably, this ministry achieved nothing, partly because Manuel was so timid that he kept the crucifix behind his ministerial desk, which contained two microphones that he did not dare disconnect.

My greatest and best surprise at Toledo was my encounter and instant friendship with José Luis Pardos, the head of the Department of International Cooperation of Felipe González's so-called socialist government. Jo Lu, as my son Eric called him, is a very warm and generous person, willing to exchange ideas on any subject, and work for any cause he believes in. While in Toledo he read my short book on the role of science and technology in development (Bunge 1980b). Jo Lu was particularly taken by my systemic vision of society. Two years later he organized a meeting, held at the old Alcalá de Henares University, to discuss my project of a HispanoAmerican Institute of Science of the Sciences. The meeting, presided over by Queen Sofía and the foreign affairs minister, Fernando Morán, went very well, but the project aborted shortly thereafter due to a political crisis. My good friend Miguel Angel Quintanilla, the philosopher of technology who had attended that meeting, refloated it successfully a few years later at the Salamanca University.

Cuba: Yes, But

Upon turning from the Toledo-Madrid meeting, I flew to La Habana, along with my family, invited by the Cuban Academy of Sciences. My task was to deliver five lectures at the academy, but in the end I gave 16 talks at a number of places. My audiences were attentive, curious, and polite. The exception was one rude woman later identified as a fellow Argentine.

Manuel, our friendly driver, whom Silvia loved, took us to a number of beautiful places, among them some unspoiled beaches. When the children caught a "bug," he drove us to a medical clinic in the middle of a wooded hill, where two physicians examined the children and prescribed a drug that was hard to find in the city's pharmacies.

We also paid a visit to Ernesto Guevara Lynch, Che's father, who told me that I was related to both of his son's parents. His two young children showed me their admirable exercise books and attractive science and math textbooks. I was so impressed, that upon our return to Montreal I wrote to Mrs Fripp, the principal of Silvia's private school, suggesting she send a member of her staff to Cuba, to learn how to teach math and science. Mrs Fripp replied that the school did not have the funds for that. Later I learned that she was a Conservative party organizer – in a province where conservatives were practically extinct.

On the eve of our return, Marta and I were invited to pay a visit to Carlos Rafael Rodríguez, then "number 3" in the Cuban government, with whom I had exchanged a letter or two four decades earlier, when each of us was editing a philosophy journal. He asked me to tell him what I had disliked, not what I liked. I replied that I was dismayed by the low status of basic scientific research and by the backwardness of philosophy in his country. Don Carlos, pointing his finger at the president of the Cuban Philosophical Society, who was sitting next to me, said: "What happens is that these copy the Soviet philosophers, who have thought nothing new in decades." When taking leave of us, he told Marta to return, because they needed mathematicians, and said to me: "Do come back, Bunge, but not too often." We did go back several times, but only as tourists. And the Spanish daily *El País* published my report on that first visit: "Cuba: Yes, but." Apparently, it displeased everybody. Objectivity is never welcomed in the political arena.

During my visit to some laboratories, I said that I disagreed with the official policy of placing technology before basic science, because modern technology makes an intensive use of basic knowledge. I understand that this defect was corrected after the collapse of the Soviet Union. But my main objections were to the lack of popular participation, for it affected more the honest critics than the American agents; to monoculture, as it forced Cubans to import Bulgarian food jars; to the absence of cooperatives due to the confusion of socialization with statization; and to the replacement of self-government with dictatorship. No one replied to my criticisms.

Philosophy Congress in Montreal

At the end of that summer of 1983, the seventh World Congress of Philosophy was held in Montreal. I participated along with my six graduate students. My presentation dealt with applied science, the link between basic science and technology. I held that the applied scientist, such as the pharmacologist, the parasitologist, and the education scientist, hunts for new knowledge, just like the basic scientist, but gives preference to knowledge of possible practical use. For example, the pharmacologist does not design new molecules just out of curiosity, but in the hope that some of them may be used to treat diseases.

One of the congress panels – composed of Quine, Putnam, and Føllesdal – dealt with the semantics of proper names. This seemed to me to be a miniproblem, hardly distinguishable from a pseudoproblem, since proper names have referents but no senses. For example, 'Francis' denotes any of the men who bear this name, but it connotes nothing. What is true is that, contrary to most other proper names, 'Francis' used to name individuals who had come from France. When I asked the panel members whether their learned papers helped to identify the referents of evolutionary biology or microeconomics, they seemed disconcerted. Perhaps they believed that semantics need not be of any use.

At the congress I met again Robert Nadeau, director of the Centre d'épistémologie appliquée of the Université du Québec à Montréal. I asked him whether he could offer me a job there, in view of the fact that my colleagues had the intention of retiring me as soon as possible. His reply was clear and quick: "*On vous redoute*. [You are feared.] None of my colleagues would support my proposal." I suppose this served me right, for having my own opinions and publishing them abundantly.

Holidays

All of the above has been about study and work. But holidays have been no less important in my life: I have always tried to organize two per year, a long one in summer and a short one in winter. I have two excuses for such apparent sloth: health and the wish to think leisurely about long term projects. I almost always do some work during holidays. For example, I wrote several chapters of these memoirs within view of Odysseus's sea.

One of the best holidays family ever had was spent at the Cala d'Or, on the beautiful island of Mallorca, in exchange for courses that Marta and I taught at the Baleares University. There I read a good many papers on cognitive neuroscience.

The most memorable but least productive holiday was the one we spent traveling in India guided by Eric, who had learned Hindi while working as an architecture apprentice near Calcutta. In Puri, on the Bay of Bengal, we visited some of the temples devoted to 33,000 Hindu gods, as well as Jaina temples and the millenary Tamil Academy. The latter used to hold yearly competitions in literature, the humanities, and religion. The many works submitted every year were evaluated using the most impartial of all methods: the manuscripts were thrown into a shallow pond. The winning works were those that floated best.

Along 2,000 years-old roads we saw milestones with inscriptions in Nepali, in memory of public works ordered by the wise emperor Ashok, the earliest peace and welfare statesman. On the border between the provinces of Tamil Nadu and Kanada, which at the time were engaged in a conflict, we waited for several hours in a wood, in uncertainty and total darkness. And one midnight, in the nation's capital, we were mobbed by 5-year old beggars. What a miserable and splendid country!

Between our return from México and the mid-1990s we spent much time in our cottage in Morin Heights, in the Laurentian hills, north of Montreal. In the summer we swam in a cold lake without motorboats, and in the winter did cross-country skiing in thick forests visited by ferocious bears, timid deer and moose, cheeky racoons and ostentatious partridges, elusive white foxes and owls, and blood-sucking black flies, the region's bane. We had finally understood how to face the ferocious Canadian winter.

One of our neighbors was an elderly but still active carpenter who lived off his generous German old-age pension. He vividly remembered the day in 1933 when, after a long period of unemployment, he got a job in the construction of a hangar: he understood that the Nazis were preparing for war. Another interesting neighbor was the friendly lady who owned "Books and Things," a quaint shop in the village, that, to our surprise, did sell some books. She also published a book of cooking recipes written by the locals, among them Marta, who just translated a recipe for a mushroom dish she had never tried. We bought groceries from the large, helpful, and cigar-chomping Vaillancourt brothers, who taught me how to bait mouse-and-mole traps. We also had to suffer disparaging comments on "cullige perfessors" from the rednecks in the café and general store, whose crossed-eyed owner hoisted the US Confederacy flag without knowing what it stood for.

I always typed a few hours a day on a heavy and noisy IBM typewriter. Before converting to the Apple faith, I wrote down every interesting idea or possibly useful datum on 5×8 cards that I kept in card-boxes. But I used them only sparingly to write papers of books, for they were just random collections. Once an unknown American scholar phoned me to announce that he was about to commit suicide because he had failed to craft a general theory of ideas out of thousands of cards that he had filled in the course of a decade. He had been a casualty of *dataism*, the idea that knowledge of anything is just a collection of bits of knowledge.

More on Systematerialism

I kept harping on systemic materialism in some papers, as well as in the books *Emergence and Convergence* (Bunge 2003d), *Über die Natur der Dinge*, with my coworker and friend the zoologist Martin Mahner (Bunge and Mahner 2004), and *Matter and Mind* (2010a). The first of them deals with systems and integrative levels, and proposes that both call for the convergene or fusion of previously separate disciplines. For example, the emergence of living things about 3.5 billion years ago prompted the fusion of evolutionary biology with citology and biochemistry.

The second book, written mainly by Martin Mahner, is a conspectus of my systemic materialism. Martin, the world authority on water fleas, spent the 1993–1996 period at my Foundations and Philosophy of Science Unit. He now directs the Wissenschaftszentrum in Rosdorf, Germany, devoted to debunking pseudoscience. I wrote *Foundations of Biophilosophy* (Mahner and Bunge 1997) with Martin.

The following diagram displays the components and causal relations of my systematic ontology:

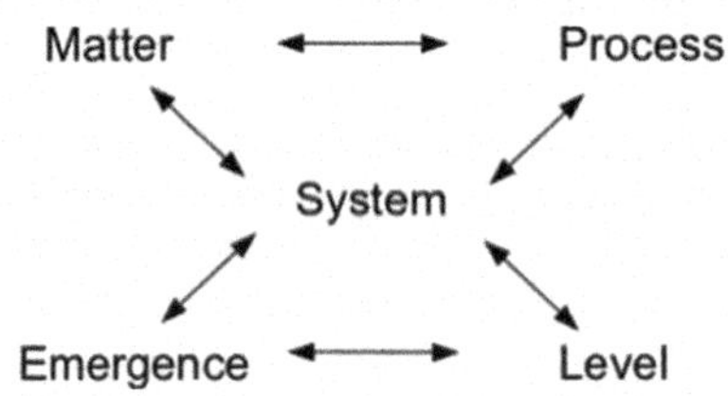

However, elaboration is so ample that it deserves a separate chapter – the next one.

My Contributions to Ontology: Summary

I believe the following to be my main contributions to ontology or metaphysics.

1. Giving precise definitions of such key concepts as those of thing, property, state, process, emergence, level, system, and causation;
2. Building mathematical and relational theories of space, time, and spacetime consistent with Einstein's theory of gravitation;
3. Crafting the fusion of materialism with systemism;

4. Emphasizing emergence, integrative levels, and the corresponding limitation of the reducionist strategy;
5. Seeking compatibility with contemporary science;
6. Bringing epistemology closer to ontology, so as to generate hylorealism;
7. Rescuing and developing some of the valuable insights of predecessors;
8. Criticizing alternative ontologies, in particular that of possible worlds.

Debunking Pseudocience

I have written and spoken against the pseudoscience swindle even before having crafted a sound view of science. Parapsychology was my earliest target, with Freudian psychoanalysis the next one. I am a member of the Committee for Skeptical Inquiry and contributor to its two magazines, *Skeptical Inquirer* and *Free Inquiry*, almost from their beginning. I have helped organize skeptical societies in Argentina, Quebec, and Spain.

In 1986, while teaching in Geneva, I was visited by Alvaro Fernández Fernández. This was a member of the Spanish engineering company that was excavating the huge tunnel for CERN. He wanted me to help him write a sort of skeptical manifesto that might give rise to organizing a skeptical society in his country. We wrote it, and when back in Spain, Alvaro had it signed by a number of influential persons, but he failed to have it published. It was rumored that the editor of the main Spanish daily had his own wizard.

Until recently, in Argentina I was known only as a critic of Saint Sigmund, and berated for that, even by professors who claim to teach philosophy of science. My first public encounter with practising psychoanalysts was a round table in Buenos Aires towards the end of the 1950s. One of my fellow panelists started to make tics so pronounced and frequent, that I recommended he see an authentic psychologist – a bad piece of advice, both because there is no known treatment for that disorder, and because at the time there were no authentic psychologists in my country. At about the same time I participated, along with the meteorologist Rolando García and the logician Gregorio Klimovsky, in a round table about pseudoscience summoned by the Sociedad Hebraica. In my talk I mentioned psychoanalysis as a clear example of bogus science. My co-panelists expressed their utter disagreement with me.

During several decades I participated in many skeptical meetings in Canada, the US, and México, organized by the indefatigable Paul Kurtz, whom I befriended when he came to Montreal to give a talk on parapsychology. I nominated him for an honorary McGill doctorate, given his distinction in the skeptical community, but my university resolved to honor a theologian instead. When James Alcock came to McGilll as a visiting professor of psychology, I urged him to write a book, which he did: it was *Parapsychology: Science or Magic?*, which I included in my Pergamon series (Alcock 1981). Another interesting character with whom I interacted is James

Randi, the stage magician who exposed Uri Geller, the spoon bender. Once Randi, fondling my favorite tie, made me believe that he was fraying it.

On several occasions I warned fellow skeptics that we ought to embrace methodological rather than absolute skepticism, if only because every criticism is relative to some body of knowledge: "I indict *A* because it contradicts our background knowledge *B*" (Bunge 2000d). Whereas the absolute or skeptics, or Pyrrhonists, reject all knowledge claims, the methodological skeptics doubt only the insufficiently substantiated ones and reject the groundless ones. Shorter: they do not reject all beliefs, but only those found to be false, and suspend belief in untested propositions.

Science Alone Does Not Immunize Against Pseudoscience

A scientific education certainly helps detect and evaluate all knowledge claims, but it is insufficient for detecting pseudoscience, as shown by the numerous cases of distinguished researchers who believe in homeopathy, psychoanalysis, or parapsychology, or who have reinvented the myths of creation from nothing, universes inaccessible from ours, or the ideas that physical space has more than three dimensions, that genome is destiny, or that humans ceased to evolve some 40,000 years ago. While scientific experience is insufficient, it is not even necessary to detect chalatanry, as shown by the cases of Mark Twain, James Randi and the proverbial Missourian who always requested: "Show me."

The lay skeptics were able to unmask charlatans because they were wary of deceit and self-deceit: they knew that, while sometimes it is enough to see in order to believe, at other times we must believe in order to see. It also helps asking what the relevant mechanisms are, from covert signals to the placebo effect, which is present in all medical and pseudomedical treatments (see Bunge 2013a). One may be skeptical in one's own research field but gullible in others. To guard against all pseudosciences one has to have a materialist worldview and a scientistic epistemology. One must begin by critically examining the relevant myths that we inherited, such as that anything is possible, that what cannot be explained in a worldly fashion calls for extraordinary (paranormal or supernatural) explanations; and that, because science is rational, it cannot explain the irrational, from taste to faith.

Every time we imagine or examine mechanisms of action we are faced with the materialism/spiritualism dilemma. How does telepathy work, how does free trade benefit agricultural countries, how do free trade and the free market help the poor, how does the dictatorship of the proletariat emancipate, and what is the mechanism of its withering away? Unless there is a proven or at least plausible mechanism, there is no satisfactory scientific explanation. This is so because, by definition, to explain X is to describe the mechanism underlying X (Bunge 1969c, 2004b). I call *mechanismic* this construal of 'explanation', which is far more demanding than the usual or "nomological model." In *Scientific Research* (1967b) I had miscalled it "interpretive."

Why Debunk Pseudoscience?

The pseudosciences have interested me for three reasons: they help define genuine sciences, evaluate the various philosophies of science, and they are symptoms of both underdevelopment and decline. Indeed, one should start by listing the traits peculiar to the recognized sciences; a philosophy that lets scientific heresies through does not deserve to be included into epistemology; and a society where pseudoscience "sells" better than science is either very immature or senile.

I have also analyzed a number of pseudosciences because they are at best false; because they divert public attention from important issues and disciplines; because they undermine trust in scientific research; and because some of them may be used to prop up socioeconomic privilege. Indeed, belief in UFOs, telepathy and horoscopes is inoffensive, whereas belief in nativism, in standard economic theory, in liberty without equality, and in equality without liberty, have been even more pernicious than the so-called alternative medicines, because they have struck entire peoples. Regrettably, most skeptics have overlooked the crucial difference between harmless and harmful pseudoscience, and have confined their attention to the oldest and less dangerous superstitions, like astrology and alchemy, perhaps because they are the easier to criticize. Let us follow the example set by Bento Spinoza, the boldest and most hated philosopher of his time, and the one whose motto, *Caute*!, he seldom practiced himself.

Coda

In this chapter I have sketched systemic materialism, the ontology first adumbrated by the Baron d'Holbach, perhaps the most enlightened of the members of the radical wing of the French Enlightenment. I came to systematerialism after struggling with the obscurities of dialectical materialism, which had seduced me in my late teens, but increasingly disappointed me as I encountered modern logic and the so-called general systems theory around 1950, and later on cognitive neuroscience and social science and technology as well. Since then I have been working on systematerialism.

The balance of these memoirs deals with the relations between that philosophy and the sciences of life, mind, and society, as well as the study of values.

Chapter 11
Biophilosophy

Early Curiosity About Living Things

Living beings interested me from a very early age, perhaps because I grew up in a small and uncomfortable shack with hardly any toys, but surrounded by a garden-orchard and by neighboring farms far more alluring than home. A bush was likely to contain a nest that my parents urged me to observe but not disturb. A hole in the ground was likely to be the home of a green lizard, a *cuy* (Guinea pig), a family of country mice, or a harmless snake. The neighbors with chicken coops complained about thefts committed by opossums. Under a tile a fat bullfrog might hibernate. In several trees and on the top of the windmill there were nests of untidy, noisy, quarrelsome, and thievish jays, as well as of quiet spinels, said to be lifelong monogamous couples. Hummingbirds were often flitting above the honey-suckle vines. The triumphal song of the *benteveo* (Great kiskadee) was often heard nearly everywhere. And the industrious ovenbird, the national bird, would always announce the end of a rainfall. How I miss them!

Farther afield I would also see clever *tero* birds (southern lapwings), singing wild doves, cheeky mocking birds, or secretive owls. Once in a while a large flock of *teros* would fly over at great heights, calling their brethren that we kept captive with half-clipped wings. And there was at least one "singing" canary in every household. I used to play classical music records to mine, but without success; perhaps it was, like its owner, only a music lover.

Feral dogs and cats were frequent visitors. Once a beautiful *barcino* cat approached my father, showed him a paw, and meowed. (How did it guess who the alpha male was?) My dad examined his paw, saw a large thorn on it, and extracted it. The cat adopted us, and I christened him Mustapha. Every time he caught a mouse he deposited it in front of the kitchen, to show that he was doing his job. I bought some waterfowl that congregated around an open faucet. There was also a pair of peacocks whose shrill call used to wake me up. The ungainly *chajás* brought from the pampas were even noisier. One day at sunrise, exasperated, I sent the dogs

M. Bunge, *Between Two Worlds*, Springer Biographies,
DOI 10.1007/978-3-319-29251-9_11

after them, but they found a sort of foam instead of meat. Of boastful people we say that they are sheer foam like the *chajá*. Occasionally my father would bring from the pampas a huge ostrich egg, enough for us all. And we all had luscious bed-covers made from furs of Patagonian *guanacos*, close relatives of the Peruvian llamas.

There were also insects of various kinds, from houseflies and horseflies to honey bees, wasps, and bumblebees to grasshoppers, which occasionally came in thick clouds and devoured all the greenery around, thus ruining the farmers nearby. We kids were urged to bang on empty kerosene tins to shy them away, but the greedy insects just hopped to a nearby bush or tree and went on chewing. Yet they never learned that the *ombú* leaves are powerful laxatives that killed them *en masse*. My father kept a few beehives, which he handled by making smoke and wearing gloves and a wirenet mask, but I do not remember ever having tasted their honey. My father would not steal the fruit of their labors. How could dull school compete with the ever surprising delights of our garden-orchard?

I got my first biology lesson at age 5. It was a summer evening, Congress was in recess, and for a change my father was relaxed and found time for me. He sat me on his lap and told me about the Madagascar lemurs, whose huge eyes allowed them to forage at night. He told me that those strange and timid animals were relatives of ours. I was puzzled, because they did not look quite like us; but I did not disbelieve him, for he had never lied to me. Still, that may have been my first experience of doubt. The second was when a teacher told us that those who do not save are dishonest. But my father did not believe in saving because he remembered the time of German hyperinflation. I used to play with the hundreds of colored banknotes for billions of marks that filled a shoebox.

At about 12 I became intrigued by the nature and origin of life. I took it for granted that life had arisen from non-living things. Once, when dissolving some copper sulfate crystals, which we used to fumigate the aphids, I noticed that beautiful tree-like things started to grow. I repeated the experiment many a rainy afternoon, admiring a somewhat different arborescent thing every time, and watched for signs of life. I did not know that I held the abiogenesis hypothesis, and that Alexander Oparin was trying to revive it at that time, though on the basis of biochemistry, since metabolism, not shape is of the essence of life.

It was only after Sputnik (1957) that the origin of life was regarded as a subject of serious research rather than free speculation. It was even a cornerstone of some space exploration expeditions. In the 1980s I was fortunate to interact with two of the most productive workers in this area: the American Melvin Calvin, who was later awarded a Nobel prize, and the Catalonian Joan Oró, who worked at NASA.

First Inkling of Biophilosophy

At school the problems of the origin and evolution of life were not even mentioned. We were not even told how organisms are born, grow, and reproduce. Nor were we told about spontaneous generation, teleology, or the popular belief that the way to cure indigestion is to attach a live frog to the stomach. The only thing that seemed to matter was the classification of organisms, which bored any normal child.

To my delight, those and other biophilosophical queries were discussed by the French biologist Marcel Prenant in his *Biologie et marxisme* (Prenant 1935), which I discussed with my father soon after it appeared. Prenant was an original investigator who joined both the Resistance and the communist party, from which he was expelled for attacking the charlatan Lysenko, the leader of the Soviet campaign against genetics and evolutionary biology.

A decade later I befriended Enrique Mathov, the foremost Argentine specialist in allergy, who became increasingly interested in biophilosophial problems, which we often discussed. With him I discussed in detail Julian Huxley's, *Evolution*, the earliest semipopular presentation of the so-called synthetic theory of evolution. This was the fusion of evolutionary biology with genetics that had been forged in the 1950s, but had not yet reached the American textbooks. But Enrique's deepest and most enduring influence on me is that he injected me with a dose of skepticism, that cured me of many a dogma taken for granted by most left-leaning intellectuals.

Since then, evolutionary biology has been greatly enriched: saltation was added to graduality, evolution was joined with development (evo-devo) – the way Ernst Haeckel wanted – as well as with ecology (evo-eco). And epigenetics, the latest arrival, has shown the sensitivity of the genome to the environment. Epigenetics has also ruined genetic determinism, the contemporary phase of nativism, by showing that identical twins may acquire different epigenomes; for instance, one of them may get cancer while the other stays healthy. All these novelties have shown that the standard philosophies of science are simplistic, since those findings resulted neither from blind data hunting nor from refuting hypotheses nor from the struggle for power.

My First Biologist Friend

The first biologist with whom I had a close, fruitful and lasting interaction was the eminent zoologist and paleontologist Osvaldo A. Reig, who had made his debut discovering and describing the first American Archaeopteryx. Osvaldo had published his first scientific paper while still in high school, and was expelled from my alma mater because of his anti-Peronist militancy. After the so-called Revolución Libertadora of 1955 he won a chair at Buenos Aires University before earning a doctorate. But in 1966 he had to migrate to Chile, where he was arrested by the Kissinger-Pinochet government, and was saved from execution thanks to the Organization of American States. In 1973 he moved to Venezuela, where he set up a laboratory of tropical biology. He complained over the extraordinary number of bat species: he seldom found more than one specimen of each of them when he checked the nets he had set up the night before.

I visited Osvaldo and his family several times at several places, and always found him delighted with the new things he was learning, as well as optimistic and enthusiastic over his research projects and students. Osvaldo returned to Argentina in 1983, and the newly elected president of the republic appointed him director of the National Museum of Natural History. When Osvaldo went to assume his new post, the director of the old regime, the anti-evolutionist and clerical José María Gallardo told him: "God put me here, and only He will remove me." So it happened: not even the president of the Republic could fight the Catholic Church in Argentina. Osvaldo kept working at the university, where he trained several researchers, until he died of cancer in 1992. God had not allowed him to direct the Museum, but allowed him to keep enriching paleontology, herpetology, and genetics for nearly a decade (Ponsa 2011).

I had first spotted Osvaldo in 1961, standing next to his wife Estela Santilli, one of my philosophy students, in the back of the classroom where I was lecturing. He drew my attention because he asked an interesting question, and his round face wore a friendly smile from ear to ear. During a visit I made to his herpetology laboratory, Osvaldo invited me to teach a seminar in the philosophy of biology to his advanced students. One of them was Jorge Rabinovich, who soon earned an international reputation as an ecologist. The subject that intrigued them at the moment was the ecological exclusion hypothesis, or Gause's law. According to it, if two populations of different species compete in the same habitat for the same resources, one of them will end up by displacing the other. Is this hypothesis related to other biological generalizations? And is it a universal law? I understand that these questions are still under study.

Mayr and Simpson

In 1963, while visiting Osvaldo Reig at Harvard, he introduced me to George Gaylord Simpson and Ernst Mayr. These two great scientists looked quite different: Simpson was short and reserved, whereas Mayr was tall and exuberant. Simpson remained silent, whereas Mayr declared in a loud voice: "I have read your *Causality*. It is a modern classic. But you are quite wrong in placing biology in the same bag with physics. All electrons are the same, whereas no two organisms are identical."

Shortly thereafter I wrote to Simpson asking him to criticize the draft on biological systematics I had written for my future *Scientific Research*. I wanted his view on the species-genus relation: was it one of membership or, as I argued, one of set inclusion? And how about phylogenetic relations? Do species emerge before their genera or, as I thought, was it the other way around? Simpson sent me his detailed and patient response, which I adopted.

Both scientists were extraordinarily prolific, and both were interested in deep philosophical questions, but I thought that Simpson was clear-headed while Mayr was muddled. Moreover, I believe that Mayr's emphasis on the uniqueness of living beings and their science, as well as his fight against essentialism, was obscurantist and possibly a result of his early exposure to German idealism. Granted, no two individuals of the same species are *identical*, but they are *equivalent* in most respects; likewise, biology and physics are quite different from each other if only because the former is at once a historical as well as a natural discipline, yet both share the scientific method and the naturalist ontology. As for essentialism, what reason is there for refusing to admit the essential/secondary and the primary/derivative dichotomies of properties? Surely the atomic number (Z) is an essential property of atoms, and metabolism of all organisms.

Stephen Jay Gould

I was of course a fan of Stephen Jay Gould's brilliant semi-popular essays in *Natural History*. But I also found that his thinking was philosophically muddled because of his reliance on the dialectics of Hegel and Engels. I attended Stephen's stimulating lectures at McGill, starting with the one he gave in 1968 while still an assistant professor at Harvard. In that lecture he asserted that paleontology is a branch of geology, perhaps because fossilization involves mineralization; in addition, geologists and prospectors had discovered many fossils. I wrote to Gould, reminding him that paleontology had begat evolutionary biology, and that the fact that geologists use paleontology to date strata is only an example of the interdependence of the sciences. Seismologists might yet invent a more accurate method of strata dating. The history of life is entwined with that of rocks, but paleontology is a part of biology because it studies organisms, whether alive or fossilized. Stephen and I exchanged a few more letters, until he told Reig that he had changed his mind about my standing as a philosopher: whereas formerly he had placed me at the top of his totem pole, lately he had demoted me to the bottom.

A decade later Stephen and I had our last friendly collision, this time at Boston University on the definition of "species", as will be seen below. But our disagreements did not prevent me from regarding him, along with his colleague Richard Lewontin and the British all-round biologist John Maynard Smith, as the leaders of the contemporary pro-evolution movement. The fact that all three were Marxists explains their commitment to evolutionary biology and its fusion with developmental biology, as well as their occasional conceptual howlers.

One of the first things I did when arriving in McGill in 1966 was to visit Bob Carroll, one of Osvaldo's friends. He told me about his work on the fossil-rich Canadian soil, comparable to Patagonia's, first explored by Darwin. Pointing to a bookshelf in his library, he said: "And that's my philosophy library." I looked and saw that it only contained works on evolutionary biology. He may have meant that as a joke, because he was soon to become the pre-eminent vertebrate paleontologist. At Harvard he had studied under the famous Alfred Romer, who dreaded being introduced to new people because every time he learned a new family name he forgot that of a fossil.

Earliest Essays on Biology

My first essay in biophilosophy was a note on Florentino Ameghino's materialist worldview (Bunge 1945b). The Florentino and Carlos Ameghino brothers were amateur paleontologists who collected, described and classified the fossils they found in the cliffs of the Luján river, a modest water course in the pampas, that turned out to be a very rich fossils trove. The Ameghinos had discovered and studied about 6,000 fossil species while minding their modest school supplies shop in La Plata city (Casinos 2012).

They were the first Argentine scientists to win public recognition, but they were criticized by some because of their Darwinianism, by others because of some of their wild conjectures, and by still others because they sold some of their collections to French scientific organizations in order to support their work. But all the innovators in a conservative society are likely to be rubbished.

My second publication in this field was a paper on the bold explorer and naturalist Alexander von Humboldt (Bunge 1969d). This was a remarkable case of transformation of a believer in superstitions into a world famous scientist, and of an aristocrat who braved obstacles that even natives skirted. For example, he was the first human to climb the monte Avila, the 2,765-m high mountain overlooking Caracas.

A decade later came a programmatic article on method in biology (Bunge 1976b), and a mathematical theory of processes combining competition with cooperation (Bunge 1976c). Even though ecologists emphasize competition, I believe that it coexists with cooperation – as the great naturalist and anarchist theorist Peter Kropotkin had suspected. Indeed, there is direct cooperation, as in the cases of symbiosis, coral colonies, bird flocks, deer herds, human families, and the "nanny" bird that helps minding the newly born birds of other conspecifics. And there is indirect cooperation, as with the animal droppings that scarabs use. Resistance to regarding all these as cases of cooperation may have to do with confusion of this word with 'collaboration'.

While on a Montreal bus ride I imagined a simple model of biological evolution, which I formulated in mathematical terms. The physicist Rafael Pérez Pascual helped me refine it while we interacted in México, and I included it in the fourth volume of my *Treatise*. In the same book I also discussed some topical subjects, such as the concepts of species and essential property.

At the 13th Wittgenstein symposium, held in the pretty Austrian village of Kirchberg, I read my paper with my former student David Blitz on a topical problem in evolutionary biology: the gradualism/saltationism dilemma (Bunge and Blitz 1989). Our thesis was that biological evolution involves both gradual changes, such as alterations in size, shape, and nourishment, and qualitative changes, such as the emergence and submergence of organs and traits through mutation. We felt that, consequently, too much fuss had been made over saltation.

Biospecies

Aristotle, the founder of marine biology, handled deftly the concepts of species and genus, and the great Linnaeus – whose amphitheatre at Uppsala University is still in use – proposed the earliest non-utilitarian classification of biospecies. However, questions about the definition of the species concept, as well as the problem of whether species are real, are still being spiritedly discussed. In fact, it is the only subject that a number of biophilosophers have ever discussed.

These issues were first tackled by Plato, and in the Middle Ages they pitted the Platonic "realists" (objective idealists) against the nominalists (or vulgar materialists) like Ockham, leaving the Aristotelian conceptualists in the middle. Roughly, the Platonists held that individuals are just instances of predicates, or members of classes, whereas the nominalists held that there are only individuals and signs thereof, and consequently species are imaginary. Contrary to both Platonists and nominalists, the conceptualists held that species are sets and therefore concepts. But, far from being conventional or arbitrary assemblages, like $\{\pi$, Darwin, USA$\}$, every well-defined taxon is determined by a bunch of precise predicates. For example, the group of Ciliata is the set of all unicellular organism endowed with cilia. Hence, although "Ciliata" is a concept, not a thing, it is a *natural* class rather than a sheer invention like the unicorns, centaurs, or the free and greedy agents of standard economic theory. The conceptualist stand about universals ought to be the ruling one because it is the one actually taken by all the scientists who introduce or use class terms, such as 'acid', 'herbivore' or 'teacher'.

However, over the last half century most philosophers of biology have repeated dogmatically Dobzansky's assertion that species are individuals, and that the relation of species to genus is an example of the part-whole relation, like the one that holds between finger and hand. They have not realized that, if this were true, a species would belong ($\in$) to its genus instead of being included ($\subseteq$) in it, as a consequence of which humans would not be primates, nor mammals, nor vertebrates, and not even organisms.

The space-time representation of genealogies, or trees of life, may have contributed to the species-as-individuals goof when read as the branching of species, while actually only populations exist and split in space and time. A similar double-talk occurs in chemistry. For instance, the synthesis of a water molecule consists in the combination of hydrogen with oxygen; but the reagents H_2 and O_2 must be interpreted as arbitrary members of the species H_2 and O_2. That is, the Platonic formulas, written in terms of species, must be interpreted in a nominalist fashion, that is, in terms of individuals.

Parenthetically, a paper where I discussed these foundational matters with the help of a technical paper by Rutherford Aris (1965) was rejected by a philosophical journal because its referee did not understand it – but I rescued it in vol. 4 of my *Treatise*. The standard practice in scientific journals is more honest: if a referee does not understand a submission, they suggest seeking a different referee.

The eminent paleontologist George Gaylord Simpson commented on a draft of the chapter of my *Scientific Research* where I analyze the concepts at stake in the biospecies controversy. I also discussed them with Osvaldo Reig, David Hull, Michael Ghiselin, Stephen Jay Gould, and Martin Mahner. I have always understood that the individual (be it elementary particle, atom, organism, artifact, or what have you) belongs ($\in$) to its species, which is included ($\subseteq$) in its genus, which in turn is the union ($\cup$) of its species.

Thus, the individuals that biologists handle are living beings or their fossil remains, whereas the taxons they invent are concepts, so they lack biological properties. By contrast, the part-whole relation is a relation between material things, hence an ontological one, not a logical one. I first learned this point from Alfred Tarski in 1964, when he scolded Abraham Robinson for confusing set inclusion with the part-whole relation.

When Michael Ghiselin, a champion of the species-are-individuals howler, came to visit McGill in the early 1970s, I tried to explain all that to him on the blackboard. But he had never seen those symbols, and refused to learn what they meant. Nor did my much-admired Stephen Jay Gould understand them. Indeed, in 1977, while speaking at the Boston Colloquium for the Philosophy of Science, Gould interrupted me, objecting to my definition of a species as the set of individuals that share certain essential properties. I immediately replied with a question: "So, Steve, which definition do you favor?" He adopted a professorial stance, unusual in him, and started: "A species is a set . . ." I interrupted: "Stop right there, Steve. That's exactly what I claim: that species are sets, not things." Steve blushed and kept quiet during the rest of my talk – a rare attitude for a New Yorker, the animal that most resembles the *porteño*.

Essentialism Pro and Con

The concept of an essence too has a long history, and there still are people who object to it just because it has often been misused. The existentialist racket started by inverting the Platonic formula "essence precedes existence." But this sentence is just as opaque as its opposite, which led Heidegger to asserting that "the essence of being is THAT itself." Common sense and the sciences use the concept of an essential property, or basic trait upon which all the others depend. For example, metabolism is essential to life, whereas reproducibility is not. And yet the eminent Ernst Mayr conducted a crusade against *essentialism*, the thesis that some properties, like metabolism in the case of organisms, are essential, whereas others are not. If all the properties were equally important to stay alive, no one would survive the accidents and diseases that deprive us of some of them, either for a while or permanently. Nor would it be possible to categorize, or conceive of such categories as varieties, species, genera, and families. We would not have the periodic table, and would embrace the definition of a well-known Wittgenstein follower: "liquid = pourable." In all cognitive fields we assume that some properties precede others in relevance or importance for some crucial role.

Still, the fiction that all the biological properties have the same weight is the guiding principle of phenetics, a fad of biology in the 1970s when the computer started to be used to construct numerical taxonomies (see Mahner and Bunge 1997). Though mindless and divorced from everything that makes organisms tick, such taxonomies may be initially helpful in the face of the mountains of data collected by field workers.

Information and Life

In 1944 Erwin Schrödinger, the physics giant, suggested that the essence of life is *information*, conceived as negentropy (Schrödinger 1944). This suggestion was exploited by the scientist Manfred Eigen and the popular writer Richard Dawkins, the father of the "selfish gene" and "meme" fantasies. According to this variant of genetic reductionism, "genome is destiny." Moreover, Dawkins held that organisms are nothing but gene funnels between generations, so that their very existence is "paradoxical." This doctrine is open to the following objections raised by various scientists.

First, what lives and dies is not the genome but the whole organism. The methodological consequence is that genetics is only one of the essential components of biology.

Second, an organism's environment is just as important as its genetic "plan," for one and the same genome is viable in some environments but not in others. Consequently, geneticism is just as wrong as its dual, environmentalism.

Third, genes are impotent without enzymes. In particular, DNA, unlike RNA, does not divide by itself. Consequently, the genes-enzyme systems should be regarded as the building blocks of cells.

Fourth, the concept of information used in genetics has nothing to do with the one used in physics, where information is the dual of entropy or collective disorder. Genetic information is also alien to the concept of channel noise or randomness prevailing in Shannon's theory of telecommunication. Indeed, in genetics, information = order of the nucleotides in the DNA molecule. When this order changes, the kind of synthesized proteins alters too. But, since 98 % of genes are "junk," that is, do not guide protein synthesis, they cannot be all-important.

Giving absolute priority to order or genetic information is tantamount to claiming that the plan of a house is more important than the materials used to build it. What distinguishes the living from the nonliving is not genetic information, which exists at the macromolecular level, but metabolism. This is the cell's essential mechanism, for it is what makes cells tick. In short, *to live is to metabolize*. And metabolism occurs on all the levels of an organism from foraging, feeding, and excreting to breaking down, separating, and assimilating.

Biological Laws?

Mayr did not believe in biological laws. He did not elaborate, but from his writings on the peculiarities of biology one gathers that his disbelief in biological laws was due to his emphasis on the uniqueness of every organism, that is, the enormous variety ("variability" in Biologese) of individuals of the same species. Indeed, not even the clones of a given individual are identical to it, although they have identical genomes.

In my view such variety only shows that in matters biological (and social) we should speak of equivalence instead identity. No two organisms are identical, but any two organisms are equivalent in a number of respects, in the sense that they share some essential properties. In other words, biological classes are equivalence classes. Hence, the universe of discourse of biology is the family of such equivalence classes.

What holds for individual organisms also holds for biological laws: they hold for all the individuals of a given taxon, but they involve only essential properties, so they make room for individual peculiarities. For instance, all birds are winged, but the wing of some species are no longer functional.

Furthermore, presumably every evolutionary biologist who believes in biological laws assumes that such laws emerged or submerged along with the corresponding taxons. And realism suggests that such laws hold even before being discovered: we realists distinguish objective laws or patterns from their conceptualizations, namely law statements, called "lawlike generalizations" in contemporary Philosophese.

If the facts of life were not lawful, they would be miraculous. And if there were "evolutionary algorithms," as Daniel Dennett has held, organisms would be artifacts. Likewise, brains would be machines if there were "neural algorithms," as Patricia Churchland believes, since algorithms are computational devices, and as such artifacts. What biologists try to discover is real *natural* patterns, that is, laws of nature, not rules for doing things. And the laws of nature, inasmuch as they are invariant relations among natural properties of natural things, are inherent in them (Bunge 1960c).

Cybernetics Explains Teleology Away

Telelogy is the doctrine that the chief biological mechanism is goal-seeking. This doctrine was central to Aristotelianism, which was criticized, and even ridiculed, by many early-modern thinkers, such as Rabelais and Bacon. Yet, it survived in many modern biologists, such as Ernst Mayr and Jacques Monod, who rechristened it *teleonomy*.

The earliest persuasive criticism of teleology came from an unexpected quarter: cybernetics, or the theory of control. This discipline, born in engineering, became a part of general systems theory when it was shown that not only artifacts like steam engines but also organisms are endowed with control mechanisms. Indeed, in their foundational paper on the subject, Wiener, Rosenblueth and Bigelow showed that what appears as goal-seeking behavior is actually nothing but a negative feedback mechanism (Wiener et al. 1943).

Negative feedback is of course the process wherein a part of the output is fed back into the input terminal to maintain the output within a suitable range – the small box in the state-space of the organism within which life is possible. This was, of course, homeostasis, or the stability of the internal milieu first postulated by Claude Bernard and later worked out by Walter Cannon. However, this solution to the teleology issue poses as many problems as it solves, for it does not say how the livable states box emerged in the course of evolution or development, or how the control of the relevant parameters is implemented. Still, these problems can be investigated scientifically instead of being kept in a dark corner of obscurantist metaphysics. Thus scientism transformed a mystery into a bundle of feasible scientific research projects, and proposed solutions that generated further problems.

My own encounter with the cybernetic dissolution of the teleology mystery was a transfomative experience. I owe this experience largely to my friend Alberto González Domínguez, the mathematician to whom a handful of mathematics and physics students, myself included, resorted in search of explanation and intellectual stimulation during the dark 1940s and 1950s. He was the first to talk to us about communication theory, cybernetics, rational choice theory, the theory of distributions, and much more.

Alberto was also the one who arranged for Garrett Birkhoff, Laurent Schwartz, Antoni Zygmund, Jean Dieudonné, and other mathematical stars to come and teach short courses at Buenos Aires University. Birkhoff's visit was particularly significant, because he came during wartime, and we gave him a sumptuous baby-beef dinner at a large restaurant, where González Domínguez surprised us all giving an eloquent speech where he praised the American-Soviet war alliance. One decade later Jean Dieudonné, the spokesman for the Bourbaki sect, shocked his audience by shouting his slogan *À bas le triangle*! (Down with the triangle!)

Alberto, who was apolitical, had learned Russian only to be able to read some volumes of the prestigious if expensive Springer yellow series, which were available in Russian translation at low prices. He disappointed some of us when he was made dean of the Faculty of Science under Peronism, and gave jobs to colleagues who had joined the ruling party. (For example, he appointed as professor of theoretical physics a former student of mine who had no publications in this field.) But Alberto was otherwise such a generous fellow, and such an inspiring teacher, that when the regime changed in 1955, his lapse was forgiven and forgotten. Back now to biology.

My Main Writings on Biology

One of my earliest writings about biology was "Is biology methodologically unique?" (Bunge 1973b). This was a criticism of the thesis, upheld by Ernst Mayr and others, that biology is radically different from all the other sciences, in that it studies unique and purposeful things. I read that paper at the Zoology Department in the Finnish university of Turku at the invitation of Kari Lagerspetz, who in 1959 had published an excellent essay on teleology (Lagerspetz 1959).

I had gone to Finland to attend the 1970 colloquium of the Institut International de Philosophie. On arrival, at the end of August, we were greeted by a blizzard. Can you blame the Finns for not being usually in a sunny mood? My host asked me to attend the start of the academic year ceremony, assuring me that the rector's speech would be brief, and sat me on the first row. The speaker spoke with great verve, as if haranging the troops, in a language utterly incomprehensible to me. After an hour that felt like a day, the audience was invited to sing the national anthem, which starts with the stirring strophes:

Oj maamme, Suomi, synnyimaa, soi, sana kultainen!
Ei laaksoa ei kukkulaa, ei vettä, rantaa rakaampaa.

That event was compensated for by a dinner offered by Kirsti, Kari's wife and a psychology professor at the local Swedish university. The charming Kirsti was a world authority on aggression in mice and children, as well as an anti-war activist. She was an effective and vehement critic of the ruling view, that our remote ancestors lived in perpetual war, as Hobbes and the sociobiologists believed – an example of bad science generated by an ideological preconception.

Another interesting Swedish Finn I met at that time was Georg Henrik von Wright, a philosopher with multiple interests whom I had befriended a decade earlier at Stanford, and who maintained a close relationship with my friends the Argentine legal philosophers Eugenio Bulygin and Carlos Alchourrón, aka Alchie. Von Wright was a member of the Finnish Academy, and lived in a beautiful and ample house with a concert piano that the enterprising Evandro Agazzi banged on uninvited but to the benign envy of many of us.

From Helsinki I flew to Göteborg, Sweden, to give a talk at the institute headed by Håkan Törnebohm. I had obtained his excellent book on the two relativities in Buenos Aires through my friend Juan Eresky, who worked at the local branch of the Swedish firm Erikson. Håkan and his sweet wife Siv gave me a typical Swedish dinner, with a ceremonious welcome speech and plenty of candles and Skoals!

My next destination was Uppsala, invited by the economist Herman Wold, who had written about mathematical models in the social sciences. He asked me to get him a picture of my lay godfather Raúl Prebisch, to know whether the Nobel committee could keep postponing Raúl's candidacy for the coveted prize, so seldom awarded to worthy economists.

From Uppsala I went to Stockholm to join Marta, who came from the international congress of mathematics held in Nice, where Gerard Debreu gave a famous paper on his general economic equilibrium theorem, that earned him the Nobel prize despite his glossing over the desequilibria that had wrecked so many lives so many times. There we met with the Argentine quantum chemist Osvaldo Goscinsky and his wife Gunilla, who made us wear Swedish clogs and gave us dinner. We took them to see a play at an old theatre famous for its *deus ex machina*, a trapdoor on the ceiling from which something was dropped unexpectedly to the delight of the audience.

Finally we flew to Oslo, in whose naval museum we admired some of the longboats that a thousand years earlier had braved the North Atlantic carrying horses and cows along with intrepid if murderous and thievish Vikings. After giving a talk at the university, we visited the philosopher Arne Naess, who was playing with a white mouse while we waited for the dinner that I mistakingly thought Arne had promised us earlier.

One decade later, when we met in Jerusalem to pay homage to Spinoza, Arne asked me why I kept ignoring God. When he noted my perplexity, he clarified: God or nature, as Spinoza said. Arne was scolding me for not taking an active part in the environmentalist movement. He himself was active in the "Deep Green" movement, which fights civilization to save the planet – a view I reject. We could save the planet in various ways, for example, by distorting the trajectory of asteroids traveling towards us, and by preventing economists from dictating economic policies.

In 1965 I was co-opted into the Académie Internationale de Philosophie des Sciences, which at that time was presided over by the mathematicians Ferdinand Gonseth and Paul Bernays, and administered by the Flemish priest Stanislas Dockx. I organized two colloquia on behalf of the Académie: one on the foundations of physics held in the Black Forest in 1966, and the other on science and metaphysics held 10 years later in Fribourg, Switzerland.

In 1969 I was made a member of the Institut International de Philosophie, limited to 100 members. And in 1992 I was inducted into the Royal Society of Canada, the Canadian academy. The formal ceremony was held in a beautiful neo-Gothic hall of the Canadian parliament building in Ottawa. It was memorable because the president's *laudatio* was exceptionally long and praiseful, and because my dear friends, the Spanish ambassador José Luis Pardos and the Argentine ambassador Lillian O'Connell Alurralde, had rounded up all the Latin American ambassadors, so I had an abundant and enthusiastic claque.

One decade later I joined the Argentine National Academy of Sciences founded by the great president Domingo Faustino Sarmiento. In 2013 I was included in the Argentine Academy of Exact, Physical, and Natural Sciences, where I gave a talk on the puzzling Aharonov-Bohm effect, that I had sketched a few weeks earlier at the Aegean island of Euboea.

Foundations of Biophilosophy

My next work in the philosophy of biology was *Foundations of Biophilosophy*, which I wrote together with Martin Mahner (Mahner and Bunge 1997). A few years earlier Martin, then a biology student and taxi driver, had written to me from West Berlin asking some interesting questions. In the summer of 1991 we met in Bolzano, where I was attending a colloquium of the Académie, at which a Polish mathematician gave his mathematical proof of God's existence. Nobody knows whether He was flattered or felt insulted. Martin explained to me cladistics, or phylogenetic systematics, then a hotly debated subject. He persuaded me that the concept of a clade or dynasty is indispensable in systematics because it deals with populations, which are real things – but he did not convince me that it should replace that of species, which is used to describe organisms and populations between branchings.

My own view of the matter is that the two approaches are mutually complementary rather than exclusive. For example, to a non-cladist the European sparrows introduced in Argentina in the 1880s count as the founders of a new population of the same species, whereas to a cladist they are the founders of a whole new species – which seems extravagant to me since the immigrants differed from their old-country cousins only in their domicile, which is not a biological trait. In any event, the cladism controversy is not yet over. Martin had gotten involved in this controversy while working on his doctoral dissertation on aquatic "bugs," such as Daphnia. The publication of his voluminous dissertation in 1983 made him the world authority on water bugs. We agreed that he would seek a German fellowship to work with me on biophilosophy.

Martin arrived in Montreal in the fall of 1993, while I was immersed in social philosophy, and he stayed 3 years. I wrote an outline for a comprehensive and deep discussion of what I took to be the most important and topical issues in the field. Martin lay out the work schedule and managed the division of labor between the two of us. He managed the project in accordance with our respective abilities and preferences. Our collaboration proceeded smoothly because Martin had adopted my philosophy, and I fulfilled the tasks that did not require a detailed knowledge of biology. Since Martin was basically an evolutionary systematist, taxonomy has the pride of place in our book. Still, I believe we tackled all the main issues, and punctured a number of balloons along the way.

Puncturing Biophilosophical Balloons

In our book Martin and I attacked a number of popular myths, among them Richard Dawkins's genetic determinism, according to which development and evolution are mutually independent – which made us precursors of the evo-devo synthesis; David Hull's contention that species are individuals – a mistake derived from the confusion of species with populations (or with clades); Daniel Dennett's fantasy of evolutionary algorithms, understandable in a computer worshipper; Ernst Mayr's denial of the existence of biological laws – as if molecular biology, genetics, and physiology could do without them; Mayr's insistence that biology is an autonomous science – as if organisms were not composed of molecules; the sociobiological myth that inheritance and sex explain everything social; and, of course, the creationist myths, in particular the intelligent design one.

Our book did not take the biophilosophical establishment by storm. Some potential readers were still enthralled to some of the balloons we punctured, and many more were put off by our logical and mathematical formulas. The author of the review in *Nature* did not care for exactness: he dismissed it as "fastidiousness." Another reviewer complained that we did not confine ourselves to his pet subject – evolution. But no one accused us of committing any howlers. Furthermore, the book was also published in German, Japanese, and Spanish. The excellent Spanish translation was the work of Mariano Moldes, a promising young biologist who died of a neglected infection.

With hindsight I realize that Martin and I did not pay enough attention to the social environment, which sometimes favors the organism's development and at other times stunts it. This defect was entirely my fault, because I was familiar with social stress thanks to my friend René Zayan, director of the laboratory of experimental ethology at the Louvain Catholic University. René, an early fan of my treatise, kept me up to date on his work on the effects of social stress on domestic animals. He had visited me in Montreal, and I visited him at Louvain, where I also lectured several times – in French at the Walloon branch and in English at the Flemish one.

Zayan had found, in particular, that the quality and quantity of the eggs laid by free-ranging chicken was significantly higher than those produced by chicken crowded in batteries. Countless other stress students since Hans Selye's pioneering work at McGill and elsewhere since the 1940s had confirmed that stress can sicken and even kill. I corrected the said imbalance in my *Medical Philosophy* (2013a).

Incidentally, the splitting of the university's assets between the two linguistic camps had been strictly impartial. For example, the odd volumes of the complete works of an author had been assigned to one of the parties, whereas the even volumes had been assigned to the other. I do not know the fate of the bilingual professors.

75th Birthday Party

Martin and my former students Roger Angel, David Blitz, Moish Bronet, Martin Cloutier, Mike Dillinger, Michael Kary, Jean-Pierre Marquis, Don McGowan, and Dan Seni, as well as our friends the ambassadors Lillian O'Connell Alurralde and José Luis Pardos, and others, attended the 2-day symposium in celebration of my 75th birthday party in 1974. This event, held at the McGill Faculty Club, was organized by my colleagues Claudio Cuello (Pharmacology), and Bernardo Dubrovsky (Psychiatry). The main speakers were my former student David Blitz, my colleague William Shea, my old friend and *Minerva* collaborator Hernán Rodríguez, and the eminent neuroscientists Albert Aguayo and Vernon Mountcastle. The latter said that reading my book *The Mind-Body Problem* (1980d) had restored his enthusiasm for research.

Other notable attendees were Joseph and Judith Agassi, Herbert Jasper, and the Right Honourable Pierre Elliott Trudeau. Trudeau was the Canadian prime minister who had earned a reputation for saying what he thought, and for keeping his promises, among them respecting the privacy of Canadians and maintaining the country's independent foreign policy. In my short thank-you speech I said that, if I appeared to be a couple of school terms younger than my actual age, it was because I abstained from poisons like tobacco, alcohol, and existentialism. Twenty years later, as I write, I can add another longevity factor: avoiding departmental meetings.

At about that time I wrote with Martin a rather long paper on religion and science (Mahner and Bunge 1996), that provoked much criticism, to which we responded (Mahner and Bunge 1996). We confined our criticisms to the negative effects of religious belief on scientific curiosity and the habit of looking for evidence, but did not elaborate on the use of religion as a tool of social control.

In 1996 Martin and I attended the colloquium on pseudoscience and antiscience organized by the New York Academy of Sciences. In my presentation I proposed that we should be intolerant of pseudoscience, for on top of lying it is a swindle (Bunge 1996f). The philosopher of science James Robert Brown wrote later that my thesis was "a recipe for disaster," but did not explain why. I keep thinking that universities should seek and teach only truth instead of giving "equal time" to truth seekers and swindlers.

Since 1999 Martin has been the executive director of the Center for Science and Critical Thinking in Rosfdorf, southern Germany. His main jobs are to answer queries on the new pseudosciences that keep arising, and to organize meetings on the same. Martin has also compiled 30 papers of mine on scientific realism, and commented on them (Mahner 2001). The same year we published a joint paper on the widely misunderstood concept of function (often confused with those of role and goal), as well as on functionalism, a popular alternative to the search for the mechanisms underlying functions (Mahner and Bunge 2001).

Martin and I met again in 2002 at the formerly sleepy but now quite vibrant Spanish city of Vigo, where Aletheia, the local philosophy group around the enterprising philosopher Avelino Muleiro, organized a colloquium on my philosophy. The meeting was also attended by the economist Alfons Barceló, the psychobiologist Ignacio Morgado, the philosophers Jesús Mosterín and Miguel Angel Quintanilla, and the physicist Héctor Vucetich. In my good-bye speech I said that the remarkable statue at the entrance hall of the Casa das Palabras, where we were meeting, and which shows a walking man looking backwards, evokes the philosophy professors who only comment on old books.

The next year the University of Salamanca, the fourth oldest in Europe, awarded me an honorary degree. I was garbed like a bishop, and addressed the *Rector Magnificus* in Latin. My old friend Quintanilla read the *laudatio*, also in Latin, and we enjoyed the company of the minute logician Mara Manzano and the tall diplomat José Luis Pardos. We held in our hands a copy of the bowdlerized Visigothic bible, and were shown the bare classroom that in cold days the impecunious students used to warm up with their bodily heat an hour before their rich classmates arrived – for a fee, of course. And we were reminded of Miguel de Unamuno's public address to General Franco's fanatics, whom he had initially supported: *Venceréis pero no convenceréis.* [You'll win, but you won't persuade.]

Biologists Visiting My Unit

The Argentine Luis Marone visited my Unit in 1984 and 1998 to discuss some methodological issues in ecology. In particular, we exchanged ideas about explaining ecological processes in terms of such mechanisms as competition and dispersion (Bunge and Marone 1998). We also discussed Robert H. Peters's popular but failed attempt to construct "operational" definitions of the key concepts of the discipline. When I visited Luis in his native city of Mendoza, at the feet of the imposing Andean cordillera, he introduced me to his father, who had attended the course I had taught in 1962 for the Asociación Física Argentina in faraway Salta.

Marone's student Rafael González del Solar visited my unit in 1999, attended one of my courses, and fulfilled all its requirements while resisting the charms of a pretty classmate, for he was in love with his fiancée Carolina. His unfinished doctoral dissertation dealt with the lifestyle of the stealthy gray fox in semi-desert regions. Since Rafael hardly ever saw those elusive creatures, a temptation for pumas, he had to confine himself to collecting and analyzing their droppings, faithful indicators of their diet. In studying their conspecifics across the Andes, he found that their diets were significantly different – contrary to the formula "You are what you eat," popularized by the nineteenth century trophologists, from Ludwig Feuerbach to the naturists around 1900.

At the beginning of 2000, Rafael and I attended the party that my colleague Raymond Klibansky gave at his home to commemorate the work of Giordano Bruno, burned alive four centuries earlier for his heresies. The boldest of these was the hypothesis, now amply confirmed, of the existence of multiple solar systems – anathema to Christians because it would require multiple creations and multiple crucifixions, long before large-scale production was invented. Since then Rafael has been a close friend, shifted to philosophy, translated several books of mine, and moved to Barcelona.

In recent years I befriended Javier López de Casenave, another of Marone's students and an ecology professor at Buenos Aires University. For the past 6 years, the two of us have been running there for 1 month a year a seminar in the philosophy of science. So far all our lecturers have been scientists who have tackled philosophical problems of no interest to the local philosophers. Four of the stalwart members of the seminar are Héctor Vucetich and his former student Gustavo Romero, the country's top cosmologist and a fan of my philosophy; the mathematician and former faculty dean Pablo Jacovkis, son of David, my close friend while I was finishing high school; and Marcelo Bosch, an agronomist who earned his philosophy doctorate at the Catholic University writing a thesis on my philosophy. Yes, Virginia, miracles do happen once in a while, but they are poorly advertised.

Canadian Students Interested in Biophilosophy

I have had a few Canadian students interested in biophilosophy, among them Suzanne Leroux, David Blitz, and Michael Kary. Suzanne had studied agriculture, which, according to my colleagues, inhabilitated her to study philosophy. They allowed her to write a good master's thesis on the biospecies controversy, but rejected her application to do a doctoral thesis under my supervision. Who had ever heard of an agriculture expert climbing to the heights of pure philosophy?

After obtaining a master's degree in philosophy, David Bitz had worked as a machinist and union organizer at a submarine factory. Later on he worked as a counselor of juvenile delinquents, nearly all of them aboriginals. In 1984, after reading my *Scientific Materialism* at the recommendation of Mr Gerald Glass, a bookseller who had graduated from the LSE, David started to study under my wing. He got enthusiastic about evolutionary biology, and wrote his doctoral dissertation on the concepts of emergence and level (Blitz 1992). Since then he has been teaching philosophy at the Central Connecticut State University. At the drop of a hat he can do hilarious impressions of any popular pseudophilosopher.

Michael Kary audited some of my courses after getting a BSc in biology. When I asked him what he had learned from the biologists, he replied: 'meticulousness' – that is, attention to detail but neglect of the big picture. In class Michael always made interesting remarks and asked hard questions. His brother Joseph, an environmental lawyer, took one of my courses, for which he wrote a superb paper on the wrong uses of probability in the law. Hoping that he would become a mathematical biologist, I advised Michael to study mathematics. This he did, writing his doctoral dissertation on the movements of severely mutilated human bodies. He also wrote some insightful papers, one of them jointly with Martin Mahner on the concepts of information. Michael did not produce much else, partly because he wasted time on two subjects typical of the excentric amateur: the "fictitious" forces, such as the centrifugal one, and the real authorship of the works attributed to Shakespeare.

Does Biology Explain Human Nature?

I call *biologism* the programmatic hypothesis that humans constitute just one more animal species, so that biology should be able to account for everything human, from language to social stratification, and from making music to making war. Biologism comes naturally to biologists and philosophical naturalists. The earliest phase of biologism is racism, which from antiquity has been used to justify slavery and imperialism (Fredrickson 2002). Other versions of it are the beliefs that intelligence, nobility, criminality, universal grammar (should it exist), and much else are not only inborn but also hereditary, i.e., that they "run in families." Since the birth of genetics, the name for this phase of biologism has been "genetic determinism," summarized in the formula "Genome is destiny." This slogan, popularized by Richard Dawkins, was made academically respectable by the Harvard professors Edward Wilson and Steven Pinker, despite the serious attack by their colleagues Stephen Jay Gould and Richard Lewontin (2000).

In 1975, in the Mexican country house we had rented for the summer, I devoured Edward Wilson's massive and beautifully illustrated *Sociobiology: The New Synthesis* (Wilson 1975). I became an instant convert. Like many others, I thought that the world's top ant expert had discovered the path to the salvation of the social sciences. I knew little about these at the time, but I had a legitimate complaint: those sciences lacked in precision and, in my opinion, failed to answer the main question: why is there social cohesion despite great diversities? In 1977 I published a very laudatory review of that book in the Mexican journal *Naturaleza*, which I now disown.

Stephen Jay Gould and Richard Lewontin attacked *Sociobiology*, and accused it of being the latest gasp of social Darwinism, the earliest attempt to justify the status quo, in particular male dominance, in genetic terms. Initially I saw their criticism of sociobiology as an invidious reaction of the Marxist fringe, which had produced no significant novelties in social science after Marx.

However, my enthusiasm for sociobiology was short-lived. Even a fleeting acquaintance with the pronounced social inequalities in Mexican society, and the widespread disregard for them among my colleagues, was enough to persuade anyone that, at least in this case, biological inadequacies were the effect, not the cause, of social marginality. In addition, the unceasing social unrest throughout the Third World should be enough to disprove the thesis that social stratification is in the human genome: what works for ants may not work for humans.

But the decisive factor for my disillusion with sociobiology was reading Richard Dawkins's *Selfish Gene* (Dawkins 1989) published the same year as Wilson's flawed but well argued genetic determinism. Indeed, I instantly diagnosed Dawkins' genetic determinism as pseudoscientific. In fact, it was not based on new research, for Dawkins was but a popularizer; it was full of howlers, such as the statements that genes duplicate by themselves (rather than under the action of enzymes), and always override the environment; that the only evolution worth talking about is the biological one, which results from mutation and natural selection; and that, since the genome is the first mover of life, and since selection would act on genomes, not entire organisms, the very existence of organisms is "paradoxical" – that is, biology is redundant in Dawkins's scheme.

As if his nonsensical genetic determinism were not enough to tar and feather Dawkins as the pseudoscientist of the day, he also invented the "meme," or unit of cultural evolution, and tried – unsuccessfully this time – to sell his memetics, or "science of memes." Dawkins' *dicta* were so simple and outrageous, and so in tune with the prevailing nativism recently reinforced by Chomsky, that he became instantly famous, even among skeptics, who care more for Dawkins's irreligiosity than for his lack of scientific credentials and his utter insensitivity to everything social. When I made fun of Dawkins's fantasies (Bunge 1999c), his fan Steven Pinker wrote me a furious letter.

Human Nature Is Quite Unnatural

It is well known that the behavior of ants, bees, and other eusocial insects, is genetically programmed: they do not enjoy the luxury of choice, hence do not have to make decisions; and their brains lack the plasticity required to invent anything, in particular devices to beat heredity. When they take a trail to go foraging, ants are driven by irresistible pheromones; and when they pile up upon one another to survive a flood, they are guided by instinct, not calculation.

Human behavior, by contrast, results from a combination of heredity, experience, and invention, which now bows under social pressure, now reacts against it. Thus, whereas the anonymous ant's future is fully determined by both its phylogenetic past and its present circumstances, a human being can change the latter, up to a point, to suits her/his needs and aspirations. This is why it has rightly been said that *man is the self-made animal*.

In shaping ourselves we transform nature in ever-new and sometimes disastrous ways. True, the lowly earthworms too alter the natural landscape, but they do it always in the same way, namely swallowing soil: they cannot start new projects, like building nests or settling in caves. Furthermore, as Marx noted, honeybees build efficient honeycombs but, unlike architects, they do not do that according to blueprints.

Imagine now an ordinary day in the life of the proverbial Joe the handyman. He cannot charge much because, not having graduated from a vocational school, Joe lacks professional accreditation. And he did not go to school because he came from a poor rural family, so he had to work in the farm since childhood. So, from birth Joe was not only at the mercy of his genetic makeup, but also at that of his society. Still, he had the will and the ability to learn some skills, such as welding, doing electrical wiring, and using power tools, all of which would have been regarded as magical in earlier times.

Much the same applies to Joe's family: his wife buys food in the supermarket instead of tending her own farm, and their children go to school and watch TV instead of hunting wild animals. In short, the plastic brains of Joe and his family members have allowed them to behave as socially plastic animals dependent on artifacts. This dependence is such that, if they were left to fend for themselves in a pristine natural environment and without the shipwreck that first endangered and then saved Robinson Crusoe's life, Joe and his folks would not survive more than a few days.

New Science of Human Nature?

All of the preceding seems obvious, yet over the past half century the "naked ape" invented by Desmond Morris (1967), the blood-thirsty primitive imagined by Napoléon Chagnon (1968), the "imperial animal" of Robin Fox and Lionel Tiger (Tiger and Fox 1971), the "lumbering robot" invented by Richard Dawkins (1976), and the "walking fossil" adapted to the Pleistocene and still stuck in it, imagined by Lea Cosmides and John Tooby (1999), have captured the popular imagination and have been consecrated by famous professors of anthropology, psychology, social science, and philosophy, to the point that they boast about "the new science of human nature."

It is hard to understand how apparently intelligent and well-educated individuals could embrace a radically naturalistic conception of human nature while at the same time admitting that one of the defining characteristics of our species is the ability to make and use artifacts, and that one of the peculiarities of contemporary humans is their inability to survive with*out* access to a large number of artifacts, from clothes, containers and axes to medicines and computers, as well as socially condoned rules of human behavior and means to enforce them.

Reading Pinker's (2003, p. 294) list of alleged accomplishments of the "new science of human nature" raises the suspicion that it is nothing but an attempt to make the New Right scientifically respectable (Bunge 2010a, p. 256). However, it may also be read as proof that higher education can coexist with plain ignorance or, if preferred, with political naiveté.

Richard Dawkins's (2013, p. 97) conjecture about the cruelty of the SS guards in Nazi concentration camps is a case in point. He reasons roughly thus: Bullying is common in British "public" (private) schools; since school bullies are children, cruelty comes with childhood; hence those camp guards may have exemplified "a sort of retention into adulthood of a psychology that is normal in children." Disregard, if you can, Dawkins's hasty generalization and logical fallacies, and focus on his blindness to the fact that violence is a tool of dominance – a *social* relation rather than a purely biological fact – to the point that it does not occur in primitive tribes or in democratic communities. Violence can be prevented or controlled through both moral and political means, from disapproval to incarceration. And, as shown by history, such practices are artifactual, hence subject to change when shown to be either ineffective or unfair.

In short, although there is such thing as human nature – just as there is feline nature – what characterizes ours is artificiality, not naturality. Indeed, inasmuch as humans are social and self-made, they are partly *un*natural. In particular, all of us are immersed in several social groups or systems, every one of which has been organized or reorganized by humans following rules, some of which are maladaptive, rather than laws of nature. Hence biologism, whether perverse like Nietzsche's *Fiat vita, pereat veritas*, and the Nazi *Blood and soil* cult, or simplistic like Feuerbach's "Man is what he eats" and Dawkins's selfish gene, is contrary to fact. What the great composer Maurice Ravel said of himself – "I am artificial by nature" – applies to all of his conspecifics. This is why the thinkers who have glossed over our partially artifactual nature have made no original contributions at all to either psychology or social science. This is also why the naturalistic philosophers have failed to raise the scientific status of the social studies. The blind spot of naturalism for everything artifactual is also a main reason for the discredit of materialism in the humanities and social sciences communities.

Continuing Interest in Biophilosophy

I have kept up my interest in biology by looking every week into *Nature* and *Science*. I have been particularly intrigued by two comparatively recent developments: the emergence of both evo-devo – the synthesis of evolutionary and developmental biology – and epigenetics, the study of heritable chemical changes in the genome triggered by the animal's experience or by environmental inputs.

I have also learned something about drugs's mechanisms of action, and the social determinants of health while researching my book on medical philosophy. And I have kept campaigning against pseudobiology and pseudomedicine. When asked to admire the purposefulness of organisms, I ask about the purpose of toenails. And when told that chiropractors cure all disorders by manipulating the spine, I ask for the mechanism involved in the process. Merely asking *How does it work*? may work miraculous conversions.

Coda

Nobody doubts that biology has come a long way since its beginnings in ancient Greece more than two millennia ago. On the other hand, biophilosophy is still in nappies, because most of its practitioners have confined their attention to a few problems, which they have tackled in an ontological void and without the requisite formal tools. It is time for a new Claude Bernard to clean up the rubble and start building an ample and solid yet expandable biophilosophy to serve as a guide to the advancement of biology, which still suffers from theory deficiency and the concomitant conceptual confusions.

Chapter 12
Psychoneural Monism

Early Infatuation with Psychology

Like countless other adolescents, I went through a phase of passion for psychology. My own was possibly aroused by interest in sexual love and fed by a few of St Sigmund's writings – which of course did not help – as well as by a number of friends intent on educating me. As told in Chap. 2, I soon learned that psychoanalysis is a branch of fantastic literature, that it does not understand love or even sex, and that there was nowhere in my country where one could study serious psychology, which at that time was only one more subject badly taught at best by psychiatrists in philosophy faculties.

So, I gave up my ambition to become a psychologist and henceforth divided my attention between physics and philosophy. I never regretted this decision, for since that time the teaching of psychology in Argentina has considerably worsened: it is now mostly perpetrated in temples dedicated to Jacques Lacan, which have the gall to call themselves psychology faculties even though no psychological research is conducted in them. It is only during the last decade that scientific psychology has begun to develop in Argentina, though of course outside those temples. I am referring to the small research teams led by Facundo Manes and Mariano Sigman, both of whom are publishing in high-impact refereed journals.

Even during the 1943–1957 period, when I regarded myself as being primarily a physicist, my interest in the mind was never extinguished. It seems to arise naturally in all of us during adolescence, even among the self-styled eliminative materialists, who deny the very existence of the mental. In the late 1940s, my friends and I discussed Pavlov's *Selected Works* in our private philosophical circle. But we were soon discouraged by what we saw as a mechanistic view of the mind, as well as by Pavlov's vehement all-round denunciation of psychologists.

Ironically, Pavlov's view of the brain as a signal processor was independently revived in the 1960s by informationist psychology, which is still going strong among philosophers of mind. Neither school has taken notice of the momentous

M. Bunge, *Between Two Worlds*, Springer Biographies,
DOI 10.1007/978-3-319-29251-9_12

novelties in neuroscience that led to the birth of cognitive neuroscience in the 1950s, in the brains of Donald Hebb, Wilder Penfield, and their coworkers. Nor have the philosophers of mind paid much attention to Jean Piaget's *épistémologie génétique*, his misleading name for developmental cognitive psychology. The controversy moderated by Massimo Piattelli-Palmarini (1980) that pitted Noam Chomsky's nativist fantasies against Piaget's solid experimental findings was particularly pathetic. It reminded one of the schoolmen who rejected Galileo's findings, as well as of the a priori pronouncements of the later *Naturphilosophen* just at the time when German science was taking off with unprecedented vigor.

My next encounter with psychology took place in 1960 at the University of Pennsylvania, then the world center of mathematical psychology. I became enthusiastic about it and befriended Eugene Galanter and R. Duncan Luce, both of whom I met several times at the faculty club and years later at Amsterdam and Montreal. But my visit with Georg von Békésy in Hawaii, in 1971, dispelled overnight my illusions about the mathematization of behaviorist psychology, which eventually seemed to me to be both brainless and mindless.

Shortly after coming to McGill, I offered a course in the philosophy of mind based on Herbert Feigl's essay *The "mental" and the "physical"* (Feigl 1967). I liked Herbert's undogmatic stand, in particular his denunciation of what he called the *hypocritical realism* of his fellow logical positivists. But his book was anything but straightforward, so I asked him point blank: "Herbert, are you a robust materialist?" His reply amazed me: "I am a neutral monist." I thought such a stance as precarious as walking on a tight rope – and just as useless in treating mental diseases, for it countenances neither word nor pill.

However, not all was lost, for an unknown student, whom I met by chance in the street, urged me to read Donald Hebb's *The Organization of Behavior* (Hebb 1949) which I did right away. This turned out to be a transformative experience despite Hebb's hostile reception of my foundations of science project, as recounted in Chap. 8. I had finally met what I had missed in my previous readings on the subject: an imaginative but plausible neural explanation of learning, creativity, and even free will. Unbeknownst to himself, Don Hebb had fleshed out my philosophy of mind all by himself.

Philosophy of Mind

The philosophy of mind is the chapter of philosophy that studies the mind as well as the disciplines that deal with mental events. It tackles, in particular, the mind-body problem and the ways to handle it. In my scientistic view, expressed in my youthful paper on the Romantic philosophy of nature (Bunge 1944d), such study, to be fruitful, had to be done through science rather than directly. That is, a legitimate philosophy of mind had to be a philosophy of scientific psychology, and philosophical psychology was to be forgotten along with Hegel's, Schelling's, and Engels's philosophies of nature, as well as Maine de Biran's, Noam Chomsky's, Hilary Putnam's, Jerry Fodor's, Daniel Dennett's, and Steven Pinker's speculations.

The central problem of the philosophy of mind is to find out what the mental is and how it relates to the bodily. This, the *mind-body problem*, is a multidisciplinary issue, as it interests neuroscience as much as psychology, medicine, and social science, as well as theology no less than philosophy – not to mention the popular and profitable psychobabbles and the newer neuro-fantasies such as neuroeconomics and neuropoetry. Mephistopheles must be overwhelmed by so much competition over the soul.

The mind-body problem was so important at the beginning of the modern era that when the University of Padua, the most advanced of its time, received a new philosophy professor, his students asked him where he stood in the matter of the immortality of the soul. The Aristotelian Cesare Cremonini, Galileo's colleague, was very popular among students for teaching that the soul dies along with the body (see Renan 1949).

Nowadays, all physicians know that the state of health of their patients depends partly on their manner of feeling, thinking, and acting. In particular, they know that optimists respond better to medical treatments than pessimists; that the disciplined are better patients than the undisciplined; and that Dr Placebo is always in attendance (Bunge 2013a). This is why physicians give advice on lifestyle along with drug prescriptions. But they know that words alone won't cure such diseases as hypertension, cancer, diabetes, schizophrenia, or depression; nor will they prevent heart attacks, cerebrovascular accidents, or degenerative diseases like Parkinson's or Alzheimer's.

Main Answers to the Mind-Body Problem

The two main solutions proposed to answer the mind-body question over the past few millennia are monism and dualism. According to monism, all existents are mental (idealism), material (materialism), or neutral. Dualism, by contrast, holds that there are two mutually irreducible substances or modes of being: the mental and the material. However, each of these genera is subdivided into several species. I proposed a more refined taxonomy (Bunge 1980a) that several authors liked so much that they copied it without giving me credit.

Idealist monism is not just an antique: it has recently been revived by the physicists who hold that there are only observations, not *of* some things, but in themselves. As a Wittgensteinian might say, this opinion is wrong because it "violates the grammar" of the verb *to observe*. A logician would say that this is a dyadic or two-place predicate: animal A observes object B. If $A = B$, there is introspection, which is useless for understanding or altering the world and anyway is never practiced by the millions who live tethered to their electronic gizmos waiting for the next trivial text. Idealist monism also runs counter to the well-known fact that sentient beings are comparatively recent arrivals – just about a couple of billion years ago. Furthermore, the idealists ignore that the sciences of the mental study facts, such as thinking, choosing, deciding, rejoicing, and deploring, that only occur in brains, whereas physics and chemistry only study nonmental processes.

Materialist Monism

Materialist monism is not single but multiple. Indeed, the family of materialist philosophies of mind includes the following species: (a) *eliminativism* or physicalism, which denies the idiosyncrasies of the mental, such as spontaneity, creativity, and free will; (b) *biologism*, according to which all mental processes are locked in the skull and thus impermeable to social stimuli; (c) *machinism*, which views the mental as information processing following algorithms; and (d) *sociologism*, which holds that consciousness is but society personified.

Physicalism is at best a promise (or a threat), for, as a matter of fact, physics knows wavelengths but not colors, pH but not bitterness, air waves but not sounds, and so on. Biologism ignores the social environment and therefore the social emotions and the ideological pressures on imagination. Machinism overlooks development and evolution, confuses law of nature with algorithm, ignores spontaneity and creativity, and is useless to treat mental diseases. And sociologism, from Marx to Vygotsky, overlooks the individual brain, and therefore, it ignores neuroscience. In short, all four versions of reductionism are oversimplifications and therefore hinder the unity and advancement of the sciences of the mental, which will concern us below. However, let us first peek at psychoneural dualism.

Psychoneural Dualism

Mind-body dualism is the most popular of the philosophies of mind. It is tacitly held by the Mexican peasant who calls the veterinarian when his cow sickens, but consults the shaman when his wife feels ill. And dualism has been defended by sophisticated philosophers like Plato, Descartes, and Popper – though not by Aristotle, Spinoza, Holbach, or Helvétius. Dualism is also tacitly adopted by the faithful of the computational doctrine of the mind when they exaggerate the hardware/software distinction, even though they buy programs sold as diskettes that are just as material as laptops. But consistency is hard to attain: some computationalists, such as Paul and Patricia Churchland, call themselves materialists just because they do not believe in immortal souls and worship computers instead, without realizing that computer programs "embody" ideas generated by live and well-trained brains.

However, the worst problem with the computational view of the mind is that it is just programmatic: it does not tell us how we compute emotions such as love and hatred, nor how is it possible to program inventions without at the same time inventing what was unknown before. It does not even explain how we compute computations proper, such as doing sums. To put it in clearer nonacademic terms: the computational program is just high-sounding hand-waving.

To summarize, there are three main varieties of psychoneural dualism: (a) *parallelism*, every mental event has a neural counterpart; (b) *animism*, the spirit, soul, or mind guides the brain as the pilot the boat (Plato, Eccles) or as software does hardware (Putnam); and (c) *interactionism*, mind and body interact, as shown by the occurrence of both psychosomatic diseases (mind → body) and their dual, which happen when the feminine reproductive system (never the masculine one) sickens the soul (body → soul). Interactionism, usually attributed to Descartes, is possibly the earliest and most popular philosophy of mind. But of course it does not suggest any mechanisms whereby the allegedly immaterial mind affects the material brain or conversely and is useless to treat mental diseases.

How to Study the Mental

There are three main currents in the study of the mental: (a) *mysterism*, or radical skepticism, the mental is and will always be mysterious; (b) *dogmatism*: the problem has been solved once and for all by theology according to the religious and by psychoanalysis or by information technology according to lay believers; and (c) *scientism*, according to which science has solved the philosophical issue of the nature of the mental, but will keep studying the myriad of special scientific problems, how we see and how we look, how we hear and how we listen, how we feel happy and how we can overcome unhappiness, how we become frightened or calm down, how we reason or imagine, how we can learn to control emotions and antisocial impulses, and so on.

Even admitting that the mental can be studied scientifically, we must choose one of the following strategies: the study of mind and behavior belongs in neuroscience, psychology, social science, or a fusion of all three, and perhaps some additional sciences, such as genetics, endocrinology, and immunology. As a matter of fact, scientists in increasing numbers are inclined to believe that the study of the mind is the task of psycho-neuro-endocrino-immuno-sociology. I share this view, for it is being increasingly corroborated and it matches both scientism and systemic materialism.

My first essay on psychoneural monism was rejected in 1976 by the Mexican journal *Crítica*, so as to leave no doubt that the linguistic philosophy that its new editor had just imported from Oxford, and which I called *oxtec*, was hostile to science. The English version of this paper was published shortly thereafter by *Neuroscience* (Bunge 1977a). In this paper, I suggested that mental properties and events are emergent, and that they can be represented in a state space of the human brain. This is the Cartesian (though not necessarily metric) space scanned by a vector of neural states, whose components are such quantities as neurotransmitter concentration and frequency of neural firings. I also noted that the dualist fantasies prevent the construction of such a space for, according to them, the mental properties would be *qualia*, that is, subjective and qualitative, hence unmeasurable traits.

It is hard to know whether that paper had any impact other than that the fee I got for it allowed us to buy Finn. This was a beautiful, cheerful, empathic, and affectionate Flat-Coated Retriever who became a valuable addition to our family. Somehow, Finn knew from the start that he could not play rough with Silvia, who was not quite 3 years old when he joined us. Finn also intuited when we were not in a playful mood. I still complain to Marta that she is less empathetic than Finn was. Another congenial animal in our life was Timber, Silvia's and Kevin's lovely Australian shepherd, who inspired in me a short story published in Argentina's oldest daily and is included in one of my books (Bunge 2012c). The same volume also includes a few other fiction pieces along with some essays on serious issues. I can distinguish fact from fiction, but do not separate them.

My second paper on the matter at hand was "The mind-body problem in an evolutionary perspective" (Bunge 1979b). I presented it at the CIBA symposium on brain and mind held in London in 1978. This meeting was presided over by the Berkeley philosopher John Searle, author of the clever Chinese Room argument against the artificial intelligence school. Searle has also held that the brain "causes" the mind. This idea, popular among the mid-nineteenth-century vulgar materialists, betrays a coarse ontology, for the causal relation can hold only between events. As the National Rifle Association fans keep repeating, guns don't kill: only firing them may have this effect.

Other heavyweights attending that symposium were Ursula Bellugi (sign language), Colin Blakemore (plasticity of the visual cortex), O. D. Creutzfeld ("mad cows"), José M. Delgado (the charging bull stopped by a radio signal to an electrode implanted in its brain), Colwyn Trevarthen (neurobiology of action), Elizabeth Warrington (semantic dementia), and Patrick Wall (pain).

The few philosophers among us had nothing to teach those stellar neuroscientists, so they spent their ammunition shooting colleagues. David Armstrong made fun of the emergence concept; and Hilary Putnam said that my criticisms of his own machinist view of the mental stemmed from my belief that present-day computers would be the last word: just you wait for the new generation of intelligent computers! Work on such a generation was in fact begun with loud fanfare in Japan only 4 years later and terminated when its failure became obvious.

My own presentation elaborated on Darwin's suggestion that, since the mental is bodily, and since the body has evolved, the mental abilities, too, must have evolved. Of course, I did not engage in the fantasies of the self-styled evolutionary psychologists who have been supplying the pseudoscience market since the 1990s (see Pievani 2014).

The eminent immunologist Peter Medawar closed the symposium with some general remarks. When I told him that he wrongly took it for granted that there were no dualists among us, he asked me where they were, and I replied that we were surrounded by them. Apparently, Sir Peter had not read his own wife's statement that blushing proved the power of mind over matter. Even Creutzfeld, famous for his study of the Creutzfeld-Jakob disease, had defended dualism. And Delgado claimed to have overcome dualism by reinventing "trialism," the Catholic trinity composed of body, mind, and soul.

My next contribution to the mind-body controversy (Bunge 1979d) was a criticism of D.M. MacKay's attempt to buttress the Christian dogma of the soul with the information-theoretical opinion that the mind consists of information flows, which supposedly require neither matter nor energy. I argued that every information flow, or transmission of coded signals, occurs in a material system capable of detecting, coding, transmitting, and decoding electromagnetic signals. Disembodied information belongs in science fiction. In any event, every time I hear it, the polysemic word "information" activates my alarm system.

Systematic Tackling of the Mind-Body Problem

I tackled the mind-body problem in a systematic fashion only when its turn came in the writing of my *Treatise*. In doing this task, I got the help of the neuroscientist Rodolfo Llinás, the psychologists Peter Milner and Dalbir Bindra, and the psychiatrist Bernard Dubrovsky. While a student, Llinás had been strongly influenced by Warren McCullough, one of the fathers of the conjecture that the brain "embodies" ideas and works as a computer. Later, Rodolfo had learned neuroelectrophysiology from John C. Eccles, who was to earn a Nobel Prize for his experimental dexterity, as well the nickname "the Ecclesiastical neuroscientist" for his defense of the Catholic version of psychoneural dualism.

In 1976, shortly after publishing a remarkable paper on individual neurons in *Scientific American*, Rodolfo phoned me in México City to chat about the nature of the mind. When I returned to Montreal, he invited me to visit his lab at New York University. We interacted intensively over the next 3 years. Rodolfo taught me quite a few things, but he had already made up his mind on the matter of mind. He never listened to my arguments on neural plasticity, let alone spontaneity and free will. He did not feel offended when I nicknamed him Machine. But he was annoyed at my insistence in asking him what he meant by "functional state," "eigenstate," and other expressions he used freely. Because of my insistence on precision, he thought I was more interested in words than in neurons. During my visits to Llinás, we cowrote a couple of papers: a criticism of the concept of a command neuron (Bunge and Llinás 1978a, b) and a discussion of Eccles's psychoneural dualism (Bunge and Llinás 1978a).

I presented the latter at the 16th International Congress of Philosophers that met in Düsseldorf on the Rhine. Eccles himself sat in the front row of the large auditorium and opened his mouth only to threaten me with retaliation the next day, which he did not do. I started my exposition by praising the magisterial experimental half of Sir John's presentation, but added that the second half – where he stated that the mind is to the brain what the pianist is to the piano – was not scientific but theological, on top of which he had disregarded the whole of biological psychology, from Broca and Wernicke to Hebb and Penfield. If only for this reason, his book with

Popper, *The Self and its Brain*, though published in 1977, could have been written three centuries earlier (Popper and Eccles 1977).

Repercussions

My criticisms of the famous scientist had an unexpected repercussion. The local daily published on its front page a detailed and faithful report on the session titled "The fight over consciousness." Neither of Eccles's partisans present at the session, among them the learned philosopher-priest Ernan McMullin and the outstanding Catholic neuroanatomist János Szentágothai, opened his mouth, as if conceding that dualism rests only on faith.

Another unexpected effect of my attack on Eccles was that Professor Vladimir S. Gott (*sic*), editor of the Soviet journal *Filosofskie Nauki*, asked me to submit a paper. I sent him "The bankruptcy of psychoneural dualism" (Bunge 1979c), which he published along with a much longer rejoinder by the Marxist philosopher David Dubrovski, who attacked me and produced a spirited defense of psychoneural dualism.

Why did a Marxist take an anti-materialist stand? Perhaps because, in his *Materialism and Empirio-Criticism* (1909), Lenin had criticized Eugen Dietzgen's statement that "thought is material." (Dietzgen had been a master tanner and self-taught philosopher whom Marx called "our philosopher.") Lenin argued that, if Dietzgen were right, materialism would not contradict idealism. At all events, my critic was only voicing the dualist philosophy of mind prevailing in the so-called socialist camp – a clear sign that Marxism as philosophy had divorced from science.

Three years later, the Hungarian journal *Magyar Filozófiai Szemle* published a translation of that paper of mine. This time, János Szentágothai was asked to comment, but he confined himself to stating that I did not know neuroscience. This was just as true as my view that he, like Eccles, had surrendered his science to his theology. (In his autobiography, Szentágothai stated that his only objection to Catholicism was its banning of birth control.) These were not isolated incidents but symptoms of the decline of Marxism. Its faithful had found that bowing the head is easier and less risky than using it.

Political Interlude: The Attempt to Break Up Canada

In the fall of 1976, soon after our return from México, Canadians experienced the first political earthquake since gaining independence over a century earlier: the Quebec separatists won the provincial election and threatened to put their program into practice. As usual with all independence movements, the Quebec nationalists, regardless of their social preferences, rallied around their flag – that of the pre-1789

French monarchy – and their leader, Lévesque, an intelligent, charismatic, and honest journalist, but also a fanatic nationalist.

Marta and I, brand-new Canadian citizens, were appalled. We admired the country's physical and cultural diversity, as well as the autonomy enjoyed by its ten provinces. But we also feared the narrow-mindedness characteristic of the radical nationalists, who wished to exact revenge for the humiliations their ancestors had suffered long ago under the conquering Anglos.

Our fears were soon confirmed: the new provincial government started by restricting linguistic freedom, limiting access to English-language schools, and demanding that all public signs be written in French. Many businesses and professionals felt threatened and left the province, which started to decline both economically and culturally. Soon, Toronto replaced Montreal as Canada's top city in everything but food, and most of McGill's new graduates left the province for greener pastures.

Marta and I too were tempted to emigrate, but we decided to fight back instead. We attended antiseparatist meetings and joined the Liberal party, the only consistently federalist and combative one. I also joined Alliance Québec, the nonpartisan association started by my colleague Storrs McCall to defend the rights of the anglophone minority, which the immigrants tended to join.

I kept lecturing on academic subjects at all the francophone universities, always wearing my Liberal lapel pin, which only provoked some angry stares. I also spoke against separatism at various local TV and radio stations and got a couple of threats over the phone. The federalist Liberal party won by a small margin the 1980 and 1995 referenda. The radical wing of the separatist movement weakened gradually. While in power, they behaved honestly and veered leftward, until becoming more progressive than the Liberals. All that sound and fury had one positive result: the Anglos stopped treating the *québécois pure laine* with arrogance and sent their children to "French immersion" schools. Some of them even attempted to learn French. The Liberals replaced their traditional middle-of-the-road social policies and became increasingly neoliberals, at which point Marta and I let our membership cards expire. We have not gone back to politics ever since. However, we keep our keen interest in international politics.

One of the negative aspects of the conflict over the Quebec separation was the conceptual indigence of all parties, as well as the superficiality of all political analysts. For one thing, no political scientist, not even the top contemporary authorities on nationalism, namely Ernest Gellner and Michael Mann, bothered to analyze the concepts of nation and nationalism. In particular, they failed to distinguish among territory (a geographical concept), people (anthropological), and nation (politological). Furthermore, nobody noticed that there are several kinds of nationalism: defensive and aggressive, biological and cultural, economic and political, and their many combinations (see Bunge 2009.) Consequently, no one realized that the pro-independence movement was far from monolithic, and that some of its proposals in favor of the francophone community were long overdue. For the record, Marta and I abhor all the aggressive nationalisms but approve of defensive economic nationalism. End of the political interlude.

Neuroscience ought to be of paramount interest to the philosophers of mind, for it studies the organ of the mind and of social behavior. As a matter of fact, over the past three decades, there have been some hints of neurophilosophy, as Patricia Smith Churchland calls this field. But much of it has regarded the central nervous system as a computer working according to algorithms (computer programs), not laws of nature. In particular, it has assumed that brains are Turing machines, hence lacking in plasticity, initiative, creativity, and free will. Indeed, one of the postulates of the theory of Turing machines is that these change in state only when receiving an input, whereas neuroscientists and psychologists know that the living brain is constantly active even when in isolation – as Hebb had proved with his famous sensory-deprivation experiment.

In addition, the philosophy of mind has been cultivated separately from the rest of philosophy and, in particular, in an ontological void, so that it employs sloppy notions of thing and property, state and process, system and mechanism, and so on. For instance, it has been said that this organ "instantiates" or "subserves" that function, and that "the brain causes the mind." In my own work (e.g., Bunge 1977a, 1979c, 1980d, 2012a), I have tried to avoid such muddles and have placed the philosophy of mind in its ontological context (see Mahner 2015).

I first peeked at the boiling cauldron of neuroscience under the guidance of the Nobel laureate Georg von Békésy. We had met at the Ernst Mach Colloquium held in Freiburg in 1966, and 5 years later, I visited him at his laboratory in the University of Hawaii. On this occasion, I made the mistake of mentioning S.S. Stevens, his erstwhile Harvard colleague, who had claimed to have found the universal psychophysical law, relating sensation to its physical stimulus, allegedly valid for all the sensory modalities. Without mentioning Stevens, von Békésy showed me an experiment about lateral inhibition – the confinement of sensation to a small circle around the stimulated point. The von Békésy experiments on lateral inhibition refute the whole of stimulus-response psychology and show that the nervous tissue has peculiar properties that distinguish neuroscience from the rest of biology. Shortly thereafter, it was discovered that the brain is also unique insofar as it synthesizes endorphins – painkillers – as well as endotoxins.

At the University of Hawaii, to which I had been invited to explore the possibility of being hired, I gave talks on a variety of topics and visited the friendly Irving Copi (né Copilovich), whose clear logic textbooks I had used at Temple University. On his desk, I saw copies of the financial pages of the *New York Times*, instead of recent issues of the *Journal of Symbolic Logic*. He drove me around that amazing island, and the chairman of the philosophy department threw a party in my honor at his spectacular house in the midst of a tropical forest. I was put up at a hotel on beautiful Waikiki beach, which catered to army officers on furlough and their wives. There was an endless string of cargo ships carrying supplies to the hapless invaders of Vietnam and environs. I liked Hawaii, but it was not for the likes of me: it was a place to sweat, swim, eat well, and earn good money all year round, but not to reflect on world riddles, much less on burning social issues.

The only thing I took back to cold Montreal is what I learned in Hans Hermes's book on model theory, which I studied on a deserted beach. It taught me that the theory in question concerns examples of abstract theories, not the specific theories one learns in factual science. The members of the "structuralist" school – in particular Suppes, Sneed, Stegmüller, Moulines, and Mosterín – confuse the two concepts designated by the word "model," much as the self-styled measurement theorists confuse "measurement" with "measure," and as Christians call God His three different Persons.

In 1977, I was invited to participate in the Winter Conference on Brain Research, held in Keystone, Colorado – a scientific-sportive-ethylic marathon that began at 7:00 a.m. and went on till 10:00 p.m. I also took long walks in the forest and swam in an outdoor swimming pool under a heavy snowfall. I learned much experimental material from lots of talks, or rather slide shows, and had stimulating talks with such interesting scientists as Rodolfo Llinás, Melvin Calvin, Norman Geschwind, and Arnold Scheibel and his wife and colleague Marian Cleeves Diamond, who had discovered the abnormally high density of glial cells in Einstein's brain. My paper on integrative levels and reduction (Bunge 1977b) was the only theoretical one.

Two years later, I was the keynote speaker at the same venue. My talk was titled "From mindless neuroscience and brainless psychology to neuropsychology" (Bunge 1981a). In the middle of my talk, I read my satirical poem on the mind-body problem, in imitation of Lewis Carroll's nonsense poem *Jabberwocky* that starts "Twas bryllig, and ye slythy toves…"

The Brighton Congress

Two decades later, I attended the 18th International Congress of Philosophy, held in Brighton, that a century earlier had been a fashionable if tacky seaside resort. Emblematically, the large clock in the railway station had stopped years earlier.

I was chosen to preside over the philosophy of mind session. As I was about to open the meeting, a giant approached me and announced in a Stentorian voice that he was The Lord Quinton (Anthony Quinton, President of Trinity College). When I asked around why that lightweight had been awarded such a grand title, I was told that he was a protégé of Prime Minister Margaret Thatcher, who had given him the task of closing down ten philosophy departments in the country. One wonders whether anyone noticed.

One of the speakers in my session, a Wittgensteinian, stated that it was absurd to talk about the mind-body problem, since *bodies*, being lifeless, cannot have mental experiences. Apparently, he only knew the military euphemism for "corpse." Other remarkable participants were a bunch of pretty Afrikaner girls who held fascist views, as well as two polite and impeccably dressed North Korean men who claimed to believe that their country's self-sufficiency slogan was a philosophy. I declined their invitation to visit their country, but treated them to tea and cake.

I also had tea with less exotic but more congenial philosophers, such as my former student David Blitz, the funny Peruvian Alberto Cordero-Lecca, and the American logician Ruth Barcan Marcus. Ruth had become famous for her work in modal logic, as well as for deflating the semantics of proper names, which she rightly regarded as meaningless tags. We got along well because we shared our contempt for academic bullshit.

Other Meetings At the Time

I have attended several Wittgenstein symposia, held in the Lower Austrian village of Kirchberg, at whose elementary school Wittgenstein had earned a reputation of fondness for physical punishment. In my keynote speech of 1986, I analyzed the relation between the science of mind and ten popular philosophies of mind, most of which owed nothing to the former (Bunge 1987c). At these meetings, one was sure to meet interesting people of different backgrounds. Once, the learned priest Józef Bocheński OP grilled me thoroughly during a long lunch under the sun to find out how good an Aristotelian I was. He was not shocked at my materialist philosophy of mind. After all, The Philosopher had taught that, far from being detachable from the body, the soul was but its "form" – a conveniently vague term.

Another memorable meeting was that of the American Philosophical Society held in Philadelphia in 1990, devoted to neuroscience. I revisited some of the places that I had first seen three decades earlier in Marta's company, listened to some remarkable papers, held interesting conversations with researchers in various fields, and read a paper on the problem of mind. I had been invited by Vernon Mountcastle, the discoverer of the neuronal columns in the cerebral cortex and a fan of mine, as he stated in *Cerebral Cortex* as well as on my 75th birthday at McGill, 4 years later. (It has been said that Sir John Eccles blocked Mountcastle's access to the Nobel Prize due to his outspoken criticism of the religious view of the mind.)

In 1979, the McGill psychologist Dalbir Bindra, aka D.B., and I organized a symposium on the mind-body problem (Bindra 1980). The main speakers at this meeting, which filled the largest classroom of the medical school, were D.B., Donald Hebb, and I. With D.B. and other colleagues, we also taught a course on the various pseudosciences, for we thought that training in a single science does not provide enough immunity against bogus science in other fields. My admired colleague and dear friend died suddenly in 1980. During his funeral service, at the request of his widow, the psychologist Jane Stewart, I read a poem by Omar Khayyám, D.B.'s favorite poet.

Aborted Book with Llinás

In 1978, during one of my visits to his lab, Rodolfo Llinás and I agreed to cowrite a book on the mind body problem. He was to emphasize the neuroscientific aspect, whereas I would supply the psychological material. I started right away to do my part, which I sent Rodolfo a year later. Shortly after receiving it, Rodolfo invited me to visit him at his home. He told me that he disagreed completely with my central hypothesis, that the higher (non-automatic) mental functions are processes in the predominantly plastic regions of the brain, that is, those whose neural connectivity changes with learning.

Llinás had stayed faithful to McCullogh's thesis, which he had learned two decades earlier, that a brain's connectivity is just as rigid as a computer's and is determined by the genome: throughout your life you feel, perceive, think, dream, and decide whatever your genome orders. This was, of course, part of genetic determinism popularized by Richard Dawkins and his followers, in particular Steven Pinker and Daniel Dennett. This view makes no more room for spontaneity, creativity, or free will than behaviorism had made.

In view of our radical discrepancies, Rodolfo and I decided to abort our project, which was a hard but wise decision, because he knew a lot about rats and the human cerebellum, but next to nothing about human psychology. In addition, Rodolfo could not admit that a mere philosopher could alter convictions that he had acquired two decades earlier, while studying medicine in Bogotá. In addition, I was exploiting recent findings on neural plasticity, which did not interest Rodolfo, and I was also using some formal tools that were absent from his toolbox.

I finished promptly the chapter on the mental included in the fourth volume of my *Treatise* (Bunge 1979c). Shortly thereafter, I expanded that chapter into my book *The Mind-Body Problem*, which Pergamon Press published in Oxford and New York (Bunge 1980d). I deplore our scholarly breakup because we had become good friends.

I believe that the subsequent development of cognitive neuroscience has amply corroborated my conjecture about the centrality of neuronal plasticity. But, of course, the computer engineering crowd and the psychologists and philosophers under its spell, such as Daniel Dennett, Steven Pinker, Hilary Putnam, Jerry Fodor, Zenon Pylyshyn, Paul and Patricia Churchland, Margaret Boden, and many others, kept repeating or embroidering La Mettrie's two-century-old opinion that man is a machine rather than a largely self-made sociable animal. And of course machines, unlike human brains, do not program themselves and do not congregate into social systems.

Noam Chomsky

One of my best students at that time was Mike Dillinger, a linguistics student of Myrna Gopnik's, as well as a Chomsky enthusiast. (A few years later, Myrna became instantly famous for discovering that dysphasia of a certain kind runs in families.) Dillinger's view on Chomsky's opinions became the focus of one of my courses. I had nothing to say about Chomsky's contribution to syntax, which in my view was hardly of interest to philosophers. But I objected to both his semantics and his psycholinguistics: I thought the former nonexistent and the latter at odds with developmental psychology and cognitive neuroscience.

I objected to Chomsky's nativism, in particular his opinions that we are born endowed with a universal grammar and a linguistic theory that allows us to master any language without having to learn it. I believe this view to be but a gut reaction to the behaviorism ruling in American psychology until about 1960, as well as a consequence of Chomsky's ignorance of Jean Piaget's work in developmental psychology, which emphasized the child's constructive (inventive) ability, denied by Chomsky as vehemently as by Skinner, the most radical of behaviorists.

I also rejected the hypothesis of innate linguistic intuition – which Chomsky called Cartesian but I saw as Kantian – which would allow anyone to judge instantly whether any given linguistic expression respects what the French call *le génie de la langue*. In addition, I held that Chomsky had produced no semantics – something that he himself ended up by conceding. Nor could I share his early enthusiasm for pychoanalysis and concomitant contempt for the experimental branches of linguistics – neurolinguistics, psycholinguistics, and sociolinguistics.

Last, but not least, I regard Chomsky's said opinions as clear cases of pseudoscience, and his recent conversion to mysterianism (Chomsky 2009) as a pathetic if long overdue confession of his disbelief in science. At the same time, I share most of Chomsky's criticisms of American foreign politics and admire his courage in making them public. But I do not share his anarchism, which I regard as reactionary, inasmuch as it proposes turning the clock back by 5000 years and in particular replacing large factories with craftsmen's shops. Small is not beautiful but inefficient. I, for one, do not miss the kerosene lamp or the pit latrine of my childhood.

Mike Dillinger's questions and critical remarks helped me hone my own views about linguistics, and we got closer as we interacted. In 1981, I put him in charge of my office, to help me handle my copious mail in five languages and fend off annoying visitors. He kept that position, along with six other part-time jobs, for 6 years, while he was completing the requirements to get a PhD in linguistics. After graduating, Mike found his niche: machine translation, a field that has matured amazingly since its beginning in the 1950s, and where he is an international authority.

In 1982, I was invited to give a paper at the 13th International Congress of Linguistics to be held in Tokyo. I was amazed to find that there were only a handful of Chomskyans among the 5000 participants hailing from around the world. Expanded versions of my paper appeared in book form in Spanish (1983a), English (1984a), and Japanese (1986a). The philosophy professor Hiroshi Kurosaki, who a decade earlier had been a postdoc student of mine, invited me to stay at his home and arranged for a lecture tour. I spoke at universities in Tokyo, Kyoto, and Yokohama. In addition, we visited the Toyota car factory, where I made inquiries about its "quality circles," where workers and supervisors meet once or twice a month to discuss technical issues that arise in the industrial process.

One of the traits that most impressed me about those amazing people is the contrast between their love of modernity and the tolerance of backwardness. For example, the Japanese are the longest living and less religious of all peoples and just as egalitarian as the Scandinavians – except in matters of gender. I experienced gender discrimination in my host's household, where the males were served by his wife and daughter, who did not even sit at the table with the men.

My audiences were attentive and respectful to the point of never voicing disagreements with me. Apparently, the word "no" is a no-no among them. I succeeded only twice in eliciting debates. The only occasion where they seemed relaxed and ready to accept my challenges was during the dinner I offered them at the end of my stay – and this only after abundant beer and sake loosened their tongues. No wonder that Japan, which excels in mathematics, natural science, and technology, is backward in the humanities and social sciences.

Psychology Meetings

In 1980, I participated in a neuroscience symposium held in Galveston, Texas. There, I met Theodore Bullock – one of the fathers of neuroethology – the Mexican neurophysiologist Pablo Rudomín, the psychiatrist and long-distance runner Arnold Mandell, and other interesting if less picturesque characters. My talk was about the many levels encountered in studying the human brain, from the individual neuron ("feature detector") capable of detecting stimuli of certain kinds to the massive neuronal systems capable of planning (Bunge 1980f). I started my talk by chiding them and other basic scientists for claiming in their grant applications that their work was of great practical importance. I held that it should suffice to write that they hoped to learn something new and interesting. We should teach statesmen and bureaucrats to distinguish farming from marketing.

My next encounter with the mind-body problem took place 3 years later in Acapulco, México, where the 23rd International Congress of Psychology was held. There, I made two new friends: the Colombian psychologist Rubén Ardila and the German medical psychologist Ernst Pöppel, both of them charming and open to debate. I also met a congenial Québecois psychologist who broke both legs when jumping from the hotel's diving board without first making sure that there was water in the pool.

Later that year, I traveled to Munich to visit Pöppel at his Institute of Medical Psychology. There he told me how he had discovered "blindsight" at about the same time as Larry Weiskrantz in Oxford. In Ernst's case, the discovery came while evaluating the visual ability of war veterans who wished to be certified as being legally blind. Ernst noticed that, although shrapnel had destroyed their primary visual area, they moved around as if they saw the obstacles in their path. It is not that their immaterial soul had not been damaged. What happened is that there are about 20 secondary visual areas distributed around the cerebral cortex, as well as a second, possibly phylogenetically older, visual pathway, all of them disconnected from the consciousness organ. So, those veterans did "see" vague shadows but were not aware of it.

Neither Pöppel nor Weiskrantz knew that the Italian neurophysiologist Eugenio Tanzi and his student Ernesto Lugaro had published in *Brain*, toward the end of the nineteenth century, on the same phenomenon in chimpanzees. It is also usually forgotten that Tanzi and Lugaro were the first to assume that the interneuronal connections are not electrical but chemical and that some of them are plastic to boot. Why were these precursors ignored? Perhaps because they had worked in backward Italy rather than in Germany or England. Only the great Cajal knew about them and mentioned their amazing findings.

Some sociologists of science have studied cases of independent simultaneous findings, but they have neglected the more frequent cases of independent successive rediscoveries. As the linguist Michel Paradis remarked to me, a case in point is the avalanche of rediscoveries made in recent years with the help of the new brain-imaging devices.

Six years later, Pöppel invited me to attend his Munich symposium on the philosophy of mind, where I discussed again some problems involving reduction and integration (Bunge 1989a). Many years later, at a gathering to honor Brenda Milner – best known for her thorough study of Henry M., the most famous amnesiac in history– I talked at length with Endel Tulving, the first to distinguish episodic from semantic memory. It turned out that this extraordinary Estonian had read and liked some writings of mine.

Encounter with Islamic Fundamentalism

It is well known that religion and pseudoscience are more popular than science and materialism: the former are easier to learn, and they enjoy the support of both tradition and the powers that be. However, new winds have been blowing in academia since the slow and difficult birth of cognitive neuroscience. My insistence in refining and spreading psychoneural monism did not inconvenience me much in North America or Europe, because I was being carried by the new wave. Indeed, most research psychologists take monism for granted, and only a few of them, like Eric Kandel, practice it while atoning in their temples. As for the philosophers of mind, most of them claim to be monists even while professing informationism, in particular the opinion that brains work like computers guided by algorithms – the fashionable version of the fable that mind guides matter.

By contrast, in the Third World materialism is still a heresy to be either ignored or stamped out in the name of the ruling religion. In 1983, at Ain Shams University, in Cairo, I was twice at the center of the age-long conflict between science and religion. Indeed, in the midst of my first lecture on this subject, a man stood up, shouted something in an irate tone, and left the vast classroom followed by a couple of hundred students.

My host, Professor Mourad Wahba – a philosopher belonging to the Copt minority – informed me that the individual in question, a solid-state physicist, had declared that all I had said was false. Wahba explained as follows the incongruence between the science and the religion of his colleague: "He must have studied physics the religious way, that is, as a set of indisputable principles and precepts, not as a search. Do not we proceed in the same manner when discussing political ideologies?"

I finished my lecture without further incidents and asked my host to invite his colleague to return the next day and debate with me in the presence of the same student body. My opponent accepted the challenge and appeared the next day, willing to engage me. The following dialogue took place:

- *Why do you believe that your science is compatible with your religion?*
- Because all our knowledge is contained in the Koran.
- *Even quantum mechanics, which was invented one millennium after the Koran was written?*
- No doubt. But, of course, one must know how to interpret the Prophet's *suras*.

I was left astonished and wordless. My opponent had won easily by K.O.

The next day, when I was to lecture on the mind-body problem, I was unable to reach the classroom because my way to it was filled with kneeling students reciting their prayers. I had to wait before my path was cleared and I was able to keep uttering heresies in the classroom. Marta's mathematics lectures, by contrast, were not interfered with – one more evidence for the thesis that pure mathematics does not deal with the real world.

It was disheartening to hear medieval superstition in 1983. But our visit to ancient pyramids and temples, from Cairo to Karnak to Luxor to Asswan, was ample compensation. Eric and Silvia had a great time climbing ruins and exploring mounds of rubbish piled up over centuries. We also had fun bargaining with the drivers of horse-driven carriages who, after having agreed on ten pounds for the trip, and as their horses were galloping at full speed in the cold night, added: "*Den bounds ber berson*."

Spain Again

Soon after the Spanish version of my *Philosophy of Psychology* (Bunge 1988a) appeared, I started to get Spanish invitations to give talks. One of my best experiences was the crash course I taught at the University of Murcia at the invitation of the biological psychologists Luis Puelles and José María Martínez-Selva. Wenceslao González, a Catholic philosopher who stoically withstood my materialist heresies, was in charge of the logistics. The psychologists gave excellent presentations of some research papers recently published by *Nature* and *Science*, which I asked them to report on.

After my lectures, when the fierce sunshine started to wane, I met with a handful of congenial colleagues at a café across the beautiful central square to drink *horchata de chufas*, a delicious beverage made from earth almonds. They told me how General Franco had ordered the bulldozing of the beautiful public baths built a millennium earlier by the Arabs. Someone said that the Arabs had been expelled from Spain for trying to bathe the Christians. But the truth is that contemporary Spaniards shower more frequently than their European visitors.

The last night, at a restaurant, I met my friend, the great journalist and publisher José (Pepe) Ortega Spottorno, the son of the famed philosopher Ortega y Gasset, and an important figure in the transition from fascism to democracy. His wife had written a cookbook that sold so well that Pepe said that he had attained the ideal of every Spanish male: to live at the expense of his wife. The next morning, I was driven to Madrid through the beautiful fruit orchards, then in full bloom, that supply much of the fruit consumed in Europe. I went straight to the apartment of my friend Alvaro Fernández Fernández, an important engineer, businessman, and fellow skeptic, who had visited me in Geneva a few years earlier to try and start a Spanish skeptic society. There, I took a nap until noon, then we lunched on delicious lamb roast and strawberries, and finally, I boarded my plane back to Montreal.

What happened to Alvaro a few years later disproved once more the myth of the immaterial soul. In fact, he suffered the ablation of a huge cerebral tumor, which had been the seat of much of his knowledge. His frontal lobes and subcortical organs were spared, so he kept his willpower and, thanks to the plasticity of his cerebral cortex, was able to relearn his basic cognitive skills. But his engineering knowledge ended up in the clinic's rubbish pail, so he had to give up his profession. When we met again, two decades later, Alvaro had graduated in anthropology, remarried, and recovered his contagious *joie de vivre*. The lancet had spared his volition and emotions, and he had rebuilt most of his cognitive systems almost *da capo*.

In later years, I took my message on mind to universities in Barcelona, Girona, Granada, Mallorca, New York, Sevilla, and Tarragona, as well as those in Genève and Fribourg. In all these centers of learning, I was listened and questioned with great curiosity, perhaps because I spice my talks with comic episodes and commonsensical remarks. For example, I would pick a man at random and ask him why he points to his head whenever there is mental trouble, from headache to madness. Does this habit not betray the belief that the brain is the seat of the mind, just as the digestive system is the organ of digestion?

Relations with Catholics

Although most Argentines call themselves Catholics, few of them live up to the doctrine they profess, and even fewer are interested in discussing it. Also, since most of the faithful who were well read in doctrinal matters were politically committed to the ruling powers, ordinary people had few opportunities to converse with them. I never had classmates or colleagues, either in physics or in philosophy, who were knowledgeable about Thomism, one of the most interesting if outdated philosophical systems. I talked only once or twice, during a philosophy congress, with Monsignor Octavio Derisi, the leading Argentine Thomist, who had been an outspoken Peronist under Perón and later an equally fervent supporter of the military who chased Perón.

My experience with European Catholics was very different. In France, Belgium, Spain, Switzerland, and Austria, I met a number of Catholic intellectuals who were willing to engage in discussions with me. I have mentioned already my friends Agazzi, Bocheński, and Weingartner. In the 1970s, I had a long conversation with the Franciscan Herman Van Breda, who had achieved notoriety for rescuing Husserl's Freiburg archive under the Nazis. One decade earlier, I had harshly criticized his paper on phenomenology at the Inter-American Congress that met in Santiago de Chile (Bunge 1951b). But he had kept no grudges and invited me to talk at the French part of the Leuven Catholic University, which was then in the process of divorcing its Flemish counterpart. Van Breda took this opportunity to assure me that chastity was the best means to reach a ripe old age, but lived to be only 63 years.

Another interesting character was Mariano Artigas, of Opus Dei. We first met at the 1987 colloquium on probability that the Académie Internationale de Philosophie des Sciences held in Vico Equense, near Naples. After trying to persuade me that his religious order was not a sinister cabal, he summarized as follows the attitude of his fellow Thomists with respect to my philosophy: they approved of my philosophical realism but rejected my philosophical materialism. Artigas assured me that the latter was worse than his own chain smoking, which was to kill him a few years later, whereas materialism has not damaged my lungs. A few years later, he teamed up with my former friend and McGill colleague William R. Shea in writing *Galileo in Rome* (Shea and Artigas 2003), which analyzes the alleged errors of that "troublesome genius" and excuses his prosecutors. They do not distinguish between error and crime.

When John Paul II died, I published a note praising his efforts for peace – a praise that did not go well with my fellow skeptics. I have said more than once that we should share meals with priests instead of trying to eat them. In 1992, just back from India, I spoke at the colloquium on cosmology held at the Lateran University, in Vatican City. In my talk, I sketched seven possible worldviews (Bunge 1993b), and during the subsequent discussion, I claimed that the Church would not have condemned Galileo if it had kept Aristotelian-Thomist realism, instead of adopting Cardinale Bellarmino's conventionalism, which I regarded as a mere tactical maneuver to avoid taking sides in the geocentrism/heliocentrism controversy. Strangely enough, conventionalism had been refloated around 1900 by Henri Poincaré and Ernst Mach's followers, whereas most scientists kept trying to find objective truths.

My Vatican talk was warmly applauded, published, and paid for, and I was put up in an austere but well appointed apartment worthy of a cardinal. Regrettably, the food served at the refectory was far from being a *bocatto di cardinale*. As I write, my fellows in the Argentine Academy of Exact Sciences are discussing a sort of manifesto I wrote in favor of Pope Francis's recent encyclical *Laudato si* on climate changes and the need to adapt to them by making radical changes in our lifestyles (Bunge 2015b).

Reception of My Philosophy of Mind in My Native Land

Until very recently, my work was ignored or attacked in my native land. But in 1985, I received an unexpected invitation: to address the first Argentine Congress of Psychological Research to be held in Rosario, on the majestic Paraná River. I swallowed the research bait but found no fish in that congress: nearly all the participants were clinical psychologists or psychiatrists, many of them psychoanalysts. After my talk, one of the organizers asked me privately not to deprive them of their means of subsistence.

After Rosario, I went to Buenos Aires sponsored by the Alfredo Thomson Foundation, devoted to supporting scientific psychiatry. I gave several talks at the Sociedad Científica Argentina and the Universidad de Belgrano, where I discussed numerous recent findings in biological psychology. Judging from the many questions that followed my talks, my audience was only interested in the fashionable logotherapies, such as the Palo Alto and primeval scream ones, of which I knew nothing. Still, I interacted with a handful of serious psychiatrists around Fernando Alvarez, the serious and kind editor of the Foundation's *Boletín Neurológico*.

Another memorable encounter was with Luis F. Leloir in the large laboratory that the Campomar Foundation had set up for him and his numerous research staff. The great man, one of the three Nobel laureates in science that my country has produced and whom I had met in Oviedo a few years earlier, received me warmly and invited me to join his laboratory. I might have worked happily under his protection, but Leloir died only 2 years later, and it is unlikely that his successors would have appreciated my work.

Unlike his colleagues, who favored white lab coats, Leloir wore a gray one. This idiosyncrasy gave rise to César Milstein's famous confusion when he visited Leloir for the first time. When entering the building, he addressed the first individual he saw, whom he mistook for an orderly because he was wearing a gray smock: "Hey you, where is Leloir?" The individual in question replied in his usual calm and unassuming tone: "I am Leloir." Although both men had attained international celebrity after receiving the Nobel Prize, they never took on airs. Leloir's scientific autobiography, published in the *Annual Review of Biochemistry for 1983*, is a model of modesty – *not* to be mistaken with humility.

That time, I also visited the cytologist Eduardo de Robertis, whom I had befriended in 1958 during my first trip to Europe and from whose chair at the medical faculty I had spoken several times. Eduardo had discovered the synaptic vesicle of the neuron, which deserved a Nobel Prize. In his laboratory, I met again with his favorite disciple, Amanda Pellegrino, an old friend of mine and now an emeritus professor, who decades earlier had shown me how to operate the microtome and had swam in El Ombú. Marta and I had often visited her and her partner, Carlos Iraldi. They constituted the most dissimilar couple imaginable: she was serious and timid, whereas he was extroverted and playful.

Carlos was a man of many parts: a pathologist, assistant to Professor Houssay, a psychoanalysis enthusiast, curious about philosophy, always ready to help, and one of the founders of Les Luthiers. This was a group of inventive and irreverent amateur musicians and comedians who played eccentric musical instruments that they themselves invented and crafted. They also wrote and sang hilarious nonsense songs, such as the *cumbia* "Epistemology." When Marta and I left our country for good, we got rid of a swarm of annoying gnats, but we also lost many good and interesting friends.

When I revisited Buenos Aires after two decades, I stayed with my lay godfather Raúl Prebisch and his wife Eliana Díaz, whom I had encountered a couple of years earlier in Toledo. From early morning on, Raúl and I were besieged by newsmen. One of the most persistent of them was Bernardo Neustadt, a gifted journalist who had supported the military regime and was now campaigning for the so-called free enterprise. When he mentioned Bach, Haydn, Mozart, and Beethoven as exemplars of successful independent entrepreneurs, I had to inform him that all four had depended on wealthy patrons. And when he asked Raúl whether he agreed that the Argentine state was hypertrophic, my godfather replied: "Yes, Sir, our state is hypertrophic in some respects, but stunted in others."

During that visit, I renewed my acquaintance with several cousins, in particular Cecilia Bunge Shaw, whom my father and I had adored as a child. I also discovered my niece Lucía Gálvez and her husband Bartolomé Tiscornia. Lucía is a prolific novelist and an old-style historian (No numbers, please!), while Bartolo is a lawyer who used to teach Law. Over the last three decades, Bartolo and Lucía have been the pillars of the Club del Progreso, the oldest club in the country and originally a Freemason coven, of which I am an honorary member.

In Calvin's Country

My dear McGill colleagues forced me to retire at age 65 because I had perpetrated the worse of academic disloyalties: year after year, I had published more than all of them put together. After retiring me, they assigned me one or two courses per semester, paying me what an instructor earned. When 1987 approached, the department's chairman – an expert in German idealism who had been disciplined for a serious sexual indiscretion – informed me that there would be no courses for me that year. This was a hard blow both because I liked teaching and because I would not be able to start collecting my pension until 5 years later. But who would care? That I had taught novel courses which none of my colleagues could teach – on semantics, scientific metaphysics, the philosophies of physics, psychology, social science, and technology – did not count, when courses on Locke, Hamann, Kant, Hegel, Nietzsche, Husserl, Heidegger, Chomsky, and Habermas could be taken.

I wrote right away to my friend Michel Paty, a nuclear physicist with a particle to his name. He immediately got me a visiting professorship at the University of Paris, but the French Ministry of Education overruled my appointment because I was 1 year older than the retirement age. Yet my good luck held: while returning from a philosophical meeting, on a train from Fribourg to Geneva, I was recognized by someone unknown to me, who informed me that the university of Geneva had just opened a position in the philosophy of science.

I wrote to my friend Pierre Moessinger, the last of Piaget's disciples, who had attended some of my lectures at McGill. Shortly thereafter, I received the formal appointment of visiting professor at the Geneva Faculty of Sciences for the 1967–1968 course. The salary that the République et Canton of Genève paid me was twice my former Canadian one.

We rented a modern apartment at Rue Le Corbusier 1, whose basement had been fitted to resist a nuclear attack, and settled there with Silvia, while Eric stayed behind to start his architecture studies. The janitors of our building were a handyman and his seamstress wife, both friendly and industrious Portuguese who had wasted their youth working in Angola. They had only sad stories from that harsh time, like the day when they were ordered to sink all their trucks in the harbor on the eve of their hasty departure. Our apartment was near a forest beyond which was the prestigious École Internationale, which Silvia attended. There, she made friends with girls from a variety of countries on three continents and was given new freedoms and responsibilities. We shopped at the two cooperative stores, and once a week, we drove over to the nearest French town to buy good food at half the price. The university cafeteria offered no meals, and we could not afford Swiss restaurants.

We drove often to nearby Nyon to visit my old friend Hernán Rodríguez-Campoamor and his charming family, our nostalgic link with our native land. Hernán worked at the UN's International Labour Office, and his family tended a beautiful garden with its own creek. We spent Christmas with the Ranchettis in Florence and at the end of the academic year, drove to Corfú, stopping in Milano just to gape at the Duomo, the largest sculpture on Earth. While crossing the vast piazza in front of that temple, we were reminded of the wonder felt by Totò and his friends when, just arrived from the Mezzogiorno, they addressed in German the two tall *vigili* in their elegant uniforms. Such tall, well-fed, and elegant gentlemen could only have been German imports!

At Geneva University, I taught a high-level course for doctoral students, professors, and two deans. Because every one of my students had done some scientific research, their questions, though sometimes naive, were never silly, let alone absurd – except for the new dean's, a friendly mathematician, "What about the revealed truths?" Years later, I wrote jointly with one of those students, the biomedical researcher Gérald Thurler (Bunge, Thurler et al. 2003), a paper on a systems-theoretic approach to disease (Bunge, Thurler et al. 2003).

Since my office was in the physics department, I had some exchanges with physicists, especially the particle physicist Henri Ruegg, a helpful host who got Eric a temporary job at an architectural studio when he came to visit us. I also attended some physics lectures, in particular two by CERN's famous Joseph Bell, of Bell inequalities fame. He devoted one of his lectures to reviewing the 20 or so different measurable manifestations of the quantum vacuum discovered after Casimir published in 1948 his pioneering paper on the subject. Since then, one knows that

Aristotle had been right in holding that nature abhors a vacuum: every corner of the universe is full of a rapidly fluctuating quantum electromagnetic field.

That year, I read proofs of my *Philosophy of Psychology* (Bunge 1987a), which was eventually translated into Spanish and Farsi. (The Farsi translator made it clear that he did not share my materialist view of the soul.) However, my main task during that year was to research the eighth and last volume of my *Treatise*, the one devoted to ethics and political philosophy. This tome contains my political project: *holotechnodemocracy*, that is, integral democracy informed by social technologies instead of dogmatic ideological slogans improvised in the heat of unprincipled and improvised struggles for power.

Moscow Congress

I presented my political project at the 8th International Congress of Logic, Methodology, and Philosophy of Science, held in Moscow in the summer of 1987. At that time, Mikhail Gorbachev was still in power, and children did not play the violin in streets for a few coins. My talk was overshadowed by a Russian philosopher who had invented no fewer than 13 dialectical "laws," a remarkable improvement on the mere three known to Hegel, Engels, and Lenin.

At that congress, I met again with my old friends Agassi, Agazzi, Alchourrón, Hao Wang, Malitza, Miró-Quesada, and Quintanilla and made two new friends, the Peruvian David Sobrevilla and the Argentine Monsignor Marcelo Sánchez-Sorondo. I also had a discussion with a couple of high-placed Soviet philosophers whom I angered when I urged them to jettison Hegel for being opaque and inimical to science. They showed no interest in political philosophy, which reminded me of Maximilian, México's imported emperor, who was writing his court's etiquette book when the revolutionaries stormed his Chapultepec castle and seized and summarily executed him and his wife.

I was given a comfortable room at a huge hotel. Since the TV set was out of order, I asked to be let into the storeroom, where after some searching, I found a set that looked new and schlepped it to my room under the impassive gaze of the floor guard. On turning on the apparatus, I discovered that it did not work. Nobody seemed to care.

Because I was a member of the Executive Council of the World Federation of Philosophical Societies, aka FISP, I was assigned an interpreter who doubled as my driver. With his diligent help, I discovered the whereabouts of the copyrights for the three books of mine published in the USSR: they were the equivalent of US$ 2000 and could have been collected after a complicated bureaucratic process. But all I could get for that money at the huge GUM department store were ugly East German rugs, Mongolian pink (!) cashmere sweaters, Caspian caviar, and vodka – none of which interested me. So, I did not spend that money, which a few years later shrank from 2000 to 2 dollars. Back now to the philosophy of mind.

My Protracted and Fruitless Feud with Psychoanalysis

People who are still seduced by psychoanalysis do not understand why I spent so many years, from the age of 18, fighting it. I have fought psychoanalysis because it is a fraud, has seriously hindered the advancement of the science of mind, and has been used as a facile and ready-made theory of everything social or cultural. In the course of over a century, psychoanalysts have not performed a single experiment to check their silly fantasies, have not made a single discovery, and have not achieved anything to alleviate the most common mental ailments – depression, psychoses, anxiety, phobias, and the various unhealthy addictions. But, because it claims to deal with sex and can be learned in a few weeks, psychoanalysis is the amateur's psychology. Yet, because logotherapy is as expensive as it is ineffective, it became nearly extinct in the USA and the UK about 1970 (see, e.g., Shorter 1997).

Psychoanalysis is of particular interest to philosophers of science because it challenges all the popular definitions of "science." For one thing, it topples Popper's refutability condition, since all the psychoanalytic fables, except the Oedipus story when protected by the repression one, are testable; moreover, when tested by scientific psychologists, they were found false. Secondly, the case of psychoanalysis shows the insufficiency of empirical testability: for the hypothesis on trial to be accepted, it is also necessary for it to be compatible with the bulk of scientific knowledge, as well as with all the ontological presuppositions of scientific research. And in postulating psychoneural dualism, which goes against the grain of cognitive neuroscience, psychoanalysis crosses the border between genuine science and bogus science.

Lecterns from Which I Have Preached My Psychological Gospel

I have lectured on the philosophy of mind at McGill, New York University, Geneva and Fribourg universities, and other places. At Fribourg, where I taught brief courses in 1987 and 1990, I had to give every lecture first in French and immediately thereafter in German because, although the Swiss recognize four official languages, few of them take advantage of such linguistic wealth. My Genevan lecture was published by the Swiss journal of psychosomatic medicine (Bunge 1987d). Sometime later, its editor wrote me that my paper had led him to renege on psychoanalysis.

New Ideas in Psychology published several papers of mine (Bunge 1983b, 1990b, 1991b, f), as well as a long interview conducted by its editor, Pierre Moessinger. Further texts of mine on psychology, both scientific and bogus, have appeared in *Annals of Theoretical Psychology*, *Behavioral and Brain Sciences*, *Science Today*, *Neuroscience*, and the *Scientific Review of Mental Health Practice*. Some of those papers were collected in my books *Seudociencia e ideología* (Bunge 1985b), *Mente y sociedad* (Bunge 1989c), *Las pseudociencias ¡vaya timo*! (Bunge 2010b), and *The Sociology-Philosophy Connection* (Bunge 1999c).

I also spoke at a number of colloquia and congresses devoted to psychology, such as those held in Acapulco (1983a, b, c, d), Rome (1987a, b, c, d), Galveston (1988a, b), the 11th Wittgenstein congress (1986a, b), Munich (1989a, b, c, d, e, f, g), Philadelphia (1991a, b, c, d, e, f), and the 38th congress of the German Psychological Society (Bunge 1993a) held in Trier. This was Marx's birthplace, not far from Unna, the seat of the German Bunges since the 1600s, and part of the region covered by the engrossing TV series *Heimat*. From Trier, I was driven to Frankfurt, where I boarded the plane that took me to Valencia, at whose Universidad Internacional Menéndez Pelayo I taught a crash course.

At the Rome symposium, organized by the Italian Society of Psychology, the journalists mobbed an arrogant philosopher who minimized the role of the brain. Silvano Chiari, the society's president, interviewed me and told me that he was about to translate my book on the mind-body problem, but he died shortly thereafter from a heart attack while riding a streetcar.

At the 13th Wittgenstein congress (Kirchberg 1988), my old friend J.M. Bocheński O.P. interrogated me thoroughly about my philosophy. Every now and then, he assented and whispered "Aristotle." When my turn came, I asked him whether he had been frightened when flying solo for the first time at the age of 70. He exclaimed: "Of course I was frightened! We descend from fish, not birds, you know." I did not find it necessary to ask him whether he believed the story of Adam's rib. I wonder whether he was ever allowed to say in mass or, like Bolzano, he had been confined to the library. At 88, Bocheński had driven a Mercedes 380 from Fribourg to Kirchberg at 200 km/h.

A few years earlier, Bocheński had driven Günter Kröber and me through the Swiss Alps at night, also at a high speed, and had amused us with political jokes at a mountain café. Kröber was a friendly East German sociologist of science who called himself a philosopher, since sociology was not recognized as a science in the Soviet bloc. He declined politely my invitation to submit a book manuscript to my Library of Exact Philosophy, because I had written that it wished to restore "the Vienna Circle spirit," that is, respect for both logic and science. Marxists only saw Ernst Mach's black cloud hanging over logical positivism, whereas I regarded it, despite its faults, as the only interesting philosophical novelty of the twentieth century. One could always engage a logical positivist in a rational debate in front of a blackboard, an exercise impossible with philosophers of other schools.

Encounters with Professional Dualists

I had my first encounter with psychoneural dualism in the sixth grade of elementary school. A classmate with a double family name announced proudly that he had been allowed to attend spiritualist séances, where the initiated communicated with their dear departed. I laughed out loud, which infuriated my classmate. At about that time, a British fortune-teller who called himself Mr Luck became an instant celebrity in Buenos Aires. His customers had to wait 1 month and pay a fee of 100 pesos, the monthly wage of a laborer. Mr Luck did not foretell the future: he confined himself to "reading" his client's past. He told my father that his father used to sniff up snuff and made the elegant gesture of taking up a pinch of the stuff from an imaginary snuff box, but spared him the disgusting expectoration on a red flannel handkerchief – a ceremony I saw once in Oxford after dining at a high table. My father and I were impressed but remained skeptical.

Half a century later, a Hare Krishna official with a PhD in physics from the prestigious Imperial College of Science and Technology invited a group of McGill professors to attend his talk on levitation, a subject that his organization claimed to teach in exchange for a few thousand Swiss francs. I interrupted his well-prepared lecture, demanding that he levitate there and then. (*Hic Rhodus, hic salta*! – as Aesop would have requested.) The charlatan excused himself, saying that at the moment he did not have the necessary equipment – presumably a thick mattress with steel springs. I got up, accused him of being a fraudster, and banged the door on leaving. None of my colleagues followed me.

I had had a similar experience shortly after coming to McGill. Invited to give a talk at the psychiatry department, which at the time was dominated by psychoanalysts, I started by stating that I knew nothing about their specialty, but was curious as to how they evaluated the efficacy of their treatments: whether they included control groups, practiced follow-ups, and so on. A senior member of the audience told me unabashedly that they needed neither controls nor follow-ups, because they trusted their therapies. I had never seen such shameless display of willful ignorance and professional irresponsibility, so I exploded and shouted: "This is not a scientific department but a church!" and left. That was my first inkling that Canada, or at least McGill, was no longer at the scientific forefront.

Like every other important metaphysical doctrine, psychoneural dualism has enormous theoretical and practical consequences. The former is of course the split of science and philosophy into two camps that cannot learn anything from one another: spiritualism and materialism. A practical consequence of dualism is that the mental is treated differently from the bodily: either as all-important or as negligible – "it's only in your mind." This is how we got separate groups of professions dealing with mental health: verbal healers and scientific specialists. And even the latter or pro-science group is split into two, clinical psychologists on the one hand and brain repairmen on the other.

Worse yet, consistent belief in the superiority of mind over body is bound to lead to the belief that the body must be punished for the spirit to flourish, or at least to the neglect of bodily health, as shown by some recent research (Fortsmann et al. 2014). In other words, mind-body dualism not only protects charlatans but it is also bad for your health, inasmuch as it encourages unhealthy lifestyles. The YMCA slogan, *Mens sana in corpore sano*, still holds good, though *Corpore sano in societas sana* might be even healthier.

The Place of Psychology

If I were a radical reductionist, I would claim that psychology should become a chapter of neurobiology. However, my reductionism is moderate: I hold that learning and doing are just as mind shaping as genetic endowment, so that psychology should be regarded as a *biosocial science* (Bunge 1988a, 1990a, b).

Thus, the reduction in question is *ontological* ("The mental is cerebral.") but not methodological, since psychologists still use some of the traditional methods of psychology, in particular the study of overt behavior, interrogation, paper-and-pencil tests, conditioning, and desensitization.

For example, the survival of failed religious prophecies and groundless politics cannot be explained in purely biological terms. To understand such maladaptive beliefs, we must invoke psychosocial factors such as the protection felt when belonging to the elite chosen by God or protected by the powers that be, or even by history. Still, the learning and justification of such beliefs are brain processes. This is why psychologists do not waste their time speculating about brains in vats the way Putnam and his faithful have done.

The fact that the mental emerges from the living (Bunge 1977a), inasmuch as neuronal assemblies have peculiar properties, from lateral inhibition and habituation to creativity and spontaneity, explains why, up to a certain point, it may be handled separately from biology. After all, some important psychological discoveries were made long before the emergence of cognitive neuroscience. For example, only a few decades ago, using just a chronometer, Anne Treisman discovered that, when perceiving something, we detect separately the color, shape, and orientation of things and that we "bind" such partial percepts into a single one only when paying close attention to them.

However, to explain both the initial separation and the ulterior union of percepts, we must uncover the underlying mechanisms, which happen to work in different though interconnected brain regions. And this calls for advanced neuroscientific research, as every mechanism is a process in some material system. In short, brainless psychology may describe a lot, but only biopsychology can explain in the sense of answering questions of the form "How does it work?"

Much of the same applies to nonhuman animals. For instance, the classical ethologists Konrad Lorenz and Niko Tinbergen had described imprinting, but only the neuroethologists endeavored to explain its underlying neural mechanism. At the other end of the spectrum, Frans de Waal discovered altruism in capuchin monkeys, but it is up to social psychologists to explain why it can overpower egoism.

In conclusion, psychology is indispensable to describe most of the mental, since it emerges from the living; but, far from being a separate and independent discipline, psychology belongs at the intersection of biology with sociology. Note that I wrote "biology," not "neurocience," because the central nervous system interacts with the endocrine and the immune systems, as suggested by the occurrence of psychosomatic phenomena, such as the placebo effect (see Bunge 2013a).

My Contributions to the Philosophy of Mind

I believe that my chief contributions to the theory and philosophy of mind have been the following:

1. An elucidation of the mind-body problem, as well as an updating of the main solutions to it proposed over the last 2500 years
2. A precise statement of the psychoneural identity hypothesis
3. The proposal of a precise and testable hypothesis about the peculiar trait of the neuronal systems (or networks) where mental processes occur, namely, neuronal plasticity
4. A set of arguments in favor of psychoneural identity and against its rivals inspired by the current scientific literature
5. Location of the mental and of the philosophy of mind in a materialist, systemic, and scientistic philosophical system
6. Location of psychology at the intersection of biology with social science
7. Ontological and methodological criticisms of the pseudosciences of the mental

The fifth point deserves comment because the science and philosophy of mind are usually cultivated in an ontological void, which contributes to imprecision and therefore to unending controversy (Mahner 2015). For example, the thesis that the mental is "supervenient" on the physical is imprecise because so is the notion of supervenience – unlike that of emergence (Bunge 2003d). Nor should one say that "brain and mind are one" because the dualists concur, but they deny that the unity in question is the same as that between an organ and its specific functions. This is one more example of the failure of the ontologies without things.

Coda

The immaterial soul is the earliest idea of the mental. Its rightful owners are the priests and shamans who invented it and make a living out of it. God and Satan have fiercely battled over souls from the beginning, and the ancient Egyptians set up the souls market, which flourished until the Enlightenment ridiculed it. Very few people believe nowadays that embalming or ordering masses for the departed will guarantee the eternal salvation of anyone's soul.

In recent times, even philosophers, the last incarnation of the priests, have lost faith in what spiritualists used to call the Great Beyond; and the scientists and technologists of the mental have come increasingly close to the materialist thesis that the mental is the specific function of the plastic region of highly developed and continually developing brains.

Still, old myths die hard: they tend to revive in fashionable garb. In particular, the immaterialist view of the mental is often offered as the belief that the mind is a set of computer programs or algorithms, which can be incarnated (or "instantiated") in people, computers, or ghosts, as the occasion requires. Since frozen minds can now be bought at computer stores, they have lost their former prestige – except when they have to be repaired. But even losing one's mind is still often regarded as deserving the attention of a soul expert rather than that of a medical specialist.

Note that the materialist conception of the mind involves an ontological reduction but does not threaten the methodological specificity of psychology, for this discipline continues to use techniques, such as conditioning and disentization, that do not belong in the biologist's toolbox. For example, the experimental study of reasoning, whether correct or faulty, calls for paper-and-pencil tests as well as for magnetic resonance imaging; and the clinical study of stress in humans requires garden-variety interrogation as well as measurements of corticoid levels.

Psychology is so many-sided and so many-leveled that one may still learn more from William James's *Principles of Psychology* (James 1890) than from many a recently published paper in a prestigious scientific journal. Clear, imaginative, and disciplined thinking about deep problems still counts more in science than mindless statistical correlation.

Psychology has long ceased to engage in isolated speculation on apparently anomalous phenomena: it has become *psycho-neuro-endocrino-immuno-sociology*. The fusion of previously separate ideas and procedures is not exclusive to psychology, but occurs in all the fields concerned with facts that happen simultaneously on several levels (Bunge 2003d). For example, the good student of economic behavior engages in social psychology, not in neuroeconomics, because he has to place economic actors in their sociopolitical context. Similar considerations disqualify neurolaw, neuromusic, neuropoetry, and even neuroneurology. Intermediate levels bite back when overlooked.

In short, contemporary psychology is a hybrid science that extends from biochemistry to neuroscience and from the latter to social science. As stated above, there has been ontological reduction (of the mental to the cerebral) along with methodological enrichment. In particular, contemporary psychology bridges biology and social science. The latter deserves a separate chapter, because human social facts involve brains but do not occur in them.

Chapter 13
Social Philosophy

The Great Depression

Like everyone else, I knew since early childhood that some people were poorer, and others richer than us. But I took such inequalities for granted, and they played no role in the soccer field or at school, where social differences were hidden under the compulsory white smock. In addition, we all spoke the Buenos Aires dialect, some plagued with more Italianisms than others. We also spoke clearly instead of mumbling. Moreover, speech was not a crucial social class indicator, the way it still is in Britain: among us the quality of speech was only an education indicator. But speaking correct Castilian was very unusual and regarded as an affectation.

It was only in 1929, when the Great Depression arrived from Wall Street with some delay, that I saw the first hungry men who called on us for work or alms. My mother and I listened to their stories, and offered them dishes of nutritious soup. And when the train approached the Retiro railways hub we saw the earliest shantytowns ('misery villages') grown almost overnight. In Chile they are called 'mushroom neighborhoods', in México 'lost towns', and in Perú 'human settlements'. Most of them are self-governing communities, where the police are neither needed nor welcome, and that thrive on cooperative trade, as my friend Larissa Adler's anthropological fieldwork has shown.

In the poorer neighbourhoods we saw idle men going around aimlessly in pajamas and espadrilles, seeking the company of fellow unfortunates and sipping *mate* tea to cheat hunger. This was often brewed from used leaves that had been layed out to dry. Jakob, Kathy's dour and silent husband, left every morning and returned at dusk as usual. We soon learned that, having lost his job, Jakob had been walking all day long looking for employment.

Enrique Santos Dicépolo's tango *Yira yira*, famously sung by Carlos Gardel, is a beautiful if haunting testimonial of that hard time, which brought not only hardship but also humiliation. Nothing is more humiliating than being told that one has become redundant. Many of the best novelists and playwrights around the world,

M. Bunge, *Between Two Worlds*, Springer Biographies,
DOI 10.1007/978-3-319-29251-9_13

especially in the USA, wrote in the 1930s about the Great Depression, a tragedy ignored by Borges and his escapist friends.

We also started to hear stories about the homeless who traveled from town to town in cargo carriages looking for work. At the same time, the chimneys of the two electric plants in Retiro released white smoke from the combustion of surplus wheat and corn. They had to burn foodstuff rather than give it for free to the hungry if the "law" of supply and demand was to hold.

The Great Depression hit big cities worse than the countryside, where most people had kitchen gardens and chicken coops. Al Capone, the most powerful and famous of gangsters, organized soup kitchens that dispensed free soup in Chicago streets. Not having read Alexis de Tocqueville, he feared that hunger would lead to a Communist revolution. In Manhattan, the philosopher Sydney Hook and his friends walked Fifth Avenue carrying posters reading "A nickel for an idea." And off-Broadway theatres showed tragedies about misery rather than the usual frivolous comedies. Nearly everywhere the dominant mood was one of gloom and doom. Not even the Daughters of the American Revolution dared mention the American Dream any longer.

USSR

There was one land, though, where everyone was working hard and optimism seemed to prevail: the Soviet Union, often referred to as the Sixth Part of the World. My father, along with all the progressive thinkers of his generation, had initially hailed the Bolshevik Revolution because it withdrew Russia from the world carnage and promised the emancipation of the working class. Most socialist parties split into two, and so did the Socialist International, although there was also an ill organized and ephemeral Two-and-a-Half International. My father sympathized with the latter and published *La hora*, a small and short lived magazine for the left wing of the Socialist party. When the Bolshevik emancipation promises were broken, most of the erstwhile Soviet sympathizers cooled down and turned their gaze to the USA, France, and Germany.

A decade later, just when the Great Depression struck, the Soviet Union started its transmutation from a backward rural nation into an industrialized power, at a pace never seen before. This feat drew the attention of everyone except for the economists, ever hiding behind their pseudomathematical formulas. This economic transmutation was being carried out in accordance with the first Five-Year Plan, started in 1929, and declared accomplished 4 years later. This success hardened the ferocious Stalinist dictatorship and prevented many people – among them my father and me – from seeing the monstrous sides of Stalinism, in particular the repeated purges, thought control, and anomie.

Even the respected social scientists Sidney and Beatrice Webb, cofounders of the London School of Economics and pillars of the Labour Party, announced to the world that a new civilization was being born. Their enthusiasm was so catchy, that my father studied all the relevant documents he could get hold of, and wrote a whole book, *El continente rojo* (1932). He visited the USSR 3 years later, taking along the young Elvio, aka Poroto, son of his friend Natalio Botana, the founder of *Crítica*, the most popular Argentine daily. My father was more impressed by the Russian students he met in trains, with whom he conversed in German, French, or English, than by the material gains. Even the Nazis admitted the success of planned economy, and set up their own Four Year plan to prepare Germany for war. This is how they cut down drastically on unemployment and rearmed speedily under the nose of the ex-enemy nations.

Only right-wing economists, such as Friedrich Hayek and Milton Friedman, denied the merits of planning, even though it was routinely practiced by big business and finance ministries. Those writers never asked themselves whether the Soviet army could have destroyed the German mincemeat machine without the Five Years Plans. Had it not been for its destruction, the skins of millions of us would have been transformed into lampshades.

In the 1980s I studied planning as a part of praxiology, or action theory. I quickly came to the conclusion that if one does not plan one's own activities, others will do it to their own benefit, and that the best plan is both participative and flexible, rather than authoritarian and rigid. By then it had become clear that Soviet-style planning led to inefficiency, fraud, and disinterest in public matters. If *they* make all the decisions, let *them* bear the burden of the state.

Political Concomitants of the Great Depression

There had been some economic crises since the seventeenth century, but they had all been regional and had had no momentous political consequences. Only Marx and Engels raised their glasses to the imminent social revolution every time the London Stock Exchange sank (Hunt 2009).

The Great Depression (1929–1939) was the first global crisis of capitalism. The communist parties gained some advantage from this circumstance, but they were so sectarian, that they kept attacking other leftist parties with as much violence as they fought the common enemy, namely fascism. By contrast, the radical rightists took full advantage of the crisis: they seized power, as in Argentina and Brazil, or mobilized the discontent with demagogic slogans such as "The Jews are to be blamed," or "We are the Third Way."

The newly formed fascist parties attracted the unemployed, as well as recently impoverished middle-class people, and were financed by industrialists, financiers, and big landowners. But in many countries, from China and Indochina to France and the USA, most intellectuals and artists veered leftward, whereas the fascist sympathizers were little more than a handful – Martin Heidegger, Giovanni Gentile, Ezra Pound, T.S. Eliot, Salvador Dalí, and Louis-Ferdinand Céline. In Argentina only a few writers, among them my uncle Manuel Gálvez, hailed the 1930 right-wing military coup. My father and other leftists and liberals, who had initially supported the new de facto government out of hatred for the elected but inept and corrupt one, repented as soon as it showed its fascist claws and its subservience to British interests.

From Juan B. Justo to Augusto Bunge to Raúl Prebisch

In the late 1890s, the neurosurgeon Juan B. Justo, aka Juan B., founded both the Argentine Socialist Party and the first cooperative. He was also the earliest Argentine sociologist (see Justo 1947). My father, who regarded Juan B. as his teacher, followed suit, and wrote a book-long essay on the economic inferiority of native Argentines compared with the European immigrants, who came as craftmen or as qualified industrial workers.

In 1920 my father taught a short course on this subject at the Economics faculty. His best student was Raúl Prebisch, a brilliant 19 year old provincial who was soon to become a close friend of the family, as well as my beloved lay godfather. One decade later Raúl set up the Central Bank and the Internal Revenue Service. In 1948 he joined the UN, and organized its Economic Commission for Latin America (CEPAL), whose mission was to assist the region's social and economic development (Dosman 2009). Raúl's political views veered to the left as a result of his firsthand study of those societies: he moved from free trade and foreign aid to protectionism and self-reliance through "importation substitution" (industrialization) and public works. Prebisch's writings were instrumental in replacing economic orthodoxy with the so-called neostructuralist theory, which emphasizes the division of the world into the dominant center and the dependent periphery. He did not hesitate to preface my attack on the ruling economic theory (Bunge 1982b).

Despite my father's interest in social studies, and his close friendship with Prebisch and Ernesto Malaccorto, Raúl's student and successor, as well as the frequent visits of a bevy of economists, I was far more interested in philosophy and astrophysics than in the social sciences. One of the factors for this bias may have been my sharing the belief of most Marxists, that the reading of their classics sufficed to understand all the world's societies without the help of new empirical research.

I started to change this absurd attitude in the mid-1950s, when I discovered that some people engaged in social research instead of commenting on old books. But, at least in my country, most of those students of social reality were marginal to both the academic establishment and to the leftist political movement.

Gino Germani and Serious Sociology

The state of social studies in Argentina changed abruptly in 1955, when Gino Germani – with the strong support of the interim rector José-Luis Romero – set up the Department of Sociology at the Faculty of Philosophy in Buenos Aires

University. The ruling military were alarmed. They need not worry, because Germani had fought the Communist fellow students with words and fists, and his agenda was not Marxist, if only because it was centered on research. In fact, Gino had a grand academic vision, namely organizing a group of original researchers who, adopting the scientistic ("positivist") viewpoint, would cultivate all the social sciences, from social psychology and social anthropology to sociology and economics. In Latin America all of these disciplines had been either neglected or, as Germani himself pointed out, severely distorted by the ruling philosophy, which was intuitionist and therefore antiscientific, in particular disdainful of quantitative data.

Germani (1911–1979), who in his native Italy had studied economics and suffered jail, came to Argentina in 1934, and studied philosophy while working at a private firm. He was a self-taught sociologist and social psychologist, the first in the country to study Talcott Parsons, the American sociologist of the hour, challenged only by Robert Merton. Gino had an enviable energy and faith in himself, and knew how to recruit collaborators and delegate power. Despite his authoritarian, arrogant, and somewhat rude manners, Gino was immediately surrounded by a swarm of collaborators and admirers – myself among others.

A decade later the military, those professional saviors of the fatherland, dismantled overnight Gino's admirable setup, and he had no choice but to move to Harvard, where, according to our common friend Karl Deutsch, he looked like a downcast beagle. When he visited us in Montreal, in 1967, I asked Gino why he had left a country where he had grown deep roots in the course of three decades. He answered with his usual frankness: "I left Argentina because I did not understand Peronism, and a social scientist who does not understand it cannot work there."

When Germani invited me to take part in the panel for the conference on science and positivism that he had organized, I assaulted positivism and thus provoked Gino's anger. I had not realized that, in that milieu, positivism was confused with scientism. This is the research strategy that assumes that everything, whether natural or social, is best studied by adopting the scientific method (Bunge 2014b).

Like his philosophy professors, Gino had swallowed that confusion but, unlike them, he sided with "positivism." I had read some of the genuine positivists, from Comte, Mach and Duhem to Reichenbach, Carnap and Philipp Frank, and had thoroughly criticized their attempt to interpret physics in anthropocentric terms, from sensation to measurement. This philosophical graft had resulted in the Copenhagen interpretation of the quantum theory, which I criticized in detail (recall Chap. 8).

That misunderstanding over the meaning of 'positivism' did not affect my collegial and personal relations with Gino. He asked me several times to take over his classes, and listened to my queries. When I asked him whether the concept of a social class played any role in his research, he answered that he preferred to use that of occupation, which is far easier to define and detect. I now think that we need both concepts and even a third one, that of social status or prestige, since occupation does not correlate strongly with either earnings or power. For example, in the USA plumbers may earn more than scientists, and Evangelist preachers wield far more political clout than either of them. In short, to analyze social structure we need the occupation-status-class triad.

When I asked Gino to recommend to me some recent sociology books, he started with Leo Festinger's *When Prophecy Fails* (Festinger 1956), which was just out, and Talcott Parsons's much older *Structure of Social Action* (Parsons 1937). The former was an amazing empirical piece of social psychology – a subject that my uncle Carlos Octavio had taught half a century earlier at the same faculty without using any data. As for Parsons's book, it disgusted me because of its holism and wordiness. But two decades later, when I discovered his *The Social System* (Parsons 1951), I learned to appreciate his work, in particular his systemism and his plea for the fusion of sociology with economics – a process that started to be institutionalized only half a century later, when the Society for Socioeconomics was founded, which I joined from its birth.

Gino was keenly interested in philosophy, and belonged to the executive committee of ARLyFC, the society for logic and enlightened philosophy I had founded in 1956 along with the logic professors Gregorio Klimovsky and Jorge Bosch, and the meteorologist Rolando García. Gino's conception of science was not as demanding as mine: he tolerated psychoanalysis and taught Moreno's psychodramas and sociodramas, precursors of social network theory and the pivots of family therapy – an amusing and profitable business. Gino's lectures on Moreno's inventions packed his classrooms but did not incite anyone to analyze any social networks, for instance the one constituted by the Argentine power elite (the big ranchers, exporters, and representatives of the foreign-owned railways and meat-packers). Such a network would have revealed an important change that happened ever since the military budget was drastically reduced in the 1980s: the military officers are no longer welcome in polite society, so their chance of marrying an heiress, in particular one bound to inherit a ranch, has decreased significantly.

Though neither a political player nor a public intellectual, Gino was an outspoken critic of the dictatorships of his time. He also taught me a capital lesson that was to become one of the pillars of my political philosophy: voluntary participation is of the essence of genuine democracy. Three decades later, I quantitated this concept and joined it with others to construct a numerical indicator of political democracy (Bunge 1985b).

First Publications in Social Science

Unbeknown to most of my philosophical readers, I have published in sociology since 1968. My first paper in this field (Bunge 1969g) was about human migration, and it includes a formula that exactifies the commonsensical hypothesis that the intensity of a migratory flux is proportional to the difference in living standards (in particular incomes) between the regions at stake. A defect of this formula is that it overlooks the strong distortions caused by migratory policies. However, these can be quantitated and included in the formula. For example, we know the large reward in free land that Canada paid to its immigrants a century ago, and the harsh terms it imposes nowadays on the would-be immigrants. In any event, the formula in question is corrigible, and scientists, being constructive, happen to prefer corrigibility to fallibility.

My next contribution to sociology was my mathematical analysis of the concept of social structure (Bunge 1974b). Peter Blau and a few others had dealt with this problem using only intuitive ideas. I started with the concept of an equivalence class, which generalizes that of equality. For example, in theory we are all equal, but in practice we are only equivalent in certain respects, such as occupation and income bracket.

Every equivalence relation induces the partition of a collection of individuals into a family of non-overlapping groups, like pizza slices. The overall structure of the original collection may then be defined as the heap of such "pizzas" cut by as many equivalence relations as desired. Once such a framework is constructed, it can be filled with social statistics: the number of people in every cell or set can be counted, and updated as time goes by. This style of theory construction is unusual in social science, which complies with the empiricist rule: data should prevail both at the beginning and at the end.

This paper served as the basis of another (Bunge and García-Sucre 1976), about social participation, marginality, and cohesion. The Venezuelan theoretical chemist Máximo García Sucre, a colleague of Kálnay's at the Instituto Venezolano de Investigación Científica – in which I lectured several times – visited often my Unit from 1974 on. Once, after one of my lectures in scientific metaphysics, I wondered aloud what holds people with different interests together in social systems of various kinds, and what does it mean that some of them remain at the margin.

The Marxist scholars were not helpful, even though they shared Rousseau's great insight, that inequality is the root of all the major social issues. With very few exceptions, they are not even familiar with the indicators of income inequality, such as Gini's, and they do not ask themselves whether participation plays any role in attaining and preserving cohesion. This very question may embarrass them because of their adherence to Marx's infamous proposal of the "dictatorship of the proletariat."

I joined the American Sociological Association in 1974, and since then have followed its two main journals, and have attended some of its meetings. In them I got to meet James Coleman, Harrison White, and Alain Touraine. At about that time I was coopted into the editorial body of *Theory and Decision*. Robert Merton, whom I had first consulted in 1955, and with whom I held an intense correspondence during the last decade of his life, recommended to me a number of publications, and praised my first book on social science (Bunge 1996a), which appeared the same year we met for the first time.

I also benefited from the advice of the Swedish socioeconomist Richard Swedberg, as well as of my McGill colleagues the economist Tom Asimakopoulos, the sociologist Axel van den Berg, the political scientist Michael Brecher, the anthropologist Bruce Trigger, and the political historian John A. Hall, who called Michael Mann, "the Weber of our time," which in my opinion was not a recommendation. I did not share Mann's (2004) definition of fascism as the rule of armed militias, for it led him to denying that Franco's regime had been fascist.

He overlooked the fact that Franco did no rely on the Falange because he had the support of the bulk of the armed forces. Since every one else regards Franco's regime as fascist, Mann should have changed his definition, adding the economic component, namely that it was the regime that most favored the ultra-wealthy. In addition, Franco's treatment of the underground union organizers was so harsh, that it provoked the rebuke of Heinrich Himmler, the ferocious SS Reichsführer.

Following John's advice, I wasted a couple of years reading Weber in German. It took me a while to realize that Weber was far more interested in the past than in his own time, in religion than in science, and in the wealthy than in the poor. I also learned that the thesis which earned him fame – that the Protestant ethic had bred capitalism – was utterly false, if only because capitalism emerged before the Reformation in Italian republics like Florence, not in Luther's or Calvin's lands. I suspect that Weber has been placed on a high pedestal only because he was taken to be superior to both Marx and Durkheim, while in fact he was a basically historical scholar, whose only merit was his plea for the fusion of sociology with economics.

I have discussed my sociological essays in Aarhus, Buenos Aires, Dubrovnik, México City, Monterrey, Montreal, Sydney, and Zürich, though with no useful feedback. I was booed at the Mexican Institute of Demography for representing the history of a social system as a trajectory in its state space, and a sociologist at the University of New South Wales accused me of being "wedded to the numerical paradigm." My writings in the philosophy of social science fared much better: they were published or reviewed in prestigious venues.

Game Theory

Game theory, invented around 1940 by John von Neumann and Oskar Morgenstern, pivots around the concepts of conflict, cooperation, and subjective utility, while ignoring those of social system, power, privilege, inequality, participation, monopoly, exploitation, poverty, violence, and solidarity – as if all the "players" were equal and free, as well as devoid of feelings, convictions, and commitments.

A typical "game" is the famous Prisoner's Dilemma, wherein an incarcerated individual is free to be either loyal or disloyal to his former accomplice, now locked up incommunicado in another cell. If the first prisoner is disloyal to his fellow prisoner – that is, if he confesses and rats on his former accomplice – he gets to gain, whereas if he is loyal he is punished with a longer sentence.

The game-theoretic enthusiasts claim that such a simple "game" applies to all social, economic, political, and cultural issues, regardless of previous commitments and foreseeable consequences. But in fact it has never explained anything. Nor can it possibly account for real events, for it deals only with knaves free to do what has to be done to maximize their expected gains or utilities, overriding moral precepts as well as social constraints and future sanctions. No wonder that it cannot explain why patriots fight harder than mercenaries. The theory overlooks all that distinguishes normal people from professional delinquents or even from automata. Maybe it should be rechristened *knave theory*.

In his famous *The War Trap*, Bruce Bueno de Mesquita claimed to explain all wars using game theory (de Mesquita 1981). To this end he made up the probabilities of winning and losing, as well as the corresponding utilities and disutilities, but did not warn his readers that his numbers were invented, not found through empirical research. I plead guilty of having preceded him with my paper on the American war in Vietnam (Bunge 1973c), where I accused the American leaders, like John McNamara, of misestimating the probabilities and utilities in question. *Mea culpa*! I hope to have corrected this mistake in my ulterior críticisms of all "rational choice" fantasies, which ignore the constraints placed on individual decisions by the power elites (Bunge 1989d, 1999a, b). But my criticisms will make no dent on the admirers of the John Nash portrayed by the movie *A beautiful mind*.

Development Theory

In 1974 I was invited to take part in the workshop on development held in Paris by UNESCO. This group was constituted by sociologists like Johann Galtung, politologists like Karl Deutsch, and a prominent Yugoslav economist. We were all interested in social indicators, a topical subject since the mid-1960s, but almost all of us were new to this field. So much so, that a young American sociologist felt obliged to teach us the ABC of the subject. He mentioned a number of clear examples but failed to define the very concept of a social indicator. I took upon myself the tasks of defining the concepts of indicator, social indicator, and development indicator. My presentation included a criticism of the ruling idea that GDP (Gross Domestic Product) was the best development indicator, but its highlight was my proposal of a less biased measure (Bunge 1974c, 1981c). In my view a good development indicator had to be just as integral or many-sided as an integral conception of development: biological, economic, political, and cultural.

An official of the Inter-American Development Bank told me that some people at the UN took notice of my proposal. In 1990 the UN finally abandoned the GDP as the development indicator, and adopted its own Human Development Index, which is a contraction of mine. The new UN index is an average of three parameters: GDP, schooling years, and longevity. It includes neither income inequality – which does not compensate for high GDP – nor political development, which I regard as voluntary participation in civic activities. The latest version of my development indicator (Bunge 2009) is a refinement of the previous ones; in particular, it includes a gross estimate of sustainability.

UNDP Seminar in México City

The UN Development Program convened a seminar on development, presided over by Gabriel Valdés, which was held in México City in 1979. There I met for the first time two pioneers of molecular biology: the Spaniard Severo Ochoa – who two decades earlier had shared the Nobel Prize with Arthur Kornberg – and the Chilean Jorge Allende, an enthusiastic promoter of science education throughout Latin America.

My presentation emphasized the critical role of science among the development mechanisms. A country without research scientists remains culturally and industrially backward; it copies what may no longer be true or useful; it does not train good teachers; and it is easy prey to superstition. The USA, Germany, and Japan became world powers in less than half a century thanks largely to their vigorous support of science and the concomitant technological advancement, while other countries, in particular the Ottoman Empire, Russia, Spain, Portugal and their former colonies remained stagnant.

My presentation was drawn from the 4th volume of my treatise, that had just been published, and it became the core of a new book: *Ciencia y desarrollo* (Bunge 1980f, 1997c, 2014c, d). This book was pirated in several Latin American countries, and translated into Portuguese. Presumably, it did nothing to immunize against the antiscience plague that was spreading at the same time. That meeting was troubled by an official complaint of the Argentine military dictatorship, which upset a well-known physicist whom I had taught in the 1940s, and who got very excited, and threatened to leave if the democratic component of development was mentioned. He feared for his daughters, who had remained in Argentina after he left for Brazil.

Bento Spinoza and Marcel Roche

At the end of the PNUD seminar I travelled to Jerusalem to attend the meeting convened by the Institut International de Philosophie to honor Spinoza, the first modern Jew. Since no one mentioned Spinoza's key identity *Deus sive natura*, I reminded my fellow symposiasts that this principle had not escaped his contemporaries. They realized clearly that its author was a naturalist, hence an atheist as well. Jonathan Israel (2001) emphasizes this point in his recent books on the radical fringe of the French Enlightenment.

Back in Montreal I typed my book *Ciencia y desarrollo* (1980b), graced by a preface written by Marcel Roche, the director of IVIC (Instituto Venezolano de Investigaciones Científicas), and one of my most interesting friends. Marcel had studied medicine in the USA, and had conducted research in endocrinology and parasitology – the latter in hopes of helping the poor – until he realized that science is insufficient to solve social problems.

Marcel had attended Catholic schools in France and in the USA, but lost his faith when becoming a scientist. In his 1996 autobiography he warned that, if he gave signs of religiosity while agonizing, they should be attributed to deficient oxygen supply to his brain.

Marcel had a wide culture, played classical music on his cello, and was a *bon vivant*. Thanks to Flor, his wife, I read J.M.G. Le Clézio long before he won a Nobel Prize. The characters of this great novelist, all humble and generous, live in exotic places like the Sahara desert, the Panamanian forest, the island of Mauritius – or the Parisian *bidonvilles*.

Roche worked hard to organize and support the Latin American scientific community. He set up an institute of biomedical research, founded the *Interciencia* journal, and presided many years over the IVIC, refuge of many Argentine scientists, among them my student and coworker Andrés Kálnay. IVIC was also the scientific basis of the theoretical chemist Máximo García-Sucre, who visited my Unit a couple of times, and with whom I wrote several papers.

Between the late 1960s and the late 1980s I talked at IVIC on various subjects. At the end of one of my talks, Roche asked me what I thought about Claude Lévi-Strauss's structuralism, very popular at the time. I confessed that I had only read his early *Tristes tropiques*, which was not a theoretical work. Several years later, when its author had renounced structuralism, I learned that its central thesis was that societies are like languages. This absurd glossocentric thesis is being currently revived, though of course not in the serious sociological publications.

Repercussion of My Thesis on Integral Development

National development is not only the goal of the UN Development Program, but also the object of the attention of the economists who equate development with economic growth as measured by increase in the GDP. One afternoon of 1979, in La Paz, Bolivia, I walked to the Centro de Estudios del Desarrollo, but only the janitor answered the bell. That failed visit was part of the mission that the UNDP had entrusted the Chilean Guillermo Ramírez and me: to visit laboratories in the Andean nations to foster their cooperation with similar organizations in the region. Ramírez, a high-ranking civil servant, was competent and friendly, but he smoked all the time, so we travelled in separate compartments.

Our first stop was Caracas, where I gave a talk to a meeting of scientists, and Ramírez and I exchanged ideas with the officials of the scientific research agency. Their boss thought that it would be faster and cheaper to import ready-made science. He did not understand that science is not a thing, like a barrel of crude oil, but an activity, like playing soccer, and that a productive science center yields not only new knowledge but also valuable subproducts such as science teachers and technical consultants to both the state and the private sector.

From Caracas we flew to Bogotá, where we met with officials of Colciencias, the state organism that, in addition to subsidizing basic research, has its own enterprise of "appropriate" (locally feasible and useful) technological innovation, Las Gaviotas. Fernando Chaparrro, the able scientific director of Colciencias at the time, summoned a group of scientists to converse with me. Among them were the mathematician Carlo Federici and the neurologist Fernando Rosas. (Federici was the last member of the once-famous Italian school of mathematical logic, starved to extinction by the fascist minister of education, the neo-Hegelian Giovanni Gentile.) I suggested they set up the Colombian Association of Philosophy of Science. This was done, as I confirmed during my second visit to Bogotá, a couple of years later. However, three decades later I found no traces of that organization – a case of sowing in the sea.

On a Sunday morning my lecture on the mind-body problem filled the auditorium of the Academy of Medicine. I reported on some recent findings in cognitive neuroscience, and entertained a number of questions. One of them was "What is love?" It took me by surprise, for the serious study of love had been initiated by Harry Harlow only two decades earlier, and even today it is not in mainstream social psychology.

In Quito, Ecuador, we visited a biology laboratory, where we met two researchers who were repeating in their homeland the experiments they had learned as doctoral students in the USA. One of them was on the effect of the curare poison on the muscles of a rodent, and the other on the foraging habits of humming birds. I don't know what impact, if any, these experiments had. I learned far more from a book on biological underdevelopment by the local biomedical researcher Varea-Terán. Ramírez and I had long arguments with the economists in charge of development, all of whom believed that it was wasteful to spend money on basic research. They attempted to bribe us by putting a car and driver at our disposal. Seeing that our arguments had made no dent in their fortress, we took the bribe and revisited some of the exotic places I had first visited with Marta in 1962, when we saw a strong street demonstration of Indians demanding the "abolition of feudalism."

Our next stop was Lima, where we shared a modest hotel with a cloud of houseflies. There we talked to the bureaucrat in charge of Peruvian science, who kept a picture of Saint Rosa of Lima on his desk. The man was more interested in showing off his acquaintance with the patrician family of my friend Francisco Miró-Quesada than in science. Still, he got us an invitation to visit the state laboratory of marine research, where we learned a lot about krill.

The day's question was the sudden crash of the sardine population and the consequent ruin of the fishing industry. The marine researchers were baffled, but none of them did more than talk about the crisis. The solution came a few years later from an unexpected quarter: the oceanographers discovered the recent deviation of the cold and nutrient-rich El Niño current in the Pacific Ocean. This showed once more the futility of the sectoral approach to systemic problems. The ocean does not take notice of the division between physical and biological oceanography.

The Universidad Mayor de San Marcos, the oldest in South America, invited me to give a talk. I spoke about Einstein and his work to a vast public, and told the scientist's reply to my request for permission to collect and translate his complete work: "That would not be worthwhile, because most of my findings have been superseded or are no longer topical." One of the attendants asked me whether I had renounced some of my own ideas. The list of my mistakes was so long, that I did not know what to reply. In addition, optimists, like myself, tend to forget their failures.

From Perú to Chile via Bolivia

Landing in La Paz, at nearly 4,000 m above sea level, is a frightening experience because its rarefied atmosphere provides only a weak buoyancy to the aircraft. We went to San Andrés University, where I had taught briefly in 1955 as a UNESCO guest. We visited the chemistry lab, whose director could not believe that chemistry had a theoretical branch in addition to the experimental one. The man was in need of a course in the history and philosophy of science, which would teach him that all chemical experiments, however modest, are designed with the help of theories.

The laboratory was full of precision instruments lovingly protected by plastic sheets. They were out of order and no one knew how to repair them. This is a common occurrence in Latin America, where inexperienced researchers, or even enterprising bureaucrats, set up laboratories without the support of maintenance and repair workshops. At one point we were joined by Marcelo Robert, an international official, who had compiled an interesting catalogue of modest but profitable technological innovations, mainly in agriculture, due to Latin Americans (Robert 1972).

The last leg of our mission was Chile, then suffocated by the ferocious Kissinger-Pinochet dictatorship imposed in 1973 with the blessing of Milton Friedman and Frederick Hayek. On landing, we were photographed by a number of policemen in civilian clothes. We were invited to visit a single laboratory, and talked only to General Manuel Pinochet, a cousin of the supreme dictator and boss of my mission colleague.

Chile had withdrawn from the Andean Pact but still belonged, at least nominally, to the Andrés Bello Treaty on cultural exchanges with the other Andean nations. I suggested to the General that his country stay in this Treaty, so as not to weaken even further the faint cultural bonds among the Andean nations. The General made no promises, but took advantage of the coffee break to express his contempt for the Caribbean *negrada* (nigger rabble). I replied: "The Caribbean nations may have all the shortcomings you mention, but at least they do not suffer military dictatorships." The General and his subordinate froze, and we hurried to say our goodbyes with icy courtesy.

Our UNDP mission did not accomplish anything: the Andean governments did not strengthen their scientific cooperation, and kept economists in charge of development – which is like asking the fox to mind the chicken coop.

Fortunately, a few unofficial physical and biological associations kept their Latin American meetings going. And some distinguished scientists, like the Mexican nuclear physicist Jorge Flores Valdés, the Chilean molecular biologist

Jorge E. Allende, and the Argentine cognitive neuroscientists Facundo Manes and Mariano Sigman, have organized a number of Latin American scientific congresses, and have brought science to the public. The sociologists of science should study these fruitful activities instead of repeating constructivist-relativist nonsense imported from France or Britain.

Opposition to Integral Development

The idea that, because big social problems are many-sided, development policies should be systemic rather than sectoral, was at odds with the two ruling dogmas in science policy: the economism that places technology behind basic science, and the anti-scientism preached by both right-wingers and the New Left that arose in the 1960 decade, first in Berkeley and then in Paris and its cultural dependencies. This obscurantist ideology claimed that basic science had to be stamped out because it was either in the service of foreign scientists or the handmaiden of the industrial-military complex.

This primitive reasoning was popular among the humanities and "soft science" chronic students, postmodern Marxists like Louis Althusser, and failed scientists like Oscar Varsavsky. The latter, whom I got to know quite well while we were colleagues at the Buenos Aires University Science Faculty, was a talented but unoriginal and ambitious dilettante who had jumped from chemistry to physics to mathematics to economics to development policy, without ever attending scientific meetings, publishing in refereed journals, or training productive investigators.

In Caracas, where he expatriated himself after the 1966 military coup, Varsavsky landed a good job at CENDES, the development studies center. There he produced two pieces. The first, written with Carlos Domingo (Domingo and Varsavsky 1967), was a mathematical model of Thomas More's utopia. It was just a *jeu d'esprit*, utterly useless to understand contemporary society. Varsavsky's coworker, Carlos Domingo – a former physics student of mine – was an enthusiastic parachutist best known for this anecdote. Once, after jumping from a high altitude, the catch of his parachute got stuck, and Carlos went into free fall for a while. Asked what he thought during that terrifying experience, he replied: "Damn! The boss will kill me if I miss my 3 o'clock lecture!"

Varsavsky's second piece was a vitriolic attack on basic science (Varsavsky 1969). The chief villain in this diatribe was a paper published in a prestigious scientific journal such as *Nature*. In the same pamphlet he also proposed a "rebel science," though he did not explain what this was, nor did he construct a sample of such a marvel. He had nothing to say about such key sociological concepts as those of social system, social structure, participation, marginality, or anomie. Nor did Varsavsky explain how "rebel science" differs from pseudoscience, let alone how it could help development. And yet such poison is compulsory reading in certain Latin American universities. Not even orthodox economists have dared mount such frontal attacks on basic science: at most, they warn – against all evidence – that it is economically unprofitable.

Dissatisfaction with Orthodox Economics

I was indifferent to economics until about 1970. Moreover, I had felt a superstitious respect for mathematical economists because John von Neumann, who seemed to use powerful mathematical tools, was among them. But I had never read any of them. No sooner did I muster the courage required to tackle mathematical economics, towards the end of the 1970s, than I was disappointed by its shallowness, sloppiness, lack of empirical evidence, and blindness to severe disequilibria such as chronic unemployment, inflation, and stagflation.

I soon realized, in fact, that contemporary orthodox economics treated its hallowed market as a structureless set of sellers and buyers of goods dropped from heaven rather than produced by workers; that it pretended that traders paid more attention to preferences and decisions than to needs and constraints; and that it assumed that the market players are free agents rather than members of systems or networks that have preceded them; that political power and moral rules played no role in market transactions; that the market is always in equilibrium (supply = demand), so that it never needs help from the state; and that the natural resources are inexhaustible.

I also found eventually that the orthodox economists use without hesitation the ill-defined yet key concepts of subjective probability, subjective utility, and risk; and that nobody, in the course of a century, had subjected the theory's assumptions to experimental tests. Could this be a scientific theory useful to manage private enterprises, cooperatives, or government agencies? The preceding doubts occurred to me while reading Keynes's *General Theory* (Keynes 1936), as well as technical articles in the journals recommended to me by my friend, the Argentine Canadian Jorge Niosi, who teaches at the Université du Québec à Montréal: journals such as *American Economic Review*, *Economic Journal*, and *Journal of Economic Literature*. I also learned from my colleague Athanasios Asimakopulos, a disciple of Dame Joan Robinson, Keynes' successor, and just as interested in the philosophy of economics as his famous mentor. My conversion from indifference and readiness to believe, to keen interest and heterodoxy was accelerated by reading a few socially insensitive economists like the much-lauded Milton Friedman, a Pinochet fan.

Economics and Philosophy

The first time I presented my half-baked objections to economic orthodoxy was at the colloquium held in 1980 at Essex University. This meeting was attended by a number of mathematical economists, among them Sydney Afriat, Geoffrey Heal, Graciela Chichilnisky, and Michio Morishima. Sydney, as witty as sharp, worked on subjective utility functions, which I regarded as fictitious, but we agreed that Kenneth Arrow's social choice theory, as well as Amartya Sen's elaborations of it, were useless; Heal was starting to try and put prices on natural blessings such as clean air and water; Graciela – who could not believe that Milton Friedman was evil – paraded her mathematical sophistication; and Morishima, always surrounded by Japanese disciples dressed in black, gave us a hint of his brave attempt to join Marx with Walras and Keynes.

It seemed to this newcomer that, aside from Morishima, nearly all the participants took economics as a pretext for doing mathematics, since none of them tackled any topical economic issues. In any case, at that meeting I did not learn any economic laws, but Sydney taught me something about Pareto's distribution, and a Turkish professor – who had read some of my work while studying at Oxford – polished my version of the law of diminishing returns.

I took this opportunity to visit the huge Norman castle of Colchester, as well as the town of Lewes. The latter disappointed me for not keeping any traces of the polymath George Henry Lewes, who one century earlier had introduced the concept of emergence, and had been George Eliot's loving common-law spouse.

Joan Robinson's *Economic Philosophy* (1962) was perhaps the finest work in this field since John Stuart Mill's 1848 *Principles of Political Economy*. Although it offered a devastating criticism of the neoclassical dogma built one century earlier by Jevons, Menger, Walras, Marshall, and Pareto, I felt that it had to be updated to cover the neoliberal counterrevolution, so I wrote *Economía y filosofía* (Bunge 1982b).

This short book was prefaced by the UN economist Raúl Prebisch, my lay godfather. In that book I criticized in particular the use of subjective utilities and probabilities in microeconomics, as well as Milton Friedman's abuse of programmatic hypotheses. These are conjectures of the form "Variable *Y* depends on variable *X*," which are vague, hence untestable. At best they invite research projects of the form "Find the precise form of the dependence of *Y* on *X*." That, is, find f in "$Y = f(X)$."

My criticism may be summarized as follows. Standard economics is built with vague concepts, it lacks empirical support, and it is useless to both business practice and economic policy-making. In particular, it cannot help weather economic crises because it postulates that every economy is always in equilibrium (Debreu's theorem). In addition, the theory's devotees are indifferent to the sufferings caused by poverty, inequality, unemployment, and crises.

Most economists would be shocked by the recent finding (OECD 2014) that income inequality is the enemy of growth – a finding ignored by the economists who preach austerity to the Greeks who are currently debating whether they can afford the euro.

When my book appeared, the Catalonian economist Alfons Barceló wrote to me making some justified criticisms, although we agreed on matters of principle. He also recommended that I read Luigi Pasinetti, who helped, as well as Piero Sraffa – whose text I found as inscrutable as boring – and Albert Hirschman, always insightful about ideas, but seldom helpful to deal with economic realities. With Alfons we also discussed some of the problems of cooperative firms, among them the one noted a century earlier by Juan B. Justo (1947/1909), the founder of both cooperativism and socialism in Argentina: the coop ceases to be self-managed if it grows beyond the point that its managers lose contact with the membership. However, to my amazement, Alfons was uninterested in coops, whereas to me they are the heart of genuine socialism. Too bad if the founder of Mondragón, the flagship of Spanish cooperativism, had been the Jesuit priest José María Arizmendiarreta, not a left-wing activist.

Prince of Asturias Prize

In the Fall of the same year 1982, Prince Philip Borbón handed me the Prince of Asturias Prize for the Humanities and Communication, the most coveted prize in the Hispanic world. To attend the ceremony, presided by the royal family, I had to get a new suit and shoes. And because I was wearing new shoes, I tripped when climbing the little staircase leading to the stage of the Campoamor theatre, where I was to shake hands with Prince Philip, a nice boy who the year before had made friends with Eric.

After the ceremony, held in the old Campoamor theatre, Marta and I chatted with a number of interesting people, among them the philosopher Ferrater Mora, the chemist Severo Ochoa, the biomedical researcher Luis Leloir, and the economist and novelist José Luis Sampedro, who amused us with some good jokes. My favorite was this: Upon arriving in heaven, a village priest asks to talk to Saint Paul. He is so insistent that finally he is taken to meet the saint, who asks him what he had in mind. "Only one thing, Your Sanctity: What did the Corinthians reply?" On another occasion, Marta asked Leloir, the eminent physiologist, why he had supported the Argentine military dictatorship. Leloir, with his usual calm, replied: "We were all culprits."

Ferrater introduced us to the excentric writer Jesús Aguirre, a former priest who had married the Duchess of Alba, an ugly British aristocrat. (What a comedown for the house of Alba, whose head under Philip II had terrorized and looted the Netherlands!) We also drove to a seaside village, where we lunched on delicious grilled fish and almond tart while listening to the sea waves crashing against the wall.

Not everything was fun that time. As soon as I received Miró's commemorative sculpture, the king and queen congratulated me and excused themselves, for they had to return to Madrid to face the last military conspiracy of their reign. There was a moment of anguish, when several hundred participants in the ceremonies congregated in the courtyard of the beautiful Reconquista hotel to listen to the latest rumors. Fortunately the conspirers, a bunch of Francoist nostalgics, turned to be as inept as reckless.

Meeting for Democracy

In April of 1983 I participated in the *Encuentro en la Democracia*, a meeting of Hispano-American politicians, intellectuals and writers that started in beautiful old Toledo and ended in Madrid. There I met, after half a century, my godfather the economist Raúl Prebisch, and his second wife, the charming Chilean lawyer Eliana Díaz. With Raúl we reminisced the sad episode of his break with my father, who had reapproched him his collaboration with the government of the day. When saying good-bye to me, Raúl had told me: "I'm not a politician. I'm just a technician." Half a century later in Toledo, he told me: "I was wrong. It is not true that economic technicians are politically neutral. We are part of the political system."

In Madrid I talked to King Juan Carlos, who showed interest in my proposal of a Hispano American Confederation. The idea of mediating in inter-American conflicts attracted him. I also had long conversations with my old friends the historian Nicolás Sánchez Albornoz and the metallurgist Jorge Sabato.

JoLu

The greatest novelty for me was to meet and instantly befriend the diplomat José Luis Pardos, director of the Cooperation Department of the Spanish Ministry of Foreign Affairs. He got enthusiastic about my little book *Ciencia y Desarrollo*, and later on about my project for a Hispano-American Institute for the Sciences of Sciences. We launched this Institute at a meeting held at Alcalá de Henares University in the presence of Queen Sofía and the Minister of Foreign Affairs. This project was supported by a number of important Spanish institutions, but it foundered suddenly after a political change. Some years later my good friend Miguel-Angel Quintanilla rescued it at Salamanca University.

We became close friends with José Luis aka JoLu and his wife Mercedes, whom we saw more often when they came to Canada as ambassadors. Shortly before their arrival, JoLu phoned me from Canberra asking for advice. I strongly advised him not to get entangled with the Spanish fishermen who were looting the dying Canadian fish banks. Sure enough, his first task was to try and free the fishermen that had been caught red-handed and were jailed in Newfoundland. When he traveled there to visit them, José Luis had to confront a crowd of angry Canadian fishermen who pelted him with eggs. By contrast, in the fishing harbors of Galicia, the homeland of the cod pirates, JoLu was hailed as a hero. Such is life nowadays for diplomats: they are seldom if ever summoned to resolve international conflicts in a civilized way.

In the fall of the same year 1983 I returned to Oviedo, taking Eric along, and this time as a member of the jury to choose the next year's Prince of Asturias winner in the Humanities and Communication. I proposed the daily *El país*, the only Spanish newspaper in the same league as *Le Monde*, *The Independent*, and *La Repubblica*, and because of its key role in the transition from fascism to democracy. My candidate won.

This time, again at the same hotel, I got acquainted with other interesting people, in particular José Ortega Spottorno, aka Pepe, son of the famous philosopher. Though trained as an agronomist, Pepe became a great journalist and editor of about 2,000 books. He was brimming with anecdotes and jokes, and we met several times. Once Pepe invited me to dine in the company of the linguist Antonio Tovar, who had acted as interpreter in the famous encounter of Franco with Hitler at Hendaya, in the Pyrennees. Later Tovar had pioneered the scientific study of the Toba aboriginals in the Chaco region, who had been neglected by the Argentine anthropologists and linguists.

Making Friends in the Social Science Community

My social studies have gained me more friends than my philosophical work. Here are some of them: the Catalonian economist Alfons Barceló and his medical wife Marimar; Amedeo Amato, director of the Instituto di Economía at Génova university, who was my host between June 1993 and March 1994; the Canadian economist Athanasios (Tom) Asimakopulos, a member of the Cambridge school, who read critically my early essays on economic theory, and with whom I shared a graduate student, Martin Cloutier, whom he received in his death bed; the eminent sociologist Robert K. Merton, the father of the sociology of science, who advised me, on and off, during half a century; the Swedish-British criminologist Per-Olof Wikström, who invited me to inaugurate the symposium held at Cambridge University in 2005 (Bunge 2007a, b); and the Taiwanese sociologist and philosopher Poe Yu-ze Wang, author of *Reframing the Social* (2011), largely devoted to my social philosophy.

Another great friend has been the German Canadian Andreas Pickel, who teaches sociology and political science at Trent University in Peterborogh, near Toronto. Andreas visited me to discuss my 1996 and 1998 books, which had led him to abandon Popper, his first hero. Since then we have interacted regularly, discussing problems, like the communism-democracy transition in Europe, that are still waiting for a satisfactory solution, perhaps because each student of them has focused on a single side of the polygon, and has neglected the national peculiarities of the historical background.

Andreas has written a flattering report on my social philosophy (Pickel 2001), and has organized two published symposia on the same subject (Pickel 2004, 2007), to which a number of European scholars have contributed. Unlike most North Americans who have written on Cuba and Venezuela, Andreas learned Spanish so as to be able to interact with nationals of both countries on their own turf. We last met in Vienna and Madrid.

Heinz Droste, another German social scientist, has written extensively on public relations, and has become interested in my philosophy to the point of writing a semi-popular book on it (Droste 2015), which I find both clear and faithful. Another good friend with whom I have exchanged ideas about the social, the political and the legal, is Antonio Martino, a pioneer in digital law, whom I first met in 1994 in

Pisa, at whose university he taught for several decades. His passion is the legal digest à la Justinian – a clear case of a conceptual system – on which he has written a book that he asked me to preface (Martino 2014).

Oscar Defante deserves a separate paragraph because he is a man of culture but does not live off it: he is a businessman who produced power tools until free-trade put him out of business. Oscar is a music lover, meets regularly with interesting intellectuals and statesmen, and has organized several talks of mine in his native Rosario despite the boycott of the local philosophers.

Last, but not least, comes the publisher Serafín Senosiain, based in traditionalist Pamplona, of fighting-bull runs and Opus Dei fame. He is publishing Spanish translations of the main works of the French Enlightenment, such as those of Holbach and his friends, which had been banned in Spain and consequently profited Dutch publishers and book smugglers. He is also the publisher of the Biblioteca Bunge and of a series of books criticizing the pseudosciences as well as postmodern nonsense. Serafín is what Spaniards call a white housefly: an independent publisher who, as Voltaire might have said, has courageously adopted the *parti pris de la vérité*.

Coda

My contributions to sociology and the philosophy of social science may be summarized as follows.

1. A defense of the scientific study of all the facets of human existence, both individual and collective.
2. A realist, materialist and systemic view of society and social issues, that transcends the individualism/holism dilemma.
3. A mathematical analysis of key sociological concepts, chiefly those of social structure, solidarity, participation, marginality, and cohesion.
4. The introduction of the concept of integral development, to replace the one-sided equation of development and economic growth favored by most economists with a handful of exceptions, such as Gunnar Myrdal, Raúl Prebisch, and Amartya Sen.
5. An elucidation of the concept of social indicator, and the proposal of special indicators, such as those of feminine emancipation, participation, democracy, and overall (or integral) social development.
6. An indictment of the ill-defined concepts of subjective utility and subjective probability, which are central to economic orthodoxy.
7. The denunciation of the aprioristic hypotheses of rational egoism, free market, and market self-correction.
8. Criticisms of some naive, confused, false, moth-eaten, or antisocial social doctrines and policies, from economic orthodoxy to rational-choice theories to Marxism.

9. Criticisms of one-sided views of human nature (biologism, psychologism, sociologism, economism, and culturalism), and a plea for the fusion of all the sciences of man.
10. A denunciation of the ideologies, from neoliberalism to communism, which are often passed off as scientific theories and have made uncounted victims the world over.

Chapter 14
Weird Ideas, Strange Places, and Freak Events

Like everyone else, I have had my share of unsettling experiences along with moments of emotional and intellectual ecstasy. In the following I will recall only a small sample of unusual experiences, such as the Montreal Ice Storm of 1998, my first exposure to postmodernism, the fall of Australian philosophers into the many worlds pit, the detection of outstanding students, and my recent exchanges with Chinese colleagues. I will also mention a few of my thrilling moments while reading about some scientific findings, or having certain philosophical and political epiphanies.

Retreat from Reason in the Humanities and Social Studies

The Second World War, just like the First, dashed many illusions and bred many delusions. Its main intellectual casualty was the Enlightenment philosophy, in particular its trust in reason as a powerful tool to discover and master reality. Indeed, in the humanities and social studies, rationality, realism, intellectual honesty, and hard work were demeaned by hordes of radical skeptics, irrationalists, psychoanalysts, and other cultural arsonists. The 1950s were propitious for the flourishing of cut-rate scholarship: the universities, formerly elitist, became massified. Overnight tens of thousands of teaching positions opened in all fields and many countries. In the humanities and social sciences, many of those jobs were assigned to half-baked individuals who had never done any serious research, so they tended to avoid serious problems. As my chairman told me when I declined to teach the history of science because I lacked expertise in this field: "But you know how to read, don't you?"

The more timid among those deviants from the scientific path confined themselves to exalting Nietzsche (the case of Foucault), misinterpreting Marx (like Althusser), imitating Heidegger (like Derrida), reheating biologism (like Dawkins), or ridiculing the aspirations of scientism (like Hayek). The bolder ones built new

M. Bunge, *Between Two Worlds*, Springer Biographies,
DOI 10.1007/978-3-319-29251-9_14

academic niches, such as feminine philosophy, pop evolutionary psychology, queer sociology, Austrian economics, memetics, and narrative medicine. The retreat from rigor, depth, and originality became particularly pronounced in the philosophy and history of science and technology, which the Bomb and Sputnik had pushed into public attention. The Dalai Lama offered his opinion about quanta, and President Clinton informed the world that the sequencing of the human genome had revealed the essence of humankind. Paul Feyerabend's slogan "Anything goes" was used to pass off nonsense for wisdom.

From Merton to Kuhn

Up until the 1950s, the study of scientific activities and communities had been a task of philosophers, sociologists, and historians of science intent on finding truths about science, that much-celebrated yet still elusive animal. Suffice it to recall the studies by Henri Poincaré, Émile Meyerson, Federigo Enriques, the Vienna Circle, Karl Popper, Morris R. Cohen, Ernest Nagel, Richard Braithwaite, Eino Kaila, Aldo Mieli, George Sarton, and Robert K. Merton.

Merton was not an amateur but a professional sociologist who had studied with the most famous sociologists of his time – Sorokin, Parsons, and Lazarsfeld – as well as with the chemist, biologist and sociologist Lawrence J. Henderson, who had rescued and spread the concept of a social system at a time when individualism ruled in the social studies. In addition, thanks to Harriet Zuckerman, his student and wife, Merton met personally with a number of Nobel laureates, who told him what had motivated them, what they had done, and how their respective scientific communities had now supported, now inhibited them. Besides, unlike most other social students, Merton was politically alert and outspoken.

In his classic 1938 paper "Science and the social order," published in the new-born journal *Philosophy of Science*, Merton argued that the peculiarities of basic science are *disinterestedness*, *universality*, *epistemic communism*, and *organized skepticism* – not the doubt of the isolated researcher but critical scrutiny by a whole community (in Merton 1949/1968). Merton was not a dilettante but the first professional sociologist of science, and he wrote for the few who could distinguish research from popularization, and science from technology.

Robert Merton was recognized as the founder of the rigorous sociology of science. He was also the most balanced of the members of the emerging science-studies community: the only one who, though not a philosophical idealist, stressed the disinterestedness of basic research; while not a positivist, Merton admitted the continuity and cumulative nature of science; and, though not a Marxist, he emphasized the social embeddedness of the scientific community, as well as the economic and political pressures to which it was subjected.

By contrast, John D. Bernal's *The Social Function of Science* (Bernal 1939) published just 1 year after Merton's classic, enjoyed a great popularity because its author was an eminent crystallographer and a professed Marxist. Like most of his readers, Bernal confused science with technology and consequently assigned it functions that it could not possibly fulfill, such as helping the neglected British civil defense system. His own scientific work, in X-ray crystallography, earned him a distinguished place in pure science, not only in his specialty but also in molecular biology, but he was not qualified to help the coming war effort. I bought Bernal's book as soon as it appeared, but did not read it then. My former mentor, the mathematics teacher Manuel Sadosky, snatched, then divulged it during three decades wherever he went. And, just as Bernal, Manuel sided with the charlatan Lysenko when the famous controversy about genetics erupted. No wonder, as Marxism lacks a philosophy of science. However, back to Robert Merton, who did have one.

The Legendary Elephant Vanishes

However, Merton's picture of basic science was only one of many. Raymond Klibansky, my senior colleague, thought that the gallery of portraits of science resembled the mutually incompatible descriptions of the elephant proposed by the proverbial blind Indian sages. Still, any of their one-sided portraits would have helped in finding an elephant escaped from a circus, for every one of them held a grain of truth. Yes, truth, the black beast of postmodernists.

Suddenly, the elephant and the blind sages vanished in the postmodern wilderness: Thomas S. Kuhn, one of many history of astronomy popularizers, published in 1962 *The Structure of Scientific Revolutions* (Kuhn 1962/1970). This book became an instant success, largely because its title contained two fashionable words, "structure" and "revolution," and it held that scientists do not seek the truth because there is no such thing – nor is there a body of knowledge that may grow and is being repaired and made ever deeper. Indeed, Kuhn's central thesis was that, once in a while, there are scientific revolutions which sweep away everything preceding them. Such radical changes would not solve long-standing scientific problems, but they would respond to alterations in the *Zeitgeist* or cultural fashion of the day. Kuhn's book matched the rebellious mood that had recently arisen in university campuses, from Berkeley to Paris. Since it claimed that science is only a matter of opinion, any amateur with enough *chutzpah* was thought to qualify for a job in one of the many "science studies" and "science and society" programs that have proliferated over the past half a century.

Kuhn's counter-revolution was so massive and so sudden, that it took the academic community by surprise and by storm. It was also a fifth-column operation, since it appeared originally as a fascicle of the *Encyclopedia of Unified Science*, until then a stronghold of scientism. I witnessed how Karl Popper, in his memorable confrontation with Kuhn at Bedford College, in the summer of 1965, adopted a defensive tactic and attempted to downplay his differences with Kuhn. Karl did not ask Kuhn to analyze any examples of revolutions out of the blue which replaced well-confirmed hypotheses that were also congruent with the bulk of antecedent knowledge. He could not resort to this tactic because of his radical skepticism – which banned the very concepts of example and confirmation – and of his contempt for the sociology of knowledge (see Stove 1982). Thus, Kuhn emerged victorious from his confrontation with Popper because of the latter's weaknesses rather than because of Kuhn's superiority.

Participation in the Science Wars

Initially I underrated the anti-science antics of Kuhn, Feyerabend, and their French counterparts, from Gaston Bachelard to Michel Foucault: I thought that they were a passing fad that had gained notoriety just because it surfed on the wave of mistrust of science, which the Berkeley rebels had just denounced as an accomplice of the warlike and anti-environmental establishment. It took me two decades to realize the seriousness of the damage those shameless amateurs had done to academia, as well as the reason for their popularity, namely that they flattered the many students and scholars who had chosen the wide door – that of cultural studies, gender studies, science marketing, and the like. Every revolution elicits similar reactions from the disempowered members of former elites.

I reacted only in 1990, when Marcel Roche urged me to respond to a paper by Hebe Vessuri, an Argentine armchair anthropologist who had reported uncritically on the pile of nonsense in *Laboratory Life*: *The Construction of Scientific Facts*, the best seller by Bruno Latour and Steve Woolgar (1979). I did so in my paper "A cartoon of science" (Bunge 1991c), which I expanded in my pieces in *Philosophy of the Social Sciences* (Bunge 1991d, 1992), which were later included in *The Sociology-Philosophy Connection* (Bunge 1999c). In these texts I examined not only the main contemporary anti-Mertonians, but also some of their precursors, from Karl Marx to Gaston Bachelard and Ludwik Fleck – the first to claim that "scientific facts" are inventions of scientific communities. It transpired later that the young Kuhn had read this outrageous thesis, which Bruno Latour and others of his ilk was to exploit with great success among people whose brains had already been mashed by Louis Althusser and Michel Foucault.

In those papers I also traced the philosophical roots of the counter-revolution in question: the subjectivism of Berkeley and Kant, as well as Marx's sociologism (or externalism) of 1859, which held that "society thinks through the individual". One century later the Spanish logician Jesús Mosterín (2006, pp. 400–401) expanded society to the universe, writing that "The Universe thinks itself through me" – a sentence that I am still trying to formalize, in an effort to understand it.

I was one of the few senior philosophers to participate in the so-called *science wars*, by criticizing the contemporary constructivist relativists from a scientistic viewpoint. The silence of Popperians, logical empiricists, and Marxists through that conflict has been deafening. Was their silence due to indifference to topical issues, or else to secret sympathies – with radical skepticism in the case of Popperians, anthropocentrism in that of empiricists, and sociologism in that of Marxists? These tacit complicities have yet to be studied seriously.

Fortunately, a few years later the physicists Alan Sokal and Jean Bricmont published their outspoken *Impostures intellectuelles* (Sokal and Bricmont 1998) for a much larger audience. And, as I write, Dominique Raynaud's (2015) excellent *Scientific Controversies* is about to be published. His well-documented study confutes the theses that "science is politics by other means," and that scientific controversies are over power, not truth, and as a result they are terminated by arbitrary political decision. True, the Galileo/Inquisition and Genetics/Stalinism controversies did involve extra scientific authorities, but only for a while: in the end, truth won out. My own experience as a frequent scientific controversialist confirms Raynaud's thesis.

Contact with European Social Scientists

In 1991 I participated in two European colloquia on social science, one near beautiful Siena, in the heart of Tuscany, and the other in the dull Paris suburb of Cachan. The former, devoted to indicators of income inequality, such as Gini's, was well organized by some Italian statisticians and sociologists, and attended, among others, by the macroeconomist Edmond Malinvaud, who had presided over the Banque de France. My own presentation dealt with genuine (well-defined) quantities and pseudoquantities in social science (Bunge 1999c). Among the latter I included the subjective utilities and probabilities occurring in all rational-choice theories, which are assigned intuitively. I thus revived Peter Sorokin's (1956) fight against what he called "quantophrenia, while avoiding his own rearguard combat against scientism.

This meeting was held at the old Carthusian monastery of Pontignano, which had been acquired by the University of Pisa, and whose two cloisters were well suited to conversing while walking. This center was situated in the midst of the Chianti region, which had been producing wine for centuries. We were served piles of unusual and delicious ravioli, that Malinvaud and other economists consumed obediently to placate the equilibrium god. The last night I had my first very painful esophageal cramps, which I have been nursing ever since. Because they come unannounced and with no obvious cause, my gastroenterologist at the time declared that "the esophagus has a mind of its own." In this Tuscan region land is valued in an objective way that economists should envy: by the number of hectoliters of wine it produces per year. The bottles of Chianti wine are lovingly protected by straw jackets and shipped to the bibulous of the world.

From Siena I traveled to Cachan to attend a meeting of FUR (Foundations of Utility and Risk), as a charter member of the editorial board of the journal *Theory and Decision*, which had been founded by Werner Leinfellner. Werner was a jovial and enterprising Austrian scholar who had emigrated from beautiful Vienna to dismal Lincoln, Nebraska, because the Austrian philosophical establishment was hostile to science-oriented philosophers.

The co-symposiasts who interested me the most were the organizer, Bertrand Munier, as well as the polymath Maurice Allais and the psychologist Daniel Kahneman, both of whom had been awarded a Nobel Prize for work that was at odds with the ruling decision theory crafted by von Neumann and Morgenstern. During the final banquet, held at a huge fortified castle, I sat next to Allais. We had fun challenging a table companion, who believed that all mathematical concepts have an empirical root, to tell us which are the percepts behind the zero and the various infinities.

Three years later I took part in a symposium on social science and its philosophy held at the Ludwig Boltzmann Institute on the philosophy of science, located in Graz and directed by Johann Götschl. Planning my journey from Montreal to Graz via Greece had been complicated by the McGill travel agent, who confused Austria with Australia, booked me to Sydney, and told me with a straight face that this is how she called the homeland of Mozart and Boltzmann. I flew early from Corfu, where I was spending the summer along with my family, and waited until evening at an Athens café for my flight to Graz via Vienna. But my patience was rewarded: whereas Athens had been hot, dirty and drowned in loud rock noise, the Graz streets were cool, clean, and graced by music students who played Mozart on violins and flutes. Clearly, these days Athena's owl nests more often in provincial Graz than in its grand birthplace.

The colloquium did not produce any breath-taking results, but was spared the other-worldly pronouncements of the Austrian school, and it consolidated many a useful personal connection. It was crowned by the reception offered us at city hall by the town's mayor, an attractive woman who, defying the traditional Austrian pomposity, spoke plainly and wore becoming shorts.

Hungary and the Philosophy of Mathematics

One of the most interesting journeys our family undertook was to Turkey, where all four of us spent most of the summer of 1993. We drove from Genoa in our much used yet still elegant Alfa Romeo. Our first destination was Budapest, where Marta and I attended the colloquium of the Académie Internationale de Philosophie des Sciences on the foundations and philosophy of mathematics. We stayed at the visitor's residence of the University of Budapest. The rooms were clean and comfortable, but all the meals were based on meat – which explains the high incidence of heart diseases in Hungary. Thanks to Silvia, who is a strict vegetarian, we patronized the vegetarian restaurant in town.

The eminent mathematician Saunders Mac Lane argued that mathematics is the study of structures, and chided the logicians who did not keep up to date with the foundations of mathematics. The logicians in attendance, particularly Charles

Parsons, felt offended when Saunders told them that they had remained in the time, more than half a century earlier, when set theory had become settled and Gödels proof was still a novelty. He informed them that the theories of categories and topos had replaced set theory as the foundation of mathematics. Saunders knew because he and Sammy Eilenberg had founded category theory half a century earlier. And I knew because Marta had studied under two of Saunders's top students.

McLane's remark confuted Imre Lakatos's thesis that mathematics has no foundations. What is true is that its foundations have sunk ever deeper, as Hilbert had said long ago. Two decades earlier, when coming to Montreal, Imre had asked me to get him invited to give a talk to the McGill's mathematicians on his extravagant opinion. No one was interested, for at that time Montreal was a world center in the foundations that Lakatos, who had no mathematical credentials, claimed to have shattered. Karl Popper welcomed this heresy because it seemed to corroborate (not "confirm", please!) his own radical skepticism. Karl had also believed Imre's assertion, that he had earned a PhD in mathematics. As he told me soon after his protégé succeeded him in his LSE chair, Imre had lied to him, for his PhD was in mathematical pedagogy. This disappointment served Karl right, for he demanded abject obsequiousness in exchange for patronage, and Imre had flattered him shamelessly – and cut him as soon as he got his chair.

My own talk at the Budapest colloquium was about the moderate mathematical fictionism that I had sketched in volume 7 of my *Treatise* (Bunge 1985a). My central thesis was that mathematical objects are just as imaginary as those of fantastic literature and theology, except that they are exact, law-abiding, and susceptible to change – though not by themselves. I believe that mathematical fictionism allows mathematicians the same creative freedom as Platonism, but deprives idealists of their favorite argument against realism and materialism. Of course mathematical objects are ideal, but they are not arbitrary; and, while they are immaterial, their creators are flesh and bone beings.

Consequently, there can be no reasonable realistic or materialist philosophy of mathematics. Realism and materialism can help only the psychologist, sociologist and historian of mathematics. In Budapest we took an interesting boat trip on the Danube – which of course is not blue. We chatted with Alberto Cordero, Paulette Février, Paco Miró-Quesada, Evandro Agazzi and his new companion, and the charming if excentric René Thom, whose brand new red Citroën was stolen the same night he arrived. Cordero was an interesting and gregarious Peruvian who taught in New York City; Paulette, a French philosopher who had been married to the physicist Jean-Louis Detouches, had claimed that quantum mechanics calls for a many-valued logic; and Paco, ever eager to keep up to date, was bravely attempting to learn category theory.

After that meeting we visited a couple of dull provincial Hungarian towns, where the only edibles were delicious pastries. We also visited the large but uninteresting Lake Balathon, and the huge but ugly Esterhazy estate. The original owner, an ennobled new rich, had hired as his *Kapellmeister* Joseph Haydn – who was made to eat in the company of the other servants.

Czech Interlude

From Hungary we drove to the Czech Republic, where we admired prettier landscapes but had even worse food problems. The square of every small town was equipped with a potent loudspeaker destined, according to one malicious traveler, to hide the unmusicality of the national language. In Prague we stayed at a gloomy bed-and-breakfast ruled by a man who did nothing but cultivate his enormous mustache. We had nearly all our meals at a pizzería, visited the huge and sinister castle immortalized by Franz Kafka, as well as the minute flat were he had written his masterpieces. Once, while waiting for a street car, we were encircled by a group of Gypsies to whom we attributed bad intentions, perhaps because of our racist prejudices. In a small hall we attended an excellent concert of classical music played by world class musicians, and dominated by an incongruent painting of a battle. Our friend Abel Posse, the Argentine ambassador and author of 14 highly original novels, among them the funny *The Queen of the Plata River*, entertained us regally. He received us in his vast official residence, rented from the famous writer and brave politician Vaclav Havel, whose existentialist musings I could not abide. Why is it that most anti-Marxists march backward instead of forward?

In the charming city of Chesky Krumlov we listened to a rock concert offered in the public square by lively Gypsies wearing blue jeans. They were said to hold regular jobs and to be on good terms with the locals. The local castle and the surrounding Italianate garden were so huge, that their upkeep must have taken thousands of serfs to work the fields. Its vast gallery showed aristocrats, looking like Spanish courtiers of the sinister Phillip II, with devout and ascetic expressions, and wearing black clothes and large white ruffs. But, since the castle contained a rather large theater, they must have enjoyed the mundane comedies of the day played by itinerant troupes like Molière's.

Turkey in Transition

In Venice we boarded a Turkish ferry that took us to Izmir (Smyrna), from which raisins and dried figs used to come, and which in the early 1920s had been the exit port for masses of Greeks and Armenians fleeing from the Turkish army intent on performing a full ethnic cleansing.

Our ship was loaded with exultant Turks on holiday who had been working hard in Germany, and were now taking their second-hand Mercedes to show off in their native towns. The friendly ship engineer invited me for talks in his office on the pilot's deck, but ignored me after the day I alluded to the atrocities that were still being committed against the Kurds, to whom the great powers had promised independence six decades before.

Along the Aegean coast we visited Greek ruins even larger than those in Greece, as well as some of the earliest planned towns. Among them stood out Ephesus,

Heraclitus's birthplace as well as the site of an important early Christian community, where Paul of Tarsus had entreated the slaves to stand with awe and fear before their owners. Ancient Ephesus had a sophisticated system of running water and sewers, and an open air toilet where ten men could sit bum-to-bum at a time. It is not known whether the saint kept preaching while sitting there.

In Olympia, on the Mediterranean coast, we found a chalet in the midst of a lovely oleander copse, that seemed adequate for spending a few weeks, for it was quiet and near a long deserted beach. Regrettably there was only one towel, the owner's, who offered us to share with us. In addition, the shower pointed straight at the toilet, though at least this was not in the far more popular Turkish style. In our relentless quest for a beautiful and quiet place with good swimming, we stumbled on Üçağiz, or Uchai, the remains of an ancient Lycian city on the shore of a narrow bay ensconced between two hills. Parts of the old city lie sunk in crystal-clear water. The ancient Lycians seemed obsessed with the after-life, as there was at least one large (family size?) stone sarcophagus in every orchard, and large tombs dotted the hillside across the bay – which made for a sad albeit beautiful landscape. All of those graves were had been looted long ago by the Phenicians, Persians, Romans, Byzantines, Venetians, and Genoese who took advantage of the peaceful Lycians.

At the beach, sitting on his boat and contemplating the bay, we met Mustapha, a friendly man who called himself a fisherman, although he never brought any fish. He offered us to rent part of his wooden house. We closed a deal and had meals with him, his wife, and their daughter, a sweet, intelligent, beautiful, sociable, and industrious 12-year-old girl who, having just finished elementary school, was about to start Koranic school – the perfect wife for an imperfect matrimony. When we left, she gave us a potted basil plant which, she assured us, would keep the mosquitoes away. Silvia carried it on her lap for hundreds of kilometers. We stayed in Uchai until finding some gasoline for our nearly empty tank. We drove straight north to beautiful Konya, in the heart of Anatolia, center of radical Islamism and of the swirling dervishes, who seemed to have escaped from insane asylums.

Tegekkür Ederim

When we visited Turkey, in the Spring of 1993, the country had just started its transition from the secular and Europhile nation founded by the progressive if brutal Mustapha Kemal Atatürk, to a backward religious one. The only openly pro-Western Turks we met were those who had spent some time in Germany as guest-workers, earning well but willing to be deported the moment they became "redundant." One of them, a mechanic who now managed the hotel where we stayed for a while, had tinned German sausages and anchovy paste for breakfast, whereas we stuck to fresh bread, feta, boiled eggs, tomatoes, and cucumbers. I suspect that the characters of Orhan. Pamuk's sad but powerful novels are only found in his birthplace, the magnificent and miserable, cosmopolitan and cat-loving Istanbul. In the heart of Anatolia we first saw very large mosques, full of worshippers, only recently built and staffed with plenty of Saudi or Iranian money. The modernizing Atatürk, who liked to dress as a Parisian, is still being officially revered in name. But Islamism is currently back in power after a century, largely in reaction to the repeated Western aggressions against the Islamic peoples who happen to sit on huge oil reserves. Let us pray to Allah that Oklahomans may soon swim in oil, so that the Arabs may finally live in peace.

We experienced Islamic fundamentalism only once, when an individual threw a pack of burning matches inside our car. But nearly everyone else went out of her/his way to help us in various ways. Some just wanted to show off their German or practice their English, but many more attempted to teach us their language. Eric and Silvia quickly learned enough Turkish to hold primitive conversations, but I only learned to say 'thanks': *Tegekkür ederim.* This phrase served me well when meeting Turks abroad. Linguistic differences are secondary by comparison with community of interests and habits. If most linguists were polyglots, glossocentrism might be eccentric.

During our Turkish journey I use to hide in secluded gardens or quiet balconies to read books and papers in social science, which I had mailed to a wholesale grain dealer in Izmir. Shortly before leaving Izmir to return to Venice, we made up several parcels with those publications, along with many novels, and took them to the central post office, to mail them back to Canada. The clerk who served us refused to take them, by virtue of an eighteenth century *fatwa* that forbade the export of foreign printed matter. After I complained and demanded to see the manager, I was immediately taken to the boss's large office ornated with sumptuous Turkish carpets, where he was sipping coffee like a pasha, surrounded by his protégés. He listened politely to my request and, invoking a different *fatwa*, authorized me to mail our parcels without even hinting that he expected a *bakshish* in return. Was this a relic of the efficient Ottoman bureaucracy?

The last leg of our journey was the Black Sea coast, utterly different from all the rest. There we saw plenty of dense and humid fir forests, visited Atatürk's museum, climbed a steep hill to visit disappointing old paintings, watched a movie spoken in Navajo with Turkish subtitles, and ate delicious meals for one dollar per person. Finally we boarded a ferry that took us to Istanbul, where we visited the amazing Aghia Triada, the enormous and beautiful Byzantine temple that withstood earthquakes, thieving Crusaders, and Ottoman depredations. We also saw the Byzantine hippodrome, which had been a political forum, and turned out to be tiny by modern standards. A long beach nearby, where we had hoped to swim, was disgustingly covered in drug syringes and plastic garbage. But the Blue Mosque, the old wooden houses with elaborate balconies, and the boat cruises in the Bosporus are unforgettable.

In Turkey, Western comforts are hard to find outside big cities, but awesome landscapes, millennia of history carved in stone, wholesome meals, and friendly people more than compensate for the inconveniences. Turkey is, indeed, one of the most interesting, beautiful, and welcoming countries on earth.

Lectures in Foreign Parts

During the last half a century I have traveled to many distant places, to lecture on social matters: to Aarhus (1972), Zürich (1973), Dubrovnik (1973), Perú (1996, 1997, 2009), México City (1975), Monterrey (1976), Madrid (1983), Buenos Aires (1996, 2001), Pisa (1993), Santa Fe (2001), and Sydney and Melbourne (2001), most of the time to large audiences. In Arequipa, Perú, the enrollment was so high, that I had to speak in the cathedral. At the Inca Garcilaso de la Vega University, in Lima, I was protected by a police officer for fear of the Shining Path, whose leader had a degree in philosophy. When I dropped the chalk or the eraser, a helping hand appeared from behind the curtain. My tall bodyguard, always armed with a big gun, accompanied me every time I went to the park for a walk.

My lectures were attended by professors and students who asked interesting questions. In Buenos Aires, when asked whether we could go ahead without Marx, I replied that Marx had led us to disaster, but without him we could go nowhere. A woman in Santa Fe City, annoyed by my statement that religion had always been used as a tool of social control, asked me what would it take for me to convert to Catholicism. I replied that a single miracle would suffice.

The National University of Trujillo, alma mater of many famous Peruvians, gave me an honorary doctorate and made me a member of the Orden del Libertador. When I asked foolishly which one, Bolívar or San Martín, they did not bother to answer: of course, it was the greater of the two, in whose honor an entire country, Bolivia, had been named. In Lima, at the Center of Higher Military Studies, attended by colonels who wished to be promoted to generals, I started by saying that the soldier's duty is not do die for his fatherland but to live for it, assisting the victims of calamities and aggression. I was presented with a plaque inscribed with the admirable saying of General Marín: "Ideas are to be expounded, not imposed."

The Cayetano Heredia, Lima, Inca Garcilaso de la Vega, San Agustín de Arequipa, and Mayor de San Marcos universities gave me honorary doctorates. The Garcilaso also gave me two large llama woolen ponchos that the moths have spared and still look new after two decades. My hosts also took me to museums full of erotic ceramics, and archaeological sites like Chan Chan, a huge earthen city in northern Perú that is a model of large-scale urban planning for a rigidly stratified society. It is not known whether they produced scholars whose task was to justify class inequality.

In Arequipa I visited what had been the huge Saint Theresa cloistered monastery of the barefooted Carmelites, where no males had been allowed. It had been a comfortable prison for the unmarriageable daughters of the wealthy, who were locked up for life along with their servants, doing nothing but pray, embroider, cook sweets, and prepare for death, which was only one step ahead. I took my greatest pleasure in chatting about philosophy and politics with my old friends Paco Miró-Quesada and David Sobrevilla, and gaining a new one, Lucas Lavado Mallqui. Lucas, founder of the Garcilaso University Press and endangereded twice his career, when he took to court two rectors of his university, accusing them of embezzlement, and of appointing relatives for well remunerated jobs that they neither deserved nor worked at. Believe it or not, Lucas won both trials while undergoing chemotherapy, teaching, and performing his administrative duties.

Lecturing in Porteñol

In 1991 I taught a crash course at the Universidade Federal de Santa Caterina, in an island in Southern Brazil, and 4 years later in Goiania, in central Brazil. In both cases I spoke *porteñol*, a mixture of Portuguese and Spanish, and talked to a number of interesting people, among them a female anthropologist who had been adopted by an Amazonian tribe, and took us for a dip in a rapid river at the bottom of a large crater made by a meteoritic impact. I also had long discussions with my former McGill student and assistant Mike Dillinger, who was now a professor of linguistics in Northern Brazil, and made his last attempt to recruit me to the Chomsky sect. His former hero had only recently published in *Mind* a long paper where he claimed that his discipline was a natural science and, more particularly, a branch of biology.

Chomsky's sudden conversion of a former idealist to the version of naturalism that makes no room for conversation, imitation, invention or convention, must have puzzled most of his admirers. These would not explain, for instance, the important linguistic changes that followed the Aryan, Norman, Mongolian, Arab, British, or American conquests. Did the Normans import into England the gene that suppressed the older Saxon gene responsible for attributing genders to inanimate things such as stones, as in *la pierre*? What biological transmutation caused the substitution of American English for Spanish in the Philippines after 1898? Of course, some of Chomsky's faithful invoked the escape clause: "But he has not reneged his politics."

Portuguese Interlude

In the spring of 2001 Marta and I drove through the colorful Portuguese countryside, swam in the cold waters of Algarve, and in Tomar we roamed through the huge Templar fortified hospital and visited the tiny synagogue. In Lisbon we stayed at the cozy "As Janelas Verdes" hotel, because it had belonged to the great novelist Eça de Queirós, who might have gotten a Nobel Prize had he written in an international language – or in Icelandic. We visited the very rich Calouste Gulbekian and Ancient Art musemus, and saw an interesting play by the rector of the local university.

One night we slept in what had been the window-less cell of the fortress guarding the majestic estuary of the Tagus river. From there, a mad Portuguese prince had taken pot shots at the stevedores who unloaded the ships packed with riches from the Portuguese colonies on three continents, and which only benefited a small class of merchants. We rode the brand-new subway, and visited the ruins of the cathedral destroyed by the famous 1755 tsunami, which the Inquisitors marked with an *auto da fe*: the public execution of all the unbelievers they could lay their hands on. That calamity produced an uncounted number of religious skeptics throughout Europe, and 4 years later Voltaire used it in his *Candide* to poke fun at the Leibnizians who believed ours to be the best of all possible worlds.

In Coimbra, Portugal's oldest learning center, we lodged at the hotel with the romantic name Garden of Tears. At the local university, where six decades earlier my teacher, Guido Beck, had taught after escaping from a Vichy camp, I gave talks to the physicists and sociologists while Marta interacted with the mathematicians. We also visited the colorful baroque library that had been commissioned by the Marquis of Pombal, the enlightened eighteenth century statesman. Apparently it contains only old theology and law books, because the Inquisition prohibited everything that came from enlightened France. The library's only modern trait is the "ecological" method used to protect its antiques. Indeed, the reading room is the home of an ancient bat family respected by the librarians because those residents protect the books by eating the worms off paper – on top of which they keep silent, as befits a good library visitor.

We were delighted by the enormous advances that had occurred in the country since our first visit at Christmas of 1965, when it was still languishing under Salazar's fascist-clerical dictatorship armed by NATO. However, I paid my due to tradition: throughout our Portuguese journey I loyally kept the St Christopher medal that Julha, our Portuguese cleaning lady, had forced on me. The same Julha explained to us why the flight from Lisbon to Montreal takes 1 h longer than the eastward trip: because, as the map shows, Montreal lies "above" Lisbon, and so the aircraft has to make a greater effort climbing up than going down. Julha, who had grown on a farm in central Portugal, had not attended elementary school because she had to work on the fields from age five. But in Canada she had learned to read the prices in the supermarket's catalogues.

Bright Spots in Dark Times

One of my most popular seminars at McGill was the informal discussion group held at the student cafeteria after my classes. It was open to all interested individuals, and we would discuss any of the subjects proposed by any of them. Some of the most frequent discussants were Moish Bronet (social work), Omar Ahmad (biology), Michael Kary (biology), and Dan A. Seni (management science). My dear colleagues deprived me of my chair between 1985 and 1990, but I was allowed to teach most years as a lowly paid instructor. I taught then courses in social philosophy, ethics, philosophy of technology, and materialism. The last two were being taught for the first time at our university.

One of my most interesting students at that time was Karim Rajani, who held a fellowship from the Aga Khan Foundation, reserved to Ismailis. Karim, who was studying political science, called my attention to Bueno de Mesquita's *The War*

Trap (1985), that had been highly praised by the political-science community. That book was a typical specimen of what Pitirim Sorokin had called *quantophrenia*, or the putting arbitrary numbers on the probabilities and utilities of events. In fact, it was Karim who introduced me to game theory and its applications to politics and political history. Karim was an open-minded and delightful fellow, always ready to discuss matters of intellectual interest over the coffee and cake we got at a small cafeteria that always played the same cassette: Antonio Vivaldi's "Four Seasons." Because his religious beliefs were so alien to current events, Karim was able to discuss them in an objective fashion, and to produce well-informed and critical papers on the mostly silly doctrines he was being taught by my political science colleagues.

Another interesting person I interacted with at the time was Frank Forman, an education economist who worked in Washington DC for the US government. Funnily enough, he was a fan of both James Buchanan, the orthodox economist and right-winger, and myself. Forman was seriously interested in deep and general issues, but he fell for the half-baked and reactionary pop philosophy of Ayn Rand, that oracle of American libertarianism and mentor of Alan Greenspan, of Federal Reserve Bank and 2008 world financial crisis fame. I never met Forman, but in the late 1970s, while I was working on ethics, we had an intense letter exchange, and he helped me get hold of a number of books in ethics and social philosophy in the days before the arrival of Amazon.com. End of interlude. Let us now return to social science.

Writing About Social Mechanisms

Every concrete system is driven by one or more mechanisms. The specific mechanism of a material system is the thing or process that makes it tick, and hence identifies it as such: energy exchange with other things if it is a physical or chemical system; metabolism if it is alive; love if it is a couple; production or trade of goods if it is an economic system; struggle and the exercise of power if it is political, and so on.

The elementary things, such as electrons and photons, are not systems, hence they have no mechanisms. Likewise abstract and symbolic entities, for they do nothing. The mechanisms of material systems are, then, just as important as their composition, environment, and structure (or set of bonds among its parts). Therefore, any suitable sketch of a material system ought to be an ordered quadruple: composition – environment – structure – mechanism. The usual definitions of 'mechanism' recently improvised by some philosophers are inadequate because they are imprecise and they omit reference to both system and materiality. I have argued that the concept of a mechanism is the clue to scientific explanation: to explain X is to show how X works (Bunge 1959a, 1967b, 1996a, 1998b, 1999c, 2003c, 2004a, b). Typically, the pseudosciences do not involve plausible mechanisms. Scientific psychology explains something mental when it exhibits the underlying brain process;

and social science explains to the extent that it finds social mechanisms such as sexual rivalry, the struggle for power, government regulation, competition, and cooperation. For example, territorial expansion explains most wars at the time when land was the most valuable resource, whereas most international conflicts over the past half a century have been over oil. And religion has been used as a pretext for many wars and colonial occupations, whereas today small-scale terrorism is the excuse for large-scale aggression.

I have mentioned or analyzed systems of various kinds and their mechanisms starting with my *Scientific Research* (Bunge 1967b) – where I analyzed the mechanism underlying the ideal gas law – but especially in my publications on social science (e.g., Bunge 1996a, 1998a, 2004a) and in my *Medical Philosophy* (Bunge 2013a). In the latter I proposed calling *platinum standard* the gold standard (randomized control trials) plus a plausible mechanism of action.

The first international colloquium on social mechanisms met in Stockholm in 1996. It was convened by the Swedish sociologists Peter Hedström and Richard Swedberg, who quoted some of my writings on the subject, in particular my *Scientific Research.* That successful meeting was attended by some of the best known social students, such as Robert Merton, Thomas Schelling, Charles Tilly, Arthur Stinchcombe, and Jon Elster. I was asked to comment on the presentation of Harvard's Aage Sørensen's. I confined myself to formalizing his hypotheses, a task for which he thanked me without enthusiasm because he should have done that himself, particularly since he complained about the backward movement of theoretical sociology in recent years. Chuck Tilly commented favorably on mine, though he started by complaining about its length. We became instant friends, and I learned much from him in the following years, for he was a socioeconomist and social historian of enormous learning and generous with his time – which, alas, turned out to be much shorter than he thought.

Ronald Burt, another interesting co-symposiast, asked me over breakfast whether I had any inkling about who the next Nobel-Prize winner in economics might be. I replied with what the physicists to whom I had spoken the day before had told me: that they were campaigning for the elimination of that prize because it had been ill-awarded in most cases. Two such cases were those of Robert Aumann and Thomas C. Schelling, both of whom participated in our meeting, and were about to be awarded the distinction in question. Aumann had studied markets with a continuum of agents – hence, with no distinct individuals; and Schelling had justified the segregation of Blacks in the USA, claiming that they had freely chosen to clump together.

The colloquium organizers put together all the papers except Tilly's and mine, in a volume published by Cambridge University Press. Neither Chuck nor I ever learned the reason for these exclusions, nor for the inclusion of papers like Schelling's, which confused mechanisms with theories about them, or Elster's, with his absurd claim that "mechanism is the antithesis of law." A number of philosophers (e.g. Craver and Darden 2013) have liked so much my ideas on mechanism and mechanismic explanation, which I had been discussing for half a century, that they have appropriated them, though failing to mention the original source. One can imagine several mechanisms for this oversight.

In the long flight back to Montreal I sat next to the owner of a homeopathic products factory. When I told him what I thought about that swindle, he asked me whether I believed in God. On hearing my negative reply, he said: "Nor do I. Belief or disbelief in homeopathy is similar." I spent the rest of the journey reading Swedberg's fascinating biography of the remarkable socioeconomist Joseph Schumpeter, of "creative destruction" fame – who, to everyone's surprise, ended up sympathizing with fascism.

Interacting with Canadian Social Scientists

I have interacted with a number of Canadian social scientists. In addition to Asimakopulos, Brecher, Niosi, and Trigger, mentioned earlier, I have had frequent conversations with the sociologists Axel van den Berg and John A. Hall, as well as with the political scientists Michael Brecher and Andreas Pickel, the historian Peter Hoffmann, and Ian Jarvie, the sociologist of film, or rather cinema audiences – which nowadays may be similar to those who in the past used to attend public executions.

Jarvie, originally a theoretical anthropologist, had come under the influence of Joseph Agassi, Popper's most lucid, loyal, and yet independent student. I first contacted Jarvie in 1973, when I submitted a paper to *Philosophy of the Social Sciences*, the journal that he founded in 1970 and that he has been managing ever since. He rejected that submission because it contained some symbols, but accepted some of my subsequent papers, asked me to referee a number of papers – most of which I had to reject – and gradually we became friends, as I did with his new wife, Jeanette Bicknell, one of the rare philosophers of music.

Ian's nice and smart son Max asked me to supervise his PhD, but I found him unteachable, for he was a radical skeptic unwilling to take anything scientific for granted – a stance ridiculed by the Australian David Stove (1982). It is seldom if ever noted that Pyrrhonism is logically untenable, for any rational attack on a proposition boils down to proving that it contradicts one or more propositions that we must assume at least for the moment.

My sociology colleague Axel Van den Berg sought me out because of our shared antipathy to constructivism-relativism. But he attacked so harshly my concepts of social system and social mechanism, that I was forced to respond to him in equally harsh tones (Bunge 2004c). The Argentine economist Arturo O'Connell, brother of my friend Lily Alurralde, helped me with the bibliography. John Hall, a nice historical sociologist, came to McGill from the UK, where he had written his scholarly yet highly readable *Powers and Liberty* (Hall 1986). Regrettably, his fire paled when he came to frigid Montreal: he got lost in trivia, and even accepted a deanship. But during his tenure he implemented my project of a postgraduate program in the social sciences – for which I got no credit.

John kindly undertook the task of improving my spotty sociological education. He started by persuading me to waste 2 years reading Max Weber in German, in particular his monumental and soporific *Wirtschaft und Gesellschaft*. I found this work more erudite and historical than original and sociological. It is also disorganized and partly inconsistent, for he does not use the "interpretive" or Verstehen method he claimed to use. His wife Marianne, a distant relative of mine, is to be blamed for some of those defects, because she built the three tomes by cobbling together a pile of her husband's papers (Richard Swedberg, personal communication, 2014.) The very young Raymond Klibansky, who eventually would become my colleague, had helped Marianne reading proofs of that work while he was her lodger soon after her husband's death.

One decade later Klibansky met Heinrich Rickert, Weber's philosophical mentor, and reproached him for having signed, along with many other German professors, an abject manifesto in Hitler's support. The same Klibansky noted the small size of the academic world at that time: he had shaken the hand of his first employer, Ferdinand Tönnies, who had gone to England to meet Engels, who in turn had Marx at hand. So, Klibansky could brag that he was separated from Marx by only three handshakes – and I, four.

Preaching in Exotic Places

My work, parts of which was translated into a dozen different languages, attracted a number of invitations outside North America. I have lectured and taught weeklong courses on different subjects, in five languages and on five continents. I have given talks in places that, though no longer exotic, still are out off the trodden trail, such as India, Japan, China, Costa Rica, Egypt, India, Japan, Nepal, Panama, and the Andean nations. Most of my presentations were well attended, but I doubt that they achieved anything more than awakening curiosity, since none of them brought me new students. But I always enjoyed the questions and the lively debates that my talks usually provoked.

In Kathmandu, in 1979, I participated in a UNESCO colloquium on development, that was unproductive because most participants were like frogs out of their ponds. Indeed, one of us specialized in ancient Greek mathematics, another in child development (hence he equated underdevelopment with childhood), a third in dialectical logic, a fourth in standard economic theory – which, because it centers on equilibrium, cannot see the great disequilibria that accompany socioeconomic development – and so on. No Nepalese scholars attended, perhaps because they were busy developing.

For me that was the chance to contemplate the highest peaks in the world, admire deep valleys, converse in signs with natives, and watch Nepalese worshipping first at a Buddhist temple and immediately thereafter a Hindu one. You never know! Talking with local professors with offices in modern buildings we soon noticed the scarcity of experts capable of guiding the development of a people divided into traditionalists and Maoists, as well as sandwiched between India and China on the one hand, and the imperial powers on the other.

On my way back I stopped in New Delhi, where I stayed at the home of my friend Dhirendra Sharma, the editor of *Philosophy and Social Action* and a professor of political science at the Jawaharlal Nehru University. His two teenaged sons, born and bred in London, wandered about tethered to their cell phones: they were disoriented and unmotivated. When asked why astrology was still popular in that hapless multination, a colleague of Dhirendra's assured me that these days astrologers were very accommodating about dates. After a long and frightening ride on a motor tricycle in the midst of crazy traffic, I talked about development at the famous All India Institute of Medical Sciences. To my amazement, nearly all the questions I was asked concerned parapsychology, which I had not even mentioned in my talk.

Two decades later I returned to India, this time with my family and just to gape at some of its many wondrous sights. We traveled in all directions to visit Hindu, Moslem, Buddhist, and Jayna temples, often meeting more monkeys than worshippers. The beautiful sculptures and bas-reliefs, many of them with erotic themes, did not erase the painful impression left by beggars, some with huge scars, who gathered in large numbers at the entrances of temples. On the old trains, ever full, we always chained our suitcases, but had lively conversations with people who were as amiable as curious about personal matters. Unlike their previous masters, the Indians did not engage in weather talk. We saw elephants working in the midst of thick forests, buffaloes swimming in wide rivers, and open-air fairs full of flowers and reeking with spices. In Cochin, a spice center and former French colony, we saw cobra snakes and visited the empty synagogue founded by Doubting Thomas two millennia earlier. In Trivandrum, the capital of the progressive state of Kerala, we watched enviously the beginning of the construction of a science museum. We swam at the southern tip of the subcontinent, and once we saw on the beach the corpse of a man in the sitting position, which would have kindled Sherlock Holmes's imagination. And in New Delhi, on our way back, we were mobbed at midnight by a swarm of child beggars.

The Great Ice Storm

Cold weather weakens social ties. Indeed, in the winter it is disagreeable and sometimes even dangerous to go out to meet people. This explains why Anglo-Canadians are rather frosty – though not why French Canadians tend to keep their cheer. There must be cultural factors that puzzle me. In 1998 we suffered the Great Ice Storm that paralyzed Montreal and other cities during a couple of weeks. In the early morning of January 4th we were awoken by the tree branches, loaded with ice, that broke and crashed loudly on the ground. A few hours later the power was cut off because the iron pylons that held the electric wires bent and fell down as if they were made of putty. The light went off just when my surgeon started to make incisions on my left hand, to repair a muscular contracture (Dupuytren's). The fact that he was a namesake of the discoverer of the circulation of the blood did not help.

Back home, we noted that it was cooling down fast for lack of heating. Marta improvised a bed for me on the floor, facing the fireplace, and lit a fire with firewood that did not last long. The pile of firewood in the garage had been burnt the previous winter. Due to my improvidence, when night came and I started to feel feverish, we were left without fire, light, and hot water. That first night we dined on leftovers. At midnight arrived my former student and good friend Moish Bronet, carrying firewood and ritual Jewish candles, the only ones he could find. The next few days Marta cooked *cous-cous*, the Algerian staple, in the fireplace. When we proposed a couple of close friends of us to share our resources, they replied that they were O.K. But they did give us a hand a few years later, when we suffered a flooding. Giving is easier than sharing, because you hand out a surplus, whereas you share a scarce resource.

Millions of Canadians experienced the same or worse. However, there was an admirable solidarity: thousands of people offered help to total strangers. To the amazement of Americans, during those weeks there were neither burglaries nor lootings. Some say that the burglars had declared a ceasefire, and others that they were busy saving their own skins. In any event, I finally understood why cave painting had been the only invention made during the last ice age, between 110,000 and 10,000 B.P.E.

The Paradise of Marsupials and Possible-Worlds Philosophers

In 2001, Marta and I traveled to the paradise of marsupials. We took an unusual route: from Montreal to Buenos Aires to Sydney via Auckland. We went as visiting professors, Marta to Macquarie University, and I to the University of New South Wales.

We stayed there from the Fall to the early summer of 2001. Sydney harbour captivated us from the start, but it took us a month to adapt to the new time zone. After much search with the help of Eric's kind and patient friend and colleague Andrew Burges, we settled in an apartment in the much-criticized new building popularly called "The toaster." It lies across the water from the iconic Harbour Bridge, alongside the Botanical Garden, near the amazing Opera House, which we regard as the Notre Dame of our time. At twilight we watched flocks of huge bats silently circling above that building and ours before alighting on gigantic wild-fig trees. We visited often the public gardens, as well as the Arts Museum, various theatres, the large Blue Mountains Forest, and Kangaroo Valley, where my generous host, Michael Matthews, had a "rough and ready" cottage full of children, his own and others, as well as a family of mice that became very fond of our soap cakes. On the grounds jumped cheeky kangaroos, while fat and timid wombats hid in their enormous burrows. And in the village we could make out the odd customer of the second-hand bookstore patronized by Michael and a few other fellows of the lost Tribe of Book Readers.

Michael had founded the *Science & Education* journal in 1992, and edited it for a quarter of a century with the assistance of nearly 1,000 referees – no mean feat for any scholarly journal. This journal reflects Michael's commitment to utilisation of history and philosophy of science in the clarification and resolution of theoretical, curricular and pedagogical issues in science education. It published a couple of articles of mine jointly with Mahner's on religion and science education (Bunge 1996d, e), as well as two symposia on my work. Michael's bêtes noires are social constructivism and the pedagogic constructivism introduced by Ernst von Glasersfeld, who held that teachers should not teach, but only give students the chance to rediscover the world by themselves. The underlying assumption is that knowledge is mostly innate.

Michael had been educated in Catholic schools, was well versed in Catholic philosophy and theology, and completed science, philosophy and education degrees at University of Sydney. He lost his faith when discovering delights that Christianity condemns, but he kept his admiration for Aquinas and his teachers. We became close friends while attending together the Australasian congress of philosophy of science, which met in Melbourne a few days after my arrival to the country. I read there my paper on the first century of quantum physics, which I had not been allowed to present at the McGill philosophy seminar (Bunge 2002a). This paper provoked many rebuttals, to which I responded (Bunge 2002b). A decade later Michael published a symposium on my philosophy (Matthews 2012).

In Melbourne I met again with David Armstrong, the last survivor of the triad of Australian materialists who had made a splash in mid-twentieth century, when they dared claim that mental events are cerebral – the "identity theory," – who I had met around 1950. Regrettably, Australian materialism had been dead for quite some time, because the gentle John Smart, the uncouth David Armstrong, and the philosopher-farmer Ullian T. Place had been repeating themselves and, with one exception, had remained confined to the philosophy of mind. The exception was David Armstrong (Armstrong 1997), whose latter metaphysics boils down to the thesis that the ultimate bricks of the universe are states in themselves, such as moving, cooling, and disintegrating, without mention of that which moves, cools down, or disintegrates. Plato might have concurred with this version of immaterialism, which renders labs unnecessary. Let us hope economists won't hear of it.

Around 1980 the Australian materialists were replaced overnight with the American David Lewis and other possible-worlds fantasists. They sound attractive because they use a single tool, namely modal logic, to work on all the miniproblems and pseudoproblems they can think of. In addition, they hide behind the hypothesis that nobody can check what they claim about the said worlds, since these are assumed to be inaccessible from one another, in particular from ours – a new version of esotericism.

As soon as we arrived in Sydney, Michael introduced us to his close friend Hans-Juan-Johnny Schneider, who overnight became our friend, guide, driver, and protector. Johnny, 5 years younger than I, was married to Christine, an artist, made a living repairing computers, and kept in shape by walking, swimming, and skiing whenever he could. His curriculum is hard to believe. Motherless, Hans acquires a loving foster mother, Otti, who extracts him from his native Berlin a few months before the Second World War starts, and takes him to Santiago de Chile.

In the new land, Hans quickly becomes Juan, integrates into Chilean society, learns by himself to repair electrical instruments to earn a living, finishes high school, graduates from the local state university, dabbles in leftist politics, and marries Cristina Cordero, a sociology student. The two of them spend the year 1963 in Beijing, where Juan works as a translator for Foreign Languages Press. Both are confined to a foreigner's ghetto, so they hear nothing about the massive purge called Cultural Revolution.

Back in Chile, Juan specializes in geography and in 1970 he earns a French fellowship, both go to France, where he earns a PhD in geography, and they suffer an automobile accident that terminates Cristina' sociology career. Three years later Juan accepts a visiting professorship at the prestigious University of Sydney, but his wife stays behind. When the Kissinger-Pinochet coup occurs, Juan decides to stay in Australia for good, and becomes Johnny. Despite all these accidents, he never lost his cheerful disposition, generosity, work ethic, and pleasure in travel and in outdoor activities. He knows how to live and help others live.

Peter Slezak was another friend we made in Sydney who was always cheerful and ready to help me resolve the conflicts I had with my computer. He had been born in a Displaced Persons's camp after his parents, born in Hungarian Transylvania, were freed from Auschwitz. Initially the Central Agency for Jewish Emigrants in Vienna refused to help the Slezaks because Peter had not been circumcised, as they discovered upon peeking into his diaper. But all three arrived safely in Australia in 1948 when Peter was 1 year old. Peter did his PhD in philosophy at Columbia University and was director of Cognitive Science Programme. He is an enthusiast of functionalist psychology, and critical of experimental work on visual imagery, although it includes the stunning neural trace on the primary visual cortex of a monkey forced to stare for a while at a set of concentric circles, which the said traces match almost perfectly. I never succeeded in persuading Peter to read any of the papers in cognitive neuroscience I gave him.

He does not go to the extreme of holding, as a coreligionist of his said some three decades ago, that brains are only good to be eaten fried. He does not doubt the legitimacy of brain studies, but does not believe that they are relevant to the science of mind: the *obiter dicta* of Hilary Putnam, Jerry Fodor and Zenon Pylyshyn suffice for him, even though they have never helped discover or explain anything mental, whether normal or pathological. It would seem that psychoneural dualism is even more firmly anchored than belief in the supernatural, particularly when dressed in Computerese.

Australian Talks

I gave talks in nearly all of the departments of my host university, always tailored to the interests of my audience. The science education audience was particularly interested in my Universidad Obrera, especially after I replaced the formal lecture with group study. In the Department of Sociology I spoke about social structure and the need to count the number of people in each social cell. This provoked the following comment from one of the listeners: "So, you are committed to the numerical paradigm." I had had a similar experience in the Mexican Institute of Demography. Acalculians of the world, unite!

I did not fare better with the psychologists, for they were more interested in computers than in brains. The explanation is obvious: the fantasies about the mental as computational are easier to learn and more general than the hypotheses of cognitive neuroscience, and they do not call for any empirical tests. Nor do they risk failing in the clinic. For example, acalculia can be "explained" as failure of the numerical computation program, instead of demanding the search for a lesion in the parietal lobe.

My worst failure was my talk to the philosophers at Macquarie University. My theme was the crisis in philosophy and the need to reconstruct it with the help of science (Bunge 2002c). To make my point, that philosophy is in poor shape, I examined the ontology and semantics of possible worlds, which had been invented to give modal logic something to do, for so far it had not helped solve any philosophical problems since its birth in 1912. Actually there are 256 possible systems of modal logic, as David Makinson told me. But, since none of them has ever been used, nobody knows which is the more useful.

My criticisms caused unusual hostility, for all my listeners had invested many years studying silly pseudoproblems, such as whether the name of a person is the same (a "rigid designator") or different in different worlds, and what would life be like on a dry Earth. Obviously, my choice of topic had been unfortunate, for what I told my audience is that they had been toiling in vain.

My research in Sydney, or rather at the Botanical Garden and the well-stocked NSW university libraries, centered on inverse problems, such as that of "inferring" (actually guessing) the intention that triggered a given behavior. This problem is the heart of my book *Chasing Reality* (Bunge 2006a). It is so unfamiliar to philosophers, that the referees for several philosophy journals recommended rejecting my paper on the subject, even though nearly all of them started by admitting that they had never met the expression 'inverse problem' before. They did not know that mathematicians keep trying to discover the assumptions entailing certain theorems, scientists keep guessing the causes of certain events, and creative engineers try to design artifacts to perform desired operations – whence the phrase 'reverse engineering'. The first international congress on this subject was to meet in Hong Kong only a year later.

While in Sydney, Marta and I walked on the coastal sidewalk, boarded ferries to various islands, and took walks in the rich Botanical Garden across the street. There we admired not only exotic plants and flowers, like the voluptuous *ceibo*, the Argentine national flower, but also Egyptian ibises that, like their ancestors, nested on the top of palm trees; majestic white cackatoos that ate beautiful flowers with enormous waste; and "gray foxes" (bats) that by day hang from trees like enormous fruits, and after sundown dined silently on huge wild-fig trees.

When Eric and Silvia came to visit, we all traveled to the Great Barrier Reef, where we swam amidst huge shoals of fish of all sizes and colors. After sunset our Long Island garden was visited by timid wallabies that kept the lawn well trimmed and fertilized. Silvia, always the first to go for a swim, was bitten by a poisonous jellyfish after swimming a few strokes. She was lucky, since the sting of some jellyfish in the same cove is deadly.

On September 12th, 2001, gigantic screens were installed in Sydney harbor to show images of 9/11, the biggest of the small-scale terrorist attacks in history – the large-scale ones being wars, of course. A few hours later Eric phoned from New York to tell us that he had watched the two towers crumbling from a bus, for a few minutes earlier he had been visiting a client in one of the towers. A decade later he and his Columbia University students left Fukushima the day before it was struck by the tsunami. Lucky Bunges!

The Flood

At the beginning of 2004, on our return from a family reunion in Puerto Escondido, México, we found that a waterfall had knocked off the complete works of Shakespeare, Lope de Vega, Balzac, Eça de Quirós, and other friends, and had flooded both the main floor and the basement of our house on the slope of Westmount. About 1,000 books, first of all the arts ones, were utterly ruined. Only one book was floating: the thick volume where the famous physical chemist Wilhelm Ostwald had expounded his energetics, the idiosyncratic ontology according to which everything is energy. He claimed that, since energy is neither material nor spiritual, energetics overcomes both materialism and spiritualism. The flotation of this book may not confirm his theory, but is a clear testimonial to the superiority of the paper and ink used in 1902 over the current ones.

The flood had been caused by the burst of the radiator under the window through which the burglar had entered: when the water froze, it dilated, thus breaking the pipe. The burglar took away everything that was or looked golden, including the gold-plated ring that Salamanca University had given me together with the honorary doctorate diploma.

Everything but the drowned books was recovered or repaired. The craftsman who came at midnight to weld the burst pipe arrived soon after I phoned him, along with the policemen and the insurance agent. The policemen could do nothing, but the insurance official compelled us to move to a hotel, because our house had been invaded by dangerous mold and fungi.

Marta and I spent the next 6 months at a hotel in the middle of the gray and cheerless old town, empty during the night. Our suite, built three centuries earlier with granite stones as a harbor deposit, was inhospitable, but we kept teaching and studying, and even hosted our good Australian friend Michael, whom we regaled with Chinese beer. We went daily to our century-old home, to inspect the extensive repair works.

Marta was happy because of the proximity of both the river and the subway, but I felt dejected by the gray old buildings and the lack of green spaces. The only serious book I remember having read during that period was the one by the Peruvian Argentines Carmen Dragonetti and Fernando Tola about the striking parallels between the Indian and European philosophies. The only fiction books I read at the time were Lindsay Davis's engrossing novels about the private investigator Marcus Didius Falco, set in Rome after AD 70. And our only interlocutors were a Korean grocer and the Greek who parked our car in the ice-covered yard. His colleague slipped on the ice and had to be hospitalized.

Beijing

In the early 1980s a delegation of the Academia Sinica caught me by surprise at my office. One of its members told me that, before the so-called Cultural Revolution (1966–1976), he had worked in theoretical physics, and at the beginning of that tragedy he was sent to a farm to distribute human waste. When freed, his mind had become too rusty for science, so he devoted himself to philosophy. Think of the nauseous chain reaction: science-political fanaticism-shit-philosophy.

Shortly thereafter, a Chinese university invited me to spend a year as a visiting professor. My daughter Silvia, about 10, was enthusiastic: "Think how that would look on my vita!" I declined the invitation because I was expected to teach 8 h a day, as well as visit several other universities, which would have wrecked my work plans.

I did not hear again from China till a day in the summer of 2011, when my former student and secretary Robert Blohm, protégé of the Nobel laureate Robert Mundell, and who was working in China as a financial counselor, called unexpectedly. He brought along his friend Jason Chung, a friendly Taiwanese American businessman who was buying big firms on behalf of the Chinese government.

Jason came lugging a suitcase filled with books of mine, which he wanted me to sign, to donate to the Peking and Tsinghua universities, the oldest and most prestigious in China. They had commissioned him to invite me to give some lectures at them. He himself would collect us in Montreal and accompany us during the whole round trip.

Nobody guessed that Jason's constant coughing was a symptom of the lung disease that would kill him a few months later. He knew it, and told us that he wished to do at least one disinterested action in life. As promised, Jason came in October and flew us first class to Beijing, where he introduced us to his secretary Amy – who was to be Marta's personal assistant – his partner, the effective and friendly Mister Hu, the photographer, and the driver, all of whom had free access to all the official buildings, including the Party's.

Next morning, my first appointment was with a journalist from the official news agency, who wished to know what I thought about Chinese socialism. I told him that so far I had seen only the hotel's breakfast hall but, from what I had read in the Western press, no traces of socialism remained in China, whereas there was some of it in Argentina, where health care and education were free. The journalist looked nonplussed and asked no further questions.

We were treated like royalty: driven to admire the usual tourist sights, like the Great Wall and the Forbidden City, and invited to sample the most delicious and wholesome meals, unknown in the West. When I tripped and hit my head against a wall, as a consequence of which I suffered a concussion and my scalp bled profusely, I was immediately ferried to a modern hospital, where I got five stitches and a CAT scan. I was also ordereded to sit on a wheelchair wherever I went. When told that I was a philosophy professor, the worker in charge of lifting wheelchairs at the Forbidden City, asked me what I thought about Chairman Mao's conflict theory. Did he doubt it?

Beijing Lectures

Marta and I gave talks to select audiences at Peking University, the oldest in the country. I also lectured at Tsinghao University, the Academy of Sciences, and the Marxist School of the ruling party. Our audiences were curious, attentive, and polite. We were welcomed with flowers galore, as well as with flattering speeches, and seen off with delicious lunches. In short, we enjoyed lavish Oriental hospitality, although I did not mince my criticisms of both the regime and its philosophy.

I dealt with different subjects in every one of my five talks, but in all of them I stressed my central message: Chinese philosophy had not moved along with its industry, technology, and science. In fact, dialectics, the heart of Marxist philosophy, has remained untouched despite being false in the best of cases, and confused, hence incapable of being discussed rationally, as well as empirically untestable, in the worse. For example, the elementary particles are not "unities of opposites," and "negation of a negation" makes no sense except as affirmation. In particular, I don't see my great grandchildren as negating the negation of poor me.

Nor is it true that conflict ("contradiction") is the mother of all change. Although there is competition, and even conflict, in everything social – in part because all the important resources are limited – cooperation is more important than competition, for it breeds systems, within which and between which conflicts happen. Moreover, the cult of conflict is politically suicidal, since a leader's main task is to resolve conflicts, not to exacerbate them. Remember that the disastrous Cultural Revolution was provoked by the Chairman's idea that, since Chinese society had resolved its main "contradictions," it risked remaining stagnant, whence it needed a kick to keep advancing. Never before had a philosophy provoked such a colossal tragedy.

Hence my exhortation: "Jettison Hegel and his dialectics, and update realism and materialism with the help of modern logic and the sciences, both natural and social. Further, admit that these disciplines have developed outside the Marxist box, and that most twentieth century Marxist philosophers have played a reactionary role inasmuch as they rejected all the major scientific breakthroughs of their time. Advance beyond Marx and Engels: replace dialectical materialism with systemic and scientific materialism. Seek guidance in the contemporary scientific literature rather than in obsolete books." Perhaps I should have added: "Stop confusing socialism with statism, start replacing dictatorship with participation and cooperation, and remember that great revolutionaries do not necessarily make wise statesmen."

My talks were received respectfully, perhaps largely because of my white head of hair. Many of the issues my listeners raised were relevant and interesting, though somewhat long-winded. However, most of them concerned authors rather than problems, and nearly all of the authors that interested my interlocutors were either lightweights like Jürgen Habermas or charlatans like Michel Foucault – the scribbler who had combined Marx with Nietzsche. Worse, none of my answers provoked debates.

My listeners seemed to admire the quickness and vehemence of my replies. Presumably, an old teacher is expected to be slow and moderate, as well as to respect icons and refrain from cracking jokes. Perhaps all of this is part of a millenary tradition without dialogue, where quoting alleged authorities replaces both empirical evidence and rational debate, and white-haired teachers are treated either as sages to be venerated or as old fools to be suffered.

I do not know what impact, if any, my criticisms and proposals have had. However, the leaders of the schools where I lectured assured me that my talks were successful, they invited me to return, and projects to translate and publish some of my books were hatched. Was this mere Oriental courtesy? *Qui vivra verra.*

Coda

All of us are subject, once in a while, to unforeseen calamities. Unless the latest of them is also the last, the best way to reconstruct one's life after a disaster is to try and resume the work one was doing when it struck. And for this to happen one needs to stick to long-term plans demanding tenacity and concentration: in this case the accident is only a temporary interruption. To paraphrase Socrates, an unplanned life is not worth living, for it risks being grabbed by someone else.

Chapter 15
Practical Philosophy

It is widely believed that philosophy is contemplative and that to take things philosophically is to resign oneself. Actually there have always been proactive philosophies and ideologies, like pragmatism and Marxism, alongside contemplative ones, like original Buddhism and Stoicism. In addition, whereas some moralities – like Confucius's, Christ's, and Kant's – have been socially conformists, others have been dissident or even rebellious, though fragmentary rather than systematic. Finally, there have also been double moralities, like the Nietzschean or fascist one, that praised the leader's "heroic" (violent) ways while exhorting "Believe, obey, fight!" to the masses.

All the extreme ethical attitudes may be regarded as vulgar versions of practical philosophy, which is a serious reflection on action and its conceptual concomitants, from moral issues arising in daily life to the general principles guiding policies, projects, and actions in engineering, medicine, education, the law, and politics. For example, the debates on gender discrimination, torture, prison, war, the distribution of wealth, the duties and limits of the state, the treatment of animals, the negative side of technological advances, and more belong in practical philosophy, not in semantics, epistemology, or ontology – the main chapters of theoretical philosophy.

Axiology

Axiology, or value theory, analyzes the general concept of value or worth and examines particular values, both personal like well-being and social-like peace. I became interested in this discipline while reading Alfred Stern's comprehensive *La Philosophie des Valeurs* (Stern 1936), published during the first period of its author's expatriation process, which took him from his native Vienna to Paris, México City, Los Angeles, and Puerto Rico. In 1944 he mailed me a copy of that book from México, along with his first submission to my journal *Minerva*.

M. Bunge, *Between Two Worlds*, Springer Biographies,
DOI 10.1007/978-3-319-29251-9_15

Stern's article was a frank and acerbic criticism of Max Scheler's defense of war – in particular of the German side in the First World War – in the name of phenomenology, which then, along with Thomism, ruled the establishment philosophy in the Western world. Stern's exposure irritated Francisco Romero and his circle, all of them admirers of Dilthey, Brentano, and Husserl. Some of them despised Anglo-Saxon philosophy because of its clarity – as Nietzsche had written – and because, as Romero told me, the Brits wrote mostly short essays instead of heavy tomes. Stern's paper also angered all those who, like the influential and politically pliable legal philosopher Carlos Cossio, hid their own confusions behind Husserl's high-sounding nonsense.

My second encounter with axiology happened around 1960, and it was a reaction to the obscurity of most value theorists, even my colleague and friend Risieri Frondizi. He had been Whitehead's doctoral student at Harvard, and neither of them had bothered telling their readers what values are, let alone how they emerge and submerge. When I tried to give the logician Jorge Bosch an impartial abstract of Frondizi's philosophy, he said: "I understand. He is at the vanguard of the rearguard."

My reaction to conventional axiology came in two parts: an analysis of the general concept of value (Bunge 1962b) and a value theory that remained unpublished, but which eventually I used in the first of my papers on approximate truth-values (Bunge 1963b). In both cases, my aim was to show the superiority of exact over intuitive philosophy. I did not attain this aim, not only because of my own shortcomings, but also because nobody except for mathematicians and physical scientists was interested in exactness at that time, which was when postmodernist vandalism began to seduce some scholars because it encourages nonsense and others because it looks leftist.

Value Systems

After those essays I became increasingly interested in bundles or systems of values (Bunge 2009, 2014d, 2014e). Indeed, values come in bundles, in the sense that no value can be upheld in isolation from others – as in the case of education, which comes together with income and health. Secondly, higher values, such as knowledge and liberty, can only be attained if such basic values as employment and health are secured at the same time. However, there is one proviso that some legal philosophers have noted, namely, that, whereas some values reinforce one another, others conflict with one another. In particular, every right is constrained by some duties – whence the immorality of libertarianism and the irrelevance of contractarianism. Likewise, duties without rights are unbearable – whence the legitimacy of rebellion against oppression. It follows that pursuing one value at the expense of others is bound to either failure or to exact painful social costs – and yet one-sidedness is the rule in both value theory and social practice, from public works to social policies.

A systemic approach would have spared many costly endeavors. I hope I have heeded this advice in my political philosophy (e.g., Bunge 1989e, 2009).

Let us now deal briefly with the nature of values. Platonists assert that values are pure ideas and that things, attitudes, and values "participate" in them – an obscure idea; by contrast, intuitionists hold that values are desiderata or tastes and, as such, not objects to be analyzed, let alone be the subject of rational dispute. In my view values are relational properties, just like position, friendship, and quality of life. For example, one says that an animal *a*, such as a person, appreciates (or not) item *m* as a means to attain goal or desideratum *g* in situation *s*: *Vamgs* (Bunge 1962b, 1989e).

Thus, one does not handle values in themselves but in quadruples: *evaluating organism, evaluated item, goal, and situation*. This analysis expands my previous one, in that it includes situations and it generalizes valuations from people to all animals. In other words, there are no values in themselves, as Platonic ideas, but *valuations* made by living things – from unicellular organisms to humans – of things, properties, or processes, that may alter with the situation that the evaluator assesses.

The explicit reference to evaluators and their circumstances does not validate the relativistic thesis that all values are personal and local or tribal. When all the members of a very large group appreciate the same thing in a similar situation, one is entitled to talk of impersonal and universal values, such as health and nourishment, coexistence and solidarity, truth and efficiency, and so on. That is, our ontological value relationism does not imply the moral relativism involved in the belief that the members of the in-group have privileges denied those of the out-groups.

Another important if obvious point is that there are many kinds of value: biological and social, technical and artistic, conceptual and moral, and so on. Hence one and the same item may be good in some respects and bad in others. For example, the nuclear bomb is at once a technological marvel and a moral and political monster; likewise rock "music," particularly in its acrobatic version, is a pricey commercial product that has seriously damaged music and poetry, not to mention our hearing organs.

These are not just personal opinions, but outcomes of objective analyses: indiscriminate mass murder is the worse crime, and uneducated noise that combines exhibitionism with vulgarity, or even brutality, is a cultural crime. True, art appreciation is a matter of taste, but taste can be educated, evaluated, and argued about: there is good taste and bad taste. For example, we all distinguish between erotica and pornography, paintings and posters, symphonic orchestras and bands, and so on. Thus, the popular saying *de gustibus non est disputandum* is empirically false. In sum, I subscribe to *moderate axiological realism*, according to which all tastes can be the object of rational debate, and some values are objective. This is why the test of time is so important. A related achievement of axiological analysis is that some values can be ordered as "tables," or rather *n*-tuples, such as satisfactory-good-excellent, true half-true-false, and pretty-beautiful-sublime.

A fourth result is that, since values are goal oriented, they call for an objective study of the best means to attain them. And, since science and technology are the best means to acquire knowledge about means-goals pairs, ethics is best built upon those fields of knowledge. Shorter: efficient ethics is scientific (Bunge 1961b). This is, of course, a scientistic thesis and as such abhorred by all antiscience philosophies. But since, as is well known, the end does not justify the means, technology – the arm of science – ought to be subjected to ethical vigilance. So, we face a difficult, albeit interesting, vicious circle: to be efficient, moral action must be guided by science and technology – which in turn may be bad unless controlled by a humanist ethics.

A fifth and far from trivial finding of axiological research, intimated above, is that values come in bundles or systems rather than one at a time. For example, diet must be combined with occupation and housing to be a health factor, and in turn scholastic achievement depends critically on health, which depends on public sanitation. Likewise liberty, equality, fraternity, and competence constitute a political package, whereas privileging one of the members of this quadruple is a recipe for injustice and failure. I have discussed this matter at length in my *Political Philosophy* (Bunge 2009).

Finally, there is the open problem of the emergence of values in the course of evolution. On an evolutionary worldview, before the emergence of life, there were no values: these emerged just a few billion years ago together with the earliest living beings. Moreover, some values have preceded others. For example, questions of the forms "What is *X* for?" and "Why are there still organs and functions without any survival value?" are about all animals, whereas "Which were the advantages and disadvantages of encephalization, bipedalism, agriculture, and urbanization?" are much younger, since they make sense only with reference to humans: they are human values. Note also the difference between these axiological problems, which can be investigated scientifically, and the teleological dogmas that everything happens for a reason and that what distinguishes organisms is goal seeking. Goals involve valuations, but the converse is false.

In short, human values are made, not found, and they come in bundles; some of them can be rationally justified; some of them must be expected to emerge in all the domains of human life, whether or not they are alethic (truth dependent); and some value gains may come together with value losses, as in the classical case of the chronic backache that resulted ultimately from the adoption of the bipedal posture. Note that all of these views about values are at variance with Platonism and were suggested by scientific findings over the past couple of centuries.

Analethic Values

Let us shall call *analethic* those activities that do not accompany the search for truth even though they cannot help using some truths. For example, domestic chores, games, sports, language, art, and religion are not centrally concerned with the acquisition of knowledge proper. Indeed, they involve noncognitive values, such as home comfort, physical fitness, communication, beauty, or spiritual elevation, and they are best learned by doing. All analethic activities are rule abiding rather than truth seeking, and they are not transferable to other fields. For example, games and religious practices have no use for one another; likewise, home economics and linguistic skills are mutually independent. By contrast, all alethic activities produce or use truths that can be exported to some or all other fields.

Another peculiarity of alethic activity is that it involves integrity. Cheats may triumph in the market, from financial speculation to alternative medicine to postmodern scribbling, but they do not go far in a scientific or technological laboratory. Of course some literary or artistic masterpieces – such as Aesop's fables, Maurits Escher's visual illusions, Isaac Asimov's science fictions, or Italo Calvino's fantastic stories – are untruthful, but not deliberate and harmful lies.

Lastly, non-alethic activities are not evaluated by their truth content but by nonalethic criteria, such as nutritional value, entertainment, political gain, or pecuniary worth. For instance, social psychologists have found that value sharing is critical in the emergence and maintenance of social groups, from families to gangs to political factions.

Actually everything social, and also much that is just biotic, involves values. Therefore, contrary to so-called received wisdom, there is no value-free social science, let alone value-free sociotechnology. Values should be evaluated in the light of reason and experience, instead of being either adopted or rejected dogmatically.

Evaluating Alethic Items

Truth, particularly if original and deep, is the whole point of research in basic science. For example, two millennia ago the polymath Eratosthenes conjectured that the Earth is not flat but spherical and devised an ingenious yet simple method to measure its circumference, and his result, 40,000 km, is a good approximation to the true value. Eratosthenes would not have bothered to carry out this investigation if he had been a radical skeptic or if he had thought that he was about to *construct* our planet instead of trying to *represent* it as faithfully as possible. In other words, he was an epistemological realist, and accordingly he was after objective truths, not philosophical mirages. But, being the highest value in science, original truth is also the hardest to attain, so much so that the classical empiricists preferred the timid expression "highly probable," the logical positivists "strongly confirmed," and Popper "so far not falsified."

One of the reasons for such epistemic prudence on the part of the abovementioned philosophers is that they had in mind only isolated hypotheses supported or undermined exclusively by the relevant empirical data. But this is not how science has worked since Newton. Indeed, ever since his 1687 *Principia,* scientists have confronted their guesses not just with relevant data but also with conjectures in the same or closely related fields, as well as with the ruling worldview. This is why we should combine empirical with theoretical truth-values, as I suggested in *Scientific Research* (Bunge 1967b).

For example, one of the reasons string theory is still in limbo after half a century of intensive investigation by thousands of able mathematicians is that it postulates extra-spatial dimensions that occur nowhere else; "intelligent design" fantasies are universally rejected because they invoke at once teleology and miracles; and dualist psychology is quickly losing ground because it is incompatible with neuroscience as well as therapeutically barren. Isolates do not make it in science. Science is just as systemic as the world it studies. For instance, a single negative experimental finding may falsify an isolated hypothesis, but not one belonging to a whole system of well-confirmed hypotheses held together by the logical relation of deducibility. This is why, when Wolfgang Pauli heard that a certain nuclear reaction violated energy conservation, he surmised that the missing energy was carried away by a hitherto unknown particle – the elusive neutrino – whose existence was later confirmed, but one of whose basic properties, mass, has been the subject of investigation and debate for half a century.

Popper's much-debated assertion that disconfirmation is more reliable than confirmation, so that scientists ought to seek to falsify their pet theories instead of trying to confirm them, is at once psychologically and methodologically false. Indeed, only masochists behave that way; disconfirmation is not more reliable than confirmation, and the only truth indicator we know and rely on is *confirmation together with compatibility with extant knowledge* (Bunge 1967b).

A proposition that fails both conditions is neither true nor false, but an untested proposition that coheres with the bulk of background knowledge is plausible and therefore promising. And, while a few Nobel Prizes have been awarded for confirming hypotheses, such as that of the existence of transuranians, most of them have rewarded discoveries, such as that of the position neurons in the hippocampus. More to the point, none have been awarded for just refuting hypotheses – except in the trivial sense that confirming H falsifies not-H. The same cognitive values occur also in the evaluation of research (or grant) proposals. The evaluator must judge whether the project is likely to yield new and nontrivial truths, besides estimating the chances that s/he will be capable of carrying it out with the means at her/his disposal.

However, the first consideration of a project's alethic evaluator is whether the project is *scientific*, particularly at a time when many genuine scientific finding come accompanied by grand pseudoscientific fantasies in hopes of attracting handsome financial support. Remember the cold fusion swindle, the unfulfilled promises of string theory, the multiverse extravaganza, and the idea that the universe started out from nothing and is bound to end the moment the Higgs boson ("the God particle") is produced? In short, legitimacy questions, all of which are about values, arise at all the stages of any scientific project (see Bunge 2004c).

So much for basic science, whose overriding goal is finding new truths about the world. Technology and the bridge between it and basic science, that is, applied science, have values of their own, to which we turn now.

Technophilosophy

I was one of the five or so youngsters who enrolled in 1938 in the Faculty of Physicomathematical Sciences of the National University of La Plata, about 100 km south of my home, to study physics: the overwhelming majority of my fellow students wished to become engineers. We all knew that technology differs from science and that the job prospects of science graduates were dismal at that time. We also took it for granted that, for the modern engineers, conceptual tools are far more important than manual skills: they are intellectuals, not craftsmen. This is why the first 2 years for both parts of the student body – those few who were just curious about the world and those who wished to alter it – had roughly the same program of studies, worked at the same labs, and were taught by the same teachers.

However, let us not confine technology to engineering. From the start I have understood technology sensu lato, as the study of human action in all fields, from agriculture and industry to education science and jurisprudence and from management science to public policymaking.

The idea that one ought to understand the world before attempting to change it is the exact obverse of the lemma engraved on Marx's tombstone, his famous Thesis XI on Feuerbach – "The philosophers have only interpreted the world in various ways; the point is to change it." I rejected this thesis even during my Marxist phase, but it was only after Che Guevara's tragic failure in Bolivia that I fully realized that replacing the scientific study of social reality with dogma and heroism can only lead to the barbaric policy summarized in the slogan "Shoot first, ask questions later."

The same mistake – acting without the requisite knowledge – also helps understand the inability of Mikhail Gorbachev's team to prevent the sinking of the Soviet Union in 1991: they did not know the ship they were attempting to steer because there was no social science in the country. And there was no such science because they believed that a handful of Marxist slogans could replace the detailed study of social reality. In my *Social Science Under Debate* (Bunge 1998a), I suggested a bundle of interrelated causes of several kinds to explain that shipwreck. Mindless repetition of Thesis XI is surely one of them, as well as a source of the anti-intellectualism that has been infecting many Marxist groups.

Modern technology is based on science but it has its own values, which we keep changing along with it. New civilian technological projects are required to be compatible with the underlying sciences, as well as leading to useful and profitable, efficient and safe, and also user-friendly things and processes. Obviously, new products of military technology are expected to embody somewhat different values, in particular maximal destructiveness. For example, automated weapons such as drones will be preferred to conventional ones because they can kill at a distance and with impunity. Actually, all aggressive artifacts involve the substitution of an underlying criminal behavior code for a humanistic one. But the point is that all deliberate human action, whether beneficial, harmful, or neutral, is value laden.

The axiological commitment of technology is particularly obvious where the objects to be manipulated are humans, as in the case of management science. I became intrigued by management science (or rather technology) the moment I saw the journal by the same name in a library and a reader on operations research (OR) in a bookstore, on my way to teaching quantum mechanics in La Plata around 1956. I learned thus that the scientific method may be used to design not only nonliving artifacts but also social operations, from warehousing to shipping under threat of enemy submarines – which reinforced my scientistic conviction. I also learned then that management science is a conglomerate of mathematical models built as problems emerge in the course of managing social organizations of all kinds with the help of the universal science – mathematics. But those models use some empirical data and are evaluated by their practical utility.

During the next decade, I attended some talks by C. West Churchman in Philadelphia, and a decade later I befriended his former student Russell Ackoff, aka Russ. His talk at my seminar at the UNAM (Universidad Nacional Autónoma de México) was about his ambitious project of concentrating all the offices of the French government in Orléans, to make Paris more livable. The Wittgenstein faithful were indignant: they saw no philosophical merit in a talk that was not about words, private languages, or similarly deep topics.

Both Russ and this teacher had had unusual careers: they had written their doctoral theses on philosophical problems, for a while had taught this subject, learned some mathematics on their own, became business consultants and leaders of the general systems movement, and ended up teaching management science at the Wharton School of Management in the University of Pennsylvania, which metamorphosed from Cold War fortress into the most innovative center for the systems-theoretic study of business and economic issues.

The Canadian Dan A. Seni had graduated from the same university and audited several of my courses just "to get some intellectual stimulation," as he put it. Eventually he asked me to supervise his doctoral dissertation about planning, which in 1993 he submitted to the Wharton School of Management. As a result, Dan, his wife Helene Gaudry, and their daughter Anne became close friends of my family. His death from cancer at 69, after a long and painful fight, was a hard blow for us all. In addition, Dan's death aborted our plan to write together a book on technophilosophy.

Around 1980 Dan's father-in-law, the chemist and university administrator Roger Gaudry, invited me to join a group of captains of industry, among them the boss of Alcan, the aluminum giant, to set up a science museum in Montreal. We gave up after meeting and consulting over about 10 years, when we realized that the provincial government was not interested. How many votes could a museum bring?

Dan's dissertation subject was politically incorrect, for that was the time when the Reagan and Thatcher governments exalted spontaneity and liberty while conspiring to destabilize the governments that did not bow to their authority. Dan had earned undergraduate degrees in economics and engineering and, being perfectly bilingual, had taught management science at the two Francophone universities in Montreal. He had also worked in Saudi Arabia. One night, when returning to his office to check some computations, he saw the Bedouin in charge of tea kneeling in front of his large computer. He was worshiping the Western boss's powerful deity.

Let us not laugh at the tribesman's superstitions, for our students too worship computers and other electronic marvels: they prefer handling them to thinking without their assistance, doing mental calculations, or dealing with other people face to face. Worse, they never find time to examining their own lives, as they are tethered to their cell phones waiting for the next message, catching the latest Facebook post, or seeking an excuse to mail the next text.

The case of the Brazilian engineer Denise Fleck was similar to Dan's: she enrolled at the McGill School of Management, took a number of courses of mine, and wrote her dissertation on the firm's growth (2001) under the joint direction of Henry Mintzberg and myself. She recently flew from Rio de Janeiro to Buenos Aires, and later to Montreal, to chat with her old professor, who was able to give her some information about recent findings in experimental economics.

Incidentally, I have learned much from Henry Mintzberg's publications on management strategies. When he joined McGill, freshly graduated from MIT, he sought me out because he had liked my book on intuition and science (Bunge 1962c). At that time I overrated the mathematical modeling of business operations, whereas Henry thought that they were too far removed from real-world businesses. I have come to appreciate his viewpoint: I find most of the current papers in mathematical economics and management science to be just mathematical exercises (Bunge 1998a). I also approve of Mintzberg's proposal to require from all graduate students in his faculty to have had some business experience. Last, but not least, we share our admiration for cooperatives and other agents in the civil society (sometimes wrongly called "civil capital").

Another doctoral student who wished to work on technophilosophy was José Félix, a friendly Basque engineer whom I engaged as a house sitter while my family and I spent the long summer of 1985 in Mallorca. He seized this opportunity to photocopy all my unpublished essays and filing cabinet cards. The external examiner, who was familiar with my work, flunked José Félix's dissertation, alleging that it had been lifted from my publications. Another student who tilled the same field was an enthusiastic but undisciplined Colombian engineer, too arrogant to plagiarize me and even to accept my observation that in English one says "sectoral," not "sectorial" as in Spanish.

A third unproductive comet was a Mexican who wished to work on the foundations of classical mechanics as understood by the so-called structuralist school. He panicked when I showed him the chapter on mechanics in the *Handbook of Physics* and immediately flew back home without taking his leave. He ended up switching to the philosophy of religion. A friendly Spanish viscount who used to exhibit his crest read a lot but wrote nothing. There was also a Goan Indian who had emigrated to Uganda, then to Brazil, fleeing the cannibal Idi Amin. Every one of these individuals kept a tiny philosophical flame flickering, but none had the discipline necessary to transform it into a roaring fire.

Fortunately, my great friend Miguel Angel Quintanilla, who has been teaching at Salamanca for decades, overcompensated for those failures when, while a guest at my unit, he decided to devote himself to technophilosophy. He has become an international authority in this field. In addition, contrary to most of us, who only use some artifacts, Miguel Angel designed and built some robots, to get a feel for robotic engineering.

Ethics and Praxiology

I was always critical of Francis Bacon's praise of science for its alleged practical fruits, as well as of the Marxist classical scholar Benjamin Farrington (1951), who opposed what he called Bacon's "philosophy of works" to the schoolmen's "philosophy of words." The confusion between the two fields is common among Marxists, and it is rooted in Marx's celebrated Thesis XI of his sketchy evaluation of Ludwig Feuerbach's virtues and shortcomings. No wonder then that the Marxist philosophers Antonio Labriola – a respected Italian professor – as well as the Spanish expatriate Adolfo Sánchez Vázquez, one of the rare Marxist moral philosophers, whom I befriended in México, claimed that Marxism should be conceived as "the philosophy of praxis," that is, as praxiology. This pragmatist view might be regarded as an isolated glitch, but it comes together with Engels's pragmatist truth criterion, summarized in the proverb "The proof of the pudding is in the eating."

In my view, collapsing philosophy into praxiology is a serious mistake, for it suggests that Marxists are equipped to face social problems without first learning some social science. The social cost of this conceit can of course be huge, as shown by Che Guevara's Bolivian adventure: he and his comrades failed not just because the CIA-propped military were far better armed, but also because the Bolivian peasants whom Che wished to emancipate were not wiling to risk their lives for an unknown foreigner's promise.

In the best of cases, Marxism might help investigate social reality and design the best strategies to alter it for the better. In other words, an efficient praxiology is not the work of philosophers, ideologists, or social activists: it is a hard discipline sandwiched between specialized scientific and technological findings on one side and a humanist ethics on the other.

The most dramatic experience that confirmed to me the differences between basic science and technology happened in August 1945, right after the American nuclear attack on Hiroshima and Nagasaki, which was at once a gigantic technological feat and a monumental war crime. At that time I was strolling in Buenos Aires in the company of Dr Guido Beck, my thesis advisor, and we speculated on the forthcoming peaceful uses of nuclear energy. In particular, we wondered whether we would see the nuclear cigarette lighter, which would last a lifetime without recharging. Although we were both working in nuclear physics, we did not even know about critical masses. Our most interesting research problem, namely, the nonclassical forces that keep together the constituents of an atomic nucleus, was scarcely of interest to the pioneers of nuclear engineering, who took the existence of such forces for granted and were interested in using them, not in understanding them. They used nuclear physics only as one of their means to get things done, or rather undone. Manhattan Project, the greatest technological enterprise in history, would not have been succeeded without the huge engineering team and sophisticated organization it involved.

The Bomb should have taught everyone that, although modern technology uses much basic science, it also contains knowledge that does not result from disinterested research, and that poses utterly novel philosophical problems, from the control of microphysical processes to the control of the social forces in charge of them – what General Eisenhower famously called "the military-industrial complex." Only sociologists like C. Wright Mills pointed their finger at an even more comprehensive system, namely, the power elite at the core of any polity in any of the handful – nowadays dozen – of nuclear powers.

The moral dimension of nuclear technology became immediately obvious to everyone. By contrast, its ontological, epistemological, and axiological problematics have remained buried under centuries of neglect and misunderstanding. This problematics called for a new approach, starting with the recognition that we were facing new problems that the philosophers of virtue, utility, or justice had never faced.

The conceptual richness and moral ambivalence of technology call for a philosophy of its own. This new branch of philosophy was unexpectedly pioneered by José Ortega y Gasset, at an inappropriate place and time (Buenos Aires 1939). But it was not generally recognized as such until 1966, in the special issue of *Technology and Culture* titled "Towards a philosophy of technology." This issue included my paper "Technology as applied science" (Bunge 1966). This title was incorrect, because applied science is not the same as technology, although it often leads to it, for it is the bridge between basic or disinterested research and technology, or the design of artifacts. The original title of my article was the one that the journal's editor, Melvin Kranzberg, used to title that special issue. Nor was my paper completely original, but a rewrite of the paper published in Chile 3 years earlier (Bunge 1963a). That paper is included in several anthologies on the subject, and so puts me among the founders of the contemporary phase of the discipline.

In sum, the philosophy of technology is a recognized discipline, with its own society that meets regularly. But, ironically, it has been hijacked by a coterie of writers who embroider on the antitechnological extravagances of the anti-philosopher Heidegger and the theologian Jacques Ellul, neither of whom had a broad and clear grasp of modern technology.

Technology is vulgarly equated with engineering. In the long chapter about technology in volume 7 of my *Treatise* (Bunge 1985a), I conceived of it in a broad sense, as the discipline centered in the design of artifacts. There I distinguish several kinds of technology, in particular biological, like the part of plant science busy with genetically modified organisms, and social, like jurisprudence. What all the technologies have in common is the design or improvement of artifacts or procedures capable of changing reality for the benefit or the harm of people.

Design is the core of original technology, but not all technologists have the chance to design new things or processes. Indeed, the vast majority of them are employed in maintenance and repair jobs. Still, in the case of mega-artifacts, like hydroelectric and thermonuclear plants, maintenance can be harder than design and installation, because of the so-called human factor. Indeed, conventional management science is not prepared to cope with unforeseen events, such as the nuclear accidents at Chernobyl (1986) and Fukushima (2011), for they require quick decisions by low-level managers. And this is banned by the hierarchical chain of command involved in traditional management strategies.

Whether original or routine, the technologists's work requires far more knowledge and intelligence than that of their radical critics. It also involves concepts typical of their field, such as those of plan and design, input and output, efficiency and control, feasibility and maneuverability, risk and utility (both personal and social), and environmental harm and safety. Because all of these concepts are extremely general, they should interest philosophers as well as technologists.

Finally, let us glimpse at the technology-ethics connection, which has often been examined since the Nobel laureate Fritz Haber invented in 1916 the gas that started chemical warfare. There are two aspects of the technology-morality connection that have been generally overlooked. The first is that ethics may be seen as the technological branch of philosophy, for its aim is to help detecting moral problems and solving them in a humane manner (Bunge 1989e). Secondly, technology can help us solve some moral problems, such as that of minimizing the hazards incurred by people employed in dangerous jobs like uranium mining. But technology is not morally self-sufficient, for it does not supply moral norms. These arise and change in social practice and in our reflection about it. For example, the typical engineer is politically conformist because s/he is primarily interested in getting things done. S/he learns to behave in a fair manner at home and in the course of daily life transactions rather than by reading moral philosophers. Still, both in business and in politics, fairness pays in the long run, for it breeds loyalty and even friendship. The converse does not hold: placing loyalty and friendship above fairness leads to incompetence, clientelism, and even mafiosism.

For some reason I was invited twice to be part of international juries of large architecture and engineering works. We were conveyed by car from Madrid to Alcántara, near the Eastern border with Portugal, and stayed at the fortified monastery of a religious military order that had fought the Moors five centuries earlier. Lugubrious Gregorian chants followed us all the time wherever we went. There I had the privilege of conversing with the leader of the team of civil engineers who had designed the Alcántara Dam on the Tagus river. It is the largest in Spain, yet also a work of art because of the illusion of lightness that its concrete components cause. I also met with the famous architects Félix Candela and Rafael Moneo – both of whom had taught my son Eric at Harvard – as well as with the Duke of Calabria, proud but not a boastful manager of the Doñana National Park.

During a lunch that took a whole afternoon, José María Oriol, the CEO of the national hydroelectric company, told me in great detail why he admired Fernando el Católico, who had ruled Spain jointly with his wife, Queen Isabella: he had died poor for investing his part of the loot stolen from the Amerindians in another criminal adventure – the pillaging and murder of the peaceful Flemish who, by manufacturing those beautiful tapestries, had gotten too wealthy for their own good.

Back to my relation to art. I love it and feel moved every time I contemplate the Parthenon or Notre Dame, or listen to a piece by Beethoven or Mozart – yet not more so than when I walk on a Patagonian trail or see a beautiful child or woman. However, aesthetics is not in my philosophy. The reasons for this exclusion are that I have yet to meet an aesthetic law, and that normative aesthetics imposes mediocrity. What we do have are artistic conventions, fashions, and trends, such as Bach's revival in the 1930s, Wagnermania around 1870, the tango's eclipse around 1950, the Beetles' overnight celebrity in the 1960s – and the perceptions of "socialist realism" under Stalin's rule. But the conventions and fashions that characterize artistic genres, styles, and schools at different places and times are subjects for social psychology, sociology, and history. They are alien to both rationality and objective truth, without which there is neither science nor philosophy.

For example, one may analyze Borges's work and note its exquisite style as well as its rich and clever imagination. But, in my opinion, Borges is frigid, escapist, and devoid of moral and social sensitivity. For example, he describes the beheading of gauchos taken prisoners in a civil war with the same matter of factness that Lawrence of Arabia – a hero in Victoria Ocampo's literary circle, which included Borges – had described with the machine-gunning of a group of Turkish prisoners of war. An analysis of this text would be a piece of literary criticism or amateur psychology, not one in the philosophy of art – whatever this may be.

One may also study artistic creation, art appreciation, the art-society connection, the artwork market, or artistic communities. But none of these studies is properly philosophical. Indeed, the first two belong in psychology and the last two in sociology. Moreover, each of these studies may be conducted either scientifically, in particular experimentally, or not. For example, in his pioneering *Aesthetics and Psychobiology* (Berlyne 1971), which he summarized in my seminar, the Anglo-Canadian psychologist David Berlyne described some clever experiments in art appreciation. Likewise, Walter Benjamin's studies on the artwork as merchandise and Antonio Berni's unpublished thoughts about the art dealer as artistic arbiter belong in the sociology of art. (Incidentally, around 1950 Berni painted a portrait of Cantarito.) And we have only anecdotes about artistic creation – such as those told cleverly by Stravinsky and Ernst Gombrich.

But *philosophy* of art? Aristotle, Kant, Hegel, Croce, and other philosophers without artistic credentials wrote on aesthetics with the same assurance that celibate priests assume when pontificating about sex and family. Did those philosophers help create, understand, or spread art? There is no evidence that they did anything of the sort. On the other hand, the philosopher José Ferrater Mora, who never published in aesthetics, had an enviable artistic experience: 6 good novels and about 50 amateur films. After telling him repeatedly that he had the duty to write an aesthetics book, he started to work on it. Regrettably a heart attack struck him down when he had completed only four chapters.

Nobody can guess how many centuries will elapse till an artist-philosopher, or philosopher-artist, will attempt to carry out this task without provoking either boredom or hilarity. Lucretius, Leonardo, and Galileo stood out in both fields, but neither of them wrote on aesthetics. Is it because aesthetics exists only in certain curricula? We won't know until someone who excels at both art and philosophy makes a serious effort to join them. In the meantime I shall go on repeating Keats's questionable dictum *A thing of beauty is a joy forever* and regarding time as the most just of art critics, as my late friend Raymond Boudon thought (Boudon 2000).

Ethics and Praxiology

My first book on ethics, *Etica y ciencia* (Bunge 1960h), is in three parts: the thesis that science involves moral norms, such as the commandment to abide by both truthfulness and the disclosure of sources and patrons; the suggestion that ethics would benefit from respecting empirical tests; and utilitarianism à la Mill.

The first thesis had become rather familiar after the Bomb, and I had learned the second from the Argentine polymath José Ingenieros, who had been a close friend of my father's and whose children Delia and Julio I had befriended independently. I have retained the thesis that moral norms ought to be empirically testable, if only because they change along with the societies that adopt them. By contrast, I have never believed in moral naturalism – the unlikely intersection of ancient Stoicism, Thomism, and the naturalism rooted in Darwinism (despite Thomas Henry Huxley's reasonable objections) – and resurrected a century later in the wake of sociobiology and evolutionary psychology. If moralities were natural, there would be only one, and it would not have to be learned, debated, reformed, and enforced. In any event, the thesis of the practical testability of moral norms, which is the core of consequentialism, deflates the fact/value division and is usually dismissed by moral intuitionists and religious moralists as the summum of the "naturalistic fallacy." I still adhere to it because I cannot accept moral dogmatism any more than epistemic dogmatism.

On the other hand, I abandoned utilitarianism long ago, as soon as I was shown it to be wrong. I owe this sudden conversion to Tom Settle's question around 1970: "Where have you left obligations?" Tom had had an unusual career: he had been ordinated as a Methodist minister and sent to Hong Kong as a missionary. There my friend Joseph Agassi, who was preaching Popper's gospel at Hong Kong University, converted Tom to philosophy. After getting his PhD, Tom got a position in Canada at the University of Guelph and tried to persuade me to join him. A short visit to the dreary town of Guelph persuaded me to stay on at McGill. Ironically, a couple of years after converting me, Tom forgot his obligation to his family, took a lover whom he proudly paraded, and started wearing a monocle – quite a metamorphosis for a Methodist.

I had first met Tom at a philosophical colloquium held in 1969 at Stony Brook University in Long Island, where I gave a talk on "Theory meets experience" (Bunge 1970c). Sidney Morgenbesser interrupted it with one of the guffaws that made him famous – his single contribution to philosophy. Tom stood up and scolded that clown with his impressive preacher's voice.

In sum, I revised my moral philosophy, eventually replacing the concept of utility with that of the good, and the utility maximization postulate with that of balancing rights with duties. Only recently I learned that the motto of the International Workingmen's Association, or First International, founded by Karl Marx in 1864, was precisely *No duties without rights, and no rights without duties.*

Every moral philosophy has a supreme norm. Mine is *Enjoy life and help others live.* This principle joins egoism, needed to survive, with altruism, needed to coexist (*convivir* in Spanish). Therefore it may be said to be *egotuist.* I call *agathonism* the ethical system it heads, for it enjoins us to seek the good (*agathon*) for ourselves and others. Agathonism is an alternative to both hedonism (the pursuit of pleasure) and eudemonism (the practice of virtue), as well as to positive utilitarianism (the maximization of utility) and negative utilitarianism (doing no harm).

Far from being separate from the rest of culture, my moral philosophy is sandwiched between axiology and praxiology, and it resorts to the special sciences and technologies to evaluate particular means-ends pairs. For example, the designers of sanitary policies who use both scientific medicine and social epidemiology practice agathonism along with realism and systemic materialism even if they are not familiar with these words.

In contrast, the health economists who do not care about public welfare will complain that the usual 6 % yearly increase in the public health budgets of the advanced nations does not translate into a comparable longevity increase. They are likely to invoke the law of diminishing returns, instead of learning that public health is only one of the components of a complex social system, so that it cannot improve unless the other components (mainly income, lodging, and education) improve at a similar pace – as do the corresponding budgets. Recall the physician in *The New Yorker* cartoon who advises people to choose the treatments they can afford instead of the ones that may cure them. Protect us, o Asclepius, from run-of-the-mill health economists!

I have expounded agathonism in volume 8 of my *Treatise: Ethics: The Good and the Right* (Bunge 1989e). By supplying me with information and criticizing some of my ideas, my correspondents Antonio Colomer Viadel, Frank Forman, Ernesto Garzón Valdés, Pierre Moessinger, Hernán Rodríguez-Campoamor, and Tom Settle helped me hone them during the 1970s.

As indicated earlier, my ethics is placed between axiology and praxiology. The latter, better known as *action theory,* is the philosophy of praxis. Its best-known champion was the Polish philosopher Tadeusz Kotarbiński – a kind man preceded by a huge mustache and surrounded by a swarm of admirers – whom I first met at the Venice International Congress of Philosophy in 1958. Kotarbiński conceived of praxiology as "the science of efficient design and action" in all fields. Actually, his praxiology is not a *theory*, or conceptual system, but a collection of common sense precepts, on top of which it lacks moral norms. The latter is a serious shortcoming, for the doctrine might be used to justify inhumane actions, such as using weapons of mass destruction as long as they are efficient. In general, as argued above, the pursuit of a single desideratum is bound to have some bad consequences.

Conceived in extremely general terms and detached from both ethics and science, praxiology has hardly any practical value. We still do not have a general theory of efficient and good actions. But at least we know that Ludwig von Mises's version of praxiology is nothing but the principle of maximization of subjective utility – a fancy version of egoism.

My *Social Science Under Debate* (Bunge 1998a) contains a chapter on praxiology. There I explain two inference patterns that everyone uses, but logicians have ignored them because they revolve around the general value concept. I have called them *modus volens* and *modus nolens.* Here they are.

Modus volens:

Law If *A* is done, then *B* happens.
Valuation Fact *B* is valuable (to someone, etc.)
Conclusion Do *A*.

Modus nolens:

Law If *A* is done, then *B* happens.
Valuation Fact *B* is disvaluable (to someone, etc.)
Conclusion Avoid *A*.

Notice that these inference patterns are not only alien to deductive logic. They also combine facts and laws (natural or social) with values, in defiance of the alleged fact/value and law/norm dichotomies. But those inference patterns do not teach us how to estimate the worth of a decision, a plan, or a course of action. Every valuation calls for specific expertise and, in some cases, moral sensitivity and an explicit morality as well.

Medical Philosophy

The philosophy of medicine is a very immature sector of philosophy except for its ethical branch. And even this part is confined to personal issues such as abortion and assisted suicide: it ignores the social context, although all social epidemiologists know that health can only be improved if sanitation, vaccination, adequate housing, and other social desiderata are attained. I knew this because my father, a physician and champion of socialized medicine, never tired of reminding us of the social context of medicine.

Medics and nurses have always philosophized even without knowing it. For one thing, they are philosophical realists: they take it for granted that their patients and nurses have existed before them and that only few of the ailments they treat are iatrogenic. Second, medics are also philosophical materialists, insofar as they believe all diseases to be "organic" rather than the effect of curses or actions by malignant spirits. Only shamans hold diseases to be purely spiritual, so that it is their professional task to either fight or bribe them. And only the social constructivists, from Ludwik Fleck and Ivan Illich to Bruno Latour, have claimed that all diseases are invented by medical establishments, so that sick people should avoid "official" (scientific) medicine.

The job of medical philosophers is to tackle the specific ontological, methodological, and ethical problems raised by medical practice, such as the nature of disease, medical diagnosis as an inverse problem (symptoms → mechanisms), the mechanisms of action of remedies, the mechanisms of the placebo effect and psychosomatic diseases, the reach and limitation of RCTs (random controlled trials), and the pitfalls of Bayesian (subjectivist) statistics. Medical philosophers and sociologists are also expected to make use of research on the effect of stress on immunity and to help find out whether medical knowledge is purely biological or has a sociological component, the effects of income inequality on drug prescription and dosage, the corruption of Big Pharma and its impact on medical diagnoses and treatments, whether health care is a private matter or concerns us all, etc.

Some of these problems, particularly that of medical diagnosis, intrigued me from the moment I started to read fragments of philosophically inclined medical classics, such as Hippocrates, Harvey, Bernard, and Selye – whose first book on allergy I had skimmed soon after its publication and with whom I held a long conversation over the phone soon after my arrival in Montreal. However, I started to tackle those problems seriously only in 1976, when the Mexican Academy of Medicine invited me to participate in a colloquium on the philosophy of medicine, or *iatrophilosophy*, as I called it. There I presented my state-space approach of the disease and healing processes (Bunge 1978b, 1997b).

It was only a quarter of a century later that I conceived of medical diagnosis as an *inverse problem*, in particular one from syndrome to disease (Bunge 2004a, 2007c). This helps explain why medical diagnosis is so hard – as hard as guessing the forces acting on a body from knowing only its trajectory. I proposed that the clue to solving inverse problems in any field is to convert them into direct problems and to invent and try out hypotheses about mechanisms of action, such as the binding of the drug in question with its specific receptor in the case of medical diagnosis and treatment. Eventually I suggested various levels of sophistication of medical praxis:

Clay standard: traditional casuistry
Silver standard: experimentally based diagnosis and therapy
Gold standard: experiment + RCT
Platinum standard: RCT + mechanism(s) of action, where RCT abbreviates "randomized controlled trial"

While visiting Buenos Aires in 2010, I was approached by Daniel Flichtentrei, a cardiologist, university lecturer, and medical journalist in charge of *ProMed,* an electronic weekly. While conversing with him, we agreed that the philosophy of medicine was in poor health and that it would be nice to have a book on the subject, which were to cover more than just biostatistics and bioethics.

We planned this project, but I raced alone because, unlike the overworked Daniel, who starts his workday at 5 in the morning, I was able to work full time on it – and also because I felt increasingly weak. I wrote a few pages that Daniel criticized and stopped for a while until inspiration struck. I wrote a new outline, and the book started flowing effortlessly, at all times encouraged and advised by Daniel, my demanding but generous reader. A few additional researchers, notably the pharmacologist A. Claudio Cuello, the biologist Pierre Deleporte, the social immunologist Carles Muntaner, and the cell biologist Nicolás Unsain, checked some fragments of the book. This appeared first in Spanish (Bunge 2012b) and a year later in English (Bunge 2013a). Since then, Daniel has been promoting it with great generosity and energy.

Medical Philosophy deals not only with individual medicine – the one that treats individuals – but also with social medicine, which tackles public health issues like vaccination, drinking-water supply, and city planning. This subject emerged on the political agenda toward the end of the nineteenth century along with the earliest shoots of the welfare state, in the wake of smallpox and yellow-fever epidemics, and when tuberculosis and syphilis scared everyone and were regarded as social diseases – the subject of my father's 1900 doctoral dissertation. That was the time when medics of different ideologies started to talk about universal free medical assistance, even in remote Argentina (Sánchez 2007). My father was of course one of the leaders of this movement.

This could happen because medical treatment was dispensed only in small private clinics and public hospitals. The latter's wards used to start the day with rounds conducted by the ward's director surrounded by students and independent physicians who refreshed and updated their medical knowledge by participating pro bono in the discussions of problems posed by the examination of the progress of day's patients. All that started to change radically in the USA from about 1970, when health care became big business conducted by large clinics and medical insurance firms. These were of course the HMOs (health maintenance organizations) which, protected by conservative legislators, advised by economists devoid of social sensibility, started to defame Canadian medicare, starve public hospitals, and employ salaried physicians who were required to process their patients as quickly and cheaply as possible. It was only then that some in the medical profession started to wonder whether the existing economic order was consistent with a healthy medicine.

Political Philosophy

One of the first works in political philosophy I read was Rousseau's 1762 *Contrat Social* in an old edition that I had inherited from my grandfather the judge. I was not impressed because of Rousseau's earlier writings against science and even rationality and because I failed to realize the originality, depth, and subversiveness of his thesis that inequality is the mother of all social evils. Yet most political and economic thinkers have neglected the economic inequality issue. There is only one world-class institute specializing in this subject: that headed by James K. Galbraith at the University of Texas, which straddles economics, political science, and political philosophy.

Political philosophy is the philosophical mate of political science: it tackles such questions as whether politics is only concerned with political struggle or also with managing the commonwealth; whether politics flows from minds, material interests, or guns; whether or not there is a private sphere that ought to be protected; and how best to combine personal interests with the public good.

My *Political Philosophy* (Bunge 2009) deals with all of the above problems and many more besides. It should be regarded as the 9th and last volume of my *Treatise* and thus proposing a political philosophy with a broad philosophical foundation. In this and only sense, it resembles the political philosophies of Plato, Aristotle, Aquinas, Hobbes, Locke, and Hegel. But it differs from them in nearly all other respects, and it was written with hindsight, after most political ideologies had been tried and found wanting. In addition, my book discusses a political philosophy that proposes a new political order, one that expands democracy from politics to all of the subsystems of society, from the family to the international community.

Some political and legal philosophers have not confined themselves to studying political and legal systems, but have advised statesmen, though seldom to the benefit of the *populo minuto*. For instance, Thomas Hobbes advised the ill-fated Charles I, the neo-Hegelian Giovanni Gentile was one of Mussolini's ministers, the legal positivist Carl Schmitt presided over the Association of National Socialist Jurists, Miguel Reale inspired Getúlio Vargas's fascist *Estado Novo*, Henry Kissinger and Samuel Huntington advised President Nixon on how best to crush "the enemy," and so on.

However, a government lacking in a solid theoretical basis may just be as bad as one with an obviously perverse one. Suffice it to recall the communist leaders whose politological knowledge did not go beyond the Leninist catechism:

Development = Soviet power + Electrification
Socialism = Statism
Government = Terror

This simplistic and murderous political catechism has no theoretical basis, does not invite to conduct research, or even suffer debate, and it is incompatible with the conception of socialism as the expansion of political democracy to all the sectors of society (see Bunge and Gabetta 2013).

A broad democratic regime demands a broad social philosophy, one that is not confined to political and economic tricks designed to attain or retain power, as well as a political technology that assigns far more importance to the management of the commonwealth than to the struggle for power. Such a technology will seek to set up and defend a rule of the law both robust and just (Bunge 2014a).

Legal and Criminological Philosophies

Jurisprudence is of course the study of the legal system, which in turn is the set of the laws of the land. It is a social technology rather than either a basic or an applied science, since it is concerned with rules or norms of social behavior. But, of course, a legal norm may be either arbitrary or based on social theories and data, as well as on ethical norms. For example, the ancient Roman right to use and abuse one's property is no longer universally enforced because it could be used to hurt individuals, in particular slaves and serfs, as well as the common good. And the death penalty, repudiated by a few eighteenth-century thinkers, has been repealed in nearly all nations, on the grounds that it is nothing but legal murder and that it is not an effective crime deterrent.

There are three main schools of legal philosophy: naturalism, legal positivism, and legal realism. Legal naturalism, which today is upheld only by Thomists, is absurd because laws are artifacts: they are made and unmade, not found. But it has the virtue that it holds that justice involves morals. The legal positivists deny this thesis: they are conformists, which is why theirs has been the legal philosophy ruling in both Nazi Germany and the Soviet Union. It is also the favorite of run-of-the-mill lawyers, who are out to make a living, not to fight for justice.

Anyone who reads the news knows that laws are not born in a political vacuum but in parliaments, boardrooms, or barracks and that they live or die in courts of law and in the street. And to a humanist, the legal writings of Hobbes, Bentham, Hegel, the first Austin, Kelsen, Schmitt, and Hart are but morally repulsive sophistry. I learned this from my friend José Juan Bruera (1945), the Argentinian jurist and philosopher, who also pointed out the contradiction of the Marxists who preached both dialectics – an ontology of universal strife – and peaceful coexistence in international affairs.

Lon Fuller's magisterial paper of 1958 (Fuller 1958) in the *Harvard Law Review* confirmed Bruera's earlier analysis of legal positivism. And David Dyzenhaus' 1997 study of the last days of the Weimar Republic documented Hans Kelsen's complicity with the murderers of the first German democratic government (Dyzenhaus 1997). So much for Kelsen's "pure" (apolitical and amoral) theory of the law.

If legal naturalism is regarded as "nonsense upon stilts" (Bentham) and legal positivism is found guilty of political abasement, then a third possibility remains. This legal realism, which admits that the law is a historical construction but adds that it is not independent from ethics, that there just as well as unjust laws, and that judges, just like parliamentarians, ought to work for legal progress (Bunge 1998a). Although legal realism has had articulate defenders, such as Oliver Wendell Holmes, Roscoe Pound, Harold Laski, Alf Ross, and other Scandinavian scholars, it is far less well known than legal conformism. You can fight City Hall, but it will cost you far more than defending it.

Let us now move over to my brief if rewarding encounter with the science and technology of crime. One day of 2004, I received an unexpected visit from Per-Olof

Wikström, the recently appointed director of the Cambridge Institute of Criminology. He came lugging a bag full of books of mine, which he had just acquired, to invite me to give the inaugural lecture at the colloquium on explanations of crime scheduled to meet the following year at his institute. When I told Per-Olof that, although I had met some delinquents, I had never studied criminology beyond Cesare Beccaria's masterpiece of 1764; he replied: "Don't worry, Bunge, I'll help you. I will mail you a batch of recent papers and book chapters, and you'll study them in light of your systemism, which I subscribe to, and then you'll decide." From the start I was intrigued by the publications that Per-Olof sent me, all of them very recent. Each of them described clearly one facet of a multifaceted problem that, obviously, was too big to be tackled only by jurists. Indeed, it also required the contribution of social psychologists (anomie and the concomitant social deviance), socioeconomists (inequality of opportunities), and even neuroscientists (immaturity of the prefrontal cortex of young criminals).

In addition, in contrast to other branches of social science, here the student could intervene and make a difference, cooperating with social workers, community leaders, and even policemen. Furthermore, one could observe in the short run the consequences of interventions like the human development project that Robert Sampson had led in Chicago. The upshot was that the problem of crime captivated me to the point that I learned a lot in a short time, and my lecture, which contained a mathematization of Per-Olof's situational theory of acts of crime, flowed quickly (Bunge 2006b).

My lecture was well received by a large and friendly audience and was commented on in detail by Sir Anthony Bottoms, the top British criminologist. Per-Olof was particularly pleased, because the participation of outsiders is unusual in meetings of specialists. He asked me to give an impromptu talk on probability. I did so on a Saturday morning and argued that probability plays no role in criminology because the acts of crime, far from being random, are deliberate and have causes, both internal and environmental, that can be discovered and manipulated. This is why increases in employment rate, decriminalization of hard drugs, and the self-policing of neighborhoods cause a decline in the crime rate.

The colloquium organizers offered us sumptuous dinners, and once they sat me next to Cambridge's counterpart of Chief Inspector Morse, who holds an easy job because scholarly crimes are not punishable by law. Professor Peter Johnstone, Marta's famous colleague, took us for supper at his College, as well as for a stroll around its beautiful gardens. At the sight of a flock of loud Canadian geese busy fertilizing the centuries-old lawn, Peter exclaimed with patriotic zeal: "They have no right to be here!" The nerve of those illegal immigrants!

I wished to take that opportunity to submit my latest book typescript to the local university press, but was unable to meet with any of its officials. Both Cambridge University Press and Oxford University Press were well protected from novelty by their philosophical advisors. They have kept me away for over half a century.

Evaluating Disciplines

This chapter opened with value theory and will end by asking about the worth of some of the main fields of knowledge, whether genuine or bogus. Any realist and materialist conception of knowledge suggests the following evaluations of both the oldest and the newest disciplines. If God is a mere idea, as atheists from Buddha to Holbach and Kant held, then it has no real referent: it belongs in the fiction shelf, with the difference that it may inspire important actions or inactions. This view will be rejected by all religious believers, but these will admit both that their beliefs rest on faith rather than research and consequently they are justified in mistrusting novelty in all fields, with the possible exception of technology. But even here neophobes may be fooled, as in Melvin Kranzberg's favorite story: when an old lady was invited to fly to a certain place, she declined and said: "Thank you, but I prefer to stick to the old ways that the good Lord taught us: I will sit by the electric fireside watching my favorite TV series."

Scientists take it for granted that scientific research is the royal road to new truths. If we concur with this article of faith, then we ought to adopt the scientific method in all alethic fields. This is the *scientistic* thesis, first proclaimed in 1783 by the mathematician and social scientist Nicholas Condorcet. This attitude was understandably rubbished by the Counter Enlightenment, from Jean-Jacques Rousseau to Wilhelm Dilthey and his followers, such as the Wittgensteinian Peter Winch and the Austrian economists and right-wing ideologists Friedrich Hayek and Ludwig von Mises (Bunge 2014b). By the same token, all pseudoscience is sheer error, whether inoffensive like ufology, or harmful like standard economic theory. The contemporary skeptical movement, in particular CSICOP (Committee for Scientific Inquiry), founded in 1976 by my friend Paul Kurtz, and to which I have belonged almost from its beginning, has done much to deflate "alternative medicines" and the natural pseudosciences, such as astrology and parapsychology.

But CSICOP has ignored all the social pseudosciences, and some of its most visible members have been taken in by Dawkins's genetic determinism, Chomsky's nativism, and the self-styled evolutionary psychologists who claim that we are walking fossils whose mental toolkit was formed in the late Pleistocene to cope with dangerous beasts, from leopards to sexual rivals.

Skepticism is insufficient to ward off secular superstitions like parapsychology or multiverse theory, for one may always hope that the next experiment will corroborate them. If the psychologist Ray Hyman had embraced the materialist hypothesis that everything mental is cerebral, he would not have wasted many years attempting to initiate a "constructive dialogue" with believers in telepathy and the like. If Popper had been a materialist, he would not have admitted the possibility of the creation of matter out of nothing, nor would he have joined John Eccles – nicknamed "the ecclesiastical neuroscientist" – to buttress the ancient myth of the immaterial mind. And the excellent journal *Skeptical Inquirer* would not waste paper ridiculing the latest reports on ghosts and UFOs, had it embraced scientism instead of publishing a long article against it inspired in Hayek's malicious caricature of it. The same journal has also failed to criticize psychoanalysis as well as the pseudoscientific pockets of social studies. Thus it practices skepticism in some sectors but tolerates it in others.

Radical skepticism, like that of Pyrrho, Sextus Empiricus, Francisco Sanches, Pierre Bayle, David Hume, Paul Feyerabend, and David Miller – Popper's last assistant – is a bulldozing stance, for it places all beliefs on the lowest level, that of opinion. In addition, Pyrrhonism is self-destructing, for the rational questioning of any proposition involves admitting other propositions, if only for the sake of argument. This is why scientific workers practice *moderate skepticism*, according to which *not* everything is possible and *not* all the received beliefs are equally questionable (Bunge 1991b, 2000d).

Serious workers in all fields practice the wise handyman's maxim: *If it ain't broke, don't fix it.* Regrettably, much of contemporary philosophy is badly broken, so it needs fixing (Bunge 2001a). Fortunately, however, most of the reconstruction tools and materials are available in vast mathematical, scientific, and philosophical depots.

Novel truth is an essential characteristic of the final products of technological and scientific work. Moreover, it is a typically modern value, alien to religion, ideology, and commerce, as well as suspect to the guardians of tradition. Whereas art, science, and technology have been yielding novelties galore since the Renaissance, most contemporary philosophers have been reluctant to admit new ideas.

In fact, the only notable philosophical novelties in the twentieth century have been existentialism, linguistic philosophy, possible world metaphysics, and logical positivism. Regrettably, the first three are not philosophies proper, for they have not solved any outstanding philosophical problems, on top of which they are utterly alien to both science and real-life issues, when not either trivial or nonsensical. In particular, existentialism serves only as the butt of jokes about nothingness, being there, anxiety, and authenticity. My favorite is this: A lady at a party opens her host's refrigerator and sees that it is filled with beer bottles plus half-a-dozen empties. Asked what the latter are for, she is told that they are for the guests who don't drink beer.

Only the logical positivists grappled with philosophical problems arising in mathematics and factual science. True, their proposed solutions did not advance knowledge because they rejected anything lying beyond the observable. For example, they attempted to conceive of probability as frequency, as well as to reduce all physical magnitudes to their observable indicators (see Bunge 1959c). Still, logical positivism was the only philosophy of its time seriously interested in science. It was certainly wrong on most counts, but its mistakes were worth discussing, as they were neither irrelevant nor silly.

If a philosophical patent bureau were established to certify both the originality and the usefulness of every new philosophical publication, it would not be exactly flooded by submissions, as nearly all such publications are just comments on theses typical of about a dozen schools.

Enjoying Life

Finally, let us jump from the useful to the pleasant, or rather to what gives me pleasure besides thinking hard and interacting with family, friends, and students. I love art or, rather, some artworks, such as many of the classical Greek and Roman sculptures in the British Museum, the Louvre, and the Galleria degli Uffizi; the paintings of the Old Masters, especially Bosch and the two Brueghels in the Amsterdam Rijksmuseum, as well as the tiny Vermeer museum in Delft; the wealthy Munich Pinakothek; Velázquez and Goya in the rich albeit chaotic Prado; the impressionists, particularly Van Gogh, Cézanne, Renoir, Monet, Berni, and Forner – but not Degas, Chagall, or Deschamps – the realists Daumier, Renoir, and Utrillo; the surrealists Picasso, Dalí, and Magritte; the clever abstracts Escher and Klee; the unique Diego Rivera, though not his wife, Frida Kahlo; and numerous jewels in New York's Metropolitan Museum of Art, its Museum of Modern Art, and the Whitney Museum and their counterparts in Boston, Philadelphia, and Washington, D.C. What a contrast to the poor art museums of Argentina and Canada!

If I had any self-respect, I would say that I also love Giotto, Leonardo, Holbein, and Rembrandt, but the truth is that I only admire them: they do not move me the way Phidias's "Elgin Marbles" or Rodin's "Burghers of Calais" do. When I climbed for the first time the great staircase of the Louvre, I was transfixed by the Winged Victory of Samothrace, whereas the Mona Lisa – the most admired portrait in the world – left me cold. What poor artistic sensibility!

Thanks to my father and Marta, both of whom had played noble instruments, I am much more appreciative of music, or better of so-called classical music, from Scarlatti to Sibelius. Since late childhood I have loved Mozart, Beethoven, and Schubert and am extremely fond of some of their predecessors, especially Bach and Haydn – but not Händel or Buxtehude. When some of their compositions bore me, I remind myself that Bach had to support a bevy of children and Haydn had to satisfy a demanding rural count. I also love much of Brahms, Mendelssohn, and Schumann, as well as Albéniz, Albinoni, Boccherini, Borodin, Bruckner, Tchaikovsky, Chausson, Chopin, Corelli, de Falla, Dvorák, Fauré, Franck, Prokofiev, Ravel, Stravinsky, and Sibelius even more than the others in the same category. But most of the compositions of Bach, Wagner, Debussy, and Delius bore me. Shostakovich, Bartók, Britten, Saint-Saens, and most of Mahler irritate me. But Marta, who studied piano and has a refined taste, assures me that, as Mark Twain said of Wagner, the music of Mahler and Bartók is really better than it sounds. I can't stand opera either, except for some arias by Wagner and Verdi.

I find jazz frivolous, despise country music, find the Beetles tacky, and hate rock, particularly when combined with acrobatics. I love the early tangos, such as "El Choclo" and "La Cumparsita," as well as those of the Depression era, in particular "Yira, Yira" and "Cambalache," but dislike Piazzolla's tuxedoed pieces. With Marta I regularly attend symphonic and chamber music concerts, but confess that sometimes she has to drag me. In short, my musical likes and dislikes are quite definite, but my repertoire is rather narrow. At 16 I had tried to learn to play the violin, my favorite instrument, but gave up because I could not stand the instrument's angry responses to my clumsy attempts.

My literary taste is far broader, perhaps because it needs no schooling. During my adolescence I wrote a few stories and essays, two novels, a drama, and a few poems, some of which appeared in ephemeral journals under the pseudonym Mario Martel. I typed them on machines that Guillermo Cavazza, my father's secretary and a close friend of the family, had discarded. Guillermo also gave me my first and last typing lesson at age 7. To this day I still type with only two fingers while looking at the keyboard, which accounts for my many typos.

I am a fan of great literature, from the towering triad made up by *Don Quijote, War and Peace,* and *La Comédie humaine*, to Chinua Achebe, Jorge Amado, Margaret Atwood, Saul Bellow, Giovanni Boccaccio, Anthony Burgess, Dino Buzzatti, Italo Calvino, Peter Carey, Alejo Carpentier, Rosario Castellanos, Miguel de Cervantes, J. M. Coetzee, Miguel Delibes, Alfred Döblin, Fiodor Dostoyevski, George Eliot, Henry Fielding, Anatole France, Carlos Fuentes, Eduardo Galeano, John Galsworthy, Robert Graves, William Henry Hudson, Ismail Kadaré, Franz Kafka, J. M. G. Le Clézio, Primo Levi, Sinclair Lewis, Naguib Mahfuz, Hilary Mantel, Rohinton Mistry, Haruki Murakami, V. S. Naipaul, R. K. Narayan, Michael Ondaatje, Orhan Pamuk, Benito Pérez Galdós, Eça de Queirós, Mordecai Richler, Romain Rolland, Philip Roth, Salman Rushdie, José Saramago, Leonardo Sciascia, Vikram Seth, Wole Soyinka, Bruno Traven, Anthony Trollope, Mario Vargas Llosa, Kurt Vonnegut, Edith Wharton, Marguerite Yourcenar, and several others – such as Jane Austen, Umberto Eco, Henry Fielding, Thomas Hardy, Manuel Mujica Láinez, Machado de Assis, Vladimir Nabokov, Horacio Quiroga, M. J. Vassanji, Voltaire, and so on.

I am no longer enthusiastic about poetry, except for Homer's *Odyssey*, Lucretius, Omar Khayyam, the Spanish *romanceros*, John Donne, Goethe, Heine, Shelley, Roberto Ledesma, Antonio Machado, Pablo Neruda, and Walt Whitman. My knowledge of Italian history is insufficient to understand Dante, and my English too poor to fully appreciate Shakespeare: I only understand his popular plays. I dislike the latter Joyce's hermeticism and T.S. Eliot for trying hard to be quotable as well as a pro-fascist English gentleman.

My favorite playwrights are Aristophanes, Athol Fugard, Nikolai Gogol, Carlo Goldoni, Henrik Ibsen, Lope de Vega, Christopher Marlowe, Arthur Miller, Molière, Eugene O'Neill, J.B. Priestley, George Bernard Shaw, Jean-Paul Sartre, Tom Stoppard, Oscar Wilde, and Tennessee Williams. But the darkness and pomposity of the ancient-Greek, French, and Spanish tragedies don't move me, and hence I fail to understand Marx's admiration for the somber Calderón de la Barca.

Finally, the cinema directors whose films most impressed me are Angelopulos, Bergman, Costa-Gavras, Eisenstein, Fassbinder, Ford, Hitchcock, the Italian realists, Kiarostami, Kubrick, Kurosawa, Fritz Lang, de Mille, Renoir, Resnais, Satyajit Ray, Spielberg, Tavernier, von Trier, von Stroheim, and Welles.

In conclusion, my artistic tastes are middlebrow. I have never been carried away by cleverness, style, or technical virtuosity alone. And I never read art critics or book reviewers, except for A. A. Alvarez, the most generous of all. My only criteria for evaluating a work of art are these: First, would I look, listen, or read it again? Second, what does Marta think of it?

Coda

Philosophy is currently stagnant, hence in need of renewal, from new insights into old but still open problems to brand new philosophical problems born in the various fields of knowledge and social life. Some of the requisite renovation materials are there, in our background knowledge; others are likely be pouring out from scientific or mathematical breakthroughs. So it will always be too early to announce the death of philosophy – as Stephen Hawking did recently. Only the philosophies that neither raise new problems nor help solve old ones can be said to be dead – or rather sleeping and stagnant.

Chapter 16
Summary

At about 16, when starting to take life seriously, I fell in love at once with philosophy and science – in that order – and have attempted to intensify their interaction ever since. As I put it in my inaugural lecture as the professor of philosophy of science at Buenos Aires University (Bunge 1957a), I have tried to philosophize scientifically and approach science philosophically. The philosophical approach to science led me to recast some scientific theories in the axiomatic format, which forces one to focus on the most important concepts and propositions of a field of study, as well as to detect possible sources of trouble. Axiomatize to understand and philosophize to do sound and useful axiomatics. And the scientific approach to philosophical problems has led me to look for both motivation and support in the science of the day. No philosophia perennis for me.

I have criticized views that seemed to me to be utterly wrong, like subjectivism or harmful like intuitionism. But I have also attempted to polish nuggets, such as realism, materialism, systemism, and humanism, and turn them from isolated opinions into precise and well-grounded components of internally and externally consistent systems (theories). I have also been a militant philosopher rather than a dispassionate onlooker and commentator, because I believe that philosophy can be beneficial or harmful and that even apparently neutral and harmless *jeux d'esprit*, such as bloodless games in linguistic analysis and in social studies, are harmful in diverting attention from burning issues. Even dangerous charlatans like Hegel and Nietzsche deserve more attention than Wittgenstein and his followers, because the former tackled, albeit wrongly, some important issues, whereas the latter only played with words. Important errors are worth more than frivolous puzzles or high-sounding nonsense. For example, Henri Bergson's intuitionism was wrong, but he grappled with important problems, wrote well, and was honest. These features of his philosophy explain why he was so popular in his time and why Bertrand Russell paid such close attention to Bergson, whereas he did not waste time criticizing Edmund Husserl and his ilk.

I was lucky not to have sought philosophical mentors. This has caused me to waste much time both reinventing the umbrella and hacking my trail independently

M. Bunge, *Between Two Worlds*, Springer Biographies,
DOI 10.1007/978-3-319-29251-9_16

and with blunt tools. To be sure, this path has been in zigzag, but it led me to fashioning a philosophy that I hope is close to the science of the day, as deep and exact as I could manage, and a system rather than a pile of disparate opinions.

My philosophy is a system, in that its various constituents support one another. For instance, my ontology is materialist because my epistemology is realist; and my political philosophy is socialist in the broad sense (as integral democracy) because it is supported by a humanist ethics, as well as protected from utopianism by its scientism.

A philosophy without ontology is invertebrate; it is acephalous without epistemology, confused without semantics, and limbless without axiology, praxeology, and ethics. Because it is systemic, my philosophy can help cultivate all the fields of knowledge and action, as well as propose constructive and plausible alternatives in all scientific controversies.

I believe that my main contribution to physics has been my book *Foundations of Physics* (Bunge 1967a), which had a strong philosophical motivation. This was my attempt to prove, not just state, that the quantum and relativistic theories are realistic (observer-free) and that their subjectivistic (observer-centered) interpretations are illegitimate philosophical grafts.

Most of the problems that the book tackled are still being discussed, often vehemently. My thesis that the typical referent of the quantum theory is sui generis and therefore deserves a new name, *quanton*, is currently under debate among the physics teachers of the upper level of the French high schools. In addition, the mathematical formalisms in that book have been updated by a bevy of physicists, starting with Guillermo Covarrubias and Héctor Vucetich and his students, such as Gustavo Romero.

As for my philosophical work since about 1950, its main traits have been the following:

1. Broad curiosity, but avoidance of dilettantism
2. Interest in big unified pictures and rather than in disparate minutiae
3. Adherence to rationality, realism, materialism, and systemism, but disinterest in miniproblems, disdain for potboilers and pseudoproblems, and denunciation of pseudoscience
4. Search for truth and fairness
5. Concern for exactness but disinterest in computation – hence in computers as well
6. Consulting with experts
7. Trying to keep up to date about basic science, by regularly skimming *Nature*, *Science*, and *American Sociological Review*
8. Ignoring most of the stones in my path: plagiarists, ignorant referees, malicious critics, and university administrators disinterested in academic excellence
9. Commitment to public interest organizations, such as learned societies, but disinterest in holding purely administrative offices, such as departmental chairs
10. Respect for legitimate authority and contempt for intellectual mercenaries

I believe that my main philosophical contributions have been to the following branches:

1. *Erotetics* or the logic, semantics, and methodology of problems: analysis of a problem into presupposition, statement, analysis, and evaluation; distinction between direct and inverse problems; and tentative solution of the latter by transforming them into bundles of direct problems.
2. *Semantics* or the study of meaning and truth: original theories of reference or denotation, sense or connotation, and representation and the correspondence theory of partial truth. I argue for the use of the latter in all fields except mathematics, which I regard as the science of fictions. I also argue for exactness (conceptual precision) in all fields.
3. *Ontology or metaphysics*: the general theories of things, systems, properties, events, and processes – in particular the philosophies of mind, such as mind-body dualism and neurosociological monism. My ontology may be called science-orientated *systemic materialism*. The only objects that stay outside this ontology are fictions, such as the mathematical and theological objects, the entities and occurrences of fantastic literature and science fiction, and the idealizations built at the start of a research project.
4. *Epistemology* or the analytical and the normative views of inquiry and its fruits, in particular the philosophy of the sciences and technologies – natural, social, and biosocial. My epistemology may be called *scientific realism*, both in that it assumes the independent and prior existence of nature and in that it shuns the fictions of idealists, empiricists, and pragmatists – such as theories without an empirical support and measurements without theories and indicators.
5. *Axiology* – or value theory – and *praxeology* or the study of both individual and collective action. My axiology and praxeology are realistic, and they support both the legitimate aspirations of the individual and just social regimes – those where duties balance rights.
6. *Ethics* or the evaluation of plans, decisions, and actions affecting other individuals as well as social systems, from couples to the international community. I call *agathonism*, or *egotuism*, my own ethics, for its supreme principle is *Enjoy life and help others live*. Such enjoyment is usually a by-product of higher purposes: it seldom derives from the hedonistic pursuit of happiness. And altruism comes together with participation.
7. *Technophilosophy* or the philosophical study of the various technologies or arts of designing artifacts, from engineering to management science. I emphasize both the peculiarities and the commonalities of both fields and point out the moral neutrality of basic science in contrast to the moral commitment of technology.

8. *Legal philosophy* or the analysis and evaluation of the legal forged in the fires of social and political movements rather than in the courts of law.
9. *Political philosophy*, in particular the analysis of the main political ideologies and their relation to both politics and the design of integral democracy – a utopian but realizable regime characterized by solidarity, participation, cooperation, and self-management.
10. *Metaphilosophy* or the philosophy of philosophy, in particular the evaluation of philosophical doctrines according to their contribution – positive, null, or negative – to the advancement of knowledge and the quality of life.

So far my work has been appreciated more outside of philosophy than within it, because, far from being inward looking, it has meddled with some topical research projects in a variety of fields, instead of embroidering on antique stuff. This is why my work has been corrected or expanded by a number of workers, such as a handful of thinkers with different backgrounds who are currently revising and expanding my legacy – as befits an ongoing research project, in contradistinction to the scripture of a sect.

Finally, allow me to blow my own trumpet: Bertrand Russell and I are the only philosophers in the Science Hall of Fame kept by the American Association for the Advancement of Science. This is the pantheon of the most famous scientists of the past 200 years, and it places me between Richard Feynman and Theodosius Dobzhansky – which may only show the mismatch between fame and merit.

Marta's Appendix: My Life with Mario

My goal in contributing to the memoirs that Mario is preparing for his 95th birthday is to supplement them by adding something new about him as a person. In order to do this I must necessarily refer to myself. Specifically, I will touch on the various aspects of our long life together in which Mario has had, and in some cases continues to have, an important influence on me. I will not dwell on Mario Bunge as a philosopher, a physicist, or the author of work that is both vast and profound.

Religion

Like most middle-class Argentinians of my generation, I was educated (so to speak) in the midst of the Catholicism professed by my parents. I spent both my elementary and secondary school years as a day student at a religious school run by Carmelite Spanish nuns, situated in the Belgrano neighborhood of Buenos Aires where we lived. The religion taught at the school was just a collection of rites and stories from the bible, which in fact was limited to the New Testament. Once I showed the nuns an edition of the Old Testament that I had found in my father's study at home. To my surprise, their reaction was one of anger – I was then ordered, without explanation, to take the book back home.

At the end of high school I found myself inclined to pursue philosophy, motivated by a course taught in its last year that included Aristotelian logic. For this reason, and since I had been forbidden by my parents to attend the University of Buenos Aires ("a focus of dangerous ideas and a bastion of Peronism"), I enrolled as a philosophy student at the National Teacher's College located near my home in Buenos Aires. The atmosphere in it was good, and the teachers, with some exceptions, were dedicated and accessible. I quickly formed part of a select group of seriously motivated girls with whom I studied beyond what was required by our professors.

M. Bunge, *Between Two Worlds*, Springer Biographies,
DOI 10.1007/978-3-319-29251-9

Particularly influential for me was the study of an excellent little book on symbolic logic by Father Bochenski. Four of us then attended, as auditing students, the course in the Philosophy of Science that the physicist and self-made philosopher Mario Bunge taught for the first time at the Faculty of Philosophy of the University of Buenos Aires. This course, which he taught several other times thereafter, became the basis of *Scientific Research*, the best textbook on the philosophy of science that I have ever encountered.

The course in question was both fascinating and hard, so much so that I had to ask Ricardo A. Cavallo, my father, a civil engineer with a solid knowledge of the basic sciences and mathematics, for help in writing up the weekly essays required for it. His help, coupled with my willingness to learn, led me to become the best student in a course of 50 or so students. It was not long before Mario took an interest in me, at first just in order to encourage me in my future career, but then in a more serious way, leading eventually to his proposing marriage. His declaration of love seemed unreal since he hardly knew me outside of the classroom and of our walking together from the school to the train station, where we would take different trains.

My interest in him was also deeper than that which a student normally has (or should have) for a much admired professor. However, among the various obstacles that I saw in accepting his proposal was that at the time I had become a practicing Catholic, influenced largely by my best friend Delia Garat, with whom I attended discussions led by Jesuit priests. This sort of Catholicism was different from that practiced in my home or in the nuns' school – it was intellectually challenging and attracted me profoundly. However, Mario was handsome and intelligent, so it was natural that I too became romantically attracted to him. Compared to Mario, my friends from "the club" (Belgrano Club) seemed to come from another world – one of a life devoted to parties, sports, and gossip. I thus came little by little to the realization that the prospect of a life with Mario was worth fighting for, and that, if I decided to marry him, I would have to go against most of the rules established by the world I lived in.

There was then still the matter of my religious beliefs. Yet, Mario's arguments against religion in general and Catholicism in particular were finally convincing. He not only discussed these matters with me extensively, but also wrote me letters and poems, one of which I reproduce in the next section. By then, the threats of excommunication that my father confessor had issued to me had become easy to dismiss. In fact, what that authoritarian priest achieved by it was to ensure that I never again stepped into a church or attended a mass except for either artistic or humanitarian reasons.

To be totally sure of having made the right decision, I wrote for myself a long memoir on the reasons that I had for it. Eventually, my agnosticism became frank atheism and I have never changed my mind about it. This was the first and perhaps the most profound influence that Mario has exerted on me during our long life together.

An Unpublished Poem: 'The Secret of Happiness' by 'Mario Bunge'

Let me show you the true secret of Happiness,
Let me sell you the infallible Magic Recipe
For baking the creamy sweet cake of Happy Life.
If you promise to pay the price of the instruction,
I'll let you use the following Definition:
Happiness equals Comfort Plus Unconcern
Plus Peace of Mind, all well-balanced And well-shaken.
The definitientia are easy to get, and cheap.
All you need to buy Comfort is some money,
And this you will earn In some way or other
If you make the stern decision
To turn it into the goal
Of every one of your actions.
Unconcern comes
Next: it is the easiest to get.
Just let things be
(Heidegger's rule for getting free)
And you'll feel far
From other people's sufferings, struggles, visions.
Once you've got cozy Comfort
And feel detached
Enough, add the third ingredient.
Your Peace of Mind
Just requires systematic
Persecution
Of doubt, fixation of belief,
And suppression
Of impractical inquiries.
Now that you know
How to become a happy human, Pay as agreed:
Give me every one and all your
Hopes and sorrows,
Even the faint, untenable
Hope that the dead hopes
May live and enliven again,
Even sorrows
Deep and bitter like the sorrow Of a lost hope.
Now that you've paid for the recipe Go, bake yourself
The dreamt, the creamy, the sweet cake Of Happiness.
(Mario Bunge, Buenos Aires, 30/11/1958)

Between Philosophy and Mathematics

During my four years of study at the National Teacher's College specializing in Philosophy, I participated in small groups of study. With Alicia Wigdorovitz and other classmates we studied symbolic logic. With Delia Garat we studied the history of philosophy, taking advantage of the well-stocked library that she had in her house. Piled up on a large table that we shared there were books by Ernst Cassirer, the Greek classics, Descartes, Kant, Hegel, and many others. I by far preferred the study of the history of philosophy (from the ancient Greeks to German philosophy and Marxism) to that of the only contemporary thinker that we were obliged to read – Martin Heidegger. None of our teachers was an original thinker. The first authentic scientist and philosopher that I had encountered so far as a teacher was Mario Bunge, who was also endowed with a vast general culture. It was only then that I became aware that to devote myself to philosophy could be something quite different from and far more rewarding than studying various authors or existing philosophical systems. However, it was Mario himself who dissuaded me from so doing without first getting a first-hand acquaintance with some scientific discipline. This is precisely what he had done himself.

The taste and facility that I seemed to have for logic led me to a career in mathematics at the University of Buenos Aires, with Mario's total support. Logic was taught by Gregorio Klimovsky. His courses and seminars were lively and fascinating since in them we learned not just logic but also its connections with algebra. I recall as particularly important, among the texts that we studied under his supervision, a work by Antonio Monteiro on filters and ideals, and the book by Roman Sikorski on Boolean algebras. Of course, in my mathematics studies I took many other courses which, due to my poor preparation from the religious school I attended for ten years, I found to be very difficult. Mario became, in fact, my first calculus tutor, beginning with trigonometry, subjects which were for me the most difficult courses, whereas linear algebra and the so-called abstract algebra seemed to me to be almost trivial. My mathematical studies were rapidly becoming, for me, a passionate and totally absorbing affair. The courses taught by Mischa Cotlar on functional analysis were as dense as his manuscripts, and better suited, perhaps, to graduate than to college studies. Before them we had studied the charming book by G. H. Hardy on real functions. I must not forget the topology courses which Mario Gutiérrez-Burzaco taught. The book with the greatest impact, which we studied in a select group under his supervision, was one by George Springer on Riemann surfaces, which later on would be useful in my mathematical research.

At the time Mario, who maintained correspondence with a number of philosophers and scientists all over the world, had received an invitation to be a visiting professor of the University of Pennsylvania. In view of this, I requested admission to graduate studies in mathematics at this Ivy League university and, to my surprise, it was granted. We spent the academic year 1960–1961 in Philadelphia, still with the idea of coming back to Buenos Aires and continuing our life there as before. This we did, but only until the year 1963, when the political atmosphere in Argentina was in turmoil and the return of military rule was seen as a dreadful possibility. We returned to Philadelphia that year on that account and so I could pursue my studies at the University of Pennsylvania (familiarly called "Penn," not to be confused with Penn State University) and obtain my doctoral degree, which occurred in June of 1966. During those years in exile, Mario was working on his books *Scientific Research*, based on the course that I had taken with him in Buenos Aires, and *Foundations of Physics*. Mario's clear thinking, along with his vast knowledge of various sciences, his strict method of work, and what he requires the reader to do in order to learn the subject, is obvious to anyone who makes the effort to study this excellent treatise.

The 1960s were especially interesting. From 1963 on we lived abroad – first in the USA, then in Germany (Freiburg im Breisgau) until 1966, and from then on in the city of Montreal in Canada, where we have resided, with one-year intervals spent in other countries during sabbaticals, until now. The idea of returning to live in Buenos Aires was sadly abandoned in view of the political uncertainty in our country of origin.

Category Theory

Until the year 1964, my intention to return to philosophy after what was meant to be a mere mathematical incursion, was still standing, but that very year something made me change my mind. During the International Congress on Logic, History and the Philosophy of Science in Jerusalem I would meet a person that would influence me almost as much as Mario Bunge in my future career. That person was F. William Lawvere, the most brilliant student at Columbia University of the famous mathematician Samuel Eilenberg ("Sammy" for the mathematicians). Lawvere, who had obtained his doctorate in 1963, was one of the few mathematicians invited to give one-hour lectures at this congress. My interest in the theory of categories, founded by Sammy Eilenberg and Saunders MacLane in 1945 in order to better present and understand algebraic topology, and with which I was already acquainted through courses given by Peter Freyd at Penn, grew even more out of conversations with Bill Lawvere in Jerusalem.

From both Freyd and Lawvere I had learned enough category theory to realize that, by making it my area of concentration, I would not have to abandon, if not philosophy, at least the foundations of mathematics. I had already become a doctoral student of Peter Freyd at Penn, but the opportunity to work also under the direction of Bill Lawvere would luckily soon present itself. On the one hand, Lawvere intended to spend a couple of years at the E.T.H. in Zürich, Switzerland, while, on the other, Mario had been awarded a generous fellowship of the Humboldt Foundation to spend a year anywhere in Germany working on the foundations of physics. At my request, Mario chose Freiburg im Breisgau, as being the nearest to Zürich. These events led me to travel weekly by train to Switzerland in order to participate in the Benno Eckman seminar at the Forschungsinstitut für Mathematik of the E.T.H., and to have long discussions with Lawvere on the subject of my thesis, which he had suggested. This alone shows the generosity and support that Mario has given me in my career.

In Freiburg, Mario had interesting interactions with physicists and, equally important perhaps, he did not have to cross paths with Martin Heidegger, whom he despised for both his empty and enigmatic philosophy and for his Nazi affiliations. What Mario could not have imagined, however, was that, by choosing Freiburg, he would be making, in Lawvere, a formidable intellectual rival. Lawvere was (and still is) a deeply convinced Marxist, but at the same time a notably original mathematician without whose contributions the theory of categories would possibly have taken quite a different path than the one it actually did as an area independent from the algebraic topology that had inspired it, changing radically as well the way to view logic and algebra as well as functional analysis and differential geometry.

In his mathematics, Lawvere employed a terminology taken from the dialectics that inspired him, but the mathematical concepts that he introduced stood on their own, and could be understood and accepted (or rejected) by anybody without any knowledge of or allegiance to Marxism. This at least is how it appeared to me. My fascination with his ideas and projects overtook all my previous interests. Mario, however, did not see it that way, and argued with me and with Lawvere, owing principally to the Hegelian impression which his mathematics gave him. What Mario did not realize was that this aspect was negligible considering the amazingly clear and concise concepts that allowed Lawvere to advance mathematics and to become the unquestionable leader of an entire generation of mathematicians, to which I was lucky to belong. From that moment on, I avoided discussing with Mario what I was doing or studying for fear of his criticism. My planned collaboration which him, which had not yet begun because, according to him, I was too immature as a scientist to even attempt it, was thus postponed indefinitely.

Although I continued to read Mario's books and to participate in his academic life, I did so mostly as his wife. The main portion of my time I was actually spending on advancing my own career, and at the same time getting away from philosophy – the latter for two different reasons. The first was not to dissent from Mario, my husband and best friend, and the second was not to dissent from Lawvere who, together with Freyd, was my thesis director. Philosophy, an area that I had envisaged dedicating my entire life to, became all of a sudden a area of disagreement, and on that account I deliberately abandoned my previous interest in it. For Mario, this became a big disappointment, at least for some time. Eventually, however, my interest in Mario's ideas came back and did so with full force. Mario's initial admiration for me also returned when he saw that I had become, in the meantime, a serious mathematician with a certain degree of professional success.

In short, Mario finally understood that what I had done in following my own inclinations was correct, and because of it, I believe, he respected me more than before. My work in mathematics since my doctoral thesis consisted in developing aspects of the theory of categories as well as in utilizing categories as a foundation for areas as varied as set theory, model theory, differential geometry and topology, theoretical computer science, algebraic topology, and functional analysis. I will not mention here my published work or the students whom I have formed because these are not relevant to a tribute to Mario, but what I will say is that he helped me in various ways throughout my career as a mathematician. For his constant faith in me, I am deeply grateful.

Politics

Born, as I was, in the year 1938 in Argentina, I lived the greatest part of my childhood and adolescence under the Peronist government, about which the only thing I was told was that it had to be opposed and never ever talked about in the presence of the domestic employees, presumably out of fear of being denounced to the police.

As I had swallowed Catholicism without questioning, I did likewise with anti-Peronism. My parents were conservative – not to say reactionary. At home, the only newspaper received was *La Prensa*, which only my father read. I was quite ignorant of world affairs, among other things. My parents had never explained Peronism to me, nor told me about Nazism, the Holocaust, or even the Second World War. At the Club Belgrano, which I used as a second home, everything was frivolous, concentrating on tennis, swimming, and the numerous parties to which we went in groups, sometimes from one to the next during a single night, so long as the midnight curfew was observed.

My first eye-opener, together with a shift to the left, I owe, as many other things, to Mario. During the period 1960–1966 our home base was Philadelphia. Our friends at the University of Pennsylvania (philosophy and mathematics) were, for the most part, liberal, with some of them on the left, but never right-wing. In addition, we chose to live in a Black neighborhood next to the university campus. Although we did not manage to get integrated into it, as we had hoped, just to see how African-Americans, decidedly poor, lived in the first state (Pennsylvania) to abolish slavery was quite shocking.

Extraordinary events succeeded one after another – the success of the Beatles, the march to Washington, DC against discrimination, of Martin Luther King Junior, the assassination of President Kennedy, the Vietnam War. These events contributed to a total change in me, not so much by themselves, but because of what was talked about in our new entourage. However, it was precisely on account of those events, in particular the unjust Vietnam War, that we started thinking that the United States of America was too complex a country to be understood completely and, partly because of that we left for Europe in 1965 without a fixed destination, until we finally came to Montreal to stay at the end of 1966. At McGill University, where we both had been employed, were arriving, a little after we did, many American academics also opposed to the Vietnam War. From that moment on my political ideas became, I think, even more radical than those of my husband without, however, our disagreeing on the essential matters. We used to read every publication that would help us find out the truth, refusing to accept without questioning what was published in the majority of newspapers, even the *New York Times*, and even less the *Globe and Mail*, the main Canadian newspaper, serious but tending to the right. I recall *The Nation* and *Le Monde Diplomatique*, among others.

For reasons not of our choice, the government of the United States of America has always been at the center of our indignation, especially on account of its interventions in Latin America, and more recently also in Asia and the Middle East. More than once I have joined marches against its various wars and interventions, both here as well as in Europe. For Mario, using the pen seemed a far more efficient method of protest, and I am sure that it was. In order to counteract the information gotten from the local and international ordinary press as well as the disinformation one gets from television news and commentaries, we now read regularly the *New York Review of Books*, the *New Yorker*, and, electronically, many excellent articles from independent and non-profit organizations such as Sin Permiso, Democracy Now, Portside, and Information Clearing House. From this point of view, we are miles away from our Montreal friends, who mostly read the *Gazette*, a provincial and conservative newspaper. I should add that *Le Devoir* is an excellent Montreal newspaper for foreign affairs as well as culture, and that I often read it online.

In Buenos Aires, every time we visit – something that we have done frequently for a month over the last few years – we glance at all the major newspapers, but with some mistrust, since neither *La Nación*, *Clarín*, nor *Página 12* consitute unbiased journalism on the country's state of affairs of the country. There are, in Argentina, quite a few friends with whom we share our ideas, which I can summarize as basically socialist, against imperialism, and for the defense of human rights. However, the possibility of discussing such ideas with other friends and relatives whose political opinions – which range from the far left to the moderate right – are for the most part dogmatic in nature, is becoming increasingly difficult.

From the onset of our life together, which coincided with the arrival of Fidel Castro and his followers in La Habana, until today, I try to be *au courant* on world events. This I can do not only on account of the selected readings I have already mentioned, but also thanks to Mario's deep knowledge of the history of humanity, which helps me to locate events in their proper context. I believe that this is where, in addition to the unconditional love we have for our children, the strength of our marriage lies. After having celebrated the 55th anniversary of our marriage, it is beyond question that we love each other dearly.

Family

Between 1967 and 1973, my life with Mario was shared with little Eric Russell, born in June of 1967. Eric was very blonde, active, lively, and sociable. He exhausted, one after the other, all the Caribbean nannies that we hired to look after him. We, on the other hand, were simply delighted with him.

We eventually noticed, however, that Eric had become somewhat selfish, and that he lacked discipline – something that we should have expected from a child whose nannies addressed him as "Master Heric" and who was the center of attention wherever we went. When Eric was just 4 years old, during a visit to Bucharest for a philosophical congress, we were tempted to accept invitations from the university to spend the academic year in Romania, hoping perhaps that a communist regime would turn Eric into a disciplined and less self-centered child. Of course, we rejected this idea when we realized that such a decision would mean moving within the political circles of Nicolae Ceaucescu.

We still had to choose a place for our first earned sabbatical from McGill University (1972–1973). On account of my mathematical interests, and since there was a standing invitation for Mario, we chose to go to Aarhus, Denmark. The latter was a model institution, particularly its Matematisk Institut. What Bucharest might have achieved in changing Eric happened almost overnight at a kindergarten in Aarhus. This, however, was due not to more discipline, but rather to far less. The children spent most of their time playing with Lego, a fact that may explain why Eric later in his life decided to become an architect. The latter he did at the McGill School of Architecture and at Harvard's Graduate School of Design, where he would meet his wife and partner, Mimi Hoang, with whom he shares a highly reputable architectural firm in Brooklyn.

Going back to the Danish experience, I remember how in the afternoons the children played freely in an empty lot, where they kept animals such as rabbits, built wooden houses, and tended gardens. The discipline that would have been the norm in Romania had been replaced by absolute freedom and lack of pressure. We believe that it was thanks to his Danish noneducation that Eric became cooperative and generous. Once (later in Zürich) Eric decided to give away his valuable collection of model cars, throwing them from a balcony of the high-rise that we occupied at the time. He thought that other children had as much right as he did to play with them. In Zürich, where we ended up after Aarhus, Eric attended the Intercommunity School and as a result he forgot the Danish that he had spoken so well in Aarhus; but he also began to learn what he had not, deliberately, been taught in Denmark.

Once our sabbatical academic year came to an end, we returned to Montreal, where we bought our first house in view of the imminent enlargement of our family by the birth in December of 1973 of our daughter Silvia Alice, nowadays a well-known cognitive neuroscientist and professor at the University of California, Berkeley. Mario was then in one of his more productive periods. The family life and tranquility that we enjoyed in our peaceful country of adoption, two children, a home, and domestic help, allowed him to devote himself entirely to his work. During the academic year 1975–1976 we decided, for various reasons, to accept positions at UNAM (Universidad Nacional Autónoma de Mexico), taking leaves of absence without pay from McGill. The time spent in Mexico was excellent in all but one respect, to wit, in what concerned our health. Our friends were interesting and numerous, and we had all the time we desired to do research, publish, and attend congresses anywhere with generous travel subsidies. Teaching was optional. Mexico seemed like Paradise on Earth, but only until we started to fall sick. Besides, some friends had warned us that children as blonde as ours were in danger of being kidnapped. With much sadness we decided not to stay in Mexico and returned to Montreal. With the proceeds of the sale of our previous house and two mortgages, we could acquire, for reasonable prices, both a large old house on the Westmount hill next to a natural reserve overlooking the St Lawrence River and a beautiful cottage in the middle of the Laurentian woods near Montreal, where we could ski in the winter, swim in the lake during the summer, and walk the year round. Thus, we spent the next 35 years in a most privileged manner. Those were very happy times.

Silvia was very different from Eric and, in several aspects, more like me than like her father or her brother Eric. She was incredibly precocious – she read and wrote English at the age of 4 (just like me, but Spanish is far easier to learn), and also spoke Spanish thanks to our year in Mexico and her Mexican nanny in Montreal. In spite of the law (at age 5 she was too young to attend school yet), we managed to enroll her in the first grade of a private English school. Just as had happened with me, being younger than her classmates often made her feel out of place. We decided then that our next sabbatical year would be spent in a place where she could profit from a change. This we did in Geneva, Switzerland, where Silvia attended the École Internationale de Génève, well known for its excellence and for the diversity of its students. When Silvia returned to her old school in Montreal, she had changed her personality almost entirely, had recovered her self-esteem, and all her previous problems had evaporated. Silvia had been interested in biology from the beginning, in particular, marine biology. Since her general education on finishing high school had seemed to us still insufficient for an early specialization, we sent her to Yale University for college, where she would get a general education. However, it was there that she began to get interested in the study of the brain and its functions. From Yale she went to Stanford, where she obtained a doctorate in neuroscience, and then to MIT for a postdoctoral fellowship in the same area. Her first job was at the University of California, where she is now full professor in both the Department of Psychology and the Helen Wills Neuroscience Institute at Berkeley. She lives in a lovely house in the Oakland hills.

I tell these seemingly irrelevant stories mostly to show how much our children have mattered to Mario and me, to the extent of our adjusting our own interests so they could have happy and productive lives. For a mother, doing so is not out of the ordinary, even for an academic mother like myself. For a father, however, this is not normally the case. As pater familias, Mario has always behaved in an admirable way. His two children with me, as well as his two older children from a previous marriage, adore him. On the one hand, Mario and I regret having induced Eric and Silvia to study and work in the United States, but on the other hand, we are proud of their achievements. Our contact with them, now enriched by the addition of our grandchildren Giao and Vi Bunge Hoang, continues to be close, even though they no longer live near us.

The Arts

By arts I here mean mainly music, painting, literature, and cinema. Although Mario and I share a great deal in our tastes for all the arts, Mario is more conservative than I am in almost all of them and therefore rejects much of what I like. Thus, when choosing for him (and thus also for me) concerts, painting exhibits, literature, or films, I must be very careful to do so "correctly"; even so I do not always hit on what he will like. Both of us have been well educated in the arts, albeit in different ways.

In music, for instance, we both like Mozart, Vivaldi, Bach, Beethoven, Schubert, Brahms, Franck, Fauré, Ravel, Sibelius, and Prokofiev, and we like both symphonies and chamber music or good soloists. In Montreal we subscribe to the OSM (Orchestre Symphonique de Montréal), currently under the direction of Maestro Ken Nagano, as well as to the LMMC (Ladies Morning Musical Club), mainly devoted to chamber music and soloists. This alone entails attending at least 20 concerts a year. Mario does not like opera, even though when he was young he used to go regularly to the Teatro Colón to hear it, as did I accompanying my maternal grandmother. Besides opera, among the composers whose styles Mario hates and I like are Bartok, Shostakovich, and Mahler. I can always listen to them at home, but attending live performances together in which any of these are part of the program is out of the question.

In painting and other plastic arts, our tastes have much in common, but there again they diverge sometimes, just as they do in music. We both admire Rembrandt, El Greco, Brueghel, Van Gogh, Goya, Cézanne, Manet, and even Salvador Dalí, Otto Dix, Diego Rivera, and several others. However, contrary to most of the people we know, Mario does not feel a great admiration for Picasso, Matisse, Kandinsky, Chagall, Kahlo, and even less for abstract art like that of Rothko. He hates the baroque in all its forms. I often agree with him on that, but baroque architecture, particularly in Sicily, fascinates me. In any case, going to art museums and visiting architectural gems in all corners of the world we visit is always a priority for us both.

We also agree in literature in our likings for certain authors, such as Cervantes, Balzac, Tolstoy, Dostoyevski, Atwood, Carey, Vargas Llosa, Rushdie, Roa Bastos, Roth, Vidal, Yourcenar, Kadaré, Sciascia, Canetti, Pamuk, Saramago, Graves, Ishiguro, Svevo, Hardy, Pérez Galdós, Eliot, Trollope, Le Clézio, and many others. However, Mario is not as impressed as I am by (all of) Borges, García Márquez, Mujica Lainez, Cortázar, Austen, Henry James, Proust, Camus, Sartre, Munroe, Joyce, Murakami, and Bolaño. We both like mystery novels and (why not admit it) also crime – organized or not. Our favorite writers in that genre are P.D. James, Lindsay Davies, Donna Leon, Andrea Camilieri, Dorothy Sayers, and Henning Mankell, among others.

The cinema is a passion for both of us. We like the great Italian and French directors, in addition to Ingmar Bergman. In past years we would never miss a film festival in Montreal, but these have lost their luster lately, with Toronto taking the center of action. Fortunately it is always possible to watch movies at home on DVD, something that we do almost every evening after dinner. The existing film clubs (to which I subscribe) are not, however, sufficient to satisfy our demand, so that I am forced to spend a lot of money to purchase what I cannot get in other ways. Doing research to locate such films takes time, but I am amply rewarded by the enjoyment they bring to Mario and I. Among the series of films that we have particularly liked are *Heimat*, on the lives of the inhabitants of a small city in Germany before, during, and after Nazism, and *Berlin, Alexander Platz*. Our love for mystery novels extends to the cinema, in particular to *Poirot*, *Miss Marple*, *Sherlock Holmes*, *Montalbano*, *Brunetti*, *Wallander*, and several lesser-known Scandinavian series. I believe that in this matter we have been influencing each other for a long time now, even more than in the other arts.

Trips

During our stay in Philadelphia, Mario invited to dinner intellectuals such as Ernst Gombrich, the famous historian. It was he who advised us on our first trip to England, France, Italy, and Greece during the summers of 1960 and 1963. From then on we traveled extensively in Europe, but on our own, equipped, every time, with one of the green Michelin guides. Following a tour of Greece, in which we visited archeological sites (Delphi, Olympia, Knossos, Athens, among others) and islands as different from each other as Crete, Rhodes, Mikonos, and Corfu, we decided to make the last one our second home. (This was before the birth of our son Eric.) In it, renting villas from villagers, we have spent several long summers. My knowledge of modern Greek, a language acquired on my own during the course of four months prior to our first visit, allowed me to interact with the locals and to easily make friends among them. Mario did not stay behind for long. His fascination for languages, of which he read and spoke several, induced him to speak Greek by imitation and some help from me. We chose this island because, contrary to most of the others in Greece, Corfu was very green thanks to the Venetian influence that showed, not only in the olive trees and cypresses that covered the island, but also in the pastel colors of its houses. In addition, its ancient history, veridical or imagined, was fascinating. The rock we could see from everywhere at Palaiokastritsa, in the northwestern part of the island, was believed to be Odysseus's petrified ship. To top it all, Corfu was incredibly beautiful and, at the time, not yet opened to mass tourism.

Our life in Corfu, where we spent the summer months over more than 40 years, was crucial for being able to endure the long winter months in Montreal. In Corfu we met several interesting foreigners, notably the British poet and novelist Lawrence Durrell, author of *The Alexandria Quartet*, among others. Mario and Durrell clashed on several issues. Whereas Durrell could not begin to understand why Mario did not drink alcohol, Mario did not share the writer's reactionary political ideas. The relationship between the two ended the day Mario asked Durrell to lend him his typewriter. Naturally, Mario needed it (in those days there were no laptops, or even electricity), but he did not think Durrell would not lend anyone his main work tool. In those days, Durrell was preparing to make a film about Odysseus, who supposedly had touched land at Palaiokastritsa on his way back from Troy. This does not explain how his ship would remain behind him, petrified, considering his ultimate safe return to Ithaca. In this film the actors were local people, including Maria, the common local young woman that Durrell and us both employed for cleaning and cooking.

The 1980s and 1990s were spent between Montreal and our extensive Mediterranean holidays, as well as in our country house in the Laurentian hills. Those summers were both productive and relaxing, with daily swimming, rowing (in the sea or the lake), and walking long distances. In addition to these long working vacations we traveled to various other countries for academic reasons or simply as tourists. Particularly interesting were our visits to Israel, México, Cuba, Egypt, India, Turkey, Italy, Russia, Australia, and China. Each one of these trips enriched our lives. Mario's curiosity for everything new and his enthusiasm and energy were surprising. In every one of our visits we tried to integrate ourselves with the local inhabitants, visited archeological, historical, and artistic sites, and we tried the local cuisine. This is not the appropriate place to narrate these experiences in detail, so I will limit myself to just a couple of comments.

In Israel we visited, among several other places, Jerusalem – before and after the Six Day War – a city we found fascinating. Mario has been to Israel three times, each for academic reasons, whereas I was there just twice, accompanying him.

The first visit, in 1964, I have mentioned above. The second visit was in December of 1974. While Mario was taking part in a philosophical congress in Haifa, I, together with Eric and little Silvia, spent a week in a luxury hotel on the shores of Lake Tiberias. Eric, who was then 7 years old, was very sociable and interacted with several of the hotel guests and employees. From these conversations, Eric became interested in the country and its problems, in particular with its vicinity to Syria, whose fighter planes we could see regularly flying over. Motivated by all this, Eric prepared a handwritten "newspaper" that he sold for a few agorot. The news that he reported in it was relevant, but mostly of his own invention. Silvia took her first steps in that same place. From Tiberias we traveled south, visiting several places mentioned in the Bible, ending in the Red Sea and Sinai. Unfortunately, and in spite of having in Tel Aviv such good and long-time friends as Joseph Agassi and Judith Buber, we stopped visiting Israel due to the "apartheid" policy adopted by its government against the Palestinian people.

Our first visit to Cuba, of the seven we made in all, was the most interesting, since Silvia came with us. For our children it would be the first and last time; having spent most of their lives in the United States, going to Cuba was not an easy option for them. Among several other personalities (but not Fidel Castro himself, who was elsewhere at a medical congress), we had an interview with Carlos Rafael Rodríguez, the only Marxist within the revolutionary group that Fidel had assembled. We were shown several Cuban films, such as *Memorias del subdesarrollo*. In La Habana we stayed in a large house most likely expropriated from some functionary of the Batista regime, but totally decayed – from a grand piano which had no sound to the water leaking from the roof onto a large dinner table adorned with silver candelabra. We were treated well by the Cubans, perhaps because we were not Americans but Canadians, and Argentinians like the Che. From the beginning we experienced a great sympathy for Cuba, and we still do, since we understand that a large part of what seems negative has been caused by the harshness with which the USA has subjected it. Our friends Ernesto Mario Bravo and his American wife Estela Bravo, who live and work in Cuba, have painted for us a picture that differs greatly from that which one gets from most of the media outside of the country. We believe, therefore, although we may very well be wrong, that once the US economic embargo is lifted and hostilities end, the Cuban regime would call for democratic elections and most likely win them. In Cuba we feel at home, although I doubt that we could endure its censorship and lack of information if, instead of spending two weeks there as we do now and then, we were to live in that country.

Epilogue

From the year 2010 on, both of us already retired from teaching at McGill but still connected to it as emeritus professors, we moved from the Westmount hill to its lower part, through the acquisition of a large apartment in one of the black towers of steel and glass known as Westmount Square, designed by the famous architect Mies van der Rohe. Literally surrounded by our well-stocked libraries, with studies for each of us, and with all the conveniences of modern living, we now watch the seasons succeed one another, each one beautiful in a different way. Twice a year we reunite with our children and grandchildren, 1 week in December, and one or two in the summer.

As I have already mentioned, we are interested in politics, literature, cinema, music, painting, and, as always, our own work. For the past four years we have been travelling to Buenos Aires for a month in the Spring, so that Mario can continue to conduct, together with Javier López de Casenave, a seminar on the philosophy of the various sciences at the Science Faculty of the University of Buenos Aires. These visits have also allowed us to reconnect, not just with our Argentinian families, but also with old friends, as well as making many new friends. To have lived for half a century away from our country makes it difficult for us to fully understand the current politics and other aspects of Argentina. But in many ways we feel it is still our country and that attracts us in a natural, effortless way.

Mario's recent incursions into the philosophy of medicine and into the theory and practice of socialism have been well received in Argentina and Spain, and this is partly why he feels he can continue working without lowering his rhythm or expectations. With no obligations which tie us up to Montreal anymore, the idea of moving, either to Buenos Aires or to Barcelona, seems often attractive, since in both cities we have more friends than we do here and since the climate is better than that of Montreal – at least in winter. However, the difficulties that such a move would carry with it (particularly regarding health care and the fact that it would take us even farther away from our "American" children) seem to be bigger than the advantages.

My life with Mario has been anything but lacking in interest, something that makes me feel really lucky. I am not going to hide the fact that Mario, who is a person incredibly disciplined in all of his actions, has been and continues to be somewhat difficult to adjust to, but this could not have been otherwise taking into account his monumental work, particularly his *magnum opus* – the *Treatise*, and what he continues to produce at the age of 95. It makes me happy to know that his having shared most of his life with me has been, rather than an obstacle to his work, conducive to it. He has always been able to count on my accompanying him on most of his academic trips. These many and varied visits have helped us feel citizens of the world. The only wish that I now have is that the state of our health will allow us to continue to enjoy, for some time, the benefits of our shared life.

Illustrations

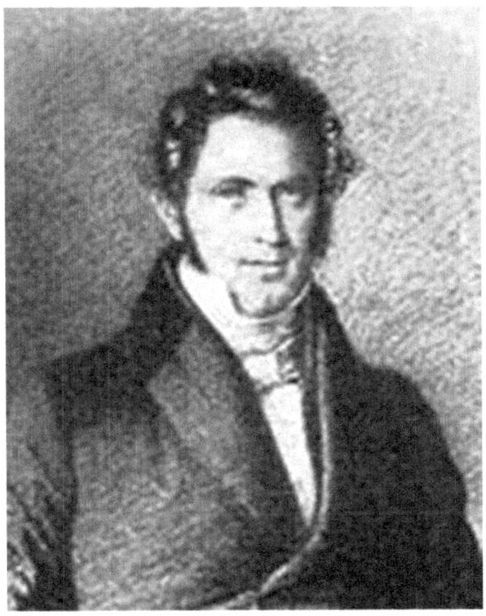

1. Carl August Bunge, Buenos Aires, ca. 1830

M. Bunge, *Between Two Worlds*, Springer Biographies,
DOI 10.1007/978-3-319-29251-9

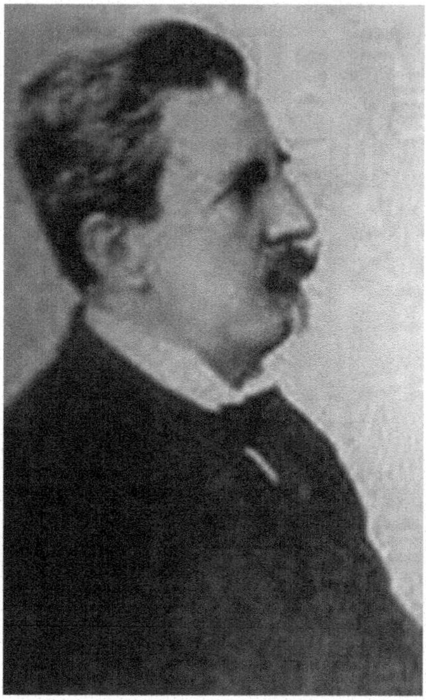

2. Octavio Bunge, Buenos Aires, ca. 1890

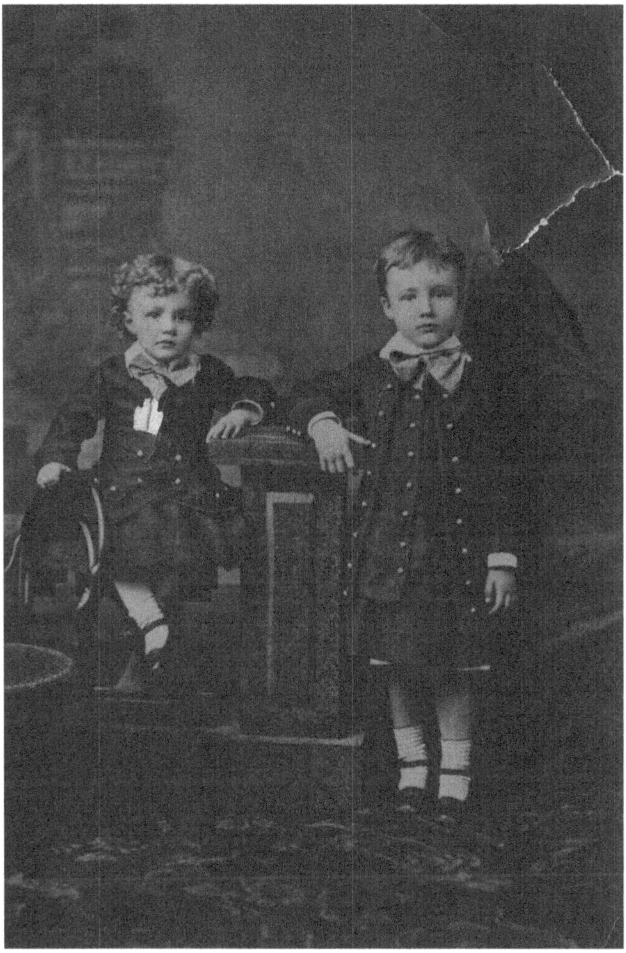

3. Auguto and Carlos Octavio Bunge, Buenos Aires, ca. 1883

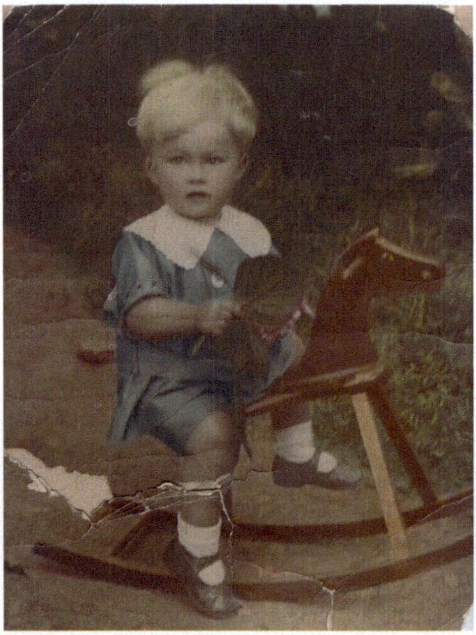

4. My first horse, 1921

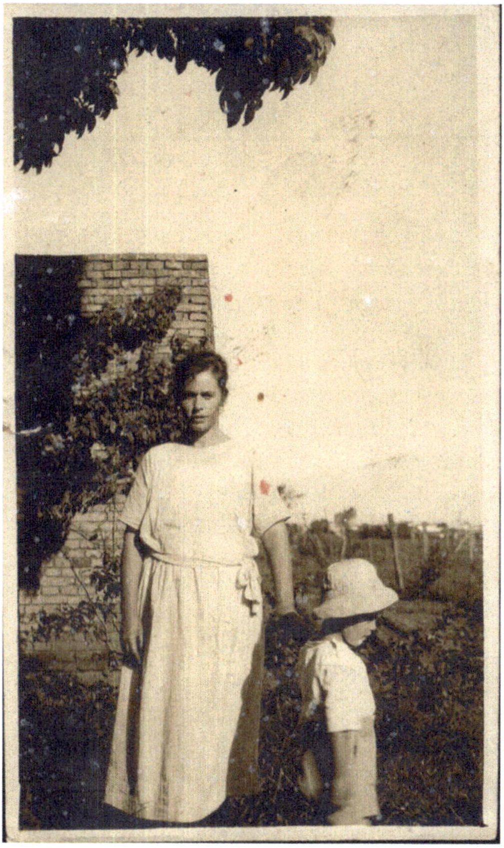

5. With mother, 1922

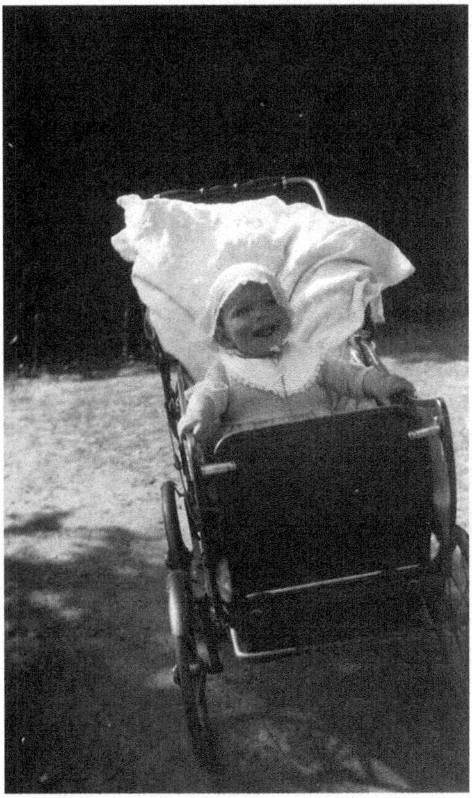

6. Eva, 1924

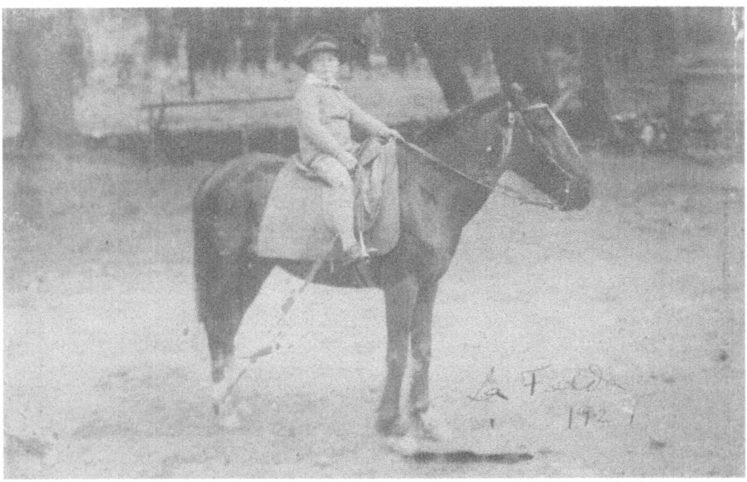

7. La Falda, 1924

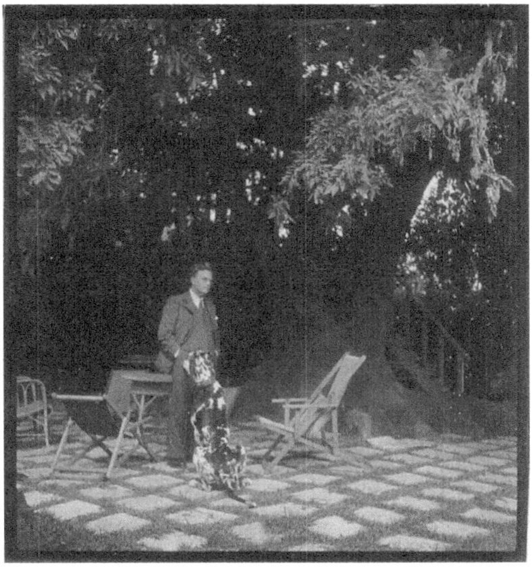

8. Augusto Bunge and Norkin under the ombú, 1935

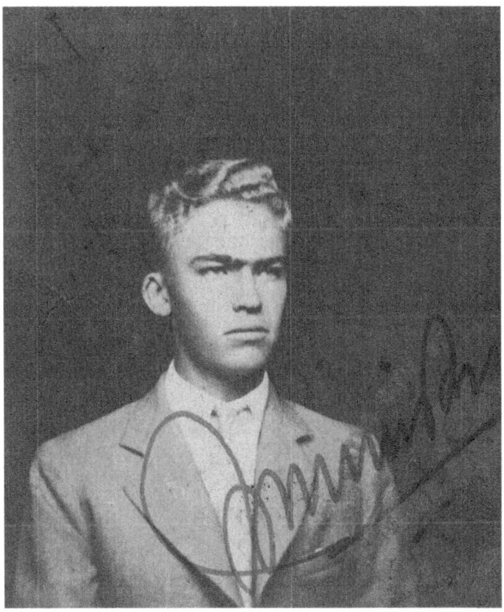

9. Military enrollment, 1937

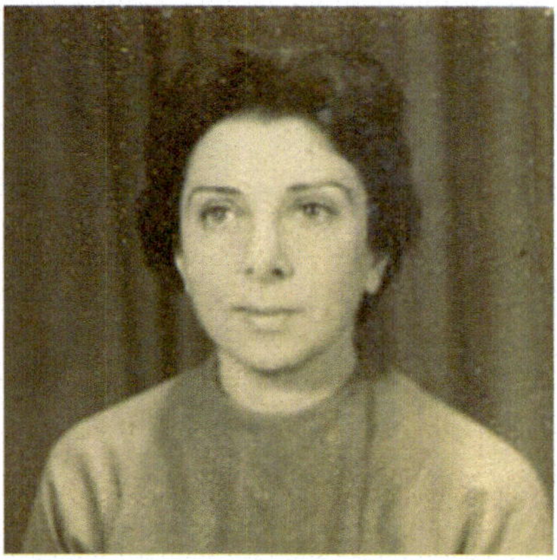

10. Julia Molina y Vedia, 1940

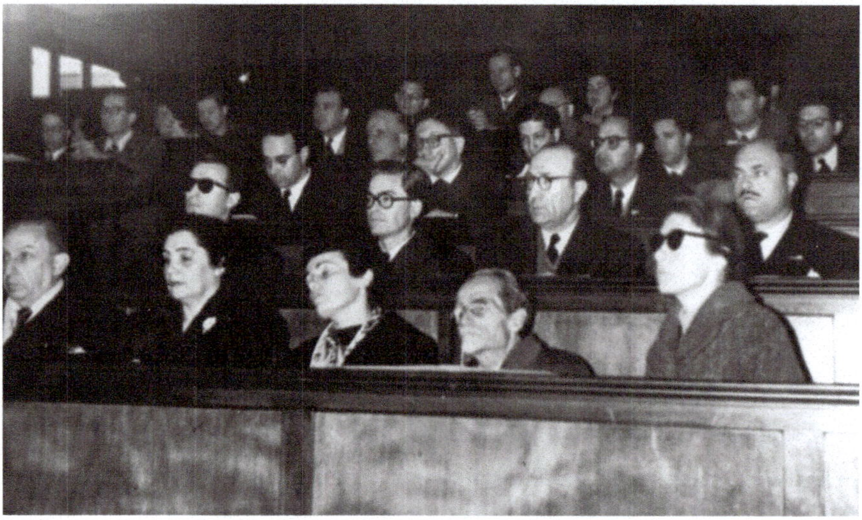

11. AFA meeting, Buenos Aires, 1949

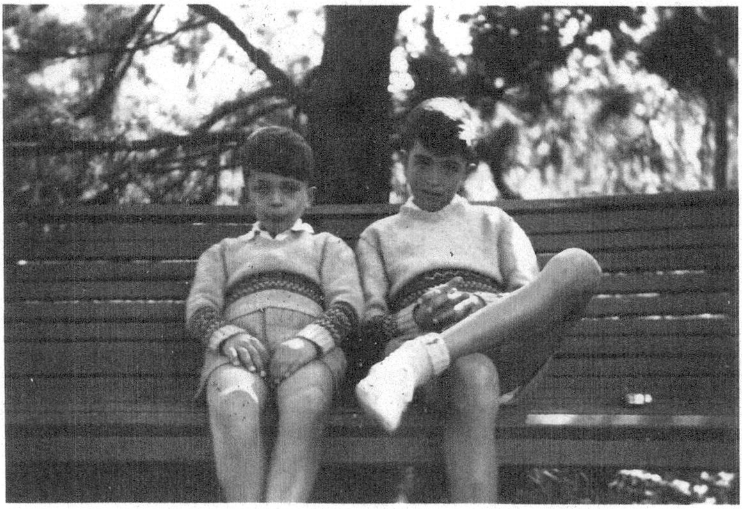

12. Cantarito and Bambi, El Ombú, 1950.

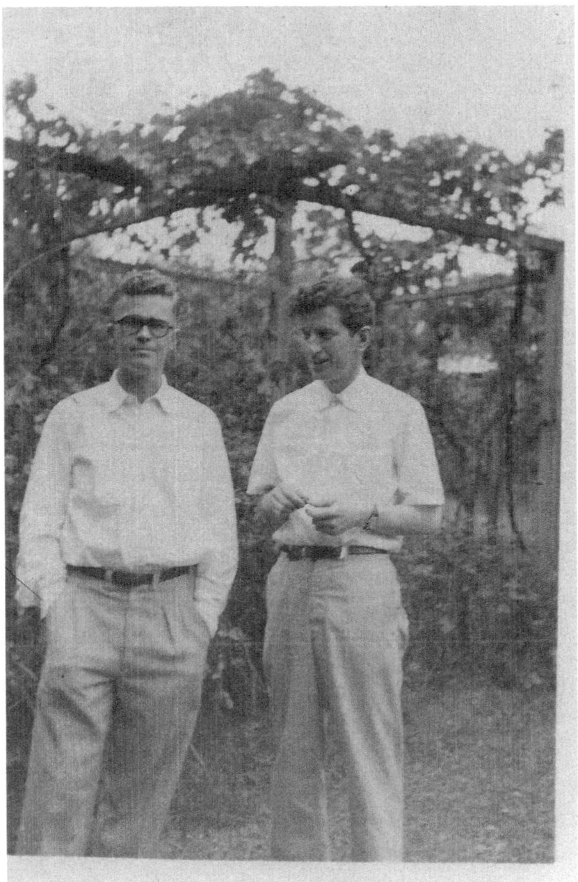

13. With David Bohm, El Ombú, 1954

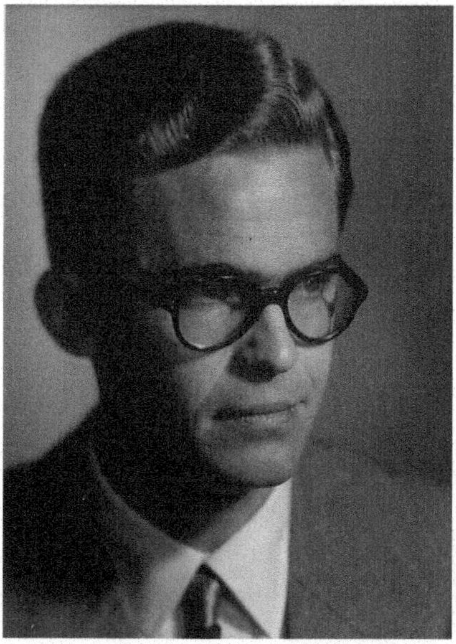

14. Buenos Aires, 1956

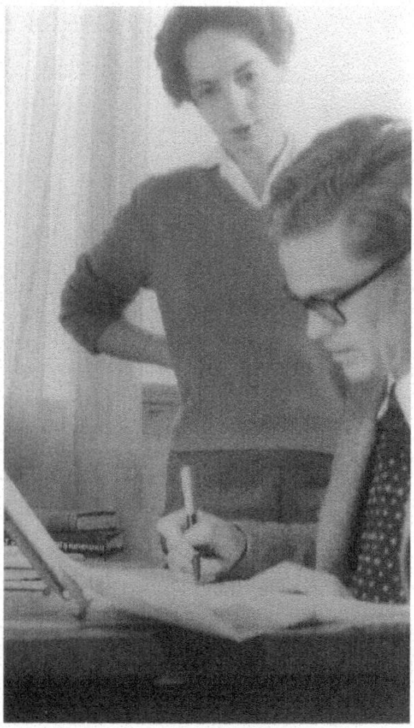

15. With Marta, El Hogar Obrero, Buenos Aires, 1959

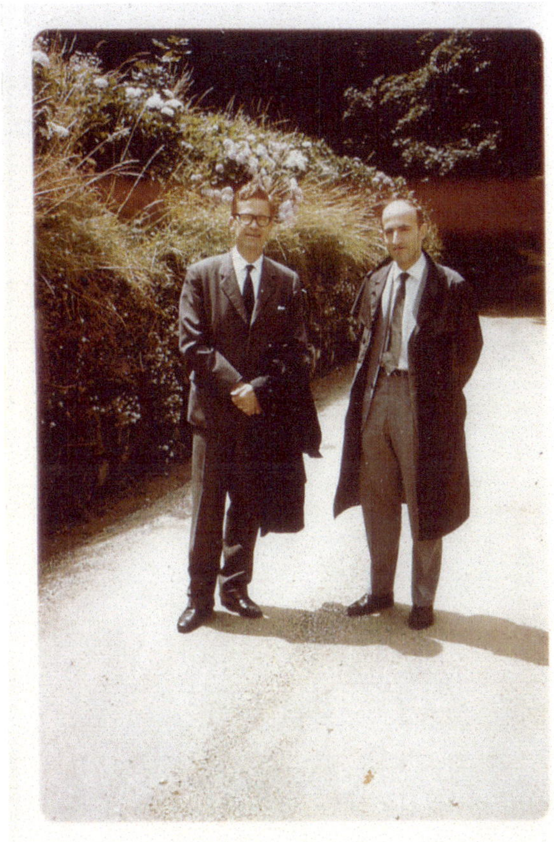

16. With Andrés Kálnay, Freiburg, Germany, 1966

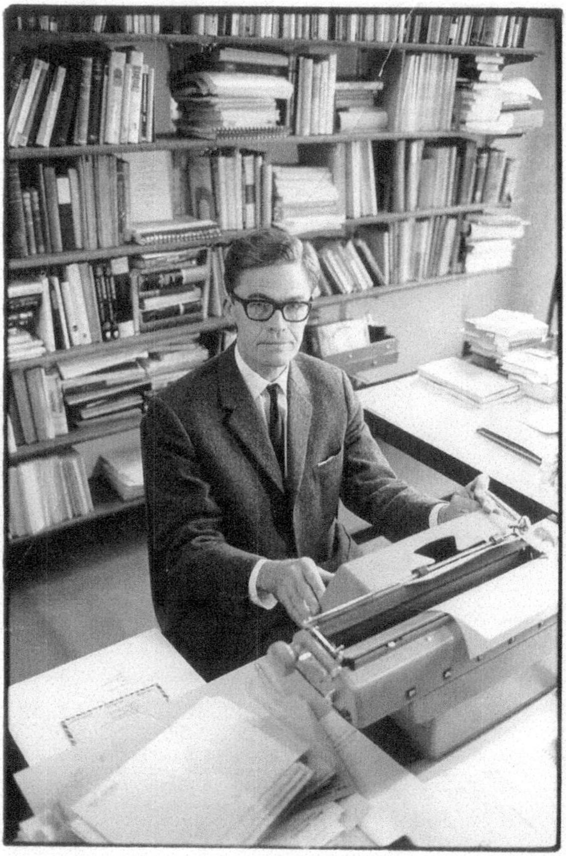

17. Just arrived at McGill, 1967

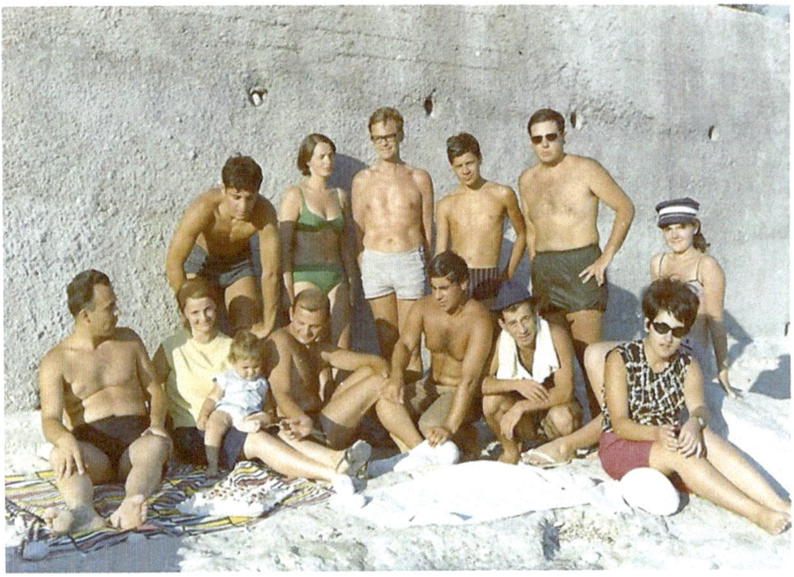

18. With Marta and our Greek friends, Corfu, 1968

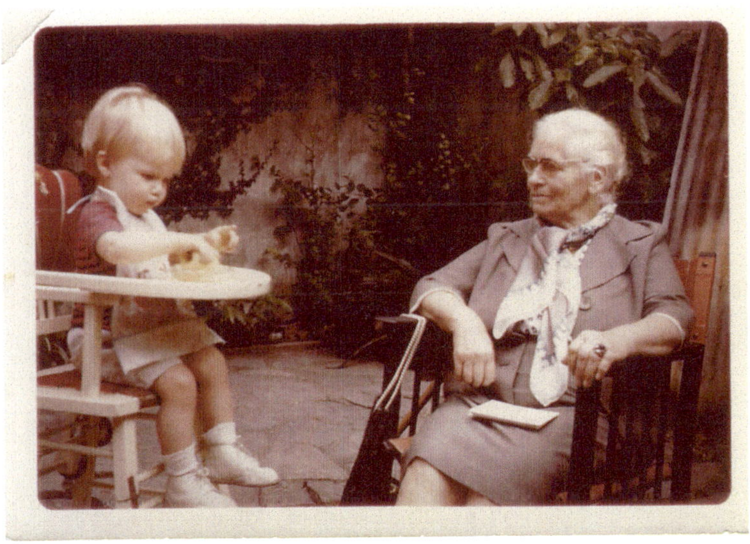

19. Mariechen with Eric, Buenos Aires 1969

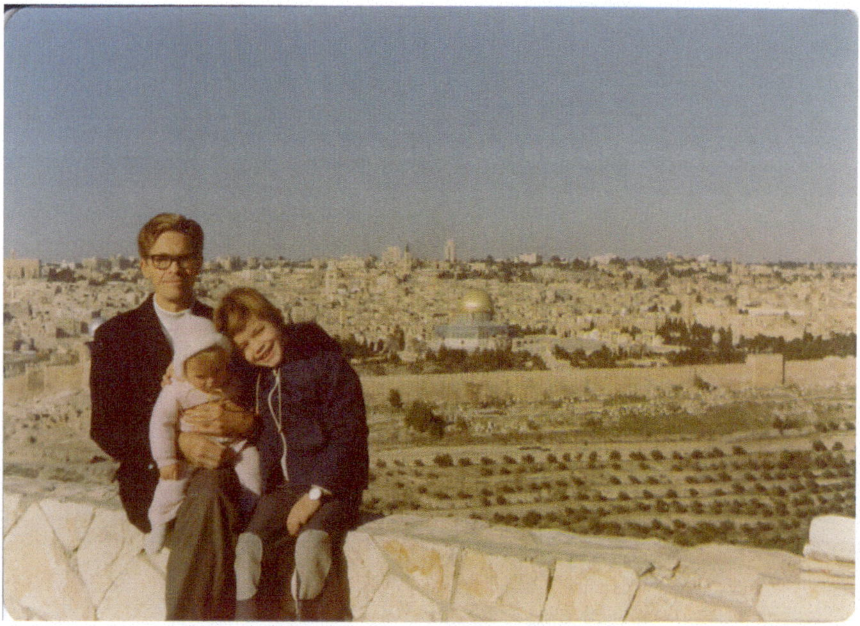

20. With Eric and Silvia, Jerusalem, 1974

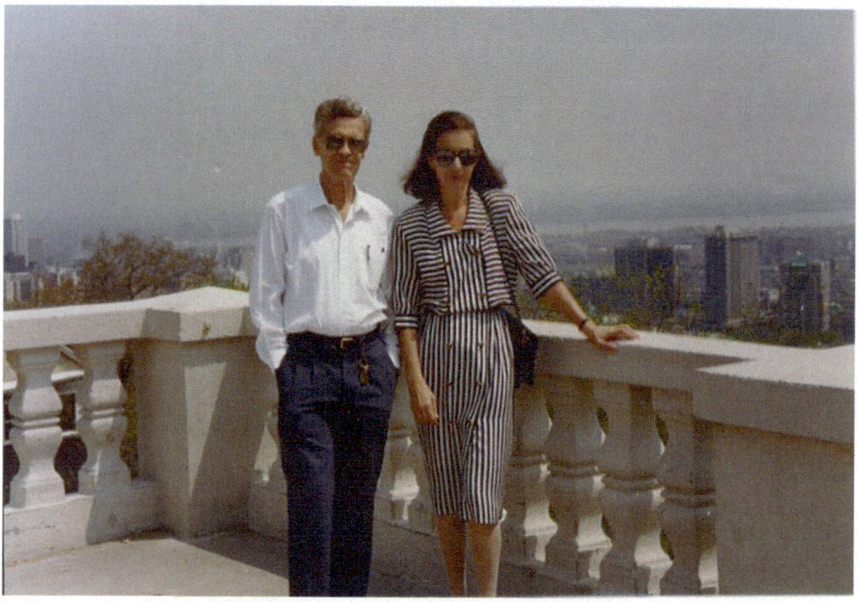

21. With Marta, Summit Park, Montreal, 1977

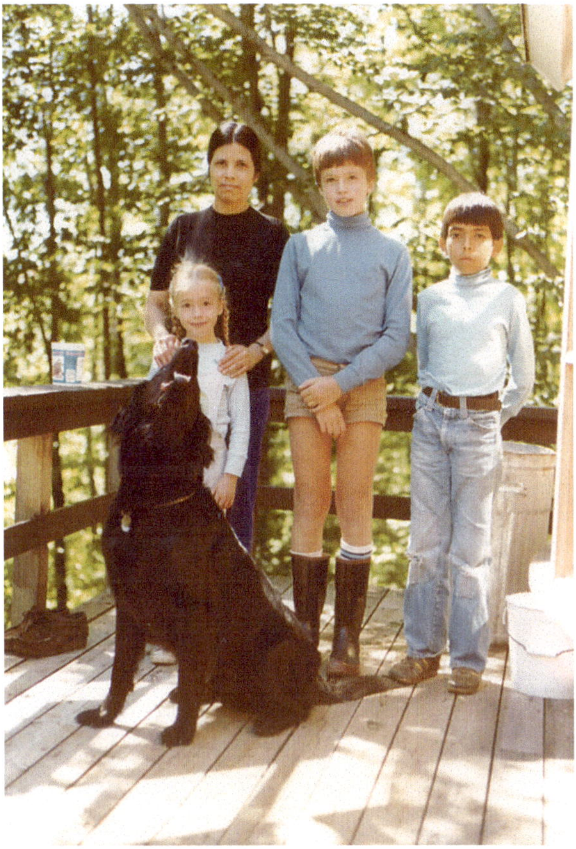

22. Eric and Silvia with Lupe, Gabriel, and Finn, Morin Heights, 1980

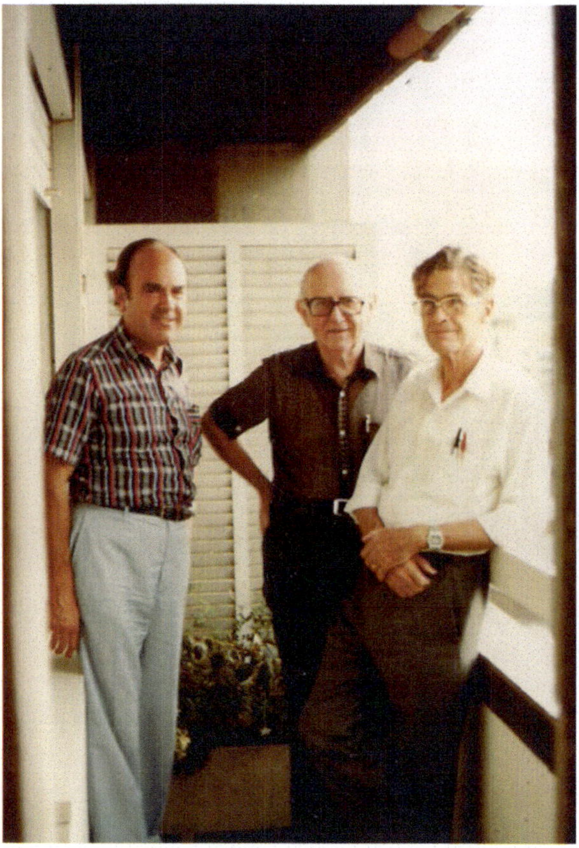

23. Prince of Asturias Prize, Oviedo, 1982

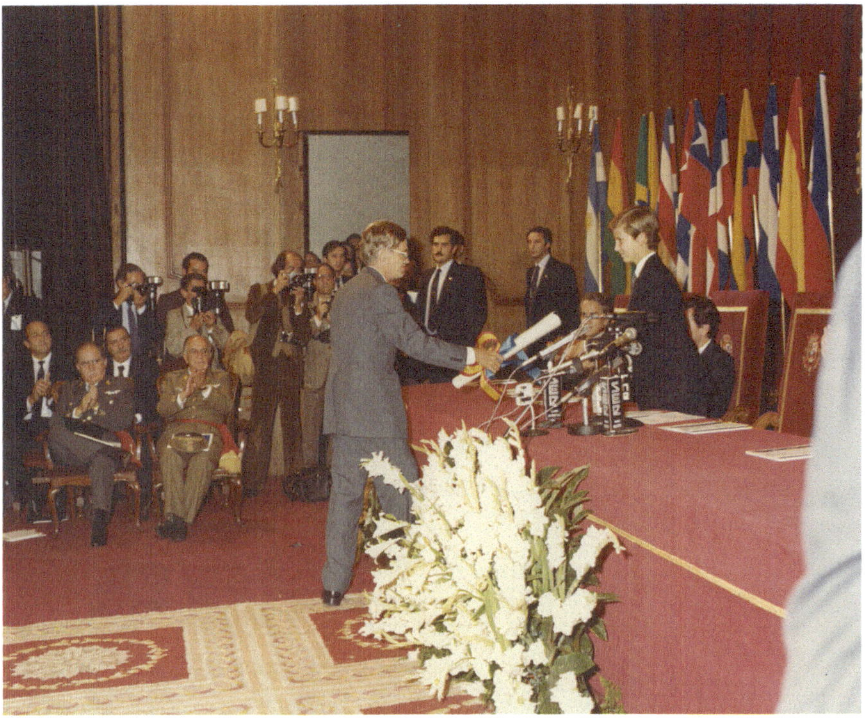

24. With Manuel Sadosky and Hernán Rodriguez, Barcelona, 1982

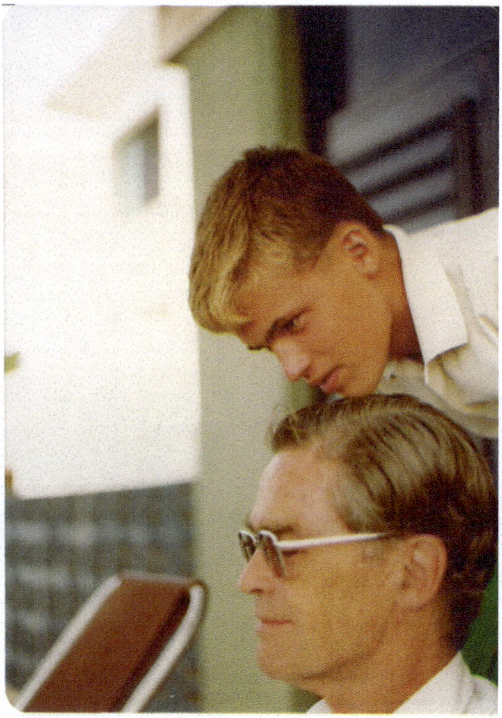

25. Teaching Eric, Corfu, 1983

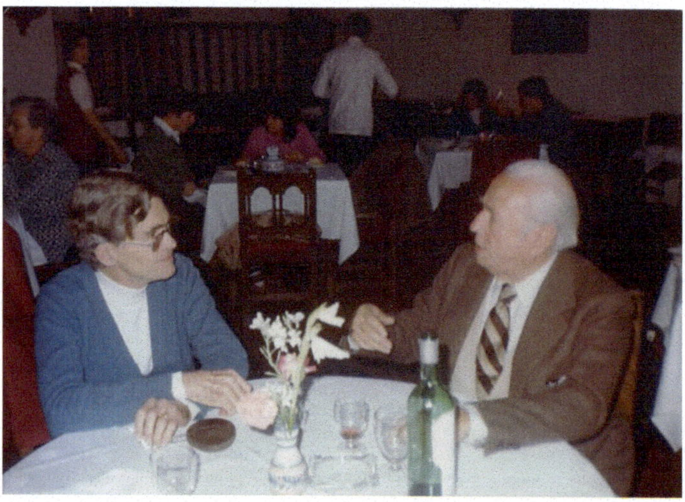

26. With Raúl Prebisch, Toledo, 1983

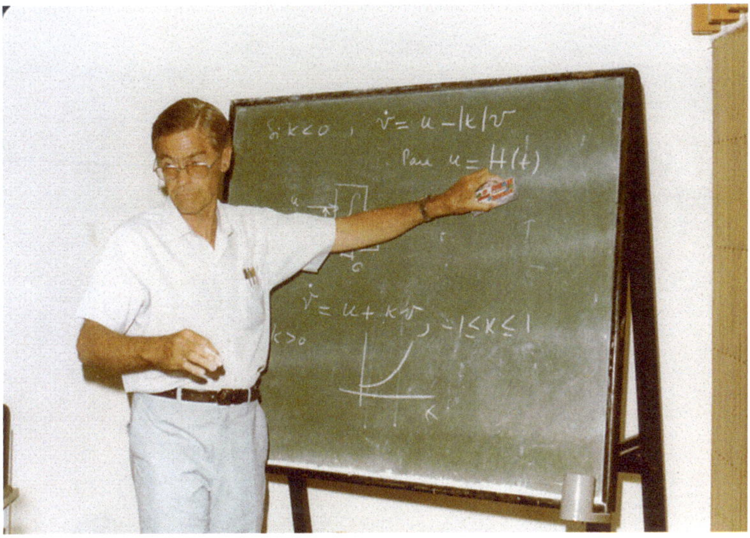

27. Santander Summer School, 1988

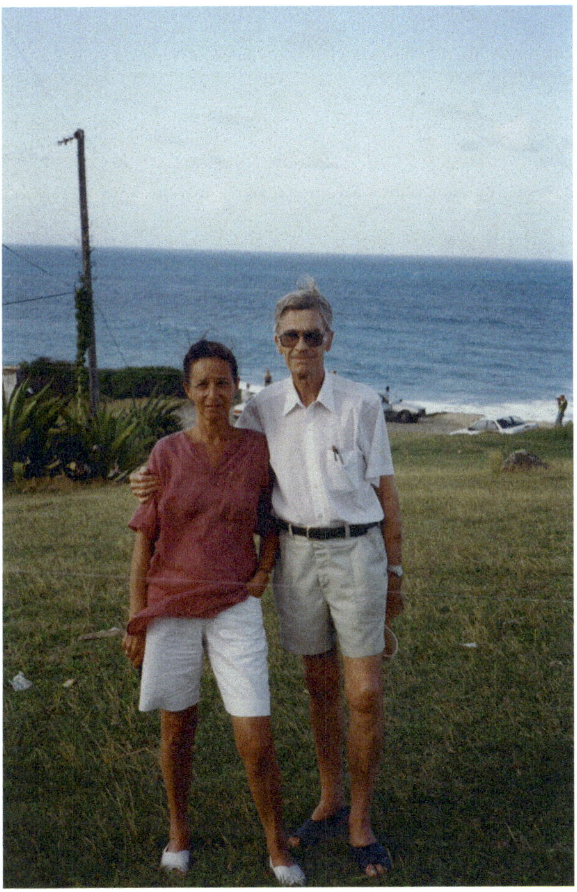

28. With Marta in Barbados, 1989

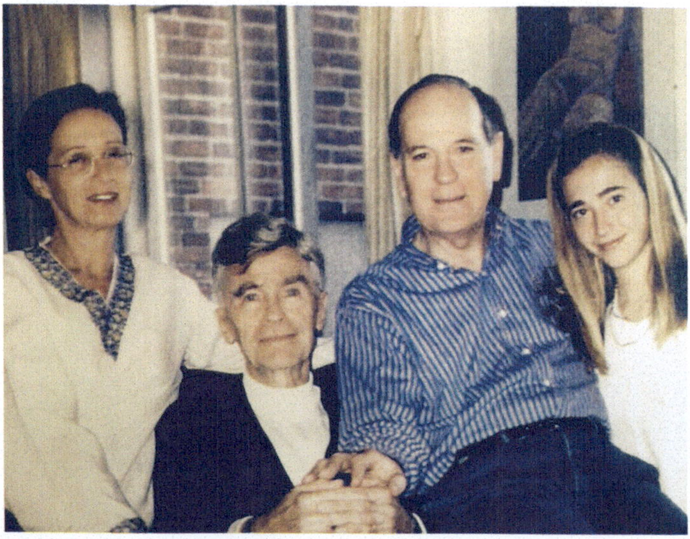

29. With José Luis Pardos, Marta and Silvia, Montreal, 1991

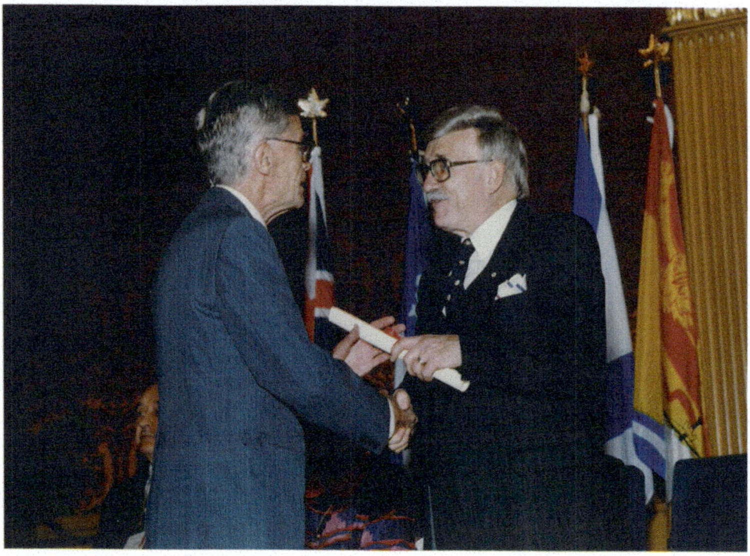

30. Welcomed by Jules Léger at the Royal Society of Canada, Ottawa, 1992

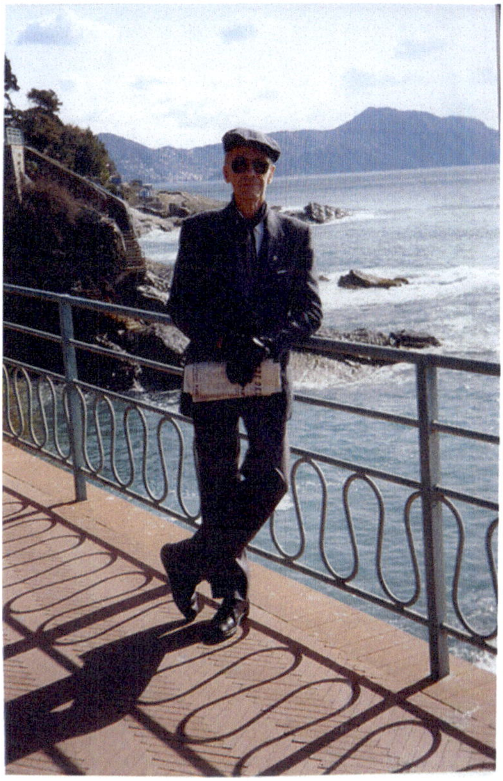

31. Nervi, Passeggiata al Mare, 1994

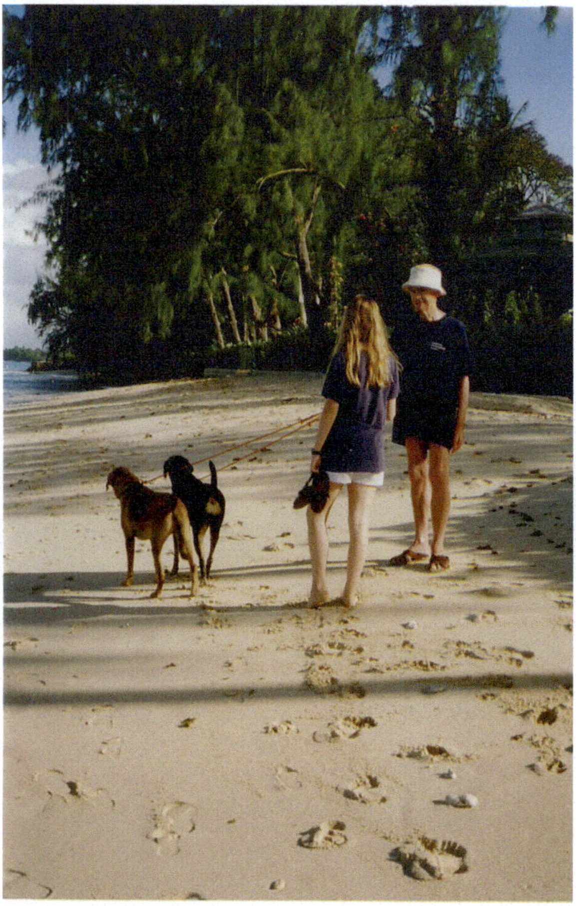

32. With Silvia, Lily and Luke, Barbados, 1994

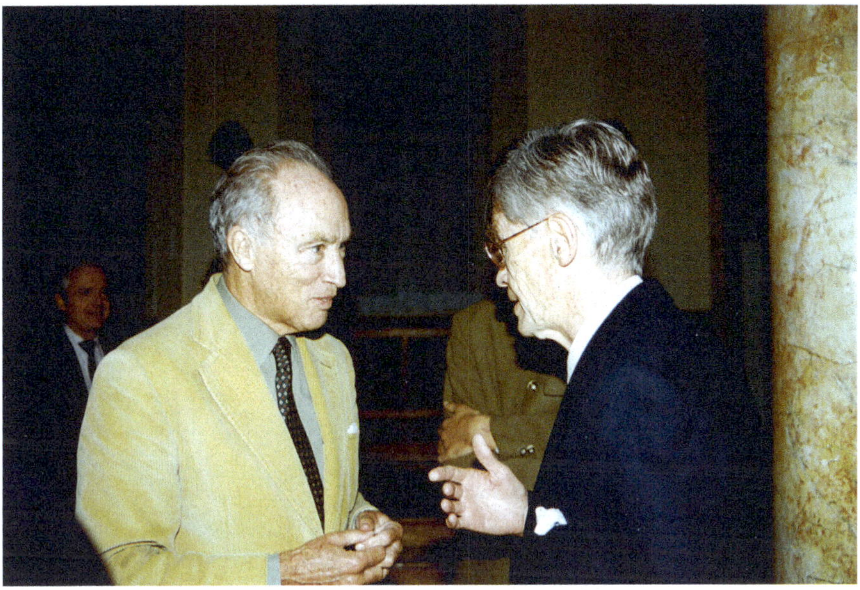

33. With Pierre Elliott Trudeau, Montreal, 1994

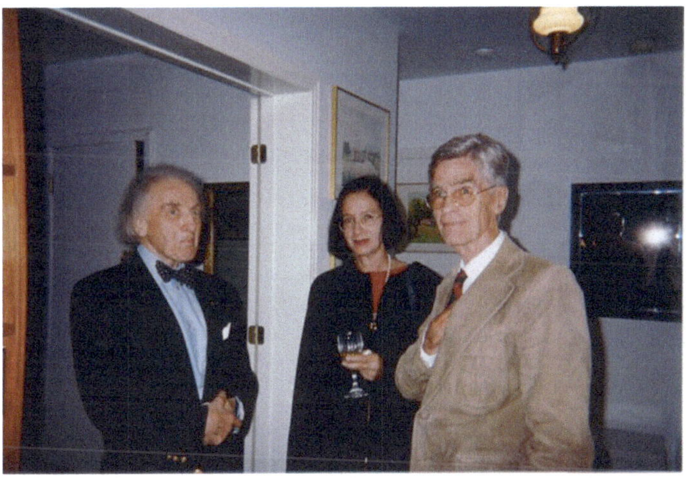

34. With Raymond Klibansky and Marta, Montreal, 1994

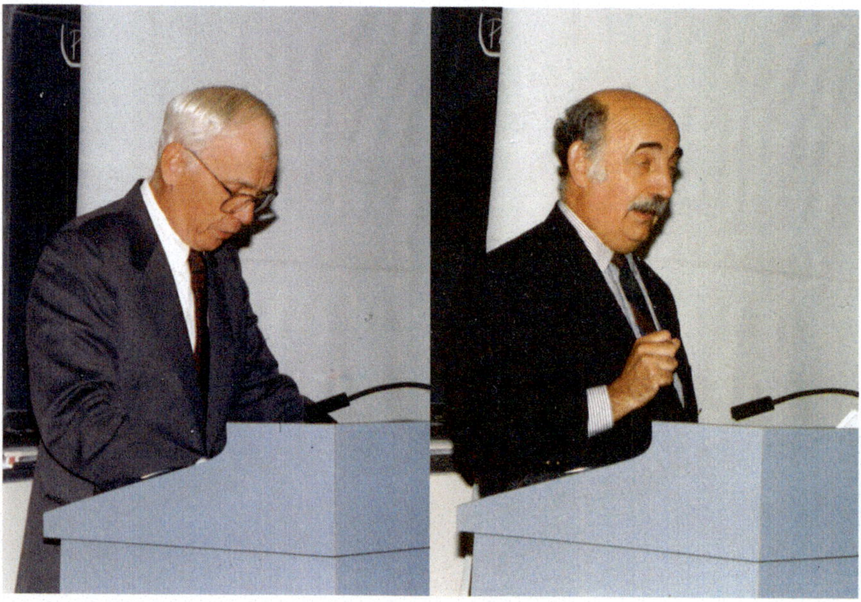

35. Vernon Mountcastle and Albert Aguayo, Montreal, 1994

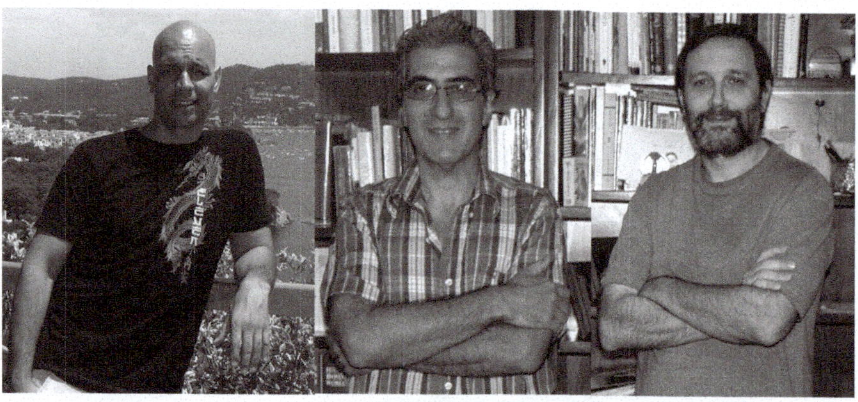

36. Rafael Gonzále del Solar, Luis Marone, and Javier López de Casenave, Buenos Aires and Barcelona, 2000

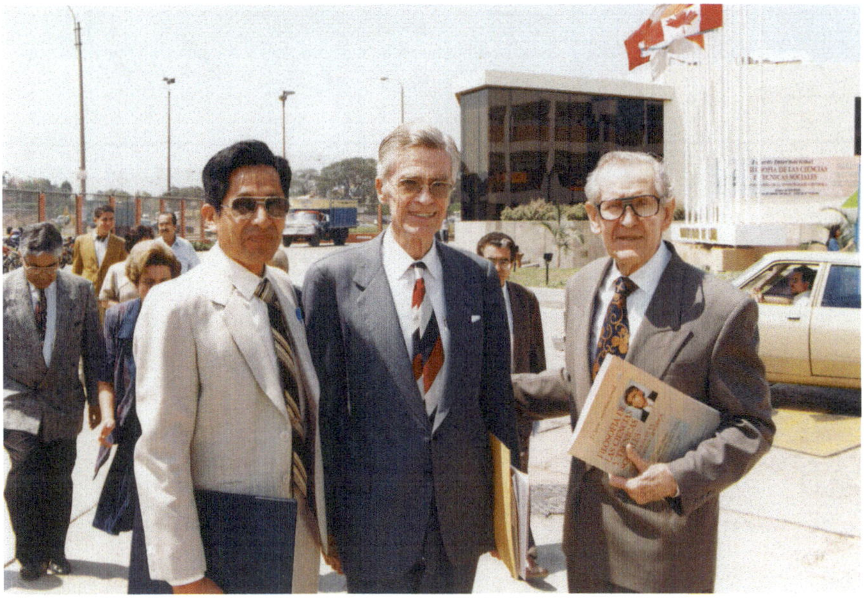

37. With Paco Miró Quesada, Lima, 1995

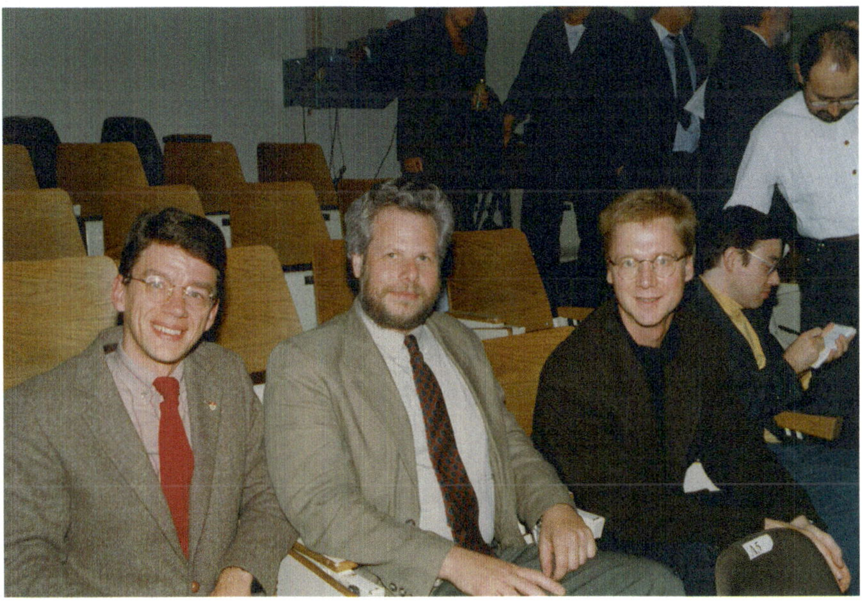

38. Mike Dillinger, David Blitz, Jean-Pierre Marquis, Michael Kary, and Martin Mahner, Montreal, 1994

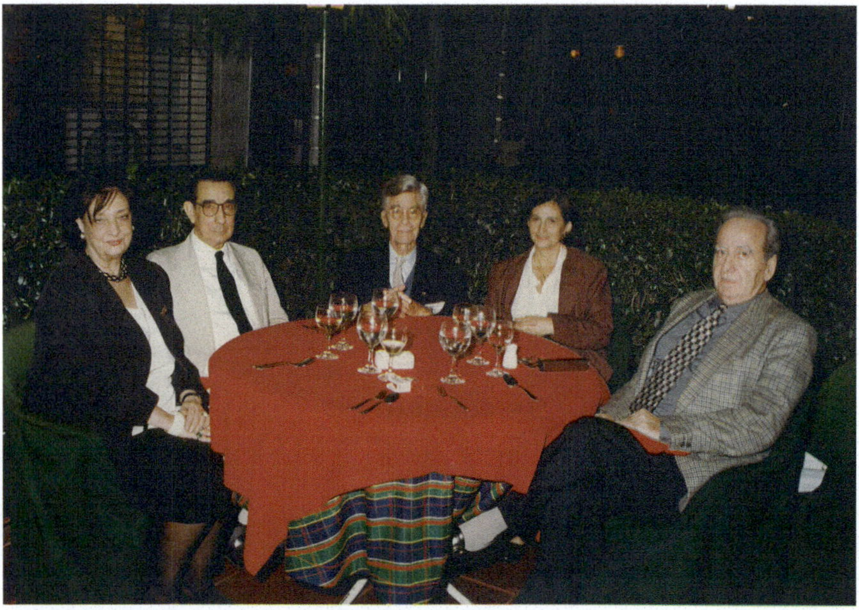

39. With the Alurraldes and the Tiscornias, Buenos Aires, 1996

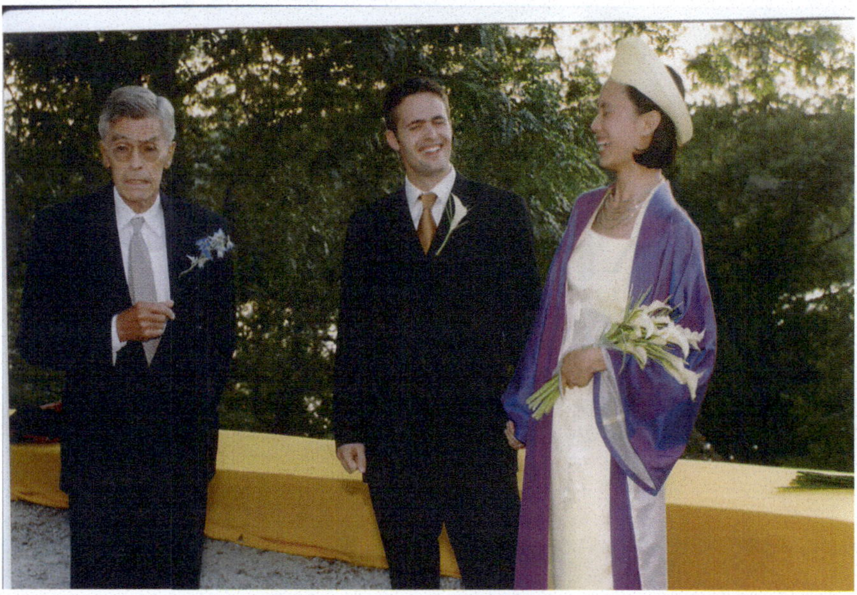

40. At the wedding of Eric and Mimi, Walden Pond, 1998

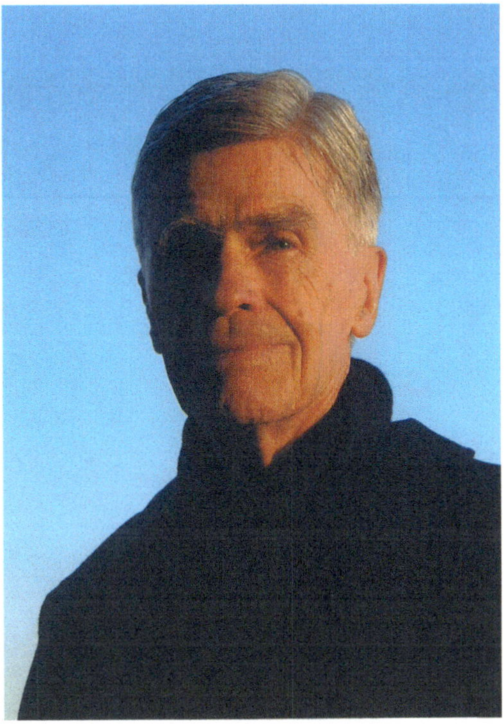

41. On the terrace at Bellevue Ave., Montreal, 2000

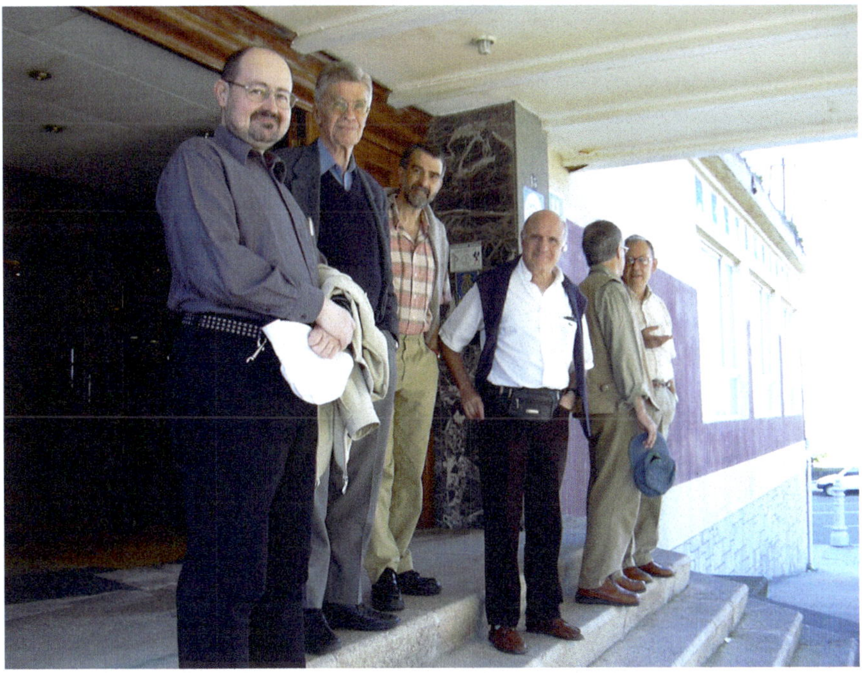

42. With Martin Mahner, Ignacio Morgado, Héctor Vucetich, Jesús Mosterín, and Alfons Barceló, Vigo, 2003

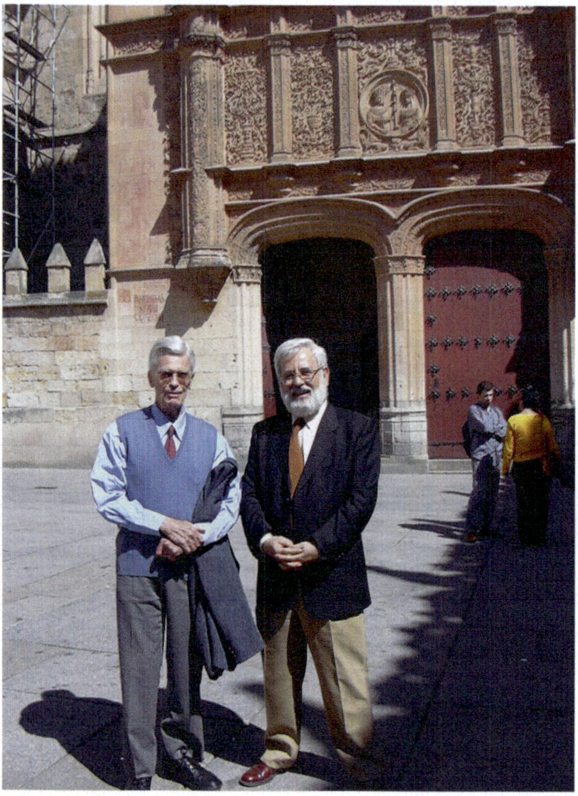

43. With Miguel Angel Quintanilla, Salamanca, 2003

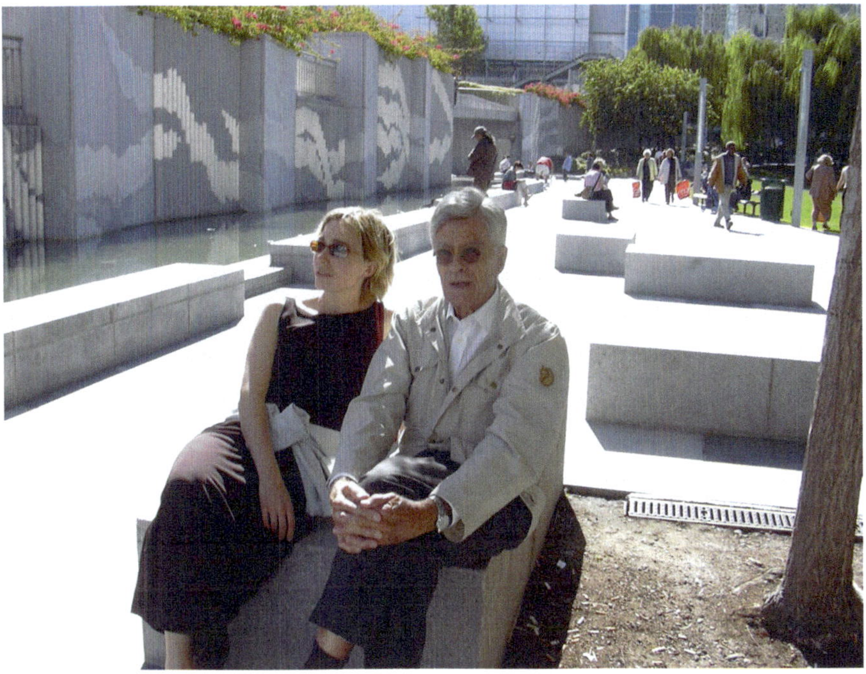

44. With Silvia, San Francisco, 2004

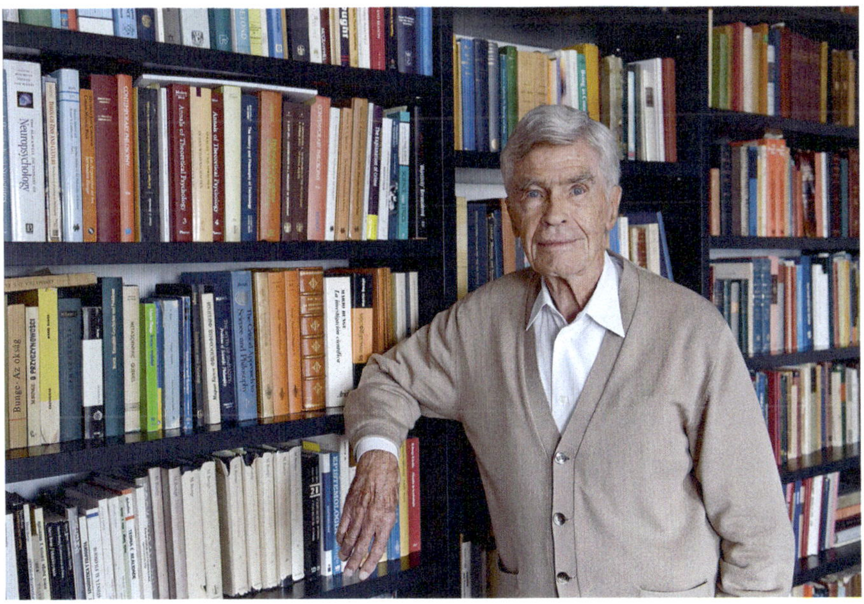

45. Mario in front of his library, Bellevue Ave., Montreal, 2007

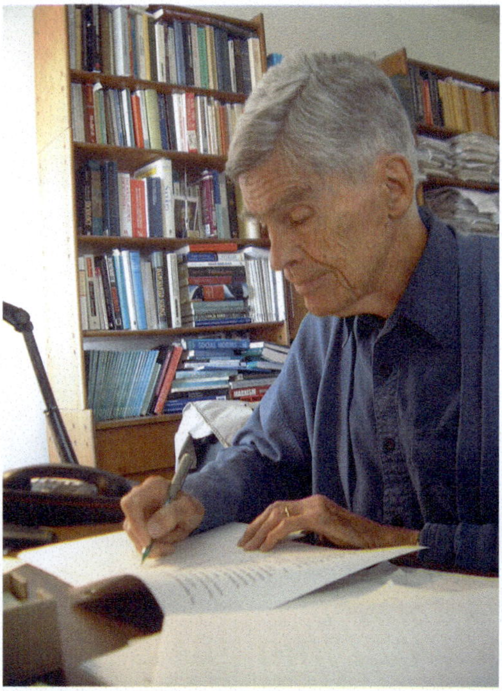

46. Mario in his study, Bellevue Ave., Montreal, 2007

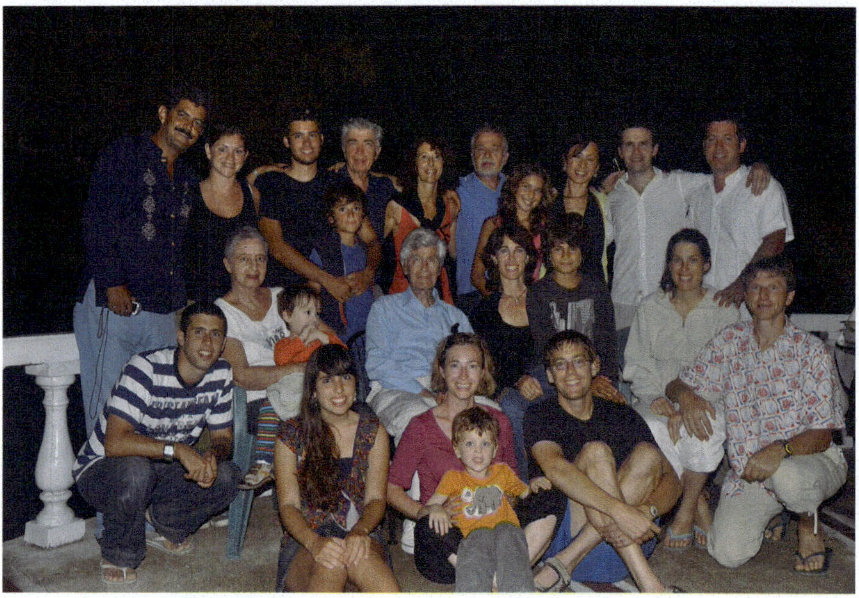

47. The Bunges, Bacalar, México, 2009

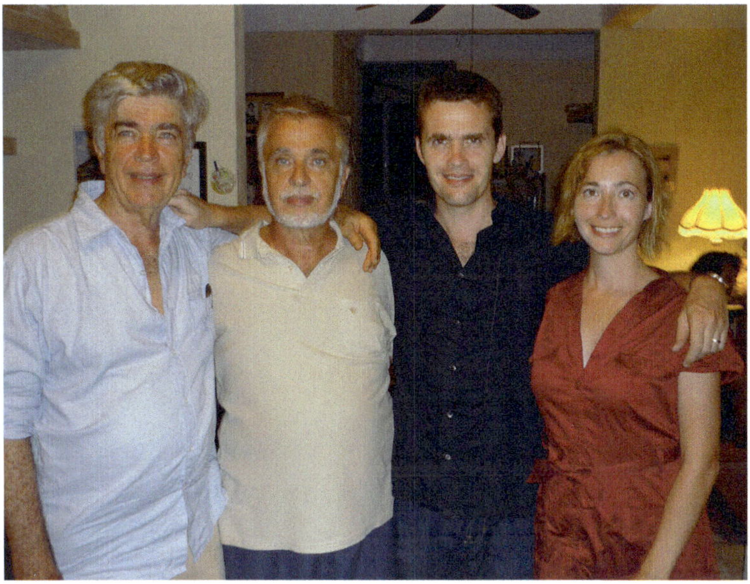

48. Carlos, Mario Jr, Eric, and Silvia, Bacalar, Mexico, 2009

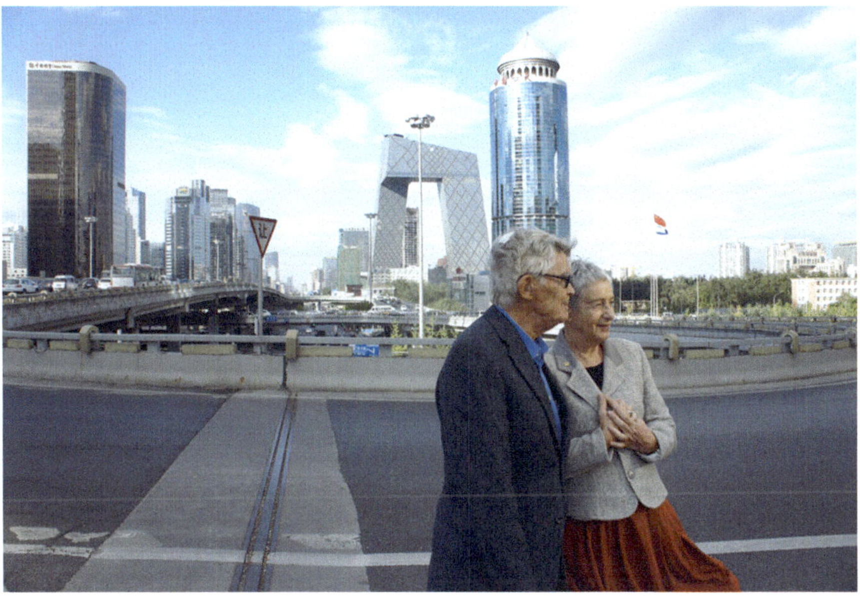

49. With Marta in Beijing, 2012

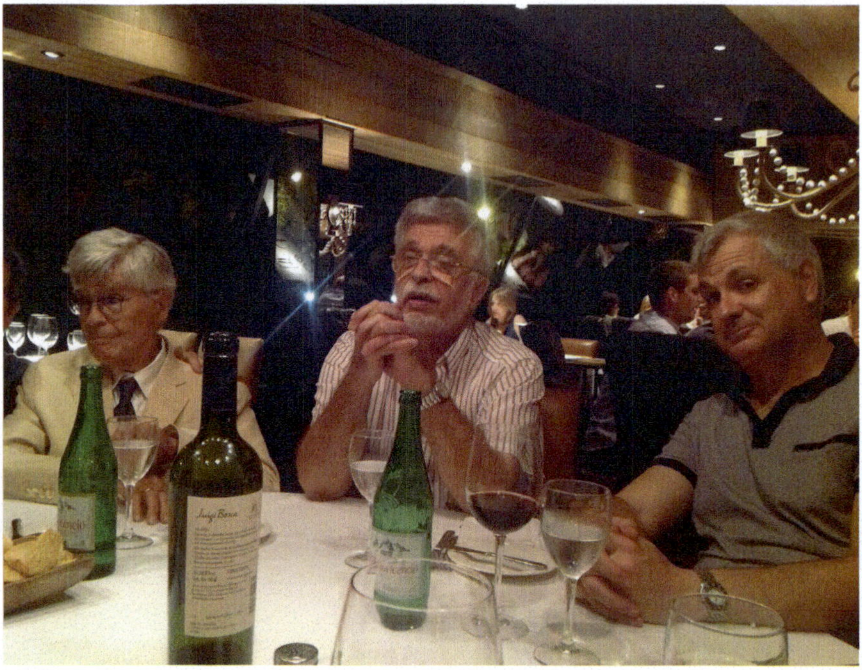

50. With Mario Bunge (junior) and Gustavo Romero, Buenos Aires, 2012

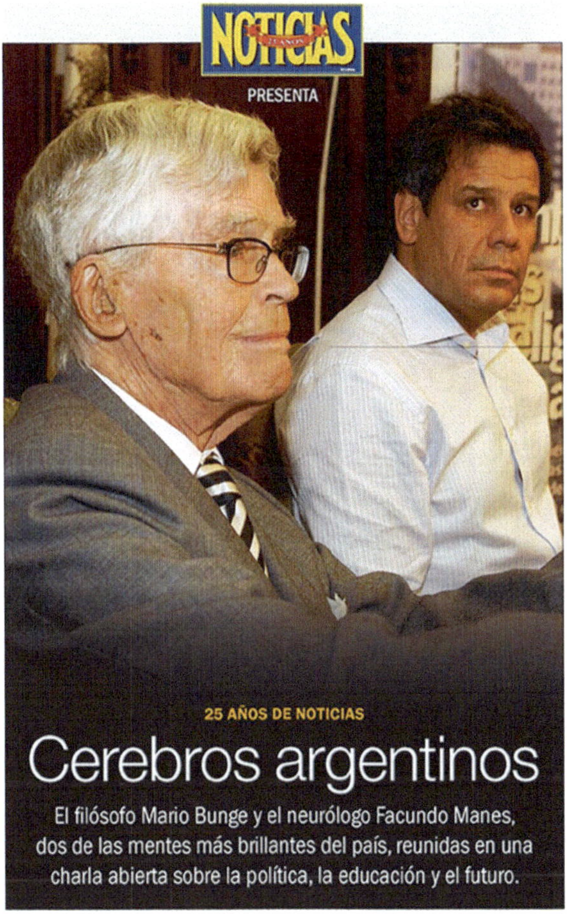

51. With Facundo Manes, Buenos Aires, 2014

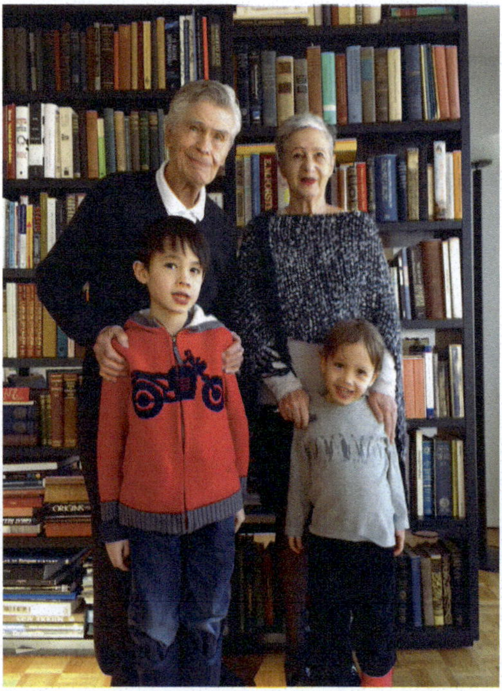

52. With Marta and our grandchildren Giao and Vi, Montreal, 2014

References

Agassi, Joseph. 1969. Review of *scientific research. Synthese* 19: 453–464.
Agassi, Joseph, and Robert S. Cohen (eds.). 1982. *Scientific philosophy today: Essays in honor of Mario Bunge*. Dordrecht/London: D. Reidel.
Alcock, J.E. 1981. *Parapsychology, science or magic?: A psychological perspective*. Oxford: Pergamon.
Anderson, Philip W. 2011. *More and different: Notes from a thoughtful curmudgeon*. Singapore: World Scientific.
Andrade, Gabriel. 2013. *El posmodernismo ¡vaya timo!* Pamplona: Laetoli.
Apostol, Pavel. 1985. About Mario Bunge's "A critical examination of Dialectics". *Journal of East European Thought* 29: 89–136.
Aris, R. 1965. Prolegomena to the rational analysis of systems of chemical reactions. *Archive for Rational Mechanics and Analysis* 19: 81–99.
Armstrong, David Malet. 1997. *A world of states of affair*. Cambridge: Cambridge University Press.
Ashby, Ross. 1956. *An introduction to cybernetics*. London: Chapman & Hall.
Barrios Medina, Ariel, and Alejandro C. Paladini (eds.). 1989. *Escritos y discursos del Dr. Bernardo A. Houssay*. Buenos Aires: Eudeba.
Bell, John S. 1964. On the Einstein Podolsky Rosen paradox. *Physics* 1: 195–200.
Benda, Julien. 2006 [1927]. *The treason of the intellectuals*. Piscataway: Transaction.
Berlyne, D.E. 1971. *Aesthetics and psychobiology*. New York: Appleton.
Bernal, John D. 1939. *The social function of science*. London: Faber & Faber.
Bernaola, Omar. 2001. *Enrique Gaviola y el Observatorio Astronómico de Córdoba*. Buenos Aires: Ediciones Saber y Tiempo.
Bignone, E. 1936. *L'Aristotele perduto e la formazione filosofica di Epicuro*. Roma: Bompiani.
Bindra, Dalbir (ed.). 1980. *The brain's mind*. New York: Gardner Press.
Blitz, David. 1992. *Emergent evolution: Qualitative novelty and the levels of reality*. Dordrecht/ Boston: Kluwer.
Bohm, David. 1952. A suggested interpretation of the quantum theory in terms of 'hidden variables'. *Physical Review* 85: 166–193.
Bohm, David. 1957. *Causality and chance in modern physics*. New York: Harper.
Bohm, David. 1980. *Wholeness and the implicate order*. London: Routledge.
Boudon, Raymond. 2000. *The origin of values*. New Brunswick: Transaction Publishers.
Bolzano, Bernard. 1972. *Theory of Science. Selection by Rolfe George*. Berkeley/Los Angeles: University of California Press.
Braithwaite, Richard B. 1953. *Scientific explanation*. Cambridge: Cambridge University Press.
Bridgman, P.W. 1922. *Dimensional analysis*. New Haven: Yale University Press.

M. Bunge, *Between Two Worlds*, Springer Biographies,
DOI 10.1007/978-3-319-29251-9

Bridgman, P.W. 1927. *The logic of modern physics*. New York: Macmillan.
Bruera, José Juan. 1945. La lógica, el Derecho y la escuela de Viena. *Minerva* 2: 170–177.
Bunge, Augusto. 1932. *El continente rojo*. Buenos Aires: J. L. Rosso.
Bunge, Augusto. 1934. *La guerra del petróleo en Argentina*. Buenos Aires: Colegio Libre de Estudios Superiores.
Bunge, Mario. 1939. Introducción al estudio de los grandes pensadores. *Conferencias (Buenos Aires) III* 105–109: 124–126.
Bunge, Mario. 1943. *Significado físico e histórico de las ecuaciones de Maxwell*. Buenos Aires: Universidad Obrera Argentina.
Bunge, Mario. 1944a. Una nueva representación de los tipos de fuerzas nucleares. *Revista de la Facultad de Ciencias Físicomatemáticas* 221–239.
Bunge, Mario. 1944b. A new representation of types of nuclear forces. *Physical Review* 65: 249.
Bunge, Mario. 1944c. ¿Qué es la epistemología? *Minerva* 1: 27-43.
Bunge, Mario. 1944d. Auge y fracaso de la filosofía de la naturaleza. *Minerva* (Buenos Aires) 1: 213-235.
Bunge, Mario. 1945a. Neutron-proton scattering at 8.8 and 13 MeV. *Nature* 156: 301.
Bunge, Mario. 1945b. Cómo veía el mundo Florentino Ameghino. *Minerva* 2: 184–185.
Bunge, Mario. 1951a. What is chance? *Science and Society* 15: 209–231.
Bunge, Mario. 1951b. La fenomenología y la ciencia. *Cuadernos Americanos*, (4): 108–122. Repr. in Bunge 2007b, pp. 265–285.
Bunge, Mario. 1951c. Mach y la teoría atómica. *Boletin del Químico Peruano* 3(16): 12–17.
Bunge, Mario. 1954. New dialogues between Hylas and Philonous. *Philosophy and Phenomenological Research* 15: 192–199.
Bunge, Mario. 1955a. A picture of the electron. *Nuovo Cimento ser. X* 1: 977–985.
Bunge, Mario. 1955b. Strife about complementarity. *British Journal for the Philosophy of Science* 6(1–12): 141–154.
Bunge, Mario. 1955c. *La edad del universo*. La Paz: Laboratorio de Física Cósmica.
Bunge, Mario. 1955d. Exposición y crítica del principio de complementaridad. *Notas del Curso Iteramericano de Física Moderna*, 27–36. La Paz: Laboratorio de Física Cósmica.
Bunge, Mario. 1955e. The philosophy of the space-time approach to the quantum theory. *Methodos* 7: 295–308.
Bunge, Mario. 1956a. Do computers think? *British Journal for the Philosophy of Science* 7: 139–148; 7: 212–219.
Bunge, Mario. 1956b. A survey of the interpretations of quantum mechanics. *American Journal of Physics* 24: 272–286.
Bunge, Mario. 1956c. La interpretación causal de la mecánica ondulatoria. *Ciencia e Investigación* 12: 448–457.
Bunge, Mario. 1957a. Filosofar científicamente y encarar la ciencia filosóficamente. *Ciencia e Investigación* 13: 244–257.
Bunge, Mario. 1957b. Lagrangian formulation and mechanical interpretation. *American Journal of Physics* 25: 211–218.
Bunge, Mario. 1958a. Review of P. W. Bridgman's *reflections of a physicist. British Journal for the Philosophy of Science* 9: 74.
Bunge, Mario. 1958b. Sobre la imagen física de la partículas de spin entero. *Ciencia e Investigación* 14: 311–315.
Bunge, Mario. 1959a. *Causality: The place of the causal principle in modern science*. Cambridge, MA: Harvard University Press.
Bunge, Mario. 1959b. Review of Popper's *The logic of scientific discovery. Ciencia e Investigación* 15: 216.
Bunge, Mario. 1959c. *Metascientific queries*. Evanston: Charles C. Thomas.
Bunge, Mario. 1960a. Levels: A semantical preliminary. *Review of Metaphysics* 13: 396–406.
Bunge, Mario. 1960b. The place of induction in science. *Philosophy of Science* 27: 262–270.
Bunge, Mario. 1960c. *La ciencia: su método y su filosofía*. Buenos Aires: Siglo Veinte. Repr: Pamplona: Laetoli, 2013.

Bunge, Mario. 1960d. *La cinemática del electrón relativista*. Tucumán: Universidad Nacional de Tucumán.
Bunge, Mario. 1960e. Probabilidad e inducción. *Ciencia y Técnica (Buenos Aires)* 129: 240.
Bunge, Mario. 1960f. On the connections among Levels. *Proceedings of the XIIth International Congress of Philosophy* VI: 63–70. Firenze: Sansoni.
Bunge, Mario. 1960g. Are there timeless entitites? *Miscelanea de Estudos a Joaquim de Carvalho. Figueira da Foz (Portugal)* 3: 290–292.
Bunge, Mario. 1960h. *Etica y ciencia*. Buenos Aires: Siglo Veinte.
Bunge, Mario. 1961a. Analyticity redefined. *Mind* 278: 239–245.
Bunge, Mario. 1961b. Ethics as science. *Philosophy and Phenomenological Research* 20: 139–152.
Bunge, Mario. 1961c. Laws of physical laws. *American Journal of Physics* 29: 518–529.
Bunge, Mario. 1962a. Cosmology and magic. *The Monist* 44: 116–141.
Bunge, Mario. 1962b. An analysis of value. *Mathematicae Notae* 18: 95–108.
Bunge, Mario. 1962c. *Intuition and science*. Englewood Cliffs: Prentice- Hall.
Bunge, Mario. 1963a. Tecnología, ciencia y filosofía. *Revista de la Universidad de Chile* 121(126): 64–92.
Bunge, Mario. 1963b. *The myth of simplicity*. Englewood Cliffs: Prentice-Hall.
Bunge, Mario. 1964. Phenomenological theories. In *The critical approach*, ed. M. Bunge, 234–254. Glencoe: Free Press.
Bunge, Mario. 1966. Technology as applied science. *Technology and Culture* 7: 329–347.
Bunge, Mario. 1967a. *Foundations of physics*. Berlin: Springer.
Bunge, Mario. 1967b. *Scientific research 2 vols*. Berlin: Springer. Repr. as *Philosophy of Science*, 2 vols. New Brunswick: Transaction Publishers.
Bunge, Mario. 1967c. Physical axiomatics. *Reviews of Modern Physics* 39: 463–474.
Bunge, Mario. 1967d. Analogy in quantum mechanics: From insight to nonsense. *British Journal for the Philosophy of Science* 18: 265–286.
Bunge, Mario. 1967e. *Delaware seminar in the philosophy of science*. Berlin: Springer-Verlag.
Bunge, Mario (ed.). 1967f. *Quantum theory and reality*, Includes an Introduction. Berlin: Springer-Verlag.
Bunge, Mario. 1967g. A ghost free axiomatization of quantum mechanics. In Bunge 1967f, 105–117
Bunge, Mario. 1968a. Physical time: The objective and relational theory. *Philosophy of Science* 35: 355–388.
Bunge, Mario. 1968b. The maturation of science. In *Problems in the philosophy of science*, ed. I. Lakatos and A. Musgrave, 120–137. Amsterdam: North Holland.
Bunge, Mario. 1969a. Corrections to *foundations of physics*: Correct and incorrect. *Synthese* 19: 443–452.
Bunge, Mario. 1969b. The metaphysics, epistemology and methodology of levels. In *Hierarchical levels*, ed. L.L. Whyte, A.G. Wilson, and D. Wilson, 17–28. New York: American Elsevier.
Bunge, Mario. 1969c. *La investigación científica*. Barcelona: Ariel. Revised edition, México, DF: Siglo xxi, 2000.
Bunge, Mario. 1969d. Alexander von Humboldt und die Philosophie. In *Alexander von Humboldt: Werk und Weltgeltung*, ed. H. Pfeiffer, 17–30. München: Piper & Co.
Bunge, Mario. 1969e. Analogy, simulation, representation. *Revue Internationale de Philosophie* 23: 16–33.
Bunge, Mario. 1969f. Models in theoretical science. *Proceedings of the XIVth International Congress of Philosophy III*: Wien: Herder. 208–217.
Bunge, Mario. 1969g. Four models of human migration: An exercise in mathematical sociology. *Archiv für Rechts- und Sozialphilosophie* 55: 451–462.
Bunge, Mario. 1970a. Problems concerning inter-theory relations. In *Induction, physics and ethics*, ed. P. Weingartner and G. Zecha, 285–315. Dordrecht: Reidel.
Bunge, Mario. 1970b. Review of Werner Heisenberg's *Der Teil und das Ganze*. *Physics Today* 23: 63–64.

Bunge, Mario. 1970c. Theory meets experience. In *Contemporary philosophic thought*, vol. 2, ed. H. Kiefer and M.K. Munitz, 138–165. Albany: State University of New York Press.

Bunge, Mario. 1971. Is scientific metaphysics possible? *Journal of Philosophy* 68: 507–520.

Bunge, Mario. 1972. A program for the semantics of science. *Journal of Philosophical Logic* 1: 317–328.

Bunge, Mario. 1973a. *Philosophy of physics*. Dordrecht: Reidel.

Bunge, Mario. 1973b. *Method, model and matter*. Dordrecht: Reidel.

Bunge, Mario. 1973c. A decision theoretic model of the American war in Vietnam. *Theory and Decision* 3: 328–338.

Bunge, Mario. 1973d. On confusing 'measurement' with 'measure' in the methodology of the behavioral sciences. In *Exact philosophy*, ed. M. Bunge, 105–122. Dordrecht: Reidel.

Bunge, Mario (ed.). 1973e. *Exact philosophy*. Dordrecht/Boston: Reidel.

Bunge, Mario. 1974a. *Philosophie de la physique*. Paris: ed. du Seuil.

Bunge, Mario. 1974b. The concept of social structure. In *Developments in the methodology of social science*, ed. W. Leinfellner and W. Köhler, 175–215. Dordrecht: Reidel.

Bunge, Mario. 1974c. *Treatise on basic philosophy*, Sense and Reference, vol. 1. Dordrecht: Reidel.

Bunge, Mario. 1974d. *Treatise on basic philosophy*, Interpretation and Truth, vol. 2. Dordrecht: Reidel.

Bunge, Mario. 1974e. The methodology of development indicators. UNESCO, Methods and Analysis Division, Dept. of Social Sciences.

Bunge, Mario. 1975a. Crítica de la noción fregeana de predicado. *Revista Latinoamericana de Filosofía* 1: 5–8.

Bunge, Mario. 1975b. *¿Hay proposiciones? Aspectos de la filosofía de W. V. Quine*, 53–68. Valencia: Teorema.

Bunge, Mario. 1975c. A critical examination of dialectics. In *Dialectics/dialectique*, ed. Ch Perelman, 66–77. The Hague: Martinus Nijhoff. Commentary by I. Narsky, "Bemerkungen über den Vortrag von Prof. Bunge", pp. 78-86, 1975. Repr.: Bunge 1981d.

Bunge, Mario. 1975d. What is a quality of life indicator? *Social Indicators Research* 2: 65–80.

Bunge, Mario. 1976a. The relevance of philosophy to social science. In *Basic issues in the philosophy of science*, ed. W. Shea, 136–155. New York: Neale Watson. Transl. in *La filosofía y las ciencias sociales*, 43–69. México: Grijalbo.

Bunge, Mario. 1976b. El método en la biología. *Naturaleza (México)* 7: 70–81.

Bunge, Mario. 1976c. A model for processes combining competition with cooperation. *Mathematical Modelling* 1: 21–23.

Bunge, Mario. 1976d. Review of Wolfgang Stegmüller's *The structure and dynamics of theories. Mathematical Reviews* 55: 333, No. 2480.

Bunge, Mario. 1977a. Emergence and the mind. *Neuroscience* 2: 501–508.

Bunge, Mario. 1977b. Levels and reduction. *American Journal of Physiology: Regulatory Integrative and Comparative Physiology* 2: 75–82.

Bunge, Mario. 1977c. *Treatise on basic philosophy*, Vol. 3: The Furniture of the World. Dordrecht/Boston: Reidel.

Bunge, Mario. 1977d. The interpretation of Heisenberg's inequalities. In *Denken und Umdenken: Zur Werk und Wirkung von Werner Heisenberg*, ed. H. Pfeiffer, 146–156. München: Piper.

Bunge, Mario. 1977e. General systems and holism. *General Systems* 12: 87–90.

Bunge, Mario. 1978a. The mind-body problem in the light of contemporary biology (with Rodolfo Llinás). *16th World Congress of Philosophy: Section Papers*, 131–133.

Bunge, Mario. 1978b. Iatrofilosofía. In *Ensayos de Yatrofilosofía*, ed. F. Alonso de Florida, 3–5. México: Academia Nacional de Medicina.

Bunge, Mario. 1978c. Quantum mechanics and measurement. *International Journal of Quantum Chemistry* 12(Supplement 1): 1–14.

Bunge, Mario. 1979a. A systems concept of society: Beyond individualism and holism. *Theory and Decision* 10: 13–30.

Bunge, Mario. 1979b. The mind-body problem in an evolutionary perspective. In *Brain and mind*, Ciba Foundation Series 69, 53–63. Amsterdam: Excerpta Medica.

Bunge, Mario. 1979c. *Treatise on basic philosophy*, A World of Systems, vol. 4. Dordrecht: Reidel.

Bunge, Mario. 1979d. The mind-body problem, information theory, and Christian dogma. *Neuroscience* 4: 453–454.

Bunge, Mario. 1979e. *Causality in modern science*, 3rd ed. New York: Dover Publications.

Bunge, Mario. 1979f. The Einstein-Bohr debate over quantum mechanics: Who was right about what? *Lecture Notes in Physics* 100: 204–219.

Bunge, Mario. 1979g. A model of secrecy. *Journal of Irreproducible Results* 25: 25–26.

Bunge, Mario. 1979h. The five buds of technophilosophy. *Technology in Society* 1: 67–74.

Bunge, Mario. 1979i. Philosophical inputs and outputs of technology. In *The history and philosophy of technology*, ed. G. Bugliarello and D.B. Donner, 262–281. Urbana: University of Illinois Press.

Bunge, Mario. 1979j. The bankruptcy of psychoneural dualism. *Filosofskie Nauki* (2): 77–87.

Bunge, Mario. 1980a. *The mind-body problem*. Oxford: Pergamon.

Bunge, Mario. 1980b. *Ciencia y desarrollo*. Buenos Aireplons: Siglo Veinte. Repr.: Buenos Aires: Sudamericana, 1997; Pamplona: Leitoli, 2014.

Bunge, Mario. 1980c. *Materialismo y ciencia*. Repr.: Barcelona: Ariel; Pamplona: Laetoli, 2013.

Bunge, Mario. 1980d. *Epistemología. Curso de actualización*. Barcelona: Ariel.

Bunge, Mario. 1980e. From neuron to behavior and mentation: An exercise in levelmanship. In *Information processing in the nervous system*, ed. H.M. Pinsker and W.D. Williams, 1–16. New York: Raven.

Bunge, Mario. 1980f. *Ciencia y desarrollo*. Buenos Aires: Siglo Veinte.

Bunge, Mario. 1981a. *From mindless neuroscience and brainless psychology to neuropsychology*. New York: Raven. . *Annals of Theoretical Psychology* 3: 115–133 (1985). Comments by de M. C. Corballis and P.C. Dodwell.

Bunge, Mario. 1981b. Half truths. In *Philosophie als Wissenschaft*, ed. E. Morscher and G. Zecha, 87–91. Bad Reichenhall: Comes Verlag.

Bunge, Mario. 1981c. Development indicators. *Social Indicators Research* 9: 369–385.

Bunge, Mario. 1981d. *Scientific materialism*. Dordrecht: Reidel.

Bunge, Mario. 1982a. Is chemistry a branch of physics? *Zeitschrift für Allgemeine Wissenschaftstheorie* 13: 209–233.

Bunge, Mario. 1982b. *Economía y filosofía*. Madrid: Tecnos.

Bunge, Mario. 1983a. *Lingüística y filosofía*. Barcelona: Ariel.

Bunge, Mario. 1983b. Speculation: Wild and sound. *New Ideas in Psychology* 1: 3–6.

Bunge, Mario. 1983c. *Treatise on basic philosophy*, Vol. 5: Exploring the World. Dordrecht: Reidel.

Bunge, Mario. 1983d. *Treatise on basic philosophy*, Vol. 6: Understanding the World. Dordrecht: Reidel.

Bunge, Mario. 1984a. Philosophical problems in linguistics. *Erkenntnis* 21: 107–173.

Bunge, Mario. 1984b. What is pseudoscience? *The Skeptical Inquirer IX* 1: 36–46.

Bunge, Mario. 1984c. Hidden variables, separability, and realism. *Revista Brasileira de Física*, special volume in homage of Mário Schenberg. 150–168.

Bunge, Mario. 1985a. *Treatise on basic philosophy, Vol. 7, Part I: Formal and Physical Sciences*. Dordrecht: Reidel.

Bunge, Mario. 1985b. *Treatise on basic philosophy, Vol. 7, Part II: Life Science, Social Science, and Technology*. Dordrecht: Reidel.

Bunge, Mario. 1985c. *Seudociencia e ideología*. Madrid: Alianza Editorial. Repr.: Pamplona: Laetoli, 2014.

Bunge, Mario. 1985d. Comment on Apostol's paper. *Studies in East European Thought* 29: 137–138.

Bunge, Mario. 1985e. Types of psychological explanation. In *Contemporary psychology: Biological processes and theoretical issues*, ed. J. McGough, 489–501. Amsterdam: North Holland.

Bunge, Mario. 1986a. *Philosophical problems in linguistics*. Tokyo: Seishin- Shobo.
Bunge, Mario. 1986b. A philosopher looks at the current debate on language acquisition. In *From models to modules*, ed. I. Gopnik and M. Gopnik, 229–239. Norwood: Ablex Publs. Co.
Bunge, Mario. 1987a. *Philosophy of psychology (with Rubén Ardila)*. New York: Springer.
Bunge, Mario. 1987b. Two controversies in evolutionary biology: Saltationism and cladism. In *Scientific inquiry in philosophical perspective*, ed. N. Rescher, 129–145. Lanham: University Press of America.
Bunge, Mario. 1987c. Ten philosophies of mind in search of a scientific sponsor. *Proceedings of the 11th International Wittgenstein Symposium*, 285–293. Wien: Hölder-Pichler-Tempsky.
Bunge, Mario. 1987d. Le problème corps-esprit. *Médecine Psychosomatique* 15: 85–94.
Bunge, Mario. 1988a. *Filosofía de la psicología*. Barcelona: Ariel.
Bunge, Mario. 1988b. Why parapsychology cannot become a science. *Behavioral and Brain Sciences* 10: 576–577.
Bunge, Mario. 1989a. Reduktion und Integration, Systeme und Niveaus, Monismus und Dualismus. In *Gehirn und Bewusstsein*, ed. E. Pöppel, 87–104. Weinheim: VCH.
Bunge, Mario. 1989b. From neuron to mind. *News in Physiological Sciences* 4: 206–209.
Bunge, Mario. 1989c. *Mente y sociedad*. Madrid: Alianza Editorial.
Bunge, Mario. 1989d. Game theory is not a useful tool for political science. *Epistemologia* 12: 195–212.
Bunge, Mario. 1989e. *Treatise on Basic Philosophy, vol. 8: The Good and the Right*. Dordrecht: Reidel.
Bunge, Mario. 1989f. Game theory is not a useful tool for the political scientist. *Epistemologia* 12: 195–212.
Bunge, Mario. 1989g. Gradualism vs. saltationism in evolutionary biology: From Darwin to Gould (with David Blitz). *Proceedings of the 13th International Wittgenstein Symposium* Wien: Hölder-Pichler-Tempsky, 297–301.
Bunge, Mario. 1990a. Des bons et mauvais usages de la philosophie. *L'enseignement de la Philosophie* 40(2): 97–110.
Bunge, Mario. 1990b. What kind of discipline is psychology? *New Ideas in Psychology* 8: 121–137.
Bunge, Mario. 1991a. A philosophical perspective on the mind-body problem. *Proceedings of the American Philosophical Society* 135: 513–523.
Bunge, Mario. 1991b. A skeptic's beliefs and disbeliefs. *New Ideas in Psychology* 9: 131–149.
Bunge, Mario. 1991c. Una caricatura de la ciencia: la novísina sociología de la ciencia. *Interciencia* 16: 69–77.
Bunge, Mario. 1991d. A critical examination of the new sociology of science, part 1. *Philosophy of the Social Sciences* 21: 524–560.
Bunge, Mario. 1991e. The power and limits of reduction. In *The problem of reductionism in science*, ed. E. Agazzi, 31–49. Dordrecht/Boston: Kluwer.
Bunge, Mario. 1991f. What is science? Does it matter to distinguish it from pseudoscience? *New Ideas in Psychology* 9: 245–283.
Bunge, Mario. 1992. A critical examination of the new sociology of science, part 2. *Philosophy of the Social Sciences* 22: 46–76.
Bunge, Mario. 1993a. Die Bedeutung der Philosophie für die Psychologie. In *Bericht über den 38. Kongress der Deutschen Gesellschaft für Psychologie in Trier 1992*, vol. 2, ed. L. Montada, 51–63. Göttingen: Hogrefe.
Bunge, Mario. 1993b. Seven cosmological paradigms: Animal, ladder, river, cloud, machine, book, and system of systems. In *Physica, Cosmologia, Naturphilosophie: Nuovi Approcci*, ed. M. Sánchez Sorondo, 115–131. Roma: Herder-Università Lateranense.
Bunge, Mario. 1993c. *Sociología de la ciencia*. Buenos Aires: Siglo Veinte.
Bunge, Mario. 1994. L'écart entre les mathématiques et le réel. In *Passion des formes*, ed. M. Porte, 165–173. Fontenay/St Cloud: E.N.S. Editions.
Bunge, Mario. 1996a. *Finding philosophy in social science*. New Haven: Yale University Press.

Bunge, Mario. 1996b. In praise of intolerance to charlatanism in Academia. *Annals of the New York Academy of Sciences* 775: 96–116.

Bunge, Mario. 1996c. The seven pillars of Popper's social philosophy. *Philosophy of the Social Sciences* 26: 528–556.

Bunge, Mario. 1996d. Is religious education compatible with science education? (with Martin Mahner). *Science & Education* 5: 101–123.

Bunge, Mario. 1996e. The incompatibility of science and religion sustained: A reply to our critics (with Martin Mahner). *Science & Education* 5: 189–199.

Bunge, Mario. 1997a. Moderate mathematical fictionism. In *Philosophy of mathematics today*, ed. E. Agazzi and G. Darwas, 51–71. Dordrecht/Boston: Kluwer.

Bunge, Mario. 1997b. *Epistemología*. México: Siglo xxi.

Bunge, Mario. 1997c. *Ciencia, técnica y desarrollo*. Rev. ed. of #52. Buenos Aires: Editorial Sudamericana.

Bunge, Mario. 1998a. *Social science under debate*. Toronto: University of Toronto Press.

Bunge, Mario. 1998b. La explicación en ecología (with Luis Marone). *Boletín de la Asociación Argentina de Ecología* 7(2): 35–37.

Bunge, Mario. 1999a. *Las ciencias sociales en discusión*. Buenos Aires: Sudamericana.

Bunge, Mario. 1999b. *The sociology-philosophy connection*. New Brunswick: Transaction.

Bunge, Mario. 1999c. The human brain and science. *Free Inquiry* 19(2): 17.

Bunge, Mario. 2000a. Ten modes of individualism – None of which works – And their alternatives. *Philosophy of the Social Sciences* 30: 384–406.

Bunge, Mario. 2000b. Systemism: The alternative to individualism and holism. *Journal of Socio-Economics* 29: 147–157.

Bunge, Mario. 2000c. *La relación entre la sociología y la filosofía*. Madrid: Edaf.

Bunge, Mario. 2000d. Absolute skepticism equals dogmatism. *Free Inquiry* 24(4): 34–36.

Bunge, Mario. 2001a. *Philosophy in crisis: The need for reconstruction*. Amherst: Prometheus Books.

Bunge, Mario. 2001b. Recuerdo de Francisco Romero. In José L. Speroni, comp., *El pensamiento de Francisco Romero*, 175–182. Buenos Aires: E. Divern.

Bunge, Mario. 2002a. Twenty-five centuries of quantum physics: From Pythagoras to us, and from subjectivism to realism. *Science & Education* 12: 445–466.

Bunge, Mario. 2002b. Quantons are quaint but basic and real. *Science & Education* 12: 587–597.

Bunge, Mario. 2002c. Introduction to Robert K. Merton, *Teoría* y *estructura sociales*, 1–8. México, D.F.: Fondo de Cultura Económica.

Bunge, Mario. 2003a. Velocity operators and time-energy relations in relativistic quantum mechanics. *International Journal of Theoretical Physics* 42: 135–142.

Bunge, Mario. 2003b. Interpretation and hypothesis in social studies. In *The European tradition in qualitative research*, vol. IV, ed. R. Boudon, M. Cherkaoui, and R. Demeulenaere, 20–40. London: Sage Publications.

Bunge, Mario. 2003c. *Philosophical dictionary*, enlarged edition. Amherst: Prometheus Books.

Bunge, Mario. 2003d. *Emergence and convergence*. Toronto: University of Toronto Press. Repr. 2014.

Bunge, Mario. 2003e. Toward a systemic approach to disease (with G. Thurler et al.). *ComPlexUs* 1: 117–122.

Bunge, Mario. 2004a. How does it work? The search for explanatory mechanisms. *Philosophy of the Social Sciences* 34: 182–210.

Bunge, Mario. 2004b. Clarifying some misunderstandings about social systems and their mechanisms. *Philosophy of the Social Sciences* 34: 371–381.

Bunge, Mario. 2004c. The pseudoscience concept, dispensable in professional practice, is required to evaluate research projects. *Scientific Review of Mental Health Practice* 2: 111–114.

Bunge, Mario. 2006a. *Chasing reality*. Toronto: University of Toronto Press. Repr: 2014.

Bunge, Mario. 2006b. A systemic perspective on crime. In *The explanation of crime*, ed. P.-O. Wikström and R. Sampson, 8–30. Cambridge: Cambridge University Press.

Bunge, Mario. 2007a. Max Weber did not practice the philosophy he preached. In *Max Weber's "Objectivity" revisited*, ed. L. McFalls, 119–134. Toronto: University of Toronto Press.
Bunge, Mario. 2007b. *Una filosofía realista para el nuevo milenio*, 2nd ed. Lima: Universidad Garcilaso de la Vega.
Bunge, Mario. 2007c. The ethics of science and the science of ethics. In *Science and ethics*, ed. Paul Kurtz, 27–40. Amherst: Prometheus Books.
Bunge, Mario. 2008a. Bayesianism: Science or pseudoscience? *International Review of Victimology* 15: 169–182.
Bunge, Mario. 2008b. *Le matérialisme scientifique*. Paris: Syllebse.
Bunge, Mario. 2009. *Political philosophy: Fact, fiction, and vision*. New Brunswick: Transaction Publishers.
Bunge, Mario. 2010a. *Matter and mind*, Boston Library in the Philosophy of Science, vol. 287. Dodrecht: Springer.
Bunge, Mario. 2010b. *Las pseudociencias ¡vaya timo!* Pamplona: Laetoli.
Bunge, Mario. 2011. Two unification strategies: Analysis or reduction, and synthesis or integration. In *Otto Neurath and the unity of science*, ed. J. Symons, O. Pombo, and J.M. Torres, 145–157. Dordrecht/Heidelberg/London/New York: Springer.
Bunge, Mario. 2012a. *Evaluating philosophies. Boston studies in the philosophy of science*, vol. 295. Dodrecht: Springer.
Bunge, Mario. 2012b. *Filosofía para médicos*. Barcelona/Buenos Aires: Gedisa.
Bunge, Mario. 2012c. *Provocaciones*. Buenos Aires: Edhasa.
Bunge, Mario. 2012d. The correspondence theory of truth. *Semiotica* 188: 65–76.
Bunge, Mario. 2012e. *Filosofía de la tecnología y otros ensayos*. Lima: Universidad Inca Garcilaso de la Vega.
Bunge, Mario. 2013a. *Medical philosophy*. Singapore: World Scientific.
Bunge, Mario. 2013b. Bruce Trigger and the philosophical matrix of scientific research. In *Human expeditions inspired by Bruce Trigger*, ed. S. Chrisomalis and A. Costopoulos, 143–159. Toronto: University of Toronto Press.
Bunge, Mario. 2014a. *Evaluando filosofías*. Barcelona/Buenos Aires: Gedisa.
Bunge, Mario. 2014b. In defense of scientism. *Free Inquiry* 35(1): 24–28.
Bunge, Mario. 2014c. Wealth and well-being, economic growth, and integral development. *International Journal of Health Services* 42(1): 65–76.
Bunge, Mario. 2014d. Big questions come in bundles, hence they should be tackled systematically. *International Journal of Health Services* 44(4): 835–844.
Bunge, Mario. 2015a. Does the Aharonov-Bohm effect occur? *Foundations of Science* 20: 129–133.
Bunge, Mario. 2015b. A systemic approach to the climate change challenge. Internal document, Academia Argentina de Ciencias Exactas, Físicas y Naturales.
Bunge, Mario. 2016a. Why axiomatize? *Foundations of Science* 21:
Bunge, Mario. 2016b. Evaluating scientific research projects – the units of science in the making.
Bunge, Mario and David Blitz. 1989. Gradualism vs. saltationism in evolutionary biology: From Darwin to Gould. *Proceedings of the 13th International Wittgenstein Symposium*, 297–301. Wien: Hölder-Pichler-Tempsky.
Bunge, Mario, and Carlos Gabetta (eds.). 2013. *¿Tiene porvenir el socialismo?* Buenos Aires: Eudeba.
Bunge, Mario, and Máximo García-Sucre. 1976. Differentiation, participation and cohesion. *Quality and Quantity* 10: 171–178.
Bunge, Mario, and Andrés J. Kálnay. 1969. A covariant position operator for the relativistic electron. *Progress of Theoretical Physics* 42: 1445–1459.
Bunge, Mario, and Andrés J. Kálnay. 1975. Welches sind die Besonderheiten der Quantenphysik gegenüber der klassischen Physik? In *Philosophie und Physik*, ed. R. Haller and J. Götschl, 25–38. Braunschweig: Vieweg.
Bunge, Mario, and Andrés J. Kálnay. 1983a. Solution to two paradoxes in the quantum theory of unstable systems. *Nuovo Cimento* B77: 1–9.

Bunge, Mario, and Andrés J. Kálnay. 1983b. Real successive measurements on unstable quantum systems take nonvanishing time intervals and do not prevent them from decaying. *Nuovo Cimento* B77: 10–18.
Bunge, Mario and Rodolfo Llinás. 1978a. The mind-body problem in the light of contemporary biology. *16th World Congress of Philosophy: Section Papers*, 131–133.
Bunge, Mario, and Rodolfo Llinás. 1978b. Restricted applicability of the concept of command in the neurosciences: Dangers of metaphors. *Behavioral and Brain Sciences* 1: 30–31.
Bunge, Mario, and Martin Mahner. 2004. *Über die Natur der Dinge. Materialismus und Wissenschaft*. Stuttgart: Hirzel-Verlag.
Bunge, Mario, and Jean-Pierre Marquis. 1992. Lógica y verdad, In D. Sobrevilla & D. García Balaúnde, eds., *Lógica, razón y humanismo: La obra filosófica de Francisco Miró Quesada C.* Lima, 359–369.
Bunge, Mario, and Arturo Sangalli. 1977. A theory of properties and kinds. *International Journal of General Systems* 3: 183–190.
Bunge, Mario and Luis Marone. 1998. *Boletín de la Asociación Argentina de Ecología* 7(2): 35–37.
Bunge, Mario, G. Thurler, et al. 2003. Toward a systemic approach to disease. *ComPlexUs* 1: 117–122.
Bunge, Mario, et al. 2004. *Honoris causa: Manuel Sadosky en Noventa años*. Buenos Aires: Libros del Zorzal.
Buchtel, Henry A. (ed.). 1982. *The conceptual nervous system: Selected papers*. Oxford: Pergamon.
Carnap, R. 1928. *Der Logische Aufbau der Welt*. Leipzig: Felix Meiner Verlag. English translation by Rolf A. George, 1967. *The logical structure of the world. Pseudoproblems in philosophy*. University of California Press.
Carnap, R. 1936a. Testability and meaning. *Philosophy of Science* 3: 419–471; 4 (1937): 1–40 (1936).
Carnap, R. 1936b. Testability and meaning. *Philosophy of Science*, III (1936) and IV (1937).
Carnap, Rudolf. 1939. *Foundations of logic and mathematics*. Chicago: University of Chicago Press.
Cárdenas, Eduardo J., and Carlos M. Payá. 1985a. *La familia de Octavio Bunge*. Buenos Aires: Sudamericana.
Cárdenas, Eduardo J., and Carlos M. Payá. 1985b. *La Argentina de los hermanos Bunge*. Buenos Aires: Sudamericana.
Cárdenas, Eduardo J., and Carlos M. Payá. 1995. *La familia de Octavio Bunge*. Buenos Aires: Sudamericana.
Cárdenas, Eduardo J., and Carlos M. Payá. 1997. *La Argentina de los Hermanos Bunge*. Buenos Aires: Sudamericana.
Cárdenas, Eduardo J., and Carlos M. Payá. 1998. *La Familia de Octavio Bunge*. Buenos Aires: Sudamericana.
Cárdenas, Eduardo J. and Carlos M. Payá. *La Argentina de los Hermanos Bung*. Buenos Aires: Sudamericana.
Casinos, Adrià. 2012. *Un evolucionista en el Plata: Florentino Ameghino*. Buenos Aires: Fundación de Historia Natural, Universidad Maimónides.
Chagnon, N. 1968. *Yanomamo: The fierce people*. New York: Holt, Rinehart & Winston.
Chomsky, Noam. 2009. The mysteries of nature: How deeply hidden? *Journal of Philosophy* 106: 167–200.
Columba, Ramón. 1947. *El Congreso que yo he visto*, 3 vols. Buenos Aires: Editorial Columba.
Cosmides, L., and J. Tooby. 1999. *Evolutionary psychology: An introduction*. London: Weidenfeld & Nicolson.
Craver, Carl F., and Lindley Darden. 2013. *In search of mechanisms*. Chicago: University of Chicago Press.
Dawkins, R. 1976. *The selfish gene*. Oxford: Oxford University Press.
Dawkins, Richard. 1989 [1976]. *The selfish gene*. New York: Oxford University Press.
Dawkins, Richard. 2013. *The making of a scientist*. New York: Ecco Press.

D'Espagnat, Bernard. 2006. *On physics and philosophy*. Princeton/Oxford: Princeton University Press.
de Abren, Alzira A. 1989. Dom Guido, físico, maestro. *Ciencia hoy* 1(2).
de Mesquita, B.B. 1981. *The war trap*. New Haven: Yale University Press.
Denegri, Guillermo (ed.). 2014. *Elogio de la sabiduría: Ensayos en Homenaje a Mario Bunge en su 95º Aniversario*. Buenos Aires: Eudeba.
Denegri, Guillermo, and Gladys E. Martínez (eds.). 2000. *Tópicos actuales en filosofía de la ciencia: Homenaje a Mario Bunge*. Mar del Plata: Universidad Nacional de Mar del Plata & Editorial Martín.
Derisi, Octavio N. 1941. *Filosofía moderna y filosofía tomista*. Buenos Aires: Sol y Luna.
Domingo, Carlos, and Oscar Varsavsky. 1967. Un modelo matemático de la utopía de Moro. *Desarrollo Económico* 7: 3–26.
Dosman, Edgar J. 2009. *The life and times of Raúl Prebisch*. Montreal/Kingston: McGill-Queen's Universities Press.
Dosne Pasqualini, Christiane. 2007. *Hice lo que quise*. Buenos Aires: Leviatán.
Droste, Heinz. 2015. *Turn of the Tide: Gezeitenwechsel: Einführung In Mario Bunges exakten Philosophie*. Aschaffenburg: Alibri Verlag.
Dyzenhaus, David. 1997. Legal theory in the collapse of Weimar: Contemporary lessons? *American Political Science Review* 91: 121–134.
Ellis, George, and Joe Silk. 2014. Scientific method: Defend the integrity of physics. *Nature* 516: 321–323.
Falk, Gottfried and Herbert Jung. 1959. Axiomatik der Thermodynamik. In *Handbuch der Physik* III/1. Berlin/Heidelberg/New York: Springer.
Farrington, Benjamin. 1951. *Francis bacon, philosopher of industrial science*. London: Lawrence & Wishart.
Feferman, Anita Burdman, and Solomon A. Feferman. 2004. *Alfred Tarski: Life and logic*. Cambridge: Cambridge University Press.
Feigl, H. 1967. *The "mental" and the "physical"; the essay and a postscript*. Minneapolis: University of Minnesota Press.
Festinger, L. 1956. *When prophecy fails: A social and psychological study of a modern group that predicted the destruction of the world*. Minneapolis: University of Minnesota Press.
Fleck, Denise. 2001. The dynamics of corporate growth. Ph. D. Thesis, Business School, McGill University.
Fondo Bunge. Archivo de Mario Bunge en la Facultad de Ciencias Exactas, Físicas y Naturales de la Universidad Nacional de Buenos Aires, Biblioteca, Ciudad Universitaria, Pabellón 2, Intendente Güiraldes 2160, Ciudad Autónoma de Buenos Aires, C1428EGA.
Fortsmann, Matthias, Pascal Burgmer, and Thomas Mussweiler. 2014. "The Mind Is Willing, but the Flesh Is Weak": The effects of mind-body dualism on health behavior. *Psychological Science*. doi:10.1177/0956797612442392.
Frank, Philip. 1946. *Foundations of physics*. Chicago: University of Chicago Press.
Fréchet, Maurice. 1955 [1946]. Les définitions courantes de la probabilité. In *Les Mathématiques et le concret, 157–204*. Paris: Presses Universitaires de France.
Fredrickson, George M. 2002. *Racism: A short history*. Princeton: Princeton University Press.
Freire Jr., Olival. 1999. *David Bohm e a controvérsia dos quanta*. Campinas: Coleçao CLE.
Freudenthal, Hans. 1970. What about foundations of physics? *Synthese* 21: 93–106.
Fry, Douglas P. (ed.). 2013. *War, peace, and human nature: Convergence of evolutionary and cultural views*. New York: Oxford University Press.
Fuller, L.L. 1958. Positivism and fidelity to law – a reply to professor Hart. *Harvard Law Review* 71(4): 630–672.
Garcia, John. 1981. Tilting against the paper mills of Academe. *American Psychologist* 36: 149–158.
Garzón Valdés, Ernesto. 2000. *El velo de la ilusión*. Buenos Aires: Sudamericana.
Gellner, Ernest. 1983. *Nations and nationalism*. Oxford: Basil Blackwell.
George, André (ed.). 1953. *Louis de Broglie physician et penseur*. Paris: Albin Michel.
Gerschenfeld, Hersch M. 2009. *Autobombo*. Buenos Aires: Zorzal.

Goethe, Johann Wolfgang. [1808] 1926. *Fausto*. Transl. in verse by Augusto Bunge.
Golan, Joanne W. 2015. The paradox of success at a no-excuses school. *Sociology of Education* 88: 103–119.
Goldstein, Herbert. 1950. *Classical mechanics*. New York: Addison-Wesley.
Goodman, Nelson. 1951. *The structure of appearance*. Cambridge, MA: Harvard University Press.
Goodman, Nelson. 1958. The test of simplicity. *Science* 128: 1064–1069.
Gracia, J.J.E. (ed.). 1980. *El hombre y su conducta/Man and his conduct*. Rio Piedras: Editorial Universitaria.
Grupo Aletheia. 2005. *Congreso-Homenaxe a Mario Bunge*. Vigo: Grupo Aletheia.
Hacohen, Malachi H. 2000. *Karl Popper: The formative years 1902–1945*. Cambridge: Cambridge University Press.
Hall, J.A. 1986. *Powers and liberties: The causes and consequences of the rise of the west*. Berkeley: University of California Press.
Halmos, Paul. 1960. *Naïve set theory*. New York: Van Nostrand Reinhold.
Harris, M. 1968. *Rise of anthropological theory: A history of theories of culture*. New York: Columbia University Press.
Hebb, Donald O. 1949. *The organization of behavior; a neuropsychological theory*. New York: Wiley.
Hebb, Donald O. 1982. In *The conceptual nervous system*, ed. Henry A. Buchtel. Oxford: Pergamon.
Hedström, Peter, and Richard Swedberg (eds.). 1998. *Social mechanisms*. Cambridge: Cambridge University Press.
Heilbron, John. 1967. *Interview with Dr. Guido Beck. Oral history*. York: American Institute of Physics.
Herzberg, Gerhard. 1944. *Atomic spectra and atomic structure*. New York: Dover Publications.
Hidalgo Tuñón, Alberto, and Gustavo Bueno Sánchez (eds.). 1982. *Actas del I Congreso de Teoría y Metodología de las Ciencias*. Oviedo: Pentalfa.
Hook, Sydney. 1936. *From Hegel to Marx*. New York: Columbia University Press.
Hobsbawm, Eric. 2002. *Interesting times: Twentieth-century life*. London: Penguin.
Høffding, H. 1912. *A brief history of modern philosophy*. New York: Macmillan.
Hume, David. 1902 [1748]. *An enquiry concerning human understanding*, 2nd ed. Clarendon Press.
Hunt, Tristram. 2009. *Marx's General: The revolutionary life of Friedrich Engels*. London: Macmillan/Metropolitan.
Husserl, Edmund. 1970 [1936]. *The crisis of European sciences and transcendental phenomenology*. Transl. David Carr. Evanston: Northwestern University Press.
Husserl, Edmund. 1977 [1931]. *Cartesianische Meditationen*. Hamburg: Felix Meiner.
Huxley, Julian. 1942. *Evolution: The modern synthesis*. London: Allen & Unwin.
Isnardi, Teófilo. 1927. Idealismo o realismo en las ciencias físicas. *Revista de Filosofía* 13: 405–421. Repr. *Revista de Enseñanza de la Física* 15: 23–31 (2002).
Israel, Jonathan. 2001. *Radical enlightenment: Philosophy and the making of modernity, 1650–1750*. Oxford: Oxford University Press.
Jacovkis, Pablo. 2013. *De Clementina al siglo XXI: Breve historia de la computación en la Facultad de Ciencias Exactas y Naturales de la Universidad de Buenos Aires*. Buenos Aires: Eudeba.
James, W. 1890. *Principles of psychology*. New York: Holt.
Justo, Juan B. 1947 [1909]. *Obras completas*, vol. iv: *Teoría y práctica de la historia*. Buenos Aires: La Vanguardia.
Kant, Immanuel. 1952 [1787]. *Kritik der reinen Venunft*. 2nd ed. Hamburg: Felix Meiner.
Kary, Michael, and Martin Mahner. 2004. Warum Shannons "Informationstheorie" keine Informationstheorie ist. *Naturwissenschaftliche Rundschau* 57(11): 609–616.
Keynes, J.M. 1936. *The general theory of employment, interest and money*. London: Palgrave Macmillan.

Kim, Jaegwon. 1978. Supervenience and nomological incommensurables. *American Philosophical Quarterly* 15: 149–156.

Kohlrausch, F. 1870, *Leitfaden der praktischen Physik.* University of Göttingen.

Kotarbiński, Tadeusz. 1965. *Praxiology: An introduction to the sciences of efficient action.* Oxford: Pergamon Press.

Kuhn, T.S. 1970a [1962]. *The structure of scientific revolutions.* Chicago: Chicago University Press.

Kuhn, T.S. 1970b. *The structure of scientific revolutions*, 2nd ed. Chicago: Chicago University Press. First edition, 1962.

Lagerspetz, K. 1959. *Teleological explanations and terms in biology: A study in theoretical biology.* Helsinki: Societas zoologica-botanica Fennica Vanamo.

Lakatos, I. 1970. Falsification and the methodology of scientific research programmes. In *Criticism and the growth of knowledge*, ed. I. Lakatos and A. Musgrave, 91–196. Cambridge: Cambridge University Press.

Latour, Bruno, and Steve Woolgar. 1979. *Laboratory life: The construction of scientific facts.* Beverly Hills: Sage.

Lenin, Vladimir. 1972 [1909]. *Materialism and empirio-criticism.* In *Collected works*, vol. 14. Moscow: Progress Publishers.

Lévy-Leblond, Jean-Marc, and Françoise Balibar. 1984. *Quantique.* Paris: Inter Editions.

Lewontin, Richard. 2000. *It ain't necessarily so.* New York: New York Review Books.

López-Dávalos, Arturo, and Norma Badino. 2000. *J. A. Balseiro: Crónica de una ilusión.* México: F.C.E.

Lovejoy, A.O. 1936. *The great chain of being: A study of the history of an idea.* Cambridge, MA: Harvard University Press.

Mahner, Martin (ed.). 2001. *Scientific realism: Selected essays of Mario Bunge.* Amherst: Prometheus Books.

Mahner, Martin. 2015. The philosophy of mind needs a better metaphysics. In *The constitution of phenomenal consciousness: Towards a science and theory*, ed. M. Miller, 294–309. Amsterdam: John Benjamin's.

Mahner, Martin & Mario Bunge. 1996. Is religious education compatible with science education? (with Martin Mahner) *Science & Education* 5:101–123.

Mahner, Martin, and Mario Bunge. 1997. *Foundations of biophilosophy.* Berlin/Heidelberg/New York: Springer.

Mahner, Martin, and Mario Bunge. 2001. Function and functionalism: A synthetic perspective. *Philosophy of Science* 68: 75–94.

Mahner, Martin, and Mario Bunge. 2004. *Ueber die Natur der Dinge.* Stuttgart: S. Hirzel.

Mann, Michael. 2004. *Fascists.* Cambridge: Cambridge University Press.

Mariscotti, Mario. 1985. *El secreto atómico de Huemul.* Buenos Aires: Sudamericana-Planeta.

Marquis, Jean-Pierre. 1991. Approximations and truth spaces. *Journal of Philosophical Logic* 20(4): 375–401.

Marquis, Jean-Pierre. 1992. Approximations and logic. *Notre Dame Journal of Formal Logic* 33(2): 184–196.

Marr, D.C. 1969/1982. *Vision.* Cambridge, MA: MIT Press.

Martino, Antonio A. 2014. *Legislación y digesto.* Buenos Aires: Eudeba.

Matthews, Michael. 2003. Mario Bunge: Physicist and philosopher. *Science & Education* 12: 431–444.

Matthews, Michael. 2012. Mario Bunge, systematic philosophy and science education: An introduction. *Science & Education* 21: 1393–1403.

Medvedev, Zhores. 1969. *The rise and fall of T.D. Lysenko.* New York: Columbia University Press.

Merton, R.K. 1968 [1949]. *Social theory and social structure.* New York: Free Press.

Merton, Robert K. 1973. *The sociology of science: Theoretical and empirical investigations.* Chicago: University of Chicago Press.

Meyerson, E. 1908. *Identité et réalité.* Paris: F. Alcan.

Montserrat. 2011. *Osvaldo Reig: Vida itinerante de un biólogo Evolucionista*. Buenos Aires: Eudeba.
Morris, D. 1967. *The naked ape; a zoologist's study of the human animal*. New York: McGraw-Hill.
Mosterín, Jesús. 2006. *La naturaleza humana*. Madrid: Espasa Calpe.
Moulines, Carlos Ulises. 1977a. Por qué no soy materialista. *Crítica* 9: 15–37.
Moulines, Carlos Ulises. 1977b. Reconstrucción estructural de las teorías físicas. *Revista Latinoamericana de Filosofía* III/2, Buenos Aires, 117–129.
North, John D. 1965. *The measure of the universe: A history of modern cosmology*. Oxford: Oxford University Press.
Novikoff, A.B. 1945. The concept of integrative levels and biology. *Science* 101: 209–215.
Nussenzveig, H. M. 1995. Beck and theoretical physics in Brazil. En *Guido Beck symposium*, H. M. Nussenzveig & A.A.P.Videira, eds., *An. Acad. Bras. Ci.* 67 (Supl. 1), 95–99.
OECD. 2014. *Trends in income inequality and its impact on economic growth*. Working Paper No. 163 by Federico Cingano.
Ortega y Gasset, José. 1939. *Meditación de la técnica*. Buenos Aires: Espasa-Calpe Argentina.
Ortiz, Eduardo L., and Héctor Rubinstein. 2009. La física en la Argentina En los dos primeros tercios del siglo veinte: algunos condicionantes exteriores a su desarrollo. *Revista Brasileira de História da Ciencia* 2(1): 40–81.
Osborne, Reuben. 1937. *Freud and Marx: A dialectical study*. London: Victor Gollancz.
Paladini, Alejandro C. 2007. *Leloir: Una mente brillante*. Buenos Aires: Eudeba.
Parsons, T. 1937. *Structure of social action*. Glencoe: Free Press.
Parsons, T. 1951. *The social system*. Glencoe: Free Press.
Pavlov, I.P. 1955. *Selected works*. Moscow: Foreign Languages Publishing House.
Peicovich, Esteban. 2007. *El palabrista: Borges visto y oído*. Buenos Aires: Marea.
Peirce, Charles Sanders. 1935. *Collected papers*, Vol. 6: Scientific metaphysics. Cambridge, MA: Harvard University Press.
Piattelli-Palmarini, Massimo (ed.). 1980. *Language and learning*. Cambridge, MA: Harvard University Press.
Pickel, Andreas. 2001. Mario Bunge's philosophy of social science. A review essay. *Society* 38(4): 71–74.
Pickel, Andreas, ed. 2004. Systems and mechanisms: A symposium on Mario Bunge's philosophy of social science. *Philosophy of the Social Sciences* 34: 169–210; 325–381.
Pickel, Andreas, ed. 2011. Rethinking systems theory. *Philosophy of the Social Sciences* 37: 391–407.
Piélou, E. 1969. *Introduction to mathematical ecology*. New York: Wiley.
Pievani, Telmo. 2014. *Evoluti e abbandonati*. Torino: Einaudi. Ponsa.
Pinker, Steven. 2003. *The blank slate*. New York: Penguin.
Prebisch, Raúl. 1981. Dialogue on Friedman and Hayek. *CEPAL Review* 15: 153–174.
Prélat, Carlos E. 1947. *Epistemología de la química*. Buenos Aires: Espasa-Calpe.
Prenant, Marcel. 1935. *Biologie et marxisme*. Paris: ESI.
Popper Karl, R., and John C. Eccles. 1977. *The self and its brain*. New York: Springer International.
Quine, Willard Van Orman. 1985. *The time of my life*. Cambridge, MA: MIT Press.
Raynaud, Dominique. 2015. *Scientific controversies*. New Brunswick: Transaction Publishers.
Reichenbach, Hans. 1959. *Modern philosophy of science: Selected essays by Hans Reichenbach*. London: Routledge & Kegan Paul.
Renan, Ernest. 1949 [1852]. *Averroès et l'averroïsme. Oeuvres Complètes*, Vol.III. Paris: Calman-Lévy.
Robert, Marcelo. 1972. *La innovación tecnológica en América Latina. Estudio de Casos*. Washington, DC: OEA. Unpublished.
Robinson, Joan. 1962. *Economic philosophy*. London: C. A. Watts.
Ross Ashby, William. 1956. *An introduction to cybernetics*. London: Chapman and Hall.
Romero, Francisco. 1952. *Teoría del hombre*. Buenos Aires: Losada.

Russell, Bertrand. 1900. *A critical exposition of the philosophy of Leibniz*. Cambridge: Cambridge University Press.

Russell, B. 1905. On denoting. *Mind* 4(56): 479–493.

Russell, B. 1912. *The problems of philosophy*. London: Home University Library.

Russell, B. 1923. Vagueness. *Australasian Journal of Philosophy and Psychology* 1: 84–92.

Russell, B. 1940. *An inquiry into meaning and truth*. New York: W. W. Norton.

Sadovnikov, Slava. 2004. Themes out of Mario Bunge's *The sociology-philosophy connection*. *Philosophy of the Social Sciences* 34: 536–587.

Salt, David. 1971. Physical axiomatics: Freudenthal vs. Bunge. *Foundations of Physics* 1: 307–313.

Sánchez, Norma Isabel. 2007. *La higiene y los higienistas en la Argentina (1880–1943)*. Buenos Aires: Sociedad Científica Argentina.

Sanguinetti, Horacio. 1987. *Los socialistas independientes*, 2 vols. Buenos Aires: Centro Editor de América Latina.

Saunders, Frances Stoner. 1999. *Who pays the piper?* London: Fontana Books.

Saunders, Stonor, and S. Frances. 1999. *Who Paid the Piper? CIA and the Cold War*. London: Granta.

Schlosshauer, Maximilian. 2007. *Decoherence and the quantum-to-classical transition*. Berlin/New York: Springer.

Schrödinger, E. 1944. *What is life?* London: Macmillan.

Seni, Dan A. 1993. *Elements of a theory of plans*. PhD thesis, Wharton School of Business. University of Pennsylvania.

Serroni-Copello, Raúl. 1989. *Encuentros con Mario Bunge*. Buenos Aires: Asociación Argentina de Investigaciones Psicológicas. 212 pp, 2nd revised ed., 2011, 262 pp.

Serroni-Copello, Raúl. 2011. *Encuentros con Mario Bunge*, 2a ed. Buenos Aires: Ediciones ADIP.

Shea, W.R., and M. Artigas. 2003. *Galileo in Rome: The rise and fall of a troublesome genius*. Oxford: Oxford University Press.

Shirer, William L. 1860. *The rise and fall of the Third Reich*. London: Simon & Schuster.

Shirer, Edward. 1960. *The rise and fall of the Third Reich*. New York: Simon and Schuster.

Shorter, Edward. 1997. *A history of psychiatry*. New York: Wiley.

Simpson, Tomás Moro. 1964. *Formas lógicas, realidad y significado*. Buenos Aires: Eudeba.

Simpson, George Gaylord. 1978. *Concession to the improbable*. New Haven: Yale University Press.

Sokal, A., and J. Bricmont. 1998. *Intellectual impostures*. London: Profile Books.

Sorokin, Pitirim. 1956. *Fads and foibles in modern sociology and related sciences*. Chicago: H. Regnery.

Speroni, José L. (ed.). 2001. *El pensamiento de Francisco Romero*. Buenos Aires: E. Divern.

Stern, Alfred. 1936. *La philosophie des valeurs*. Paris: Herman.

Stoll, Robert S. 1963. *Set theory and logic*. San Francisco/London: W.H. Freeman.

Stove, David. 1982. *Popper and after: Four modern irrationalists*. Oxford: Pergamon Press.

Strauss, M. 1969. Corrections to Bunge's *Foundations of physics*. *Synthese* 19: 433–442.

Tarcus, Horacio. 2007. *Marx en Argentina*. Buenos Aires: Siglo xxi.

Tarski, Alfred. 1944. The Semantical concept of truth and the foundations of semantics. *Philosophy and Phenomenological Research* 4: 341–375.

Thom, René. 1972. *Stabilité structurelle et morphogenèse*. San Francisco: Benjamin-Cummings.

Tiger, L., and R. Fox. 1971. *The imperial animal*. New York: Holt, Rinehart & Winston.

Trigger, Bruce G. 2003. *Artifacts and ideas*. New Brunswick: Transaction Publishers.

Truesdell, Clifford. 1984. *An idiot's fugitive essays on science*. New York: Springer.

Vacher, Laurent-Michel. 1993. *Rencontres avec Mario Bunge*. Montreal: Liber.

Vaihinger, Hans. 1920 [1911]. *Die Philosophie des als ob*, 4th ed. Leipzig: Meiner.

Van den Berg, Axel. The social sciences according to Bunge. *Philosophy of the Social Sciences* 31: 83–103.

Varea-Terán, José. 1976. *El subdesarrollo biológico*. Quito: author's edition.

Varsavsky, Oscar. 1969. *Ciencia, política y cientificismo*. Buenos Aires: Centro Editor de América Latina.

Voltaire, F.-M.A. 1759/1959, *Candide, ou l'Optimisme*, [*Candide, or All for the Best*]. Translated by Lowell Bair, Bantam Dell, New York.

von Bertalanffy, Ludwig. 1950. An outline of general systems theory. *British Journal for the Philosophy of Science* 1: 139–164.

Weingartner, Paul, and Georg J.W. Dorn (eds.). 1990. *Studies on Mario Bunge's treatise*. Amsterdam/Atlanta: Rodopi.

Westerkamp, José F. 1975. *La evolución de las ciencias en la República Argentina, 1923–1972*. Buenos Aires: Sociedad Científica Argentina.

Whitehead, Alfred North. 1929. *Process and reality: An essay in cosmology*. New York: Macmillan.

Wiener, H. 1948. *Cybernetics: Or control and communication in the animal and the machine*. Cambridge, MA: MIT Press, 2nd revised ed. 1961.

Wiener, Norbert, Arturo Rosenblueth, and Julian Bigelow. 1943. Behavior, purpose, and teleology. *Philosophy of Science* 10: 18–24.

Wilson, E.O. 1975. *Sociobiology: The new synthesis*. Cambridge MA: Belknap.

Wittgenstein, Ludwig. 1961 [1921]. *Tractactus Logico-Philosophicus*. Transl. D. F. Pears & B.F. McGuinness. London: Routledge & Kegan Paul.

Woodger, J.H. 1929. *Biological principles I*. London: K. Paul, Trench.

Žižek, Slavoj. 2014. *Absolute recoil: Towards a new foundation of dialectical materialism*. London: Verso.

Name Index

M. Bunge, *Between Two Worlds*, Springer Biographies,
DOI 10.1007/978-3-319-29251-9

C

D

Subject Index

M. Bunge, *Between Two Worlds*, Springer Biographies,
DOI 10.1007/978-3-319-29251-9

T

The manufacturer's authorised representative in the EU is Springer Nature Customer Service Centre GmbH, Europaplatz 3, 69115 Heidelberg, Germany. If you have any concerns regarding our products, please contact ProductSafety@springernature.com

Printed and bound by CPI Group (UK) Ltd, Croydon, CR0 4YY
25/09/2026
02222847-0005